Laboratory Exercises
MICROBIOLOGY

Jaime S. Colomé
A. Mark Kubinski

Raúl J. Cano
David V. Grady

California Polytechnic State University
San Luis Obispo

WEST PUBLISHING COMPANY

St. Paul

New York

Los Angeles

San Francisco

Photo credits: Figures I-1, 4-3, 4-5: Courtesy of Olympus Corporation, Lake Success, NY. Figure 3-1b: American Society for Microbiology. Figure 3-1a: P. W. Johnson and J. M. Sieburth, Univ. of Rhode Is./BPS. Figure 4-2: Centers for Disease Control, Atlanta, GA. Figure 10-1: T. J. Beveridge, Univ. of Guelph/BPS. Figure 10-3b: S. C. Holt, Univ. of Texas Health Science Center, San Antonio/BPS. Figure 15-1: Courtesy of Bausch & Lomb. Figure 21-1: H. W. Jannasch, Woods Hole Oceanographic Inst. Figures 23-3, IX-1b, 37-1, 48-2, 59-1, 64-2, 64-3, 67-2a, 67-2b, 71-7: Courtesy of the authors. Figure IX-1c: J. J. Cardamone, Jr. and B. A. Phillips, Univ. of Pittsburg/BPS. Figure IX-2: R. Humbert, Stanford Univ./BPS. Figure 60-4: Brieflet 1060—Vermont Extension Service, H. V. Atherton and W. A. Dodge (deceased). Figure 61-1: by permission. Standard Methods for the Examination of Dairy Products, 14th edition. American Public Health Association: Washington, D.C., 1978.

Cover art: *Circles in a Circle* (1923) by Wassily Kandinski. Philadelphia Museum of Art: Louise and Walter Arensberg Collection.

CONTENTS

PREFACE

Laboratory Exercises in Microbiology is for undergraduate students from diverse majors such as allied health, architecture, agriculture, biology, business, chemistry, computer sciences, engineering, environmental engineering, food sciences, physical education, predental, premedical, preveterinarian, and nursing.

This laboratory manual provides the student with a detailed explanation of the theory behind each exercise. In addition, the instructions are more than a *cookbook* list of steps. The instructions are carefully presented and contain helpful hints so that a procedure will *work well*. The explanations that accompany the instructions help the student understand *why* something is being done so that the lab manual is not used blindly. Numerous clear illustrations complement the introductions and the instructions.

This laboratory manual provides the student and the instructor with up-to-date exercises that demonstrate the importance of microorganisms in the environment (Microbial Ecology, Water Sanitary Analysis, Agricultural Microbiology), in the food industry (Food Microbiology, Dairy Microbiology), and in medicine (Medical Microbiology).

Because of the potential risk of infection that is inherent in microbiology laboratories, students are expected to adhere to the safety regulations outlined in the section at the end of the preface. Even innocuous microbes sometimes cause infections if improper procedures are followed or if careless attitudes are the norm.

The laboratory manual is divided into 16 parts. Each is introduced by a section that unifies the exercises within the part, that provides fundamental concepts, and that emphasizes the importance of each part. The first seven parts of the laboratory manual should be included in all laboratory curricula, because they provide the student with the basic techniques and fundamental understanding required to successfully carry out and understand laboratory exercises in the many specialized areas of microbiology. Laboratory exercises in specialized areas (agricultural, environmental, dairy, food, genetics, medical, water, etc.) are found in the latter sections of the laboratory manual.

The exercises in Part I (Microscopic Techniques) show the student how to use bright field light microscopes correctly and effectively. The discussion clearly explains the theory behind light microscopy and its limitations. Dark field, phase contrast, and fluorescence microscopy are also considered because of their importance and widespread use in biology.

In Part II (Cultivation of Microorganisms), the student learns the fundamental aseptic techniques needed to safely and effectively manipulate and propagate microorganisms. The principles of microbial nutrition are also considered.

The exercises in Part III (Staining) introduce staining procedures used to characterize microorganisms and help the student learn to appreciate the diversity of cells.

In Part IV (Enumeration of Microorganisms), the student is introduced to methods used to enumerate microorganisms.

The exercises in Part V (Environmental Factors Affecting Growth) show the student how various environmental factors (such as temperature, atmospheric conditions, pH, etc.) influence the proliferation of microorganisms.

Several methods used to measure the effectiveness of various physical and chemical agents in inhibiting or killing microorganisms are illustrated in Part VI (Control of Microbial Growth).

In Part VII (Metabolic Activities of Microorganisms) the student is introduced to a variety of procedures for the detection of specific metabolic activities in microbial cultures. These exercises illustrate the extensive biochemical variability among microorganisms.

Often, it is desirable to study and compare closely related microorganisms. To this end, the laboratory manual provides the student with exercises that study various important groups of microorganisms: viruses, bacteria, fungi, algae, etc.

Part VIII (Survey of Eukaryotes and Cyanobacteria) provides the student with the opportunity to learn about major groups of microorganisms and some of the techniques used to study them. There are exercises concerned with the fungi, protozoa, eukaryotic algae, and cyanobacteria.

Part IX (The Viruses) contains experiments that show the student how to isolate and grow viruses. The student also learns about the biology of viruses and how they can be used to identify certain microorganisms.

There are numerous areas of microbiology that provide fascinating careers for many microbiology students. The last seven parts of the manual cover some of the many fields that are open to those interested in microbiology as a career.

Part X (Genetics of Microorganisms) introduces basic concepts and procedures for the study of microbial genetics that will be valuable to any person interested in biological research.

Parts XI–XVI (Microbial Ecology, Water Sanitary Analysis, Food Microbiology, Dairy Microbiology, Agricultural Microbiology, and Medical Microbiology) provide numerous exercises that expose the student to specific laboratory procedures of practical importance to these fields of applied microbiology.

The 72 exercises cover a sufficient breadth of topics in microbiology to allow the manual to be used in almost any type of microbiology laboratory. Instructors can pick and choose those exercises that are best suited for the types of students taking the course. The exercises included in this manual have been tested repeatedly by the authors to ascertain that they "work" and that the students can set the experiments up within 2–3 hours and complete most of them within one to two periods.

The student should pay special attention to the "Procedure" section of each exercise. In many of the steps, additional information such as precautionary or explanatory notes are added to aid the student in understanding the reason or reasons behind each step. This extra information is highlighted by a bullet (●).

Most of the exercises in Parts I–VII should be performed individually so that each student has as much practice as possible using the basic techniques. The latter exercises may be performed in pairs. It is recommended that groups greater than four students be avoided because such groups usually include one or more "observers."

At the end of each exercise are questions that help the student realize what is important in each section, summarize the information, and challenge their understanding of the material and procedures.

Laboratory report forms for each of the 72 exercises are provided at the end of the manual to encourage the student to record and evaluate the data gathered. The use of report forms also helps the student realize what is important in each section.

The information in the appendixes should help the instructor set up, organize, and coordinate the laboratory exercises. Appendix A provides information on how to prepare dilutions and calculate the number of organisms per ml in experiments that involve the enumeration of microorganisms. Appendix B provides information on pH indicators. Appendix C provides a detailed list of organisms specified for the exercises in this manual, commercial sources, and (in certain instances) special characteristics of the organisms. Appendix D is a comprehensive listing of media used in the exercises. The constituents of each medium are indicated as well as helpful hints as to how to make it and its uses. Appendix E lists the stains, indicators, and reagents used in the laboratory manual. The constituents of each are indicated and special instructions for their preparation and use are presented.

A comprehensive index enables the student to find references to a particular topic while information in the appendixes provides additional help and understanding.

The authors have developed this laboratory manual in the hope that it will provide students with correct laboratory skills and meaningful insights into some of the many roles microorganisms play in our lives.

SAFETY REGULATIONS FOR THE MICROBIOLOGY LABORATORY

Certain rules and regulations related to laboratory safety are unique to specific institutions. The instructor will present this special information to you. Those rules listed below are designed to protect you and all others who have contact with the laboratory and should be "religiously" observed.

1. Disinfect your work area each period before you begin and after you complete the laboratory exercises.

2. Wash your hands before you begin the laboratory and just before you leave the laboratory.

3. Do not smoke, drink or eat in the laboratory and do not put objects, such as pencils, slide labels, or pipets in your mouth.

4. Wear shoes in the laboratory at all times. Lab coats or aprons are recommended.

5. Tie back long hair so that it does not hang free. Persons with long hair around gas burner flames are

at risk. In addition, hair may be a source of contamination for cultures or it may become contaminated.

6. Store clothing and books that are not being used in a designated area so that the work tables are not cluttered. This makes for a safer work area and eliminates the possibility of contaminating personal belongings.

7. Never discard microorganisms or contaminated materials into wastepaper baskets or into sinks. All objects (pipets, plates, tubes, paper, etc.) that come in contact with microbial cultures must be disposed of in appropriate containers.

8. Do not throw broken glass into wastepaper baskets.

9. Place contaminated reusable pipets (usually tip-down) in cylindrical pipet holders that contain disinfectant.

10. Do not stack Petri dishes more than three high on incubator shelves. These stacks are a hazard if they should topple in the incubator or when the incubator is opened.

11. Cover spilled microorganisms with paper towels saturated with disinfectant. The disinfectant should be allowed to soak the area for at least 15 minutes before cleanup. The contaminated area should be cleaned up and the contaminated materials disposed of in appropriate containers. The contaminated area should subsequently be disinfected.

12. Notify your instructor immediately of any accidents involving release of cultures to the environment. Your instructor should be made aware of body contamination (mouth, eyes, nose, etc.) and of any cuts or burns.

13. Label all cultures with (a) the name of the microorganism, (b) your name, (c) date of incubation, and (d) other appropriate details of the exercise.

PART I
MICROSCOPIC TECHNIQUES

Microscopes are instruments used to magnify objects that cannot be seen by the unaided eye (which is unable to distinguish objects with a diameter less than 0.1 mm). There are two categories of microscopes: **light microscopes** and **electron microscopes.** Light microscopes have glass lenses that bend light from the object so that an enlarged image of the object can be created. Light microscopes are used to produce images of objects about 1,000 times larger than the object. Images larger than 1,000 times can be produced but they are not clearly defined. Electron microscopes use electrons (rather than light) and magnetic coils in order to create an image of the object. One type of electron microscope can create images that are 100,000 times larger than the object.

Light microscopes generally are used to observe microorganisms and sections of plant and animal tissue. The large organelles of eukaryotic cells (such as nuclei, mitochondria, chloroplasts, and vacuoles) can also be seen with the aid of the light microscope. One type of electron microscope, called the **transmission electron microscope** makes it possible to observe in ultrathin sections the substructure of cells and their organelles. Another type of electron microscope, called the **scanning electron microscope,** is used to see surface details of organisms, cells, and organelles.

THE LIGHT MICROSCOPE

Light microscopes may be **simple** or **compound.** A simple microscope uses only one lens between the eye and the object to achieve magnification. A simple microscope is like a hand-held magnifying glass. In contrast, a compound microscope is one that combines two or more lenses in series between the eye and the object.

In the compound microscope, the lens (or lens system) nearest the eye is referred to as the **ocular lens** while the lens (or lens system) nearest the object is called the **objective lens** (fig. I-1). Most compound microscopes give you a choice of objective lenses: one that magnifies 10× (**low power**), one that magnifies 40× (**high-dry**), and one that magnifies 100× (**oil immersion**). The 100× objective is referred to as the oil immersion lens because it must *always* be immersed in special oil when in use. In contrast, the 10× and 40× lenses are called dry lenses because they must always be free of oil and water. Most compound microscopes have an ocular lens system that magnifies 10×.

Compound microscopes that have two eye pieces so that you use both eyes to see the image are called **binocular** microscopes, while those which have only one eye piece are referred to as **monocular** microscopes.

Magnification

Magnification is the property of a lens system that allows it to produce an enlarged image of an object. The total magnification of a compound microscope is the product of the magnifications of the objective lens and the ocular lens. Thus, if the 100× objective is in use and the microscope has a 10× ocular lens, the total magnification of the system is 100 × 10 = 1000×. How much a lens magnifies the dimensions of an object viewed through it depends upon the lens's degree of curvature. A lens that is highly curved magnifies more than a flatter lens (fig. I-2). Lenses become smaller as their magnification increases. This is because the diameter becomes smaller as the lens becomes more highly curved. The 10×, 40×, and 100× objectives have diameters of approximately 4.0 mm, 2.5 mm, and 1.4 mm respectively.

The **focal point** of the lens is the point at which parallel light rays from the object converge or are focused. The **focal length** is the distance from the focal

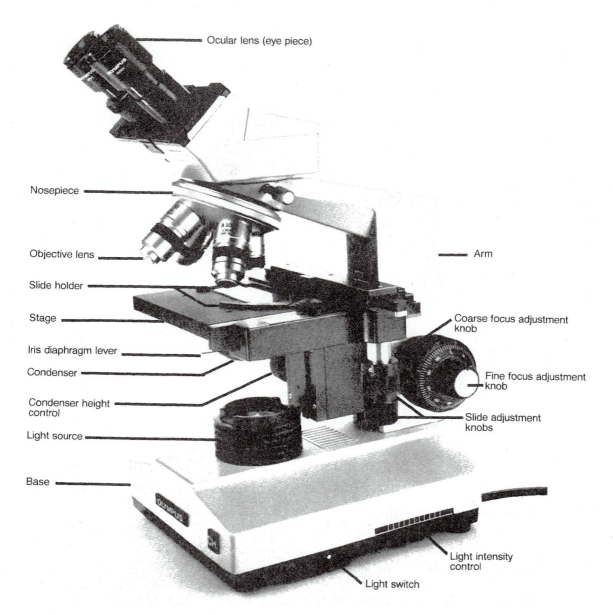

Figure I-1. **A Compound Microscope** The microscope illustrated
is also known as a bright field light microscope.

point to the optical center of the lens (fig. I-2). Notice
that the greater the curvature of the lens (the smaller
the lens), the shorter the focal length becomes. The
focal length is important since a magnified **real image**
is produced only if the object to be magnified is positioned between the focal point (f) and twice the focal
length (2f). If an object is placed more than 2f from
the lens, the image formed will be smaller than the
object. On the other hand, if the object is placed between the focal point and the lens, a **virtual image**
(an image that cannot be projected onto a screen) is
created rather than a real image.

In the compound microscope, a real image is created in the tube of the microscope and from this real
image a virtual image is created by the ocular lens.
The eye is placed close to the ocular lens so that the
virtual image can be seen (fig. I-3).

Resolution

The degree to which an image can be magnified and
seen depends upon the size of the lens and the intensity of light reaching the eye, respectively. As a lens
becomes much smaller than the 100× objective, light

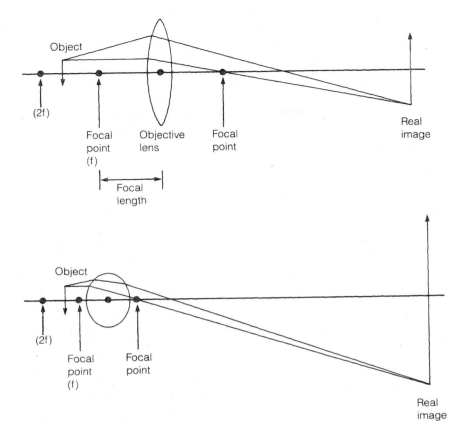

Figure I-2. Effect of Lens Shape on Magnification

going through it is spread out (**diffracted**) and no longer produces a sharp image (fig. I-4). Consequently, two points, which you might expect to be clearly separated using lenses with magnifications greater than 100×, become more diffuse and run into each other as lenses with higher magnification are used. The **limit of resolution** of a lens is the minimum distance between two closely spaced points in an object that can be seen as distinct points in the image. The limit of resolution (R) is determined by parameters of the lens system such as the wavelength of light (λ) used and the amount of light gathered from the specimen.

$$R = (0.61\lambda)/NA$$

The amount of light gathered from the specimen is expressed as the lens's **numerical aperture** (NA), which depends upon the **index of refraction** (n) of the medium between the specimen and the lens as well as the **radius** (r) and **working distance** (w) of the lens. Since the radius and the working distance are related to the angle (θ) illustrated in fig. I-5,

$$NA = n(r/\sqrt{r^2 + w^2}) = n\ (\sin\ \theta)$$

When light passes through the glass slide into air, some of it is **refracted** and is not gathered by the lens

(fig. I-5). If immersion oil with an index of refraction identical to that of the glass slide is placed between the specimen and the objective lens, more light is gathered by the lens because the light is not refracted. Since oil (n = 1.5) has a greater index of refraction than air (n = 1), the oil immersion lens has a larger numerical aperture if oil is used (NA = n sin θ) than if it is not. The larger the numerical aperture (NA) becomes, the smaller the limit of resolution (R) is, and the better the resolution, or **resolving power,** of the microscope.

The resolving power of a light microscope can be maximized by using light with as short a wavelength as possible. Consequently, blue filters are sometimes used between the light source and the specimen in order to remove light with longer wavelengths.

The effects that the wavelength of light and indices of refraction of the lenses and the medium between the object and the lenses have on the limit of resolution (resolving power) of a light microscope can be illustrated with a calculation. In a situation where red light measured in nanometers (650 nm) is used, immersion oil is not used (n = 1), and sin θ is 0.9, the limit of resolution in micrometers is approximately 0.44 μm.

$$R = \frac{0.61 \times 0.650\ \mu m}{1 \times 0.9} = 0.44\ \mu m$$

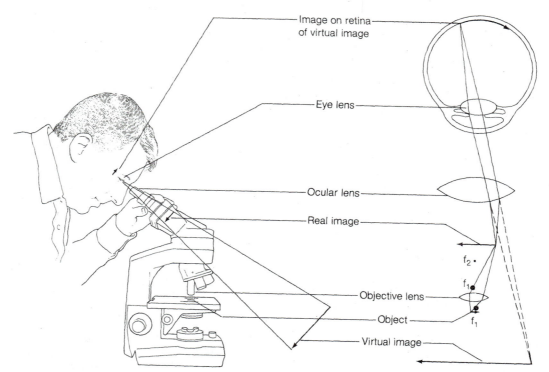

Figure I-3. How the Bright Field Light Microscope Works. Light from the specimen is bent by the objective lens so that an inverted, magnified real image is created between the objective lens and the ocular lens. The image is magnified further by the ocular lens, so that an inverted, magnified virtual image is created. The lens in the eye focuses the inverted magnified virtual image on the retina so that the specimen can be seen.

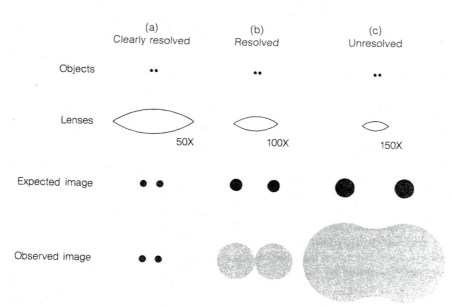

Figure I-4. Resolution and the Wave Nature of Light. (a) Very little diffraction of the light is produced by the 50× objective lens. (b) The 100× objective lens causes some diffraction because of its size. Consequently, the image is more diffuse and blurred than expected. Nevertheless, the objects can be resolved. (c) A 150× objective lens results in a severe diffraction of the light, so that the objects cannot be resolved. The light from the objects is so diffuse that it overlaps, producing a single, blurred image.

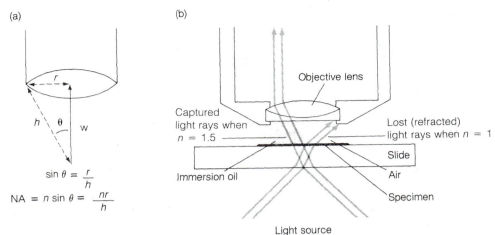

$$\sin \theta = \frac{r}{h}$$

$$NA = n \sin \theta = \frac{nr}{h}$$

Figure I-5. How Oil Affects the Limit of Resolution. *(a)* The limit of resolution of the bright field microscope is proportional to the wavelength of light used (in nanometers) and inversely proportional to the numerical aperture of the lens. The value of the NA is dependent upon the index of refraction (*n*) of the material between the specimen (object) and the objective lens as well as the radius (*r*) of the lens and the working distance (*w*). *(b)* The resolving power of the oil immersion lens can be improved by placing oil between the specimen and the objective lens. The oil eliminates refraction of light so that more light enters the lens to give definition to the specimen.

If blue light (480 nm) and immersion oil (1.5) are used, the resolvable distance is approximately 0.22 μm.

$$R = \frac{0.61 \times 0.480 \ \mu m}{1.5 \times 0.9} = 0.22 \ \mu m$$

Clearly, the resolving power of a light microscope is greater when blue light and oil immersion are used because it is possible to distinguish as separate entities two points that are much closer together (0.22 μm apart).

ELECTRON MICROSCOPES

Electron microscopes use electrons rather than light, and magnetic lenses rather than glass lenses, to create an enlarged image of an object. The transmission and scanning electron microscopes work very differently (fig. I-6). In the transmission electron microscope, electrons may pass through ultrathin sections of the specimen or be reflected or absorbed by the specimen. The pattern and intensity of electrons that pass through the specimen are used to create an image of the object on a screen. In the scanning electron microscope, the electron beam is focused onto a point on a metal coating that covers the specimen. The electron beam knocks electrons from the metal coating. The scattered electrons from the metal coating, known as **secondary electrons,** hit a detector and create an electric current that is used to control the brightness of a point on a screen. The point on the screen corresponds to the point on the object that gave up secondary electrons. An image of the object is created by scanning the entire surface of the object with the electron beam. The scanning occurs rapidly so that an image can be created out of the 200,000 or more bright points on the screen. The brightness of each point is related to the number of secondary electrons picked up by the detector.

Magnification

The magnetic lenses of the most powerful transmission electron microscopes are able to produce a clear image that is more than 100,000 times the actual size of an object. The maximum useful magnification of the electron microscope is more than 100 times greater than that of the light microscope.

Resolution

The limit of resolution of the transmission electron microscope depends upon the wavelength of the electrons and the characteristics of the magnetic lens systems. Consequently, the mathematical equation for the limit of resolution is similar to that of the light microscope. Since the structure and operation of the electron and the light microscopes are very different, however, the constants and values for the numerical aperture differ as indicated in the formula below.

$$R = \lambda/2NA = \lambda/0.002$$

The wavelength of a beam of electrons is determined by the voltage (v) used to create the beam. The higher

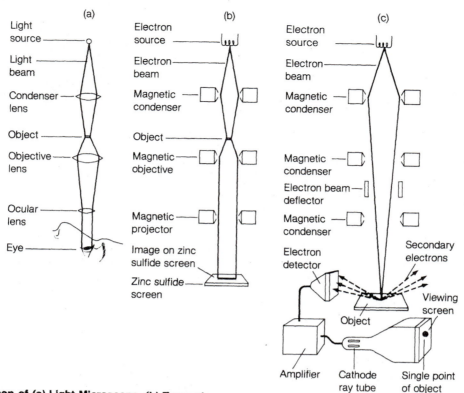

Figure I-6. Comparison of (a) Light Microscope, (b) Transmission Electron Microscope, and (c) Scanning Electron Microscope

the voltage is, the shorter the wavelength becomes. Typically, the use of 50,000 volts yields a wavelength of approximately 0.0055 nm.

$$\lambda = 1.23 \text{ nm}/\sqrt{v}$$
$$\lambda = 1.23 \text{ nm}/\sqrt{50,000} = 0.0055 \text{ nm}$$

The resolution of the electron microscope is approximately 2.7 nm when the electron beam has a wavelength of 0.0055 nm.

$$R = \lambda/0.002$$
$$R = 0.0055 \text{ nm}/0.002 = 2.7 \text{ nm}$$

EXERCISE 1
USE AND CARE OF THE MICROSCOPE

A. USING THE BRIGHT FIELD LIGHT MICROSCOPE

A **bright field light microscope** (fig. I-1) is one in which the light from the light source shines directly into the objective lens. Consequently, the field of view is generally bright. The specimen may be darker or brighter than the background depending upon how the condenser is adjusted. In subsequent sections of this laboratory manual, you will be introduced to other light microscopes such as the dark field microscope, phase contrast microscope, and fluorescence microscope.

Some students taking microbiology have never used a light microscope, and others have encountered the problems involved in using the oil immersion lens. Thus, at least one laboratory period should be devoted to learning how to use the light microscope properly and understanding the peculiarities of the microscopes you will be using. Many of your laboratory studies will be successful only if you develop technical skill in the use of the light microscope.

Modern microscopes are expensive precision instruments that must be used with care if they are to be of any value in laboratory work. *It is essential that you keep the microscope clean and in good working condition.* You should always clean (using only lens paper) the low and high-dry objectives and wipe the oil from the oil immersion (100×) objective before putting away the microscope. Oil and grit obscure and damage the lens systems and should not be allowed to dry and accumulate on the lenses. Oil or grease spilled on any part of the microscope must be removed before putting away the microscope. Microscopes that previous users have left dirty, oily, or in any state of disrepair should be reported to the instructor.

The following list of rules should help you maintain your microscope in good working order.

1. Store the microscope in a cabinet or under a cover to protect it from dust and moisture in the air.

2. When carrying a microscope, hold the arm (handle) of the microscope upright with one hand and support the base with the palm of the other hand. If the microscope is tilted backwards, ocular lenses may slide out and crash onto the floor.

3. In order to avoid scratching the lenses, clean them only with lens paper. Do not use your fingers, paper towels, rags, or facial tissue to clean the lenses. These materials are too abrasive and will scratch the lenses. Clean only the external surfaces of lenses. Do not attempt to remove the ocular or objective lens from the microscope. If dirt appears to be on the internal surfaces of lenses, notify your instructor.

4. The stage, condenser, and other parts of the microscope can be cleaned with facial tissue and soft lint-free rags.

5. After extensive use of the microscope, dried oil and other dirt may accumulate on the lenses and decrease their effectiveness. In this case, the lenses can be cleaned by using a small amount of xylene (or a special lens cleaning solution) on lens paper, which dissolves oil and removes dirt. Do not use other solvents such as alcohol or acetone unless the manufacturer indicates that it is safe. The cements that are used to hold some lenses in place are sensitive to these solvents.

6. Before storing a microscope, always thoroughly clean it. In particular, wipe all oil from the oil immersion lens so that it will not gather grit and dry. Oil and grit on a lens can cloud it and cause scratching. These effects decrease the resolving power of the lens. A slide left on the stage is a clear sign of negligence.

7. Before storing a microscope, center the stage, increase the distance between the stage and the objective lenses, rotate the low power objective into place, lower the condenser, turn the light intensity control to low, and turn off the light switch.

Materials

Prepared slides of representative microorganisms: bacteria, algae, protozoans, fungi, and microscopic animals such as nematodes and water fleas

Light microscope

Lens paper

Immersion oil

Procedure

1. Place your microscope on a firm table or bench away from sinks and gas burners.

2. Check your microscope to see whether it was put away properly.
• If it was not thoroughly cleaned, report the problem to your instructor.

3. Compare figure I-1 with your microscope to familiarize yourself with the various parts of the microscope.

4. Place a prepared slide of one of the larger microorganisms (fungi, protozoans, algae) on the stage so that it is held in place by the slide holder.

• Position the slide over the light hole in the stage so that light coming through the stage from below hits the specimen. The slide holder is positioned by using the slide adjustment knobs attached to the stage and just below it.

5. Plug in the microscope and turn on the light switch.

6. Set the light intensity control to the upper-middle range.

• The light intensity is controlled by a rheostat (sliding control or rotating knob) at the base of the microscope. An intensity (6–7) in the upper-middle range is adequate for most observations. High intensities (9–10) shorten the light bulb's lifetime and should be avoided when possible. As a rule, the background lighting in your field of view should be a diffuse white without glare or shadow.

7. Adjust the condenser with the condenser height control so that it is just below the slide.

• The structure found just below the stage, called the **condenser,** is used to focus light on the object. The condenser must be adjusted correctly for each specimen. If the condenser is not properly adjusted, you may not be able to find the specimen or it may be difficult to see any detail.

8. Open the condenser diaphragm so that the field is completely illuminated.

• When you are studying unstained specimens, the condenser diaphragm should be almost closed but when studying stained preparations, the diaphragm should be well opened (fig. 1-1). The size of the light beam through the condenser and the angle at which it hits the object is controlled by the **iris diaphragm lever** that opens and closes the condenser diaphragm. You must, of course, always have the diaphragm open to some degree in order to see an image of the object, but the amount of light you let through the condenser can mean the difference between seeing the object or not. Often, when the diaphragm is wide open, too much light hits the object and there is little contrast between the image and the background. When this occurs you may not be able to see the specimen. By closing the diaphragm somewhat, contrast between the object and the background can be increased. It is often a good idea to start with the diaphragm almost closed and then adjust the opening for the best contrast.

9. Rotate the lens turret (nosepiece) so that the lower power (10×) objective is locked into position.

• The 10× and 40× objectives generally are used to view large microorganisms and tissue sections, while the 100× oil immersion objective is used to view small microorganisms such as yeasts and bacteria as well as cellular detail within large microorganisms.

10. Adjust the distance between the 10× lens and the slide using the coarse focus adjustment so that the lens is about 0.5 centimeter (½ cm)—less than ¼ inch— above the specimen. Always position the lens by viewing the procedure from the side.

• Each of the objectives must be positioned at a different distance from the specimen in order to bring the image into focus. The 10× objective is positioned about 0.5 cm from the specimen (table 1-1). The 10× objective must never touch oil, water, or the specimen.

11. Before looking through the microscope, make sure you know which way to turn the fine focus adjustment so that the distance between the specimen and the lens increases when you turn the fine focus adjustment.

• WHEN LOOKING THROUGH THE MICROSCOPE, AVOID SMASHING THE LENS INTO THE SLIDE BY ALWAYS INCREASING THE DISTANCE BETWEEN THE SLIDE AND THE LENS.

12. Look through the ocular lens and focus the image using the coarse focus knob by increasing the distance between the specimen and the objective lens. Use the fine focus knob to sharpen the image.

13. Adjust the opening of the condenser diaphragm to improve the contrast between the organisms and the background.

14. Scan the field by moving the slide and draw what you see.

15. To use the high-dry objective lens (40×), simply rotate it into position so that it locks in place.

• It is best to center part of the specimen image in the low power (10×) field before shifting to high-dry (40×) or oil immersion (100×). Since the lenses of most microscopes are set up to be **parfocal,** the specimen will be almost in focus. A specimen that is in focus under one objective is also in focus under the other objectives in a parfocal microscope. A sharp image can be achieved by slowly rotating the fine adjustment knob clockwise or counterclockwise, a fraction of a turn. The working distance for the 40× objective is less than 0.5 millimeter (½ mm) from the specimen. The 40× objective must never touch oil, water, or the specimen.

16. Adjust the opening of the condenser diaphragm to improve the contrast between the organisms and the background.

17. Scan the field and draw what you see.

18. To use the oil immersion objective lens (100×),

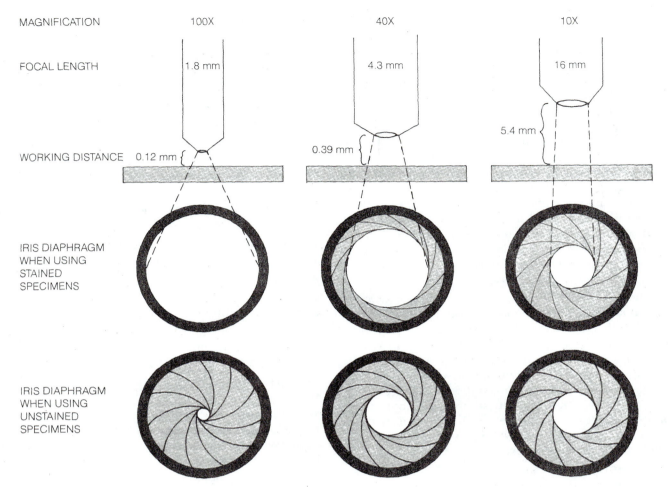

| MAGNIFICATION | 100X | 40X | 10X |

FOCAL LENGTH: 1.8 mm / 4.3 mm / 16 mm

5.4 mm

WORKING DISTANCE: 0.12 mm / 0.39 mm

IRIS DIAPHRAGM WHEN USING STAINED SPECIMENS

IRIS DIAPHRAGM WHEN USING UNSTAINED SPECIMENS

Figure 1-1. The Relationship between the Working Distance of an Objective Lens and the Adjustment of the Iris Diaphragm

rotate the high-dry (or low) power lens out of the way and place a small drop of immersion oil on the area that you wish to study. Then rotate the oil immersion lens into position. Fine focusing is generally required.

• Before using the oil immersion objective (100×), position the image of your specimen in the center of the field of the 40× objective. Clean the 100× objective lens with lens paper. *It is a good idea to clean the lens each time you look at a new slide.* The working distance for the 100× objective is approximately 0.1 mm (table 1-1). The 100× objective should always be immersed in oil but should not touch the specimen. When using the oil immersion lens, you must position the condenser just below the slide by using the condenser height adjustment on the stage support (fig. I-1). When the condenser almost touches the bottom of the slide, all the light will be focused on a very small area of the object so that most of the light from the object enters the extremely small 100× objective lens. If you do not have the condenser raised, the light will hit a much larger area of the object, and relatively little light from the object will enter the oil immersion lens. The result is a very dim image and a reduction in your ability to see detail. Looking through the microscope, see if an unfocused image of the specimen is visible. If it is, move the fine focus adjustment very

TABLE 1-1.
CHARACTERISTICS OF OBJECTIVE LENSES

OBJECTIVE	MAGNIFICATION	FOCAL LENGTH (mm)	WORKING DISTANCE (mm)	NUMERICAL APERATURE	RESOLUTION (μm)
Low power	10×	16	5.4	0.25	1.3
High dry	40×	4.3	0.39	0.65	0.52
Oil immersion	100×	1.8	0.12	1.30	0.26

slowly so that a sharp image is formed. If no image is visible when you switch to oil immersion, move the stage upward slowly using the fine focus adjustment knob so that the 100× objective is closer to the specimen. YOU MUST AVOID RAMMING THE SLIDE INTO THE LENS BY WATCHING THE APPROACH OF THE TWO FROM THE SIDE. The lens should be completely in the oil but it should not be touching the slide (or coverslip). Also, make sure that you know which way to rotate the fine focus adjustment knob so that the stage will move away from the lens when you begin focusing. As you look through the eyepiece of the microscope, move the fine focus adjustment very slowly so that the stage moves away from the lens. Within a turn of the fine focus adjustment knob, you should focus onto the object. The lens will still be in the oil. If it is not, you have gone too far.

19. Scan the field and draw what you see.

• There are a number of reasons you may not be able to find anything under oil immersion. In order of frequency, the reasons for failure are: too much oil, moving the fine focus adjustment too rapidly, condenser not adjusted correctly (usually the iris is opened or closed too much), dirty lens, too few organisms on the slide, specimen image not centered before shifting to oil, 10× objective was focused on extraneous stain on the bottom of the slide rather than on the stained smear, or 10× objective was focused on a stained smear that was upside down.

20. Study prepared slides of bacteria, algae, protozoans, fungi, and microscopic animals using the different objectives. Draw what you see.

Questions

1. Why should you always use lens paper to clean lenses? What materials should not be used to clean lenses?

2. Locate, identify, and explain the use of each major part of the bright field light microscope.

3. When do you generally use the 10×, 40×, and 100× objectives?

4. What does it mean when the objective lenses are described as parfocal? Why is this a useful characteristic?

5. Of what importance is the condenser in the light microscope?

6. Explain how the condenser should be adjusted (position and diaphragm opening) when studying unstained or specimens using the 100× objective.

7. If a bacterium is one μm in diameter, how large will its image appear in millimeters if you are using a compound microscope and the 40× objective is in place? Can you see the image? Can you see the image

of the bacterium if the 100× objective is in place? Explain.

8. Can you see the images of unstained bacterial flagella that are 5 μm long and 0.02 μm in diameter with the light microscope? Explain.

9. What is the difference between a simple and compound microscope?

10. What characteristic of a glass lens is responsible for its magnification?

11. What is the maximum useful magnification of light microscopes?

12. What problem arises when objective lenses that magnify more than 100× are used?

13. What is meant by the limit of resolution?

14. What factors determine the limit of resolution of a light microscope?

15. Explain how changes in the wavelength of light affect the resolution of a microscope.

16. Determine the limit of resolution of a light microscope if the wavelength equals 600 nm and the numerical aperture equals 1.25.

17. What is meant by the resolving power of a microscope?

18. Why it is necessary to have immersion oil between the oil immersion objective and the specimen on the slide? What happens if no oil is used with the oil immersion lens?

19. What is the difference between transmission and scanning electron microscopes?

20. What is the maximum useful magnification of a typical transmission electron microscope?

21. Determine the limit of resolution of a transmission electron microscope if the voltage equals 100,000 volts and the numerical aperture equals 0.002.

22. What is the limit of resolution of the unaided human eye?

23. How much do you have to magnify a bacterium 1 μm in diameter in order to just barely see it?

B. USING THE OIL IMMERSION LENS DIRECTLY

When studying bacteria, it is often more efficient to begin with the oil immersion lens and omit preliminary discoveries with the lower power objectives. Some instructors prefer that only experienced students omit the use of the 10× and 40× objectives before turning to the oil immersion lens. The following procedure should be followed when using the oil immersion directly.

Materials

Prepared slides of representative microorganisms: bacteria, algae, protozoans, fungi, and microscopic animals such as nematodes and water fleas

Light microscope

Lens paper

Immersion oil

Procedure

1. Clean the oil immersion lens each time you look at a new slide.

2. Position the specimen on the slide in the center of the light hole just under where the oil immersion lens will be.

3. Place a small drop of immersion oil on the specimen.

4. Make sure that there will be plenty of room between the oil and the 100× objective before you rotate the lens over the specimen.

5. Adjust the condenser so that it is just below the specimen slide. Close the condenser diaphragm almost all the way if looking at a living specimen (wet mount) but open the condenser diaphragm if looking at a stained specimen.

6. With the coarse focus adjustment, reduce the distance between the oil immersion lens and the specimen until the lens is immersed in the oil and almost touches the specimen.

- The working distance for the 100× objective is about 0.1 mm. YOU MUST AVOID RAMMING THE SLIDE INTO THE LENS BY WATCHING THE APPROACH OF THE TWO FROM THE SIDE AND ALWAYS FOCUSING SO THAT THE DISTANCE BETWEEN THE STAGE AND THE LENS INCREASES.

7. Now, look through the microscope and find the specimen image by increasing the distance between the stage and lens.

- Make sure you know which way to turn the fine focus adjustment knob to increase the distance between the stage and the lens.

8. Within a turn or so the image should come into focus. Move the slide back and forth ever so slightly while you are focusing. Moving images are often easier to see than stationary ones.

- If you do not find the image after three to five turns or if the lens is out of the oil, you have gone too far. Repeat steps 6 and 7. There are a number of reasons you may not be able to find anything under oil immersion. In order of frequency, the reasons for failure are: too much oil, moving the fine focus adjustment too rapidly, improper adjustment of condenser, dirty lens, too few microorganisms on the slide, and an upside-down slide.

EXERCISE 2
MEASURING CELLS USING THE MICROSCOPE

In studying and identifying microorganisms, it is often necessary to know their dimensions. Such measurements are readily made with a calibrated ruler known as an ocular micrometer. Generally, measurements are made using the metric system (table 2-1).

An **ocular micrometer** is a miniature ruler engraved on a flat circular glass disc. The ocular micrometer is placed just under the ocular lens (fig. 2-1) and the ruler can be seen when looking through the eye piece. Because the ocular micrometer is not magnified by the objective lens, the scale of the ruler has different values for each objective lens used. Also, because each different make of microscope may magnify to a slightly different extent and each make of ocular micrometer may not have identical rulings,

micrometers should be calibrated on each microscope. With the low power objective (10×) the smallest unit of the ruler represents a length of about 10 μm, while with the 40× objective the smallest unit of the ruler equals a length of approximately 2.5 μm. With the oil immersion objective (100×) the smallest unit of the ruler corresponds to a length of 1 μm.

In order to calibrate (assign specific values to) the ocular micrometer, a **stage micrometer** is used. A stage micrometer is a miniature ruler with units in μm engraved on a glass slide. The stage micrometer is placed on the stage and the ruler is brought into focus. By superimposing the image of the ocular scale upon the known scale of the stage micrometer, it is possible to determine the value of an ocular unit.

TABLE 2-1.
THE METRIC SYSTEM

Kilometer	km	1000 m	10^3 m
Meter	m	m	m
Decimeter	dm	0.1 m	10^{-1} m
Centimeter	cm	0.01 m	10^{-2} m
Millimeter	mm	0.001 m	10^{-3} m
Micrometer	μm	0.000001 m	10^{-6} m
Nanometer	nm	0.000000001 m	10^{-9} m
Angstrom	Å	0.0000000001 m	10^{-10} m

The dimensions of cells in dried, fixed, and stained smears tend to be reduced by as much as 10% to 20% from the dimensions of the living cells. Consequently, if the actual dimensions of an organism are required, measurements should be made on microorganisms in a wet mount or hanging drop mount (see exercise 3).

Materials

Prepared slides of microorganisms

Ocular micrometer

Stage micrometer

Procedure

1. If there is no ocular micrometer visible, unscrew one of the ocular lenses and insert an ocular micrometer disk on the rim within the ocular tube.
- Make sure that the engraved side of the ocular micrometer is on the lower surface. Replace the ocular lens.

2. Place the stage micrometer slide on the stage and focus on one part of the scale using the 10× objective lens.

- You should be able to see both the stage micrometer ruler and the ocular micrometer ruler.

3. Move the stage micrometer and the ocular micrometer so that the scales are parallel to each other.
- You may have to rotate the ocular lens so that the ocular micrometer scale is parallel to the stage micrometer scale.

4. Manipulate the stage micrometer slide so that a line on the ocular micrometer rulings coincides with one of the lines on the stage micrometer.

5. Locate the point where another pair of lines from the two rulers coincide.

6. Determine the number of ocular micrometer units (X OU) and the number of stage micrometer units (Y SU) that fit between two pairs of coincident lines.

$$(X)\ (\text{OU}) = (Y)\ (\text{SU})$$

Since the smallest stage unit is equal to 10 μm (SU = 10 μm), the equation

$$(X)\ (\text{OU}) = (Y)\ (10\ \mu\text{m})$$

can be used to determine the value of each ocular unit. Suppose that 49 ocular units correspond to 51 stage units. Fitting these numbers into the equation, it can be determined that an ocular unit = 10.5 μm:

$$(49)\ (\text{OU}) = (51)\ (10\mu\text{m})$$

7. Repeat steps 1 through 6 with the high-dry objective (40×) and the oil objective (100×). Record the values for the ocular units at each magnification.

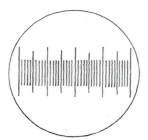

Ocular Micrometer
(O.M.)

Place the O.M. in the eyepiece. The divisions vary depending upon the magnification.

Stage Micrometer
(S.M.)

Place the S.M. on the stage. Each division equals 10 micrometers.

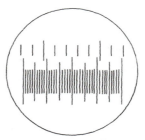

Stage Micrometer parallel to the Ocular Micrometer

Determine the value of an ocular unit (O.U.).

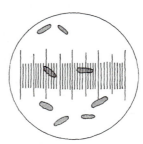

Ocular Micrometer superimposed on a slide of microorganisms

Determine the dimensions of the microorganism.

Figure 2-1. Standardization of the Ocular Micrometer and Its Use in Measuring Bacterial Cells

• Now it is possible to determine the dimensions of a microorganism at each magnification by measuring them with the calibrated ocular micrometer.

8. Remove the stage micrometer and replace it with a prepared slide. Find the specimen and line it up along the ocular micrometer. Then use the ocular micrometer like a ruler.

9. Determine the specimen's dimensions by multiplying the number of ocular micrometer units by the conversion factors you calculated (for example, 10 μm for the 10×, 2.5 μm for the 40×, and 1 μm for the 100×).

Questions

1. What are the dimensions (in μm) of some typical bacteria?

2. What are the dimensions of some typical algae?

3. What are the dimensions of some typical yeasts and filamentous fungi?

4. What are the dimensions of some typical protozoans?

5. Why is it necessary to calibrate the ocular micrometer (give values to the ocular micrometer units) with each objective and on different makes of microscopes?

6. When calibrating the ocular micrometer for use with the oil immersion lens, you find that 10 SU = 60 OU. What is the length of each ocular unit?

7. Using the microscope calibrated in question 6, you find an organism is 2 OU in diameter and 20 OU in length. What is the diameter and length of the organism in μm?

EXERCISE 3
OBSERVATION OF LIVING MICROORGANISMS

The observation of living cells can give you important information about their biology. In addition, characteristics that can be seen only in living organisms might be essential for their identification and accurate classification. For example, how microorganisms obtain their nutrients, engulf and digest food, move, and expel water from their cytoplasm can be observed only in living cells. Films have been made of algae showing how the mitochondria and chloroplasts move, divide, and fuse with each other. Such observations help to explain how these organelles propagate and exchange genetic information. In this exercise we will be concerned primarily with the type of **motility** (**vital motion** or self-directed movement) characteristic of various microorganisms.

Many different types of bacteria show no vital motion and are said to be nonmotile. However, in an aqueous environment, these nonmotile bacteria jiggle continuously and appear to be motile to the uninitiated. This motion, known as **Brownian motion** (random vibratory movement), is due to the constant and unequal bombardment of water molecules against the bacteria. The smaller the bacteria, the more pronounced is the Brownian motion. Many of the very large bacteria show little or no Brownian motion.

Different mechanisms of self-propulsion (motility) have been recognized among motile bacteria. Bacteria with **flagella** (fig. 3-1) may wiggle or dart from place to place or tumble in their aqueous environment. These organisms are said to demonstrate **flagellar motion.** Helically shaped bacteria called spirochetes have **axial fibrils** (modified flagella that wind around the cell) that form an **axial filament.** These organisms exhibit a **corkscrew motion** through their environment. Some of these bacteria also are capable of bending and so their movement is usually a combination of a corkscrew motion and a **bending motion.** Some bacteria that have neither flagella nor axial filaments show a slow gliding type of motility over moist surfaces. The snail-like movement of these gliding bacteria is known as **gliding motion.** The mechanism by which gliding bacteria propel themselves is unknown. There are numerous hypotheses, however. Some researchers believe they have discovered flagellar stubs in the wall that act like paddle wheels. A few scientists propose that the gliding bacteria give off a slime that expands when it comes into contact with water. This expansion of the slime supposedly propels the bacteria. Other scientists suspect that the gliding bacteria may be expelling hydrogen ions that induce a negative charge on the surface under the cells. When the hydrogen ions dissipate, the induced negative surface charge repels the many negative charges associated with the bacterial cell wall.

Many of the higher microorganisms (some algae and many protozoans) also demonstrate vital motion.

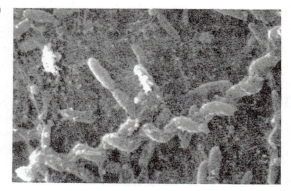

(a)

(b)

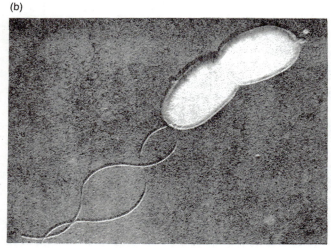

Figure 3-1. Arrangement of Bacterial Flagella. *(a)* Scanning electron micrograph of a spiral bacterium. *(b)* This transmission electron micrograph of *Pseudomonas fluorescens*, shadowed with a heavy metal, shows that it has two flagella at one pole.

They may be propelled by flagella or shorter analogous structures known as **cilia** and are said to show flagellar motion or **ciliar motion,** respectively. The flagella found in eukaryotes are structurally very different from those found in the prokaryotes (bacteria). In addition, the two types of flagella are powered by very different mechanisms. Many motile organisms like the amoeba have neither flagella nor cilia. They propel themselves by temporarily attaching to a surface and then retracting or expanding their plasma membrane in a particular direction. This type of motion is known as **amoeboid motion** and is distinct from the gliding motion seen in the bacteria. Some algae, like the diatoms, are also believed to propel themselves in much the same way as the amoeba. Since the membrane only touches the surface through a long thin opening, however, the algae appear to glide over surfaces. Consequently, their motion is known as gliding motion. Gliding motion in the higher algae looks very similar to that of bacteria but the two are believed to be powered by very different mechanisms.

A. MAKING A WET MOUNT

Wet mounts are the most common preparations used to view living microorganisms. Wet mounts are made by placing a small drop of an aqueous suspension of the specimen on a clean slide and covering it with a cover slip. Although wet mounts can be made rapidly, they dry up after five or ten minutes and so are not very useful if long-term observations are required. Hanging drop mounts are used if long-term observations are required.

Materials

Pond water, rich in gliding bacteria and spirochetes

Dental plaque or tartar rich in spirochetes

Pond water rich in algae (in particular diatoms) and protozoans

Pepper infusion held at room temperature at least one week

Bright field light microscope

Lens paper

Immersion oil

Clean slides and cover slips

Pasteur (capillary) pipets

Procedure

1. Place a small drop of the specimen on a clean slide using a Pasteur pipet.

• Many of the interesting microorganisms are associated with debris in the pond water, so be sure to include a little debris in the drop (fig. 3-2).

2. Cover the drop with a cover slip.

• Excessive water absorbs light and creates a deep field, making it difficult to see microorganisms. Remove the excess water with tissue so that the cover slip does not float down the slide when you tilt it. If there is too little water, however, the preparation will dry quickly and the organisms will stick to the slide and cover slip and show little or no movement.

3. Position the wet mount on the stage so that the specimen is over the light hole.

4. Adjust the condenser so that a minimum amount of light hits the object.

5. Revolve the nose piece so that the 10× objective is locked into position.

• The 10× and 40× objectives are used to view motile microorganisms. The 10× is used to study algae and protozoans while the 40× objective is used to view bacteria. The 100× oil immersion objective is

(a) Cover slip / Drop of culture / Glass slide

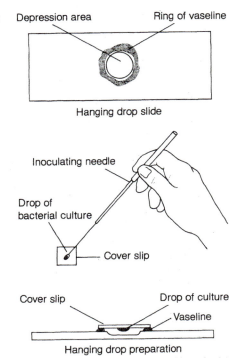

(b) Depression area / Ring of vaseline

Hanging drop slide

Inoculating needle

Drop of bacterial culture

Cover slip

Cover slip / Drop of culture / Vaseline

Hanging drop preparation

Figure 3-2. Wet Mount and Hanging Drop Mount. *(a)* The wet mount is prepared by placing a small drop of culture onto a clean glass slide and then covering the drop with a coverslip. *(b)* The hanging drop mount is prepared by placing a small drop of culture onto a coverslip and then placing the inverted coverslip onto a ring of vaseline that surrounds the depression in a depression slide.

usually not used to determine whether or not an organism is motile because of the difficulty in using oil when a cover slip is used.

6. Close the distance between the 10× lens and the coverslip using the coarse focus adjustment so that the lens is within the working distance.

7. Look through the 10× ocular lens and focus the image with the coarse focus by increasing the distance between the specimen and the objective lens.

• Use the fine focus knob to sharpen the image.

8. Scan the field by moving the slide and draw what you see.

• Label your drawings to show identifiable cellular structures. Also, record observations of any cellular movement (type and rate) or other activity.

9. Rotate the 40× objective into position and scan the field. Draw what you see.

Questions

1. Explain how the following motions are distinguished: flagellar motion, corkscrew motion, gliding motion, and Brownian motion.

2. After looking at algae and protozoans that show flagellar motion, ciliary motion, amoeboid motion, and gliding motion, explain how the motions differ. Which organisms move most rapidly?

3. Determine the velocity (μm/sec) of the different types of organisms.

4. Clearly differentiate between vital motion and Brownian motion. What causes each type of motion?

B. MAKING A HANGING DROP MOUNT

Hanging drop mounts are used instead of wet mounts when long-term observations are required (fig. 3-2). A ring of vaseline around the edge of the cover slip keeps the hanging drop from drying out.

Materials

Pond water rich in gliding bacteria and spirochetes

Dental plaque or tartar rich in spirochetes

Pond water rich in algae (in particular diatoms) and protozoans

Pepper infusion held at room temperature at least one week

Bright field light microscope

Lens paper

Immersion oil

Clean depression slide and cover slips

Vaseline

Pasteur (capillary) pipets

Procedure

1. Place four small mountains of vaseline on the edge of a clean depression slide.

• The mountains of vaseline are to support the cover slip so that the drop hanging from the cover slip does not touch the bottom of the depression. If an extended period of observation is required, the vaseline should go all around the edge of the depression. This creates a seal so that the hanging drop does not evaporate.

2. Place a small drop of the specimen on a clean cover slip.

• The drop must be large enough so that it does not evaporate before your observations are complete. But do not use too large a drop because it will vibrate and make it very difficult to see and follow organisms in the drop. Also, too large a drop may fall into or touch the depression.

3. Invert the cover slip and place it on the vaseline support so that the drop hangs into the depression but does not touch the bottom of the depression slide.

4. Position the hanging drop mount on the stage so that the specimen is over the light hole.

5. Adjust the condenser so that a minimum amount of light strikes the object.

6. Revolve the nose piece so that the 10× objective is locked into position and focus on the edge of the hanging drop. Close the distance between the lens and the cover slip using the coarse focus adjustment so that the lens is just above the specimen.

• The edge will appear as a distinct bright curved line passing across the field of view.

7. Rotate the nose piece so that the high-dry (40×) or the oil immersion (100×) objective is locked into position.

8. Before looking through the microscope, make sure you know which way to turn the fine focus adjustment

so that the distance between the specimen and the lens increases when you turn the fine focus adjustment.

• WHEN LOOKING THROUGH THE MICROSCOPE, AVOID SMASHING THE LENS INTO THE COVERSLIP AND SLIDE BY ALWAYS INCREASING THE DISTANCE BETWEEN THE SLIDE AND THE LENS.

9. Look through the ocular lens and focus the image with the fine focus adjustment by slowly increasing the distance between the specimen and the objective lens.

10. Scan the specimen slowly by moving the slide and draw what you see.

• Label your drawings and be sure to describe your findings regarding motility.

Questions

1. What are the advantages of using a hanging drop mount rather than a wet mount?

2. Why do you have to reduce the amount of light with the condenser diaphragm in order to see microorganisms in a wet mount or in a hanging drop mount?

EXERCISE 4
SPECIALIZED LIGHT MICROSCOPES

A. DARK FIELD LIGHT MICROSCOPE

A **dark field light microscope** is very similar to the bright field microscope except that the light from the light source does not shine directly into the objective lens (fig. 4-1). Light is prevented from reaching the objective lens directly by a **dark field stop** located in the condenser. The dark field stop allows a hollow cone of light, instead of a solid cone, to shine on the specimen. The only light that reaches the objective lens is light reflected from the specimen. Consequently, the field of view is dark while objects that reflect light are bright.

Dark field microscopy is commonly employed in the examination of unstained living microorganisms, microorganisms that are difficult to stain, and organisms that are too thin to be easily resolved by the bright field microscope. Dark field microscopy is frequently used to verify the presence of the spirochete that causes syphilis, *Treponema pallidum*, in exudate

from syphilitic sores (fig. 4-2). This organism is difficult to see with the bright field microscope because it is only 0.1 to 0.2 μm in diameter. It is possible to see objects just below the limit of resolution with dark field microscopes because reflected light from these objects in a dark field is sufficiently intense. If the field were not dark, it would be impossible to see such thin specimens. Although dark field microscopy makes it possible to see very thin microorganisms, it is not generally useful for visualizing internal detail within larger organisms because the more intense light reflected from their surface obscures the less intense light from internal organelles.

Materials

Sabouraud glucose broth culture of *Saccharomyces* (yeast)

Pond water

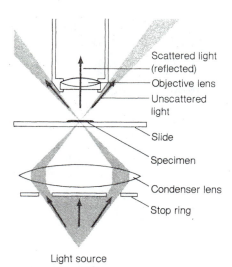

Figure 4-1. How the Dark Field Microscope Works. The dark field microscope is very similar to the bright field microscope, except that an opaque disk eliminates all of the light in the center of the beam that hits the specimen. The only light that reaches the specimen comes in at an angle. Consequently, no direct light reaches the objective lens. The only light that reaches the objective lens is light that is scattered from the specimen. Thus, the specimen and/or objects within it appear bright, while the background looks dark.

Pepper infusion

Saliva or tartar suspended in a drop of water

Dark field light microscope

Lens paper

Immersion oil

Clean slides and cover slips

Prepared slide of the spirochete *Borrelia recurrentis*

Procedure

1. Place a drop of immersion oil on the condenser lens.

2. Position a prepared slide on the stage so that the specimen is over the light opening.

3. Raise the condenser with the condenser height control so that the oil on the condenser lens comes into contact with the slide.

4. Revolve the nose piece so that the 10× objective is locked into position.

5. Close the distance between the 10× and the slide using the coarse focus adjustment so that the lens is well within the working distance of the lens (just above the specimen will do). Focus the image with the coarse adjustment knob. Use the fine focus knob to sharpen the image.

6. Rotate the nose piece to center the high-dry (40×) into position over the slide. Refocus, if necessary, with the fine adjustment knob.

7. To use the oil immersion objective lens (100×), place a small drop of immersion oil on the area that you wish to study and rotate the oil immersion lens into position. Refocus, if necessary, with the fine adjustment knob. WHEN LOOKING THROUGH THE MICROSCOPE, AVOID SMASHING THE LENS INTO THE SLIDE BY ALWAYS INCREASING THE DISTANCE BETWEEN THE SLIDE AND THE LENS.

• The 100× objective should always be immersed in oil but should not touch the specimen. Before using the 100× objective, make sure that you know which way to turn the fine focus adjustment so that the

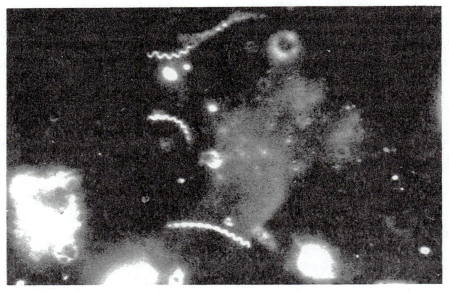

Figure 4-2. Dark Field View of *Treponema pallidum*. The helical bacteria can be seen amidst blood cells.

distance between the specimen and the lens increases when you turn the fine focus adjustment.

8. Scan the specimen slowly by moving the slide and draw what you see. Compare the image you get with a dark field light microscope with what you get with a bright field microscope.

9. Be sure to remove the immersion oil from the condenser as well as the 100× objective before storing the dark field microscope.

Questions

1. Can you see any internal detail in the bacteria, algae, or protozoan, when using the dark field microscope?

2. Are the dimensions of an organism the same when using the dark field microscope and the bright field microscope? Explain your answer.

3. Why is the field dark and the image of the specimen bright when a dark field light microscope is used to examine a specimen?

4. Sometimes, when you use the bright field light microscope, the background is darker than the microorganisms in the specimen. Why is this?

B. PHASE CONTRAST LIGHT MICROSCOPE

Organelles and other details within living cells are often impossible to see by ordinary bright-field light microscopy because they do not **absorb** (take up), **reflect** (bounce off a surface), **refract** (bend), or **diffract** (spread out) sufficient light so that they contrast with the rest of the cell. Also, many bacteria are almost invisible in wet mounts or hanging drop mounts. Cells and organelles are only visible when they absorb, reflect, refract, and diffract significantly more light than their environment. It is common practice to use dyes of various kinds in cytological studies because they produce contrast. The advantage of contrast achieved by staining, however, is offset somewhat by the distortion that results when cells are killed during fixing and staining. The **phase contrast microscope** (fig. 4-3) allows the visualization of living unstained cells and their organelles which would be invisible with the bright and dark field microscopes.

Wherever there is a **boundary** (or difference in refractive index), light is diffracted (fig. I-4). A small translucent structure diffracts proportionally more light than a larger translucent structure. The phase contrast microscope takes advantage of the small size of many translucent objects and translucent boundaries created by slight differences in the **refractive index** of the water, cytoplasm, and structures (organelles,

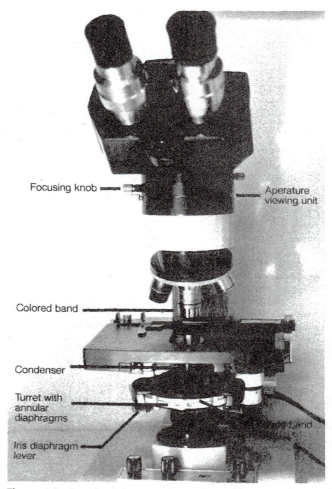

Figure 4-3. American Optical Corporation Phase Contrast Microscope. The colored band (or magnification) on the lens must match the colored band (or magnification) on the turret. The turret contains a specific annular diaphragm for each of the objective lenses. The aperature viewing unit is rotated into position in order to observe the alignment of the phase plates and the corresponding annular diaphragms.

Labels on figure: Focusing knob · Aperature viewing unit · Colored band · Condenser · Turret with annular diaphragms · Iris diaphragm lever

etc.) within a cell to increase the contrast between them.

In the late 1930s, Fredrick Zernicke developed a phase contrast microscope that uses diffracted light, which is retarded approximately one-quarter of a wave length with respect to the undiffracted light, to create contrasts (fig. 4-4). The undiffracted light is passed through a thickened portion of a **phase plate,** located just above the objective lens, that retards it one-quarter of a wavelength and also absorbs it partially. When the diffracted light and undiffracted light are recombined to produce the image, the resultant light intensity is increased because the wave fronts are in phase. Consequently, organelles, the cytoplasm, and boundaries that diffract significant light appear bright compared to the background.

The construction of the microscope described above

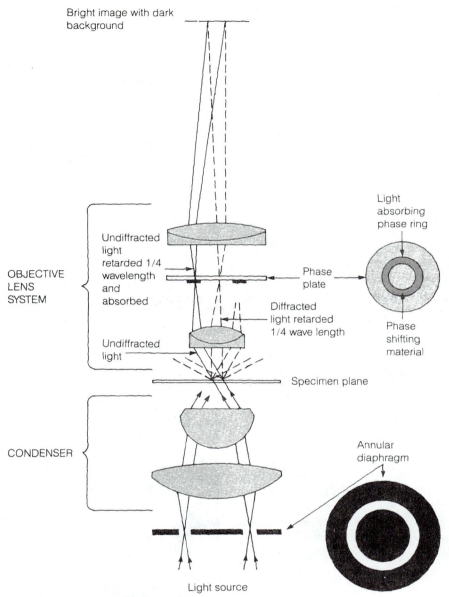

Figure 4-4. Image Formation by Phase Contrast

results in objects that are brighter than the background. This type of microscopy is known as **bright phase contrast.** If a phase plate is used in which the undiffracted light is absorbed and the diffracted light is retarded another one-quarter of a wavelength, the objects will be darker than the background because the wavefronts are out of phase by one-half of a wavelength. This type of microscopy is known as **dark phase contrast.**

Materials

Pond water

Blood

Pepper infusion

Living cultures of various higher microorganisms

Phase contrast light microscope

Clean slides and coverslips

Procedure

1. Place a slide preparation on the stage so that the specimen is over the light hole.

2. Rotate into place the $10\times$ objective.

3. Rotate into place the annular diaphragm below the

substage condenser corresponding to the 10× objective.

• The operation of the phase contrast microscope requires that the cone of light produced by the **annular diaphragm** below the condenser be centered exactly with the phase plate of the objective. Consequently, there are three different annular diaphragms to match the three phase-contrast objectives (10×, 40×, 100×) that contain the different phase plates. The substage unit contains a wheel that is rotated in order to place in position the appropriate annular diaphragm.

4. Find the image and focus with the 10× objective in place.

5. Rotate the nose piece to bring the higher power lenses into position over the specimen. Be sure to rotate the appropriate annular diaphragm into position each time you change objectives.

6. Compare the images produced by bright field, dark field, and phase contrast microscopes.

Questions

1. Define absorption, reflection, refraction, diffraction, phase shifting, constructive interference, and destructive interference.

2. What does the annular diaphragm do?

3. What happens to the phase of diffracted light in comparison to undiffracted light?

4. Explain how the phase plate works in a phase contrast microscope that produces bright objects (with respect to the background).

5. Explain how the phase plate works in a phase contrast microscope that produces dark objects (with respect to the background).

6. Of what practical use is the phase contrast microscope?

C. FLUORESCENCE MICROSCOPE

Certain substances and some microorganisms, when exposed to wavelengths of light in the ultraviolet (UV) range, emit light in the visible range. If the light is emitted only during exposure to the exciting ultraviolet light, the substance or organism is said to **fluoresce.** On the other hand, if light is given off from an object even after it is no longer stimulated with ultraviolet light, it is said to **phosphoresce.** Materials and organisms showing no natural fluorescence when exposed to UV may be made fluorescent by attaching **fluorochromes** (fluorescent dyes) to them.

A **dark field fluorescence microscope** is similar to the ordinary dark field microscope except that invisible ultraviolet light is used to excite the fluorescent portions of a specimen to release visible light (fig. 4-5). Ultraviolet light, which can cause severe retinal burns and blindness, is prevented from entering the objective lens directly by a **dark field stop** located in the **dark field condenser** (fig. 4-6). In addition, there is a UV filter (**barrier filter**) in the microscope tube to keep any reflected UV within the body tube from reaching the eye. THERE IS A DANGER OF DAMAGE TO THE EYES FROM EXPOSURE TO DIRECT AND REFLECTED LIGHT THAT HITS THE SPECIMEN. CONSEQUENTLY, YOU SHOULD AVOID LOOKING AT THE SPECIMEN WITHOUT EYE PROTECTION. When using the fluorescent microscope always wear protective glasses while focusing because this often involves looking directly at the specimen.

The dark field stop allows a hollow cone of ultraviolet light, instead of a solid cone, to shine on the specimen from below the stage. The only light that reaches the objective lens is visible **fluorescent light** from the specimen and reflected invisible ultraviolet light. Consequently, the field of view is dark while objects that fluoresce are bright and colored according to the wavelengths of light characteristically emitted by the fluorescent substance.

Fluorescent microscopy is commonly employed in the rapid detection and identification of many disease-causing organisms and in locating specific cellular constituents. A specimen can be quickly

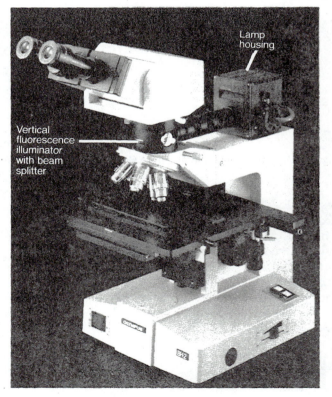

Figure 4-5. Fluorescence Microscope

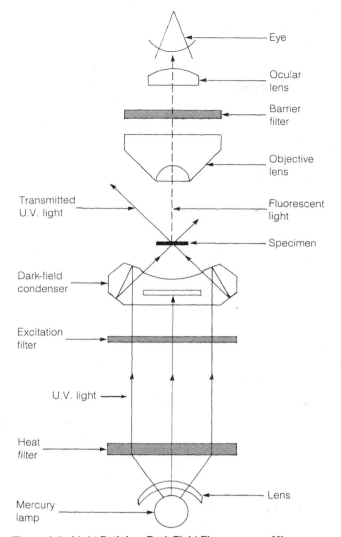

Labels on figure:
- Eye
- Ocular lens
- Barrier filter
- Objective lens
- Transmitted U.V. light
- Fluorescent light
- Specimen
- Dark-field condenser
- Excitation filter
- U.V. light
- Heat filter
- Lens
- Mercury lamp

Figure 4-6. Light Path in a Dark Field Fluorescence Microscope

screened for a particular organism by staining it with a fluorescent dye that binds specifically to the organism of interest. Only the stained organisms are visible when the specimen is viewed in dark field fluorescence microscopy and so they are rapidly discovered. Fluorescence microscopy is being used in many areas of biology. For example, fluorescent stains are being used to identify and locate various proteins of the cytoskeleton in higher cells.

An **epifluorescence microscope** differs somewhat from the dark field fluorescence microscope in that the ultraviolet light shines from above onto the specimen (fig. 4-7) and a dark field stop is not used in the condenser.

Materials

Pond water with different types of algae

Fluorescence microscope

Lens paper

Low-fluorescing immersion oil

Clean slides and cover slips

Protective glasses that filter UV light

Prepared slides of known bacteria stained with a fluorescent dye or fluorescent antibody

Procedure

1. Turn on the UV light source 30 minutes before you intend to use the microscope.

• A mercury vapor arc lamp produces the ultraviolet light in the fluorescent microscope. It takes approximately 30 minutes for the lamp to produce the optimum of invisible ultraviolet light. The mercury vapor arc lamps develop high pressures and because of this should be considered dangerous. Never inspect the lamp while it is hot and never use the microscope near sinks where water might splash on the hot lamp. The lamps occasionally explode! NEVER LOOK AT THE LIGHT FROM A MERCURY VAPOR ARC LAMP WITHOUT SPECIAL GLASSES. RETINAL BURNS AND BLINDNESS CAN OCCUR. Make sure your excitation filter and barrier filter are matched for the type of fluorescence expected and are in place.

2. Place a drop of oil on the lens of the condenser.

• In order to achieve optimum illumination, it is recommended that oil be used between the condenser and the slide. Ordinary immersion oil cannot be used because it fluoresces. A special low-fluorescing immersion oil should be used. Do not use oil on the condenser of the epifluorescence microscope.

3. Position the slide on the stage so that the specimen is over the light opening on the stage. Raise the condenser so that the oil fuses with the bottom of the slide.

4. After warming the mercury vapor arc lamp, turn on the regular tungsten filament light source and switch to this light source in order to focus on the specimen.

5. Use the 10×, 40×, or 100× objective to find and focus the specimen. Start with the low power objective.

• When using the 100× objective, place a drop of low-fluorescing immersion oil on the specimen.

6. After finding and focusing your specimen, switch to the mercury vapor arc.

7. Compare what you see in the bright field microscope with what you see in the fluorescent microscope. Can you explain any differences?

8. Be sure to clean the immersion oil off of the condenser lens and oil immersion lens before storing your microscope.

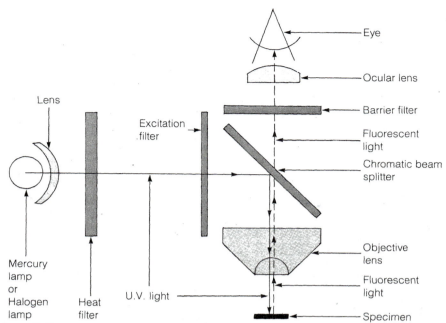

Figure 4-7. Light Path in an Epifluorescence Microscope

Questions

1. If an organism is not self-fluorescing, how can it be made visible in the fluorescence microscope?

2. What kind of light is used to make objects fluoresce?

3. What dangers are associated with the use of fluorescence microscopes?

4. Can you see any detail within the fluorescing algae?

5. Do any of the bacteria show autofluorescence (fluoresce without the use of fluorochromes)?

6. How do the following structures help prevent UV light from reaching your eyes?

   a. dark field condenser

   b. barrier filter

7. Is the ultraviolet light hitting the specimen and escaping from the objective lens dangerous to your eyes or skin? Why?

8. What is the difference between an epifluorescence microscope and a dark field fluorescence microscope?

PART II
CULTIVATION OF
MICROORGANISMS

The study of microorganisms in the laboratory is an important aspect of studying microbial life. Laboratory studies of microorganisms allow microbiologists to understand better the role microbes play in disease processes, spoilage of foods, food and antibiotic production, and detoxification of pollutants. Additionally, much of what we have learned about how genes work and about metabolic processes have come from studies of microorganisms cultivated in the laboratory. The laboratory culture of microorganisms in pure form permits scientists to determine the reproductive and physiological characteristics of microorganisms, and hence to determine what role(s), if any, each microorganism has on biological or industrial processes. Also, microorganisms are isolated from nature and studied in the laboratory so that we can understand their roles in natural environments. Many businesses, such as the petroleum, agricultural, and food industries, capitalize on the knowledge obtained from laboratory studies of microorganisms to make their products cheaper or better by involving microorganisms in the process. The techniques for growing microorganisms have numerous applications. For example, in the hospital laboratory these techniques are routinely used for isolating and identifying microorganisms suspected of causing disease.

In this section you will be introduced to the basic techniques employed in microbiological laboratories to prepare culture media, to isolate microorganisms in pure culture from various types of specimens, and to culture them in the laboratory. You will also be introduced to some of the sources of potential contamination of your cultures.

EXERCISE 5
PREPARATION OF CULTURE MEDIA

The cultivation of microorganisms in the laboratory requires that the needed nutrients and suitable environmental conditions be provided. Nutrients are those raw materials that are used to sustain living organisms, promote growth, replace cellular constituents, and provide energy for metabolic reactions and cell movement. In the laboratory, nutrients are supplied to microorganisms in culture media. A **culture medium** is an aqueous solution of the nutrients required by microorganisms. It generally contains water, an energy source, and nutritionally suitable sources of carbon, nitrogen, sulfur, phosphorus, oxygen, hydrogen, and various metal ions (trace elements). To these basic constituents, organic growth factors (e.g., amino acids, vitamins, or nucleosides) and other ingredients may be added to the medium in order to grow the desired microorganisms.

The culture medium may be in a liquid or in a solid gel form. Solid media are usually made by the addition of a polysaccharide called **agar** derived from some red algae. Agar is useful as a jelling agent for several reasons: It is not readily degraded by microbes; it remains jelled over a wide range of temperatures (approximately 1° to 95°C), thus allowing the cultivation of microorganisms with a wide variety of different temperature requirements; and it liquefies at the boiling temperature of water but does not jell until it cools to approximately 42°C. This latter property allows microbiologists to make suspensions of microorganisms in liquefied agar at around 45°C without killing them. In order to prepare agar media, a **broth** or a liquid culture medium is mixed with 1.5 to 2% (w/v) agar and brought to a boil to dissolve the agar.

When the culture medium has been prepared, it is not yet ready for use, since microorganisms commonly associated with the nutrients, water, glassware, media-maker, and air are likely to have contaminated it. It is necessary to sterilize the medium before it is used. If a nonsterile medium is left standing for several days, the contaminating microbes are virtually certain to find the nutrients suitable for growth and consume them. A microbiologist using this medium could not be sure which microbe caused the noted changes and hence could not make any conclusions. The process of **sterilization** is intended to *remove or kill all living microorganisms* in the culture medium. After sterilization, the medium can be used.

Molten agar media in tubes can be placed in a tilted position and when the agar jells, it will solidify at a slant, which provides a large surface area on which to grow microorganisms. Additionally, agar-containing media can be poured into special covered dishes called **Petri dishes,** which provide an even larger surface area on which to grow microorganisms. Sterilized media must be capped or plugged with sterile closures in order to avoid contamination.

Laboratory sterilization procedures generally are carried out using autoclaves (fig. 5-1). Autoclaves sterilize by means of steam under pressure. Routine laboratory sterilizations are carried out for 15 to 20 minutes at a pressure of 15 pounds per square inch above normal atmospheric pressure, which allows the steam temperature to reach 121°C. Heat-sensitive solutions are commonly sterilized using special filters that have pores with diameters of 0.45 μm. These filters prevent microorganisms from passing into the collecting chamber.

The nutrients used to make a culture medium largely determine which types of microorganisms can grow in it. The careful selection and preparation of ingredients for use in specific culture media provides a microbiologist with powerful tools for isolating microorganisms in pure form so that their distinctive nutritional, physiologic, and genetic properties can be studied.

Two categories of culture media based upon their formulation are routinely used in the laboratory. These categories are chemically-defined and chemically-complex (undefined) media. A **chemically-defined** medium is formulated so that the concentration and chemical nature of each added compound is known. The ingredients used to prepare chemically-defined media are highly purified ("reagent" grade) inorganic salts and simple organic compounds. Chemically-defined media are often necessary in the study of the physiology and genetics of microorganisms. However, if a defined medium calls for many ingredients, it can be tedious to prepare. Moreover, reagent grade chemicals are often expensive. **Chemically-complex** culture media are prepared using natural products of unknown chemical composition. Common sources of complex nutrients are beef extracts, partially digested proteins called **peptones,** yeast extracts, and boiled vegetable and animal materials (infusions). Some culture media are further supplemented with blood, serum, vitamins, amino acids, and nucleosides in order to support the growth of special groups of microorganisms. Obviously, the exact concentration and chem-

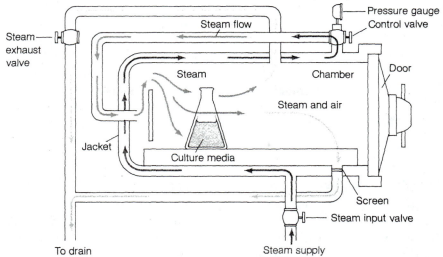

Figure 5-1. Autoclave. *Autoclaves are used to sterilize heat-stable media and equipment. Steam initially enters the steam jacket around the autoclave chamber and heats up the autoclave. An object is placed in the chamber, the door is shut securely, and steam from the jacket is allowed to enter the chamber. Air is forced out of the chamber by the incoming steam until only pure steam is being forced out. Then the air exit is closed by the high temperature of the steam,* and the steam pressure builds to 15 pounds per square inch. Eventually, the temperature rises to 121°C. When the sterilization is completed, the steam exhaust valve opens and the steam flows out of the chamber. When liquid media are being sterilized, the steam must be cooled and released slowly in order to avoid the evaporation and boiling over of the liquid media.

ical composition of each nutrient in complex media is not known. Complex media are very commonly used in the microbiology laboratory because they are easy to prepare from relatively inexpensive ingredients and can support the growth of a wide variety of micro-organisms.

Materials for A, B, C, D, and E

Erlenmeyer flasks, 250-500 ml, with closures

Graduated cylinders, 100-250 ml

Distilled water

Pipets, 1 ml and 10 ml

Weighing paper

Glucose

Ammonium chloride (NH_4Cl)

Potassium phosphate (K_2HPO_4)

Magnesium sulfate ($MgSO_4 \cdot 7H_2O$)

Ferric chloride ($FeCl_3 \cdot 6H_2O$)

Sodium chloride (NaCl)

Agar

Beef extract

Peptone

Yeast extract

Balance

Heat-proof gloves or tongs

Stirring rod

Hot plate or bunsen burner with ringstand

Test tubes with closures

Sterile Petri dishes

Autoclave

Water bath set at 50°C

A. PROCEDURE FOR PREPARING A GLUCOSE-MINIMAL SALTS BROTH (A CHEMICALLY-DEFINED MEDIUM)

1. Prepare 100 ml of glucose-mineral salts broth using the recipe outlined in table 5-1.

• To a 250 ml Erlenmeyer flask add about 75 ml of distilled water. Add each ingredient to the water in the order listed and stir after each addition until the ingredient is completely dissolved. Add the remaining 25 ml of water to wash the inner wall of the flask.

2. Dispense 10 ml of glucose-mineral salts broth into each of 10 test tubes using a 10 ml pipet and then add a closure to the tubes.

TABLE 5-1.
A CHEMICALLY-DEFINED MEDIUM

INGREDIENT	QUANTITY
Glucose	0.2 g
Potassium phosphate K_2HPO_4	0.1 g
Ammonium chloride NH_4Cl	0.05 g
Magnesium sulfate $MgSO_4 \cdot 7H_2O$	0.02 g
Ferric chloride $FeCl_2 \cdot 6H_2O$	0.0005 g
Distilled water	100 ml

- If the closure is a screw cap, leave the cap about a quarter turn loose; a tightly capped tube may explode in the autoclave during sterilization. Place the tubes of culture media in a test tube rack or basket that has been labeled with your name, date, and the name of the culture medium. Place the test tube rack or basket in the autoclave. Follow your instructor's directions to operate the autoclave properly.

B. PROCEDURE FOR PREPARING A NUTRIENT BROTH OR AGAR (A CHEMICALLY-COMPLEX MEDIA)

1. Prepare 200 ml of nutrient broth by following the recipe outlined in table 5-2.
- To a 250 ml Erlenmeyer flask add about 150 ml of distilled water. With stirring, add the three ingredients individually and mix after each addition. Add the remaining 50 ml of water, rinsing off the walls of the flask, and mix well.

2. Dispense 10 ml of the nutrient broth into each of ten clean test tubes and then plug or close with a cap.

3. To prepare nutrient agar, add 1.5 g of agar to the remaining 100 ml of nutrient broth.

4. Heat the contents of the flask and gradually bring to a boil.
- Stir the agar frequently to avoid scorching. Do not allow the agar to boil over. To avoid this remove the flask from the burner or hot plate as soon as it begins to boil. Heat the agar until it is completely melted. Alternatively, you may cover the mouth of the flask with aluminum foil and autoclave the contents before dissolving the agar beforehand. The agar will dissolve in the nutrient broth during the autoclaving process.

NOTE: If there is a double boiler available you may heat the agar-containing culture medium using the double boiler instead of heating it with a burner or a hot plate.

5. Cover the flask with aluminum foil, label properly, and place in the autoclave.

C. PROCEDURE FOR AUTOCLAVING THE MEDIUM

1. The instructor will demonstrate the operation of the autoclave.

TABLE 5-2.
A CHEMICALLY-COMPLEX MEDIUM

INGREDIENT	QUANTITY
Peptone	1.0 g
Sodium chloride (NaCl)	1.6 g
Beef extract	0.6 g
Distilled water	200 ml

- The following procedure is commonly used to operate an automatic autoclave. If your laboratory autoclave is a manual one or a pressure cooker, follow the instructor's directions carefully because the autoclave operates with steam under pressure, which may cause an explosion.

2. Load the autoclave with the freshly prepared culture media.
- CAUTION: The autoclave chamber and door may be hot, so avoid touching the sides of the autoclave or its door.

3. Close and lock the autoclave door.

4. Set the autoclave time for 15 to 20 minutes and select a slow rate of steam exhaust.
- Liquids inside the autoclave will boil over if the pressure drops rapidly at the end of the autoclaving cycle.

5. Adjust the autoclave to a temperature of 121°C.

6. Start the autoclave cycle by pushing the start button or twisting the operation knob to point to the start position.
- At this point, the sterilization cycle begins. Steam first fills the autoclave jacket (fig. 5-1) until 15 pounds of pressure is reached. After this, steam starts flowing into the chamber and displacing the air. Eventually, the temperature in the chamber reaches 121°C and the sterilization cycle begins, continuing for 15 to 20 minutes. After the period of sterilization is completed, the steam in the chamber is slowly exhausted until the chamber pressure is equal to atmospheric pressure again.

7. Using heat-proof gloves, open the autoclave door carefully ONLY AFTER THE STEAM PRESSURE IN THE CHAMBER IS ZERO. Remove the contents and place them on a cart or bench.
- BE CAREFUL! THE AUTOCLAVE AND ITS CONTENTS ARE EXTREMELY HOT AND CAN CAUSE SEVERE BURNS. Avoid shaking the contents of the flask when just removed from the autoclave because superheated liquids will boil over when shaken.

D. PROCEDURE FOR USING THE CULTURE MEDIUM

1. To remove dust and possible contaminants, clean lab benches with a sponge dipped in a disinfectant solution.

2. Remove the tubes of glucose-mineral salts broth and the flasks of nutrient agar from the autoclave when your instructor tells you it is time to do so.
- Handle these materials with heat-proof gloves or tongs because they are extremely hot.

3. Allow the tubes of glucose-salts broth to cool to room temperature on the lab bench.

4. Use the broth tubes in exercise 6 or as suggested by your instructor.

E. PROCEDURE FOR PREPARING AGAR PLATES

1. Using heat-proof gloves, place the flasks of nutrient agar upright in a water bath set at 50°C. Allow the medium in the flasks to cool to 50°C.

• This usually requires 10 to 15 minutes. Cooling is necessary to avoid excessive condensation of moisture on the lid of the Petri dish after the agar is poured. It also facilitates handling of the flasks. Make sure that the water level in the bath is as high as the culture medium in the flask. Also make sure that there are baskets or racks in the water bath to prevent the flasks from tipping over and into the water bath. Avoid contaminating the lip of the flask with your hands or with water from the water bath.

2. Label five sterile Petri dishes (glass or plastic that have been presterilized) with your name, the date, and the type of culture medium.

• Labels should be placed on the bottom of the Petri dishes rather than on the lids.

3. Remove the cooled flasks from the water bath and pour approximately 20 ml of the liquified nutrient agar into each of the five plates.

• Closely follow the procedure outlined in figure 5-2. It is important that the lip of the flask be flamed before pouring in order to avoid contamination. Be careful not to splash. Make sure that the entire bottom surface of the plate is covered with agar by swirling the agar in the Petri dish bottom gently. Replace the lid as soon as the agar is poured.

4. Allow nutrient agar plates to solidify on a cool, level surface.

5. Use plates in exercise 6 or as indicated by your instructor.

Questions

1. What is a culture medium?

2. Differentiate between a chemically-defined and a chemically-complex medium.

3. Why are culture media sterilized before they are used?

4. What is an autoclave?

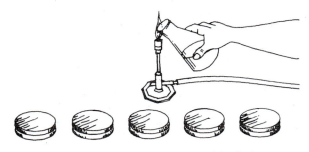

1. Remove the closure and flame the mouth of the flask.

2. Remove the cover from the Petri plate and pour about 20 ml of the cooled nutrient agar into the plate bottom.

3. If bubbles appear on the surface, flame the surface of the nutrient agar.

Figure 5-2. Pouring Agar Plates

5. Why is an agar medium cooled before it is poured into plates?

6. Answer the following questions:

    a. what is the melting point of agar?

    b. what is the solidifying temperature of agar?

    c. where does agar come from?

    d. why do most microbes fail to use agar as a carbon source?

7. Where do laboratories purchase purified chemicals and complex bacteriological media?

EXERCISE 6
MICROORGANISMS IN
THE LABORATORY ENVIRONMENT

Microorganisms are widely distributed in the laboratory environment. They are found in the air, on the work bench, on glassware, and on the clothing, hair, and hands of people working in the laboratory. All these represent potential sources of microorganisms that may contaminate cultures. The contamination occurs when these unwanted microorganisms enter a culture medium in which a pure culture of microorganism is growing.

Microorganisms growing on agar media form masses of cells called **colonies.** Colony morphology may be used as an aid to the identification of microorganisms. Although colony morphology cannot be used as the sole identifying criterion, it is used to recognize many common types of microorganisms.

Six parameters are normally used to describe microbial colonies. These are: overall colony appearance, colony margin (edge), colony elevation, colony size, colony pigmentation, and colony consistency. Figure 6-1 illustrates some of these parameters.

Materials

Tubes containing approximately 20 ml of nutrient agar (NA)

Sterile Petri dishes

Tubes containing 1 to 2 ml of sterile water

Cotton swabs

Ruler graduated in mm

Boiling water bath

Water bath set at 50°C

Incubator set at 35°C

Procedure

First Period

1. Heat three tubes of sterile nutrient agar in a boiling water bath until the agar is completely melted.
- Boiling for 10 minutes is usually sufficient to melt 10 to 20 ml of tubed culture media containing 1.5% agar. It may be necessary to heat for a longer period of time if more than 20 ml of medium are used. It is important that the nutrient agar be completely melted before pouring into plates, otherwise there will be lumps.

2. Cool the melted medium in a water bath set at 50°C for 5 to 10 minutes.
- Cooling the medium before pouring into plates reduces the amount of condensation developing on the lid of the Petri dish, and hence reduces the chances of disruption of colonies by drops of condensate falling onto the agar surface.

3. Label three sterile Petri dishes with your name, laboratory section, type of culture medium, temperature of incubation and one of the following: (a) air; (b) countertop; and (c) hair, fingers, or cough.

4. Pour the contents of one tube (or approximately 20 ml) of nutrient agar into each of the three Petri dishes.
- The proper procedure for pouring plates is illustrated in figure 5-2. Gently rock the plates so that the agar is dispersed over the entire bottom surface of the plate. If bubbles form on the surface of the agar, you may remove them by gently flaming the surface of the agar (while it is still liquid) for a few seconds, using a bunsen burner. Be careful not to burn yourself or to heat the edge of the plate. Plastic Petri dishes promptly melt when heated.

5. Allow the nutrient agar to solidify on a cool, level surface for about 10 minutes.

6. Remove the lid of the plate labeled "air" and place the bottom part containing the agar medium somewhere on the work area for the duration of this exercise (not less than 15 minutes). At the end of the period, replace the lid on the exposed agar plate and incubate as indicated in step 9.

7. Soak a cotton swab in a tube containing sterile water. Squeeze excess water from the cotton swab by pressing the swab against the walls of the tube. Swab 1 square foot of the workbench surface and then spread the contents of the swab over the entire surface of the agar plate labeled "countertop." Replace the lid of the Petri dish and incubate as indicated in step 9.

8. Cough, shake your hair, or smear your fingers on the remaining nutrient agar plate. Replace the lid and incubate as indicated in step 9.

9. Place the plates in an inverted position on the in-

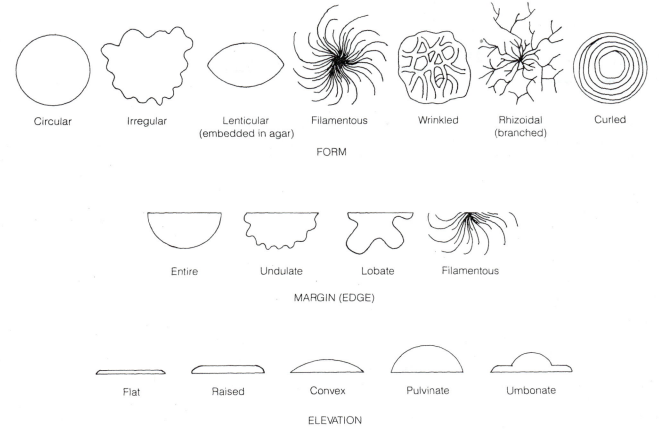

Circular Irregular Lenticular (embedded in agar) Filamentous Wrinkled Rhizoidal (branched) Curled

FORM

Entire Undulate Lobate Filamentous

MARGIN (EDGE)

Flat Raised Convex Pulvinate Umbonate

ELEVATION

Figure 6-1. Representative Colony Characteristics

cubator shelf and incubate the plates at 35°C until next laboratory period.

• Incubating plates in an inverted position reduces the amount of condensation forming on the inside surface of the Petri dish lid.

Second Period

1. Examine the colonies appearing on the plates of nutrient agar. Describe at least four different colony types using figure 6-1 as a reference.

• You may gram stain (see exercise 9) representative samples of the colonies and note the gram staining reaction and cell shape and arrangement.

2. Record your observations.

3. Discard all your cultures when finished with them in the manner prescribed by your instructor.

Questions

1. What can you conclude from the results obtained?

2. List five possible sources of contamination of cultures.

3. Explain two means of preventing colony disruption by droplets of condensate water.

4. How do bacterial colonies differ from fungal colonies?

EXERCISE 7
PURE CULTURE TECHNIQUES

In order to study and characterize a microorganism, it is first necessary to isolate it from all others and maintain it in pure form. This is usually accomplished by spreading a sample containing microorganisms on a solid medium so that a single cell occupies an isolated portion of the agar surface. The repeated multiplication of the single cell will eventually produce a colony that is made up of similar cells. Cultures consisting of a single species (or strain) are considered to be **pure cultures.**

One technique for isolating pure cultures from mixtures involves the **streak plate technique** (fig. 7-1). The streaking technique is performed by spreading a small amount of culture with an **inoculating loop** over an agar surface. As the microorganisms are spread over the surface, the concentration of microorganisms on the inoculating loop decreases. Eventually, a single microbial cell will be deposited on an area of the agar medium, multiply, and form a colony. This colony, derived from a single cell, can then be picked with a bacteriological loop (or a needle) and subcultured onto fresh culture medium. In this way, a pure culture of the isolated colony can be obtained and maintained for study.

The study of pure cultures often requires the transfer of cultures to fresh media without introducing foreign microorganisms. This transfer of microorganisms requires **aseptic** (without infection) **techniques.** These techniques are the procedures necessary to maintain the purity of a given culture during the many manipulations performed in the study of microorganisms.

Microorganisms may be propagated in either a liquid (broth) or on a solid medium. Broth cultures generally are used when fresh organisms or large quantities of organisms are required. In addition, microorganisms in broth may exhibit various patterns of growth (fig. 7-2). **Turbidity** or cloudiness results when the microorganisms reproduce throughout the broth. Large numbers (more than 5 million cells/ml) of microorganisms are required to make the cultures turbid. **Sediments** result when microbial growth accumulates at or settles to the bottom of the tube. **Pellicles** result when microorganisms reproduce on the surface of the broth, forming a film over the entire surface of the liquid. Various combinations of these three basic patterns can also occur.

The way microorganisms grow in broth cultures sometimes reflects their cellular and metabolic characteristics. For example, obligately aerobic organisms multiply only on the uppermost regions of the broth, where there are high levels of oxygen. These types of organisms sometimes form pellicles on broth. Other bacteria that form filaments, or that adhere to each other by their capsules or pili, also may form surface films on broth cultures. Motile bacteria, on the other hand, may swim throughout the broth culture and make the broth turbid. Nonmotile bacteria may settle to the bottom and form sediments. These growth patterns are sometimes characteristics of certain groups of microorganisms and are used as a criterion in identification.

Microorganisms also exhibit characteristic patterns of growth on solid media such as slants and plates. Some of these growth patterns are illustrated in figures 7-3 and 6-1, respectively. Agar slants usually are used to maintain stock cultures while agar plates generally are used to purify and/or verify the purity of a culture.

Materials for A, B, and C

Mixed culture of *Escherichia coli, Bacillus subtilis,* and *Staphylococcus epidermidis*

Pure cultures of *Escherichia coli, Bacillus subtilis,* and *Staphylococcus epidermidis*

Tubes containing approximately 20 ml of trypticase soy agar (TSA) or nutrient agar

Tubes containing approximately 10 ml of trypticase soy broth or nutrient broth

Slants of trypticase soy agar or nutrient agar

Sterile Petri dishes

Inoculating loop

Test tube rack

Bunsen burner

Boiling water bath

Water bath set at 50°C

Incubators

A. PROCEDURE FOR MAKING A STREAK PLATE

First Period

1. Pour four plates of trypticase soy agar (or nutrient agar) and label them with your name, date, and type of culture medium.

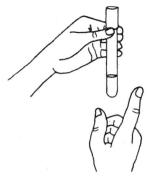

(a) Shake the tube to suspend the organisms as demonstrated by your instructor.

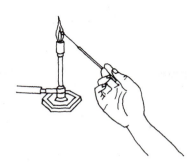

(b) Flame the entire loop and wire until it becomes red-hot. Begin heating at the middle of the wire and proceed slowly to the loop.

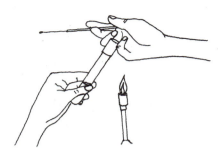

(c) Remove the plug or cap (closure), then flame the mouth of the tube. Don't contaminate the closure by placing it on the table.

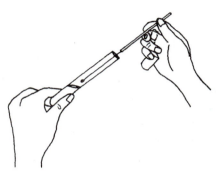

(d) After allowing the loop to cool, remove a loopful of organisms. Do not touch the sides of the tube.

(e) Flame the mouth of the tube again.

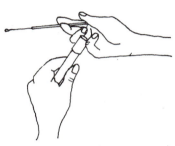

(f) Return the closure to the tube. Place the tube in a rack.

(g) Streak the plate as shown on the right or as illustrated by the instructor.

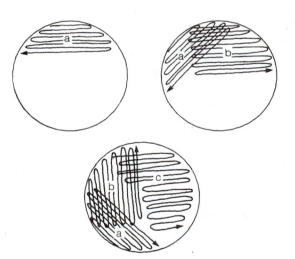

Streak plate procedure for pure culture isolation of bacteria. The direction of streaking is indicated by the arrows. Between each section, the loop is sterilized and reinoculated with a fraction of the bacteria by going back across part of the previous section. The plate is also rotated so the direction of the streak is always the same.

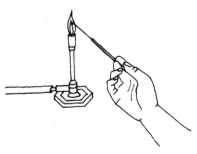

(h) Flame the loop before placing it down.

Figure 7-1. Streak Plate Technique

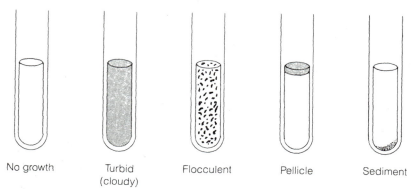

No growth — Turbid (cloudy) — Flocculent — Pellicle — Sediment

Figure 7-2. Growth Patterns of Bacteria Growing in Broth Cultures

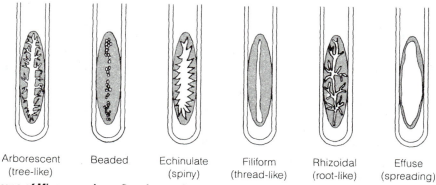

Arborescent (tree-like) — Beaded — Echinulate (spiny) — Filiform (thread-like) — Rhizoidal (root-like) — Effuse (spreading)

Figure 7-3. Growth Patterns of Microorganisms Growing on Agar Slants

• See figure 5-2 for details of the procedure for pouring agar plates. Allow the plates to jell for about 20 minutes on a cool, level surface. It is necessary that the agar be well-hardened to reduce the chances of gouging the surface of the agar with the inoculating loop during streaking. To budget your time better, you may let the covered plates stand while conducting other exercises and then return to this one.

2. Sterilize the inoculating loop in the flame of a bunsen burner.

• Heat the entire wire loop until it glows red. Let the inoculating loop cool for a few seconds before touching the culture. Inserting a red hot loop into a culture will cause spattering and create potentially dangerous aerosols of microorganisms, as well as kill those microorganisms in close proximity to the wire.

3. Remove the top or plug from the tube containing the **mixed culture** and flame the rim of the tube as illustrated in figure 7-1.

4. Pick out a loopful of the broth culture (or a small portion of a colony) with the sterile inoculating loop, replace the closure on the culture tube, place the closed tube in the test tube rack, and spread the culture back and forth over an area of the plate.

5. Flame the inoculating loop and let it cool for a few seconds. Streak the microorganisms over a second portion of the plate, passing the loop a few times over the initial streak area as a source of inoculum (fig. 7-1g).

6. Flame the loop again and streak a third section of the plate as indicated.

7. Streak the three other plates in a similar fashion; one with *E. coli*, one with *B. subtilis*, and one with *S. epidermidis*.

8. Incubate the plates in an inverted position at 35°C until the next laboratory period.

Second Period

1. Examine the plates for the presence of isolated colonies of *Escherichia coli*, *Bacillus subtilis*, and *Staphylococcus epidermidis*.

2. Record your results on the report sheet. Be careful to record the colony characteristics for each organism.

3. Discard all cultures in a safe manner as indicated by your instructor.

Questions

1. Why was the loop heated each time streaking direction was changed?

2. Why was the agar allowed to harden before streaking?

3. Which portion of the plate had the largest number of isolated colonies? Explain your observations briefly.

4. Cite two things that can be done in order to ascertain the purity of a colony.

5. Why is it dangerous to insert the red hot loop into a culture of disease-causing organisms?

6. Would you do anything different if the source of microorganisms were a colony rather than a broth culture? Explain.

7. Cite two things that may occur as a direct consequence of streaking a semisolid rather than a solid agar plate.

B. PROCEDURE FOR SUBCULTURING A PURE CULTURE INTO BROTH

First Period

1. Label three tubes of trypticase soy broth (or nutrient broth) with your name, date, culture medium used, incubation temperature, and one of the following: (a) *Escherichia coli*, (b) *Bacillus subtilis*, or (c) *Staphylococcus epidermidis*.

2. Sterilize the inoculating loop in the flame of a bunsen burner. Allow the loop to cool for a few seconds before inserting it into a culture.

3. Remove the closure from the culture tube of *Escherichia coli*. Flame the lip of the tube, then remove a loopful of the culture. Flame the lip of the tube again, return the closure, and place the tube in the test tube rack.

4. Remove the closure of the sterile broth labeled *Escherichia coli*, flame the lip of the tube, and introduce the loopful of culture into the sterile broth as illustrated in figure 7-4. Flame the lip of the culture tube once again and replace the tube closure.

5. Repeat steps 1 through 4 using *Bacillus subtilis* and *Staphylococcus epidermidis*.

6. Incubate the broth cultures at 35°C until the next laboratory period.

Second Period

1. Examine the broth cultures for turbidity, pellicles, and sediments. Determine the type of growth that took place in each of the cultures.

2. Record your observations on the report sheet provided or in your laboratory notebook.

3. Discard all cultures in a safe manner as indicated by your instructor.

Questions

1. Give two instances in which broth cultures of microorganisms are desirable.

2. Give two instances in which broth cultures are undesirable.

3. How could you use broth cultures as a criterion in the identification of bacteria? Explain briefly.

4. Could anaerobic bacteria form pellicles if they are cultivated under anaerobic conditions? Could facultative anaerobes? Explain briefly.

5. Can you use broth cultures to obtain pure cultures? Explain briefly.

C. PROCEDURE FOR SUBCULTURING A PURE CULTURE ONTO AN AGAR SLANT

First Period

1. Label three slants of trypticase soy agar (or nutrient agar) with your name, date, culture medium used, and the name of the bacterium to be cultured.

2. Sterilize the inoculating loop in the flame of a bunsen burner.

• Allow the loop to cool for a few seconds before inserting it into a culture.

3. Inoculate the slants of TSA with their corresponding culture by making a wavy line from the bottom of the slant upward using the same transfering techniques practiced in part B of this exercise, or by the technique illustrated in figure 7-5.

4. Incubate the slant cultures at 35°C until the next laboratory period.

Second Period

1. Examine the slant cultures for evidence of growth. Determine the type of growth that took place for each of the cultures.

• Describe the growth pattern using figure 7-3 as an aid.

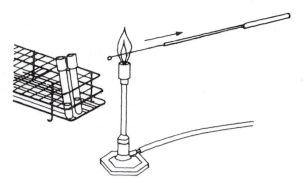

1. Flame the loop. Begin flaming at the center of the wire and proceed toward the loop.

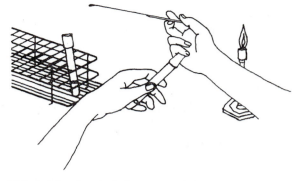

2. While holding the sterile loop, remove the closure from the tube as demonstrated by your instructor, or as illustrated.

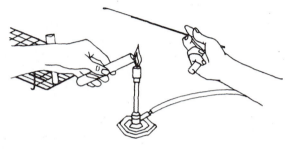

3. Heat the mouth of the tube by briefly holding it in the flame of a Bunsen burner.

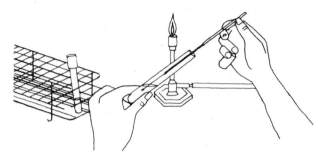

4. Remove a loopful of culture, reheat the mouth of the tube, and replace the plug.

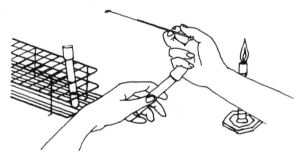

5. Pick up the sterile broth tube from test tube rack, remove closure, and flame the tip.

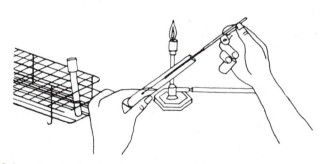

6. Introduce the loopful of bacterial culture to the sterile broth as shown. Shake the loop gently a few times to ensure a transfer of culture to the broth. Reheat the mouth of the tube, and replace the plug.

7-5. Similar techniques may be used in subculturing from slants to broth, from broth to slants, and from slants to slants.

Figure 7-4. Procedure for Subculturing Organisms Grown in Tubes. The instructor may wish to show the student the correct method for holding two tubes simultaneously as illustrated in figure

2. Record your observations on the report sheet.

3. Discard all cultures in a safe manner as indicated by your instructor.

Questions

1. Give an instance in which slant cultures of microorganisms are desirable over broth cultures. Also cite a situation in which slants are more suitable than plate cultures.

2. Give two instances in which slant cultures are undesirable.

3. How could you use growth characteristics on agar slants as a criterion in the identification of bacteria? Explain briefly.

4. Could you streak a slant to obtain a pure culture of an anaerobe? Is this a desirable procedure? Explain briefly.

5. Can you use slants to obtain pure cultures? Explain briefly.

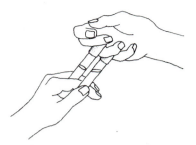

1. In one hand, hold the tubes as illustrated, or as shown by your instructor. With the other hand, flame the inoculating loop.

2. Remove closures as illustrated. Flame the mouths of the tubes.

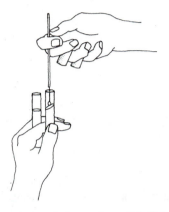

3. Transfer a loopful of culture from one tube to the other. Flame the mouths of the tubes again and replace the closures. Flame the inoculating loop before setting it down.

Figure 7-5. Procedure for Transfering Cultures from One Tube to Another

D. PROCEDURE FOR ASEPTIC TRANSFERS USING PIPETS

Materials

Flask with sterile water

Tubes of nutrient broth

Propipets

1 ml and 10 ml pipets

Incubator

First Period

1. Using a 1 ml sterile pipet, transfer 0.1 ml of sterile water into each of three tubes of nutrient broth. Use figure 7-6 as a guide.

• To bring the fluid into the pipet, place the mouth of the sterile pipet between your lips and apply gentle suction. Watch the level of fluid come up to a little above the desired level. Remove the pipet from your lips and promptly place your index finger over the opening to seal it and prevent the fluid from draining out of the pipet. Pay special attention to the way the pipet is held. The shaft of the pipet should be held with the thumb, middle, and ring finger while the mouth portion of the pipet should be held with the index finger (not the thumb). To regulate the flow of fluid release (or increase) the pressure of the index finger on the opening of the pipet. Your instructor will demonstrate this procedure. NOTE: This procedure is used only when the fluids to be transfered consist of liquids that are free of microorganisms. You must never mouth pipet liquids that contain microorganisms. Use a pipetting device instead.

2. Using a 10 ml pipet, transfer 1.0 ml of sterile water into each of three tubes of nutrient broth. In this case use a propipet (fig. 7-7) instead of your mouth to suck the fluid into the pipet.

• Notice the three "buttons" on the propipet:

a. The "button" at the top is to let the air out of the bulb. Squeeze it with your thumb and the side of your index finger while you compress the bulb with the other three fingers and the palm of your hand. This creates a vacuum that can be used to suck liquids into an attached pipet.

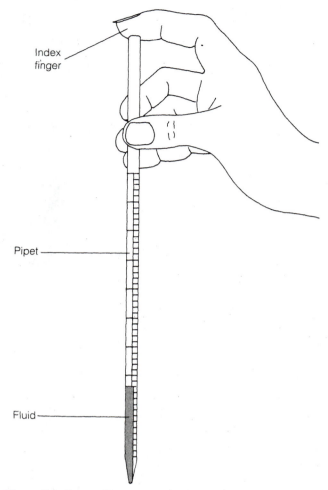

Index finger

Pipet

Fluid

Figure 7-6. Proper Method for Holding a Pipet. Pressure applied by the index finger is used to regulate the rate of flow of liquid material. However, mouth pipetting is not recommended to pipet suspensions containing living organisms.

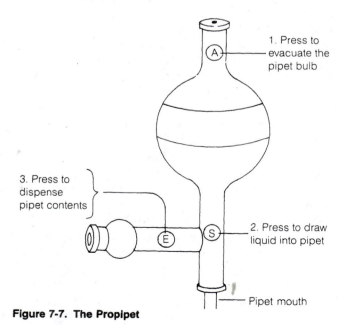

1. Press to evacuate the pipet bulb

A

3. Press to dispense pipet contents

E

S

2. Press to draw liquid into pipet

Pipet mouth

Figure 7-7. The Propipet

b. The second "button" is directly beneath the bulb, between it and the housing for the pipet. Squeeze it gently to apply suction. The strength with which you press on this button will determine the rate of flow of the fluid into the pipet. Be careful not to apply excessive suction because the fluid may flow into the pipet bulb and contaminate the bulb, rendering it useless (this is particularly true when you are pipetting cultures of microorganisms). Watch the level of fluid in the pipet at all times to prevent fluid from getting into the bulb.

c. The "button" on the side arm is to release the fluid from the pipet. Again, the amount of pressure exerted on the button determines the rate of flow of fluid out of the pipet.

3. Discard the pipets in the appropriate containers by placing the pipets, vertically, with the tip down to prevent aerosols that are sometimes created when pipet contents are squeezed through the narrow opening of the tip.

• Receptacles for used pipets are usually tall, cylindrical containers with a soap or disinfectant solution. Your instructor will indicate which containers are to be used to dispose of pipets.

4. Incubate the inoculated tubes at 35°C until the next laboratory period.

Second Period

1. Examine the nutrient broth tubes for evidence of growth.

• Any growth in the broth indicates that the tubes were contaminated during the pipetting procedure. Pipetting is an important technique that is routinely used in microbiology laboratories. Practice it often so that you become proficient and confident of your technique. Other laboratory exercises in this manual will require that you have mastered the pipetting technique.

2. Discard the broth tubes in a safe manner as indicated by your instructor.

Questions

1. Cite five uses for pipetting in the microbiology laboratory.

2. Explain briefly why mouth pipetting is forbidden when transfering liquids containing microorganisms.

3. Suppose that after incubation of the pipetted samples, growth was evident. Describe briefly what may have caused this to happen and cite three steps you may take to prevent this from occurring.

E. SELECTIVE AND DIFFERENTIAL MEDIA

Many special media have been developed by microbiologists in order to select for certain organisms and to differentiate among them. These media are called **selective** or **differential** media.

Selective media are culture media containing at least one ingredient that inhibits the reproduction of unwanted organisms, but permits the reproduction of specific microorganisms. The ingredients are called **selective agents.** Such ingredients are usually antibiotics such as streptomycin and penicillin, chemicals such as sodium azide and sodium chloride, or dyes such as crystal violet and malachite green.

Differential media are special formulations designed to differentiate among microorganisms or groups of organisms that are growing in the medium. Differential media usually contain a chemical that is utilized or altered by some organisms but not by others. When different microorganisms grow on the medium, these changes can be used to differentiate one group from another by the appearance of the culture medium or the colony growing on the medium.

In this exercise we will consider three commonly used selective and/or differential media: **mannitol-salts agar,** used to select for *Staphylococcus* and differentiate between *S. aureus* and *S. epidermidis*; **eosin methylene blue agar,** used to select for Gram-negative enteric bacteria and to differentiate between lactose fermenters and nonfermenters; and **blood agar,** used to grow nutritionally demanding organisms and differentiate between those that lyse red blood cells and those that do not.

Mannitol-salts agar is both a selective and differential medium (see appendix for formulation). The selective agent is sodium chloride, which is present at a concentration of 7.5%. It inhibits many bacteria except those in the genus *Staphylococcus*. The medium also contains 0.5% mannitol and the pH indicator phenol red. When *S. aureus* grows on this medium, it ferments the mannitol, releasing acidic byproducts which cause the pH indicator surrounding the colonies to change from red to yellow. *Staphylococcus epidermidis*, however, does not ferment mannitol and hence does not change the color of the culture medium. Therefore, you can differentiate between *S. aureus* and *S. epidermidis* by noting the color change that takes place around the colonies of *S. aureus*.

Eosin methylene blue (EMB) is also both a selective and a differential medium. The medium contains 0.065 g/l methylene blue and 0.4 g/l of eosin, dyes that at these concentrations inhibit most Gram-positive bacteria (see exercise 9). EMB differentiates between Gram-negative organisms that can ferment lactose (e.g., *Escherichia, Enterobacter,* and *Klebsiella*) and those

that cannot (e.g., *Salmonella, Shigella,* and *Proteus*). Lactose fermenters will form deep purple colonies or pinkish colonies with dark centers. Those species that produce high levels of acid during fermentation (e.g., *Escherichia coli*) will precipitate the dyes, acquiring a metallic green sheen over the purple colony. Low acid producers will precipitate only a small portion of the dyes and will develop colonies that are pinkish with a dark center.

Blood agar is a differential medium that supports the growth of a wide variety of bacteria. The medium is composed of a nutritionally rich agar base (see appendix) with 5% sheep red blood cells. Blood agar is used widely in the clinical laboratory to differentiate between bacteria that lyse blood (**hemolytic**) and those that do not (**nonhemolytic**). The breakdown of the red blood cells is seen as a clearing of the medium around the colony. When the red blod cells are lysed completely and the medium is completely cleared around the colony, the type of hemolysis is called **beta hemolysis.** If the red blood cells are partly lysed and the medium acquires a greenish hue, the hemolysis is said to be **alpha hemolysis.** Nonhemolytic bacteria cause no visible changes in the media and are said to be **gamma hemolytic.** Bacteria that are commonly differentiated based on their hemolytic activities are streptococci and staphylococci. Hemolysis is due to enzymes release by the microorganisms.

Materials

Cultures of *Staphylococcus aureus, Staphylococcus epidermidis, Escherichia coli, Proteus vulgaris, Streptococcus lactis, Streptococcus mitis,* and *Streptococcus faecalis*

Plates of mannitol-salts agar

Plates of eosin methylene blue agar

Plates of blood agar (5% sheep red blood cells)

a. The Use of Mannitol Salts Agar Plates

First Period

1. Label two plates of mannitol-salts agar with your name, the date, the type of culture medium, and the name of the organism used.

2. Using a different plate for each organism, streak the plates with *S. aureus* and *S. epidermidis*.

3. Incubate the plates in an inverted position at 35°C for 25 to 48 hours.

Second Period

1. Describe the changes, if any, that occurred on the mannitol-salts agar plates and explain what the results mean on the report form.

2. Discard all cultures in a safe manner as indicated by your instructor.

b. The Use of Eosin Methylene Blue Agar (EMB)

First Period

1. Label two plates of EMB agar with your name, the date, the type of culture medium, and the name of the organism used.

2. Using a different plate for each organism, streak *E. coli* and *P. vulgaris* onto EMB plates.

3. Incubate the plates in an inverted position at 35°C for 24 to 48 hours.

Second Period

1. Describe the changes, if any, on the EMB plates and explain what the results mean on the report form.

2. Discard all cultures in a safe manner as indicated by your instructor.

c. The Use of Blood Agar Plates

First Period

1. Label three plates of blood agar with your name, the date, the type of culture medium, and the name of the organism used.

2. Using a different plate for each organism, streak *Streptococcus lactis*, *S. mitis*, and *S. faecalis* onto blood agar plates.

3. Incubate the plates in an inverted position at 35° for 25 to 48 hours.

Second Period

1. Describe the changes, if any, on the blood plates and explain what the results mean in the report form or in your laboratory notebook.

2. Discard all cultures in a safe manner as indicated by your instructor.

Questions

1. Explain how and why the following media are selective: mannitol-salts agar and EMB.

2. Explain how and why the following media are differential: EMB, blood agar, and mannitol-salts agar.

3. What causes alpha hemolysis and beta hemolysis on blood agar?

4. Outline a procedure which would enable you to detect the presence of *Escherichia coli* from a sample which contains a variety of Gram-positive bacteria and a few Gram-negative bacteria that produce low concentrations of acid from lactose. Explain your reasons for selecting the procedure.

PART III
STAINING

Many microorganisms are difficult to see with the light microscope because they do not absorb, reflect, refract, or diffract much light. Because of this, **stains,** alcoholic or aqueous solutions of colored compounds called **dyes,** are often used to color microorganisms and/or their background. Dyes in the stains absorb and reflect light so that difficult-to-see objects become highly visible. The use of stains often makes it possible to see structural components such as endospores and flagella as well as inclusions such as starch and phosphate granules. Stains can be used to detect specialized parts of the cell. A procedure in which specific dyes are used to detect the presence or absence of specialized parts of a cell are known as **structural stains. Differential stains,** on the other hand, are procedures that enable one to distinguish between cell types. The Gram stain and the acid-fast stain are differential staining techniques that allow bacteria to be divided into major groups.

Dyes are often divided into two types; **basic dyes** and **acidic dyes.** Basic dyes are those in which the colored portion or **chromophore** is positively charged while acidic dyes are those in which the colored portion is negatively charged (fig. III-1). Basic dyes are used most often because microorganisms tend to have an excess of negatively charged groups associated with their wall, membrane, and cytoplasm under the conditions used for staining. The positively charged dyes are strongly attracted to the many negatively charged groups in cells, resulting in efficient staining. The negatively charged acidic dyes frequently are used to stain the background rather than the microorganisms themselves. Acidic dyes generally do not stain cells because they are repelled by the many negative charges associated with the cells. Nevertheless, acidic dyes occasionally are used to stain positively charged structures within cells. It is important to recognize that the charge and binding properties of dyes and cell structure can be changed by changing the pH of the environment.

Whenever a staining technique results in the staining of the objects, the technique is described as a **positive stain** whether or not a basic (positive charge) or acidic dye (negative charge) has been used. Similarly, whenever the background is stained to create contrast with the unstained object, the stain technique is called a **negative stain.**

Some staining procedures involve a combination of positive and negative staining techniques. For example, Maneval's capsule stain technique or procedure (described in a later section) is a combination of a negative stain that dyes the background one color and a positive stain that colors the organism (other than the capsule) another color. In this procedure, the capsule remains unstained and is seen as a bright area between the organisms and the background.

BASIC DYES

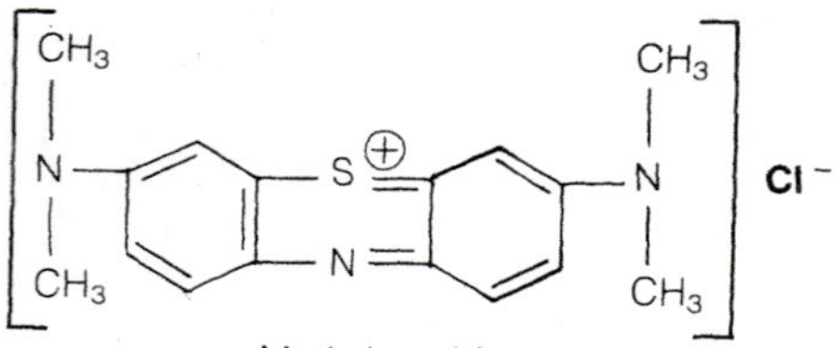

Methylene blue

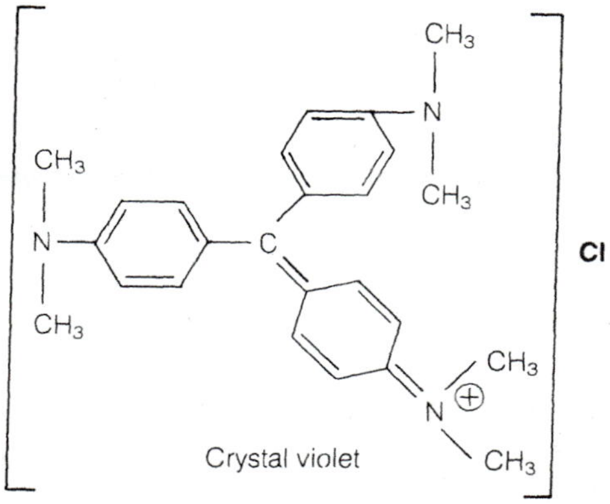

Crystal violet

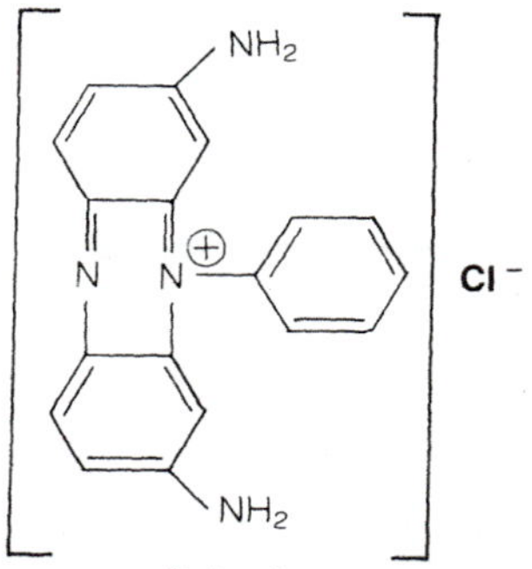

Safranin

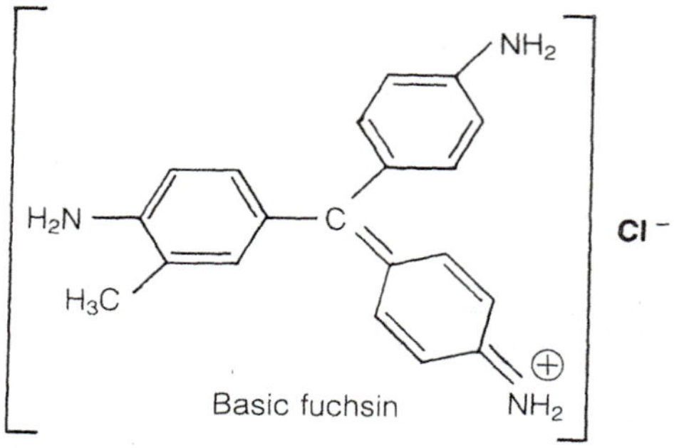

Basic fuchsin

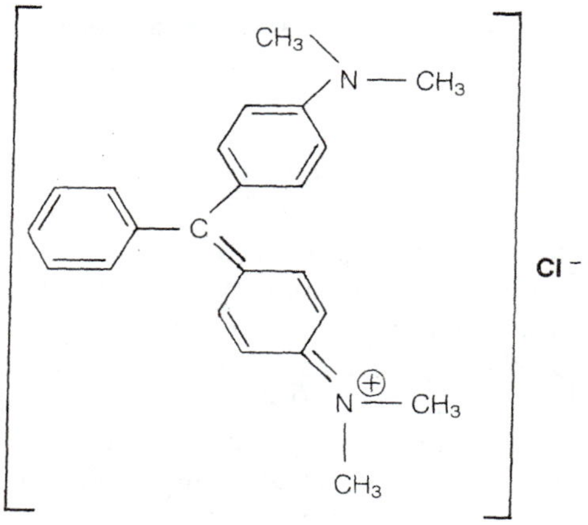

Malachite green

ACIDIC DYES

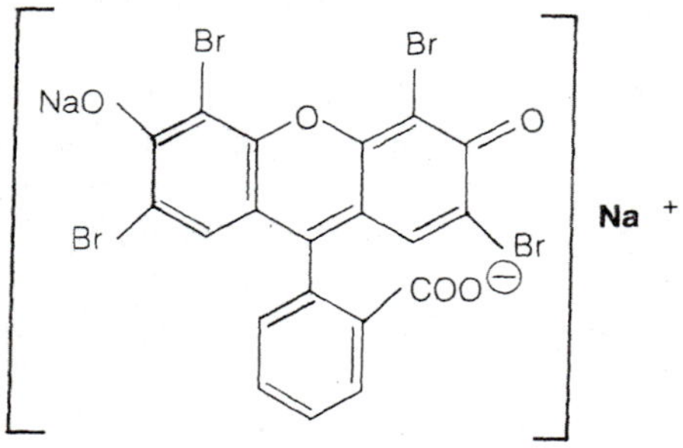

Eosine

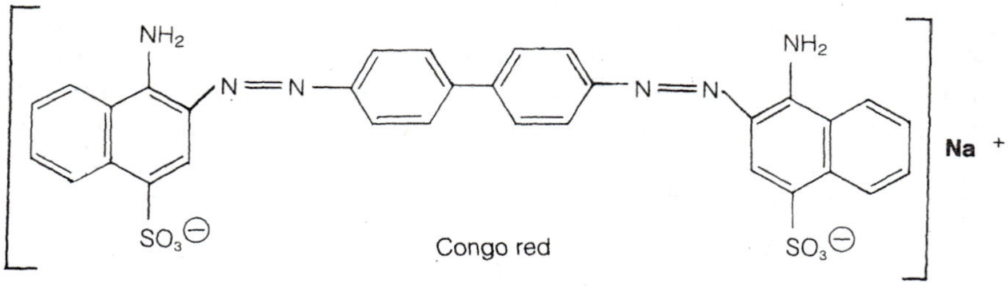

Congo red

FAT SOLUBLE DYES

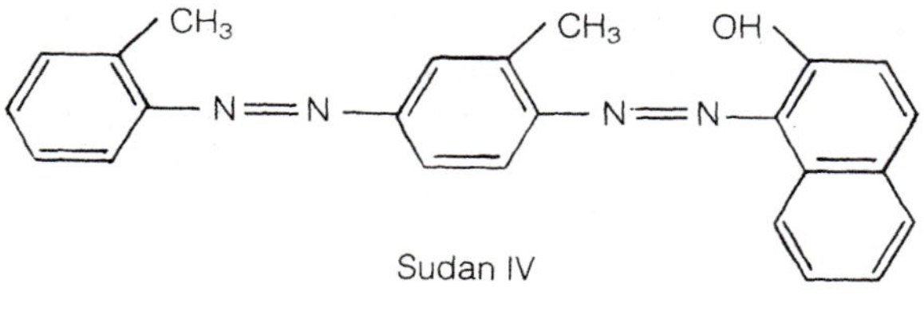

Sudan IV

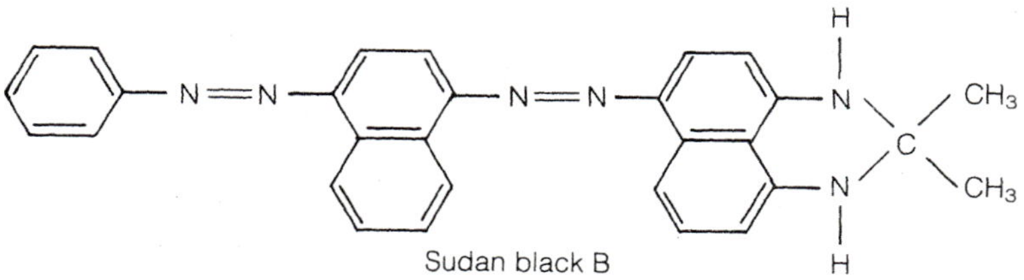

Sudan black B

Figure III-1. Some Commonly Used Dyes

EXERCISE 8
POSITIVE AND NEGATIVE STAINING

A **simple stain** is a procedure in which only one stain is used to create a contrast between the specimen and its background. Generally, simple stains involve the use of basic dyes such as crystal violet, methylene blue, basic carbolfuchsin (red), safranin (red), or malachite green. Infrequently, simple stains involve the use of acidic dyes such as nigrosin and congo red.

Since simple staining procedures are rapid and easy to carry out, they are often used when information about cell shape, size, and arrangement is desired (fig. 8-1). A simple stain helps the microscopist determine the dimensions and shape of cells. Bacteria can generally be characterized as spheres (coccus, plural cocci), rods (bacillus, plural bacilli), spirals (spirillum, plural spirilla), helices (spirochete, plural spirochetes), or branched organisms. In addition, many organisms form very distinctive arrangements that can be used to identify them. For example, bacteria such as the streptococci (strepto- = chain of) form chains of cells, the staphylococci (staphylo- = bunch of grapes)

develop in grape-like clumps, the neisseriae exist as pairs or diplococci (diplo- = pair of), and some micrococci and sarcinae (sarcina = a package) are typically found in packets of four or eight.

Usually, a simple stain involves the staining of a dried preparation of cells on a glass slide (fig. 8-2). The dried preparation of cells is known as a **smear.** Smears can be prepared from cells in a liquid culture or from growth on an agar plate or slant. When using a liquid suspension, one to several loopfuls are smeared onto a glass slide and then allowed to air dry. The cells in the dried smear are attached or **fixed** to the slide by briefly heating the slide over a gas burner flame. This procedure is known as **heat fixation.** When using colonies or growth from a semisolid medium, a loopful of water is placed on the slide and a very small amount of material is mixed with the water to separate and suspend the cells. The suspension is then spread out, air dried, and heat fixed. In a good smear, individual organisms are visible microscopically and organisms

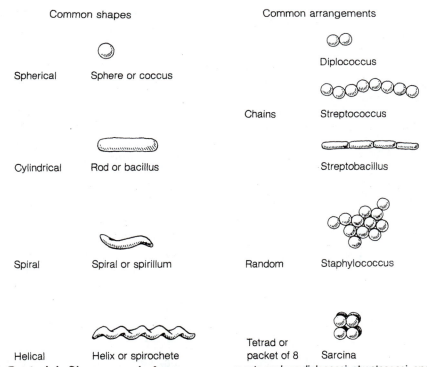

Common shapes

Spherical — Sphere or coccus

Cylindrical — Rod or bacillus

Spiral — Spiral or spirillum

Helical — Helix or spirochete

Common arrangements

Diplococcus

Chains — Streptococcus

Streptobacillus

Random — Staphylococcus

Tetrad or packet of 8 — Sarcina

Figure 8-1. Common Bacterial Shapes and Arrangements. Bacterial shapes are often useful in identifying organisms. Often cocci remain attached to each other and form typical arrangements such as diplococci, streptococci, and staphylococci. Rod-shaped bacteria in chains are sometimes called streptobacilli.

1. Place several large loopfuls of a liquid culture on a clean slide. If the culture is on an agar plate or slant, place several loopfuls of water on the slide and then mix in a very small amount of the solid growth (e.g. part of a colony).

2. Spread the liquid culture or mixture over the slide so as to create a thin film.

3. Air dry the smear. Use a warming plate if possible to speed up the drying, but do not use a gas burner.

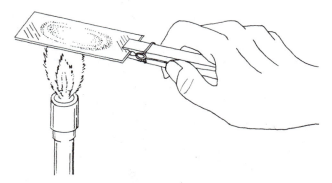

4. Heat fix the dried smear by placing the bottom of the slide over the gas burner flame for three seconds. An alternate method is to pass the slide through the flame three times.

Figure 8-2. Procedure for Positive Staining

are not piled on top of each other. In order to insure well-dispersed organisms, the drop used to make the smear should be only slightly turbid (cloudy).

In this exercise, you will learn the proper procedure for preparing a smear, how to do a simple stain, and how to perform simple positive and negative staining procedures.

Materials for A and B

Cultures (24 to 48 hours old) of *Micrococcus luteus*, *Bacillus subtilis*, *Escherichia coli*, and *Saccharomyces cerevisiae* (yeast)

Tartar (from students' teeth)

Solutions of crystal violet, methylene blue, basic carbolfuchsin, safranin, nigrosin, and India ink

Inoculating loop

Sterile toothpicks

Clean glass slides

Clothespins

Gas burner

Sink (with tap water)

A. PROCEDURE FOR POSITIVE STAINING

1. If the organisms are on an agar plate or slant, place a couple of loopfuls of water on a clean glass slide and suspend a small amount of the organism in the water. Only mix in enough material so that the drop becomes slightly turbid. If the organisms are in a liquid culture, place one to several loopfuls of the suspension on a clean glass slide (fig. 8-2).

2. Air dry the mixture on a warm hot-plate and then heat fix the organisms over a burner flame. Heat fix for only 3 to 5 seconds.

• Overheating a smear or heating a still-wet smear can badly distort the shape of cells. Clothespins can be used to hold your slide when heat fixing and staining.

3. Cover the cooled, heat-fixed smear with one of the basic dyes. Crystal violet is a good choice. After staining for 30 seconds wash off the crystal violet gently with tap water or dip the slide into standing water.

• Drain the excess water from the slide.

4. Carefully blot the slide dry with bibulous paper or a paper towel. When you hold the slide up to the light, you should be able to see your stained smear.

5. Locate the smear with the low power lens, then place a small drop of oil on the stained smear and view the organisms with the oil immersion lens (100 × objective). The condenser diaphragm should be open so that as much light as possible is captured by the oil objective lens.

• If you have carried out the simple staining procedure correctly, the microorganisms should be colored (dark blue-violet if you used crystal violet) and the background should be clear and bright.

6. Draw, label, and describe what you see. Examine several oil immersion fields in different areas of the smear to obtain a representative view of the specimen as a whole.

Questions

1. Compare and contrast basic and acidic dyes. Give some examples of commonly used basic and acidic dyes.

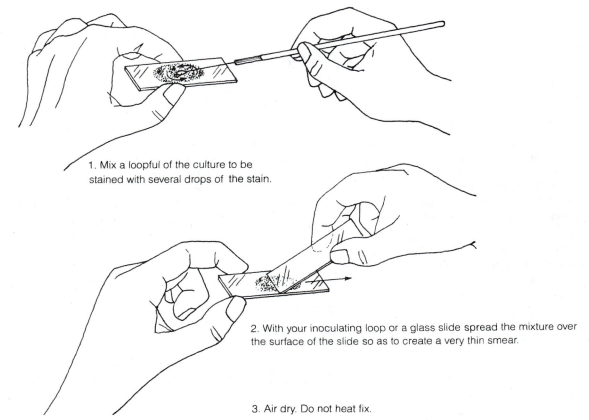

1. Mix a loopful of the culture to be stained with several drops of the stain.

2. With your inoculating loop or a glass slide spread the mixture over the surface of the slide so as to create a very thin smear.

3. Air dry. Do not heat fix.

Figure 8-3. Procedure for Negative Staining

2. What is a chromophore?

3. What is a simple stain?

4. What is a structural stain and what is a differential stain?

B. PROCEDURE FOR NEGATIVE STAINING

1. Place one to several loopfuls of an acidic stain at one end of a clean glass slide. The use of nigrosin is suggested.

• If the microorganisms are suspended in a liquid, mix a loopful of the suspension with the drop of acidic stain. If the microorganisms are on a plate or slant, use aseptic technique to obtain a very small sample of the material with an inoculating loop and mix it in the stain. Do not mix in too much of the specimen. If you do, the cells will be piled on top of each other and it will be difficult to see individual organisms.

2. Spread the mixture over the slide by drawing a second glass slide into the drop and then pulling the drop back along the slide (fig. 8-3). Try to make as thin a smear as possible.

3. Air dry the mixture on a warm hot plate but DO NOT HEAT FIX.

4. Locate the smear with the low power lens (10 ×), then place a small drop of oil on the stained smear and view the organisms with the oil immersion lens (100 ×). The condenser diaphragm should be open so that as much light as possible is captured by the objective lens.

• If you have carried out the staining procedure correctly, the microorganisms should be clear or lightly stained while the background should be heavily stained.

5. Draw, label, and describe what you see.

Questions

1. Compare and contrast positive and negative stain techniques.

2. Why are acidic dyes not commonly used to color bacteria?

EXERCISE 9
DIFFERENTIAL STAINING

A. GRAM STAIN

Differential stains are very useful in microbiology because they can be used to distinguish between groups of bacteria. An important differential stain is the Gram stain. The Gram stain is one of the most important steps in the characterization and identification of bacteria.

The Gram stain separates bacteria into one of two large groups: the **Gram-positive** bacteria that retain the color of the first stain used (crystal violet), and the **Gram-negative** bacteria that assume the color of a second stain (safranin). Bacteria stain differentially because of differences in the structure and chemical composition of their cell walls.

The Gram-positive bacteria have a thick cell wall that consists primarily of **peptidoglycan** (fig. 9-1). Many Gram-positive bacteria, however, have polymers called **teichoic acids** in their cell walls, which may account for as much as 50% of the wall's weight. The peptidoglycan layer of the cell wall consists of polysaccharides, made up of alternating N-acetylglucosamine and

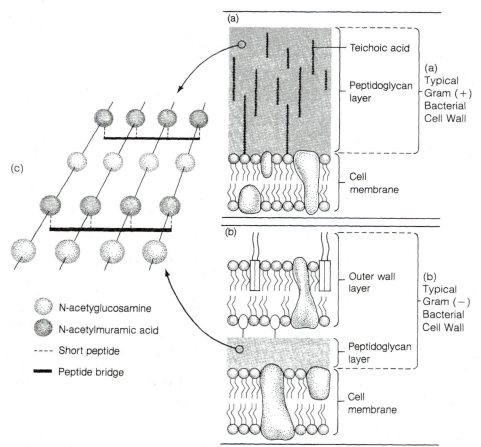

Figure 9-1. Cell Wall of Gram Positive and Gram Negative Bacteria. *(a)* The cell wall and cell membrane of a Gram positive bacterium are illustrated. The cell wall consists of peptidoglycan, wall teichoic acids, and membrane teichoic acids. The wall teichoic acids are covalently linked to the muramic acid units of the peptidoglycan layer, while the membrane teichoic acids are linked to glycolipids. The molecules found in the membrane include various proteins, phospholipids, glycolipids, and phosphatidyl glycolipids. *(b)* The cell wall and cell membrane of a Gram negative bacterium are illustrated.

The cell wall consists of a layer of peptidoglycan and a lipid bilayer called the outer envelope. The outer envelope contains lipopolysaccharides on the outer surface, which are called endotoxins because they cause allergic reactions in animals. The cell membrane of Gram negative bacteria is similar to the cell membrane in Gram positive bacteria. *(c)* Most bacteria have a peptidoglycan layer in their cell wall. The peptidoglycan consists of a polysaccharide made up of alternating N-acetylglucosamine and N-acetylmuramic acid and short peptides attached to the N-acetylmuramic acid.

N-acetylmuramic acid units, cross-linked by short peptides, while teichoic acids are long polymers of alternating phosphates and carbohydrates (e.g., ribitol or glycerol). The cell wall of Gram-positive bacteria is generally between 20 and 80 nanometers (nm = 10^{-9} meter) thick.

The walls of Gram-negative bacteria have much less peptidoglycan than those of Gram-positive bacteria. The peptidoglycan layer, usually about 2 nm thick, is surrounded by a complex lipid bilayer called the outer membrane. Some microbiologists regard this outer membrane as part of the cell wall while other scientists consider the outer membrane to be a separate and distinct **envelope.** No teichoic acids are associated with the cell wall of Gram-negative bacteria. The

STEPS		COMMENTS
1. Crystal violet	60 sec.	Primary dye. **(Primary stain)**
2. Water rinse	brief	
3. Gram's iodine	60 sec.	Mordant. Gram's iodine functions as a **mordant,** that is, it forms a chemical complex with the crystal violet that helps the dye attach to the charged groups in the wall, membrane, and cytoplasm.
4. Water rinse	thorough	
5. 95% ethanol	10 sec.	Decolorizing agent. The ethanol functions as a **decolorizing agent** that draws the iodine-crystal violet complexes from the gram-positive and gram-negative cells. Since the gram-negative cell has a much thinner wall than the gram-positive cell, the iodine-crystal violet complexes are more rapidly removed from the gram-negative cells.
6. Water rinse	thorough	
7. Safranin	60 sec.	Secondary dye. The safranin functions as a **counterstain** to color the clear gram-negative bacteria red. It is believed that the gram-positive bacteria pick up very little safranin because most of the charged groups are still occupied by the crystal violet.
8. Water rinse	brief	
9. Drain and blot dry		

Gram stain procedure illustrated in figure 9-2 consists of nine steps.

In order to obtain a reliable Gram stain, it is necessary to use a young culture of organisms, no older than 24 to 48 hours. If older cultures are used, there is the possibility of ambiguous results. As cultures age, progressively more cells sustain damage to their cell walls. This damage apparently allows dyes to be more easily leached from the cells by decolorizing agents. Thus, Gram-positive cells with damaged cell walls tend to lose their ability to retain the crystal violet-Gram's iodine complex and consequently stain as if they were Gram-negative. Damaged cells in a pure culture of Gram-positive bacteria are sometimes the reason a mixture of blue-violet and red cells is seen in Gram stained smears of pure microorganisms.

Materials

Young cultures (24 to 48 hours old) of the Gram-positive bacteria *Staphylococcus epidermidis* and *Bacillus subtilis*

Young cultures (24 to 48 hours old) of the Gram-negative bacteria *Escherichia coli* and *Neisseria subflava*

Solutions of crystal violet, Gram's iodine, 95% ethanol, and safranin

Clean glass slides

Inoculating loops

Clothespins

Gas burner

Sink with tap water

Procedure

1. Prepare a thin smear of the bacteria provided. You may wish to prepare individual or mixed smears. When making a mixed smear mix together on a clean slide a Gram-positive rod *(Bacillus)* and a Gram-negative coccus *(Neisseria)*. On a second slide mix together a Gram-positive coccus *(Staphylococcus)* and a Gram-negative rod *(Escherichia)*.

• When using organisms from a plate or slant, mix in only enough bacteria to make the drop slightly turbid.

2. Air dry the smears on a hot plate and then heat fix the bacteria over a burner flame. Heat fix for only 3 to 5 seconds.

3. Cover the heat fixed smear with crystal violet and stain for 60 seconds. Rinse the slide gently and briefly with water. Allow the excess fluid to drain from the slide.

4. Cover the smear with Gram's iodine. After 60 seconds, thoroughly rinse the slide.

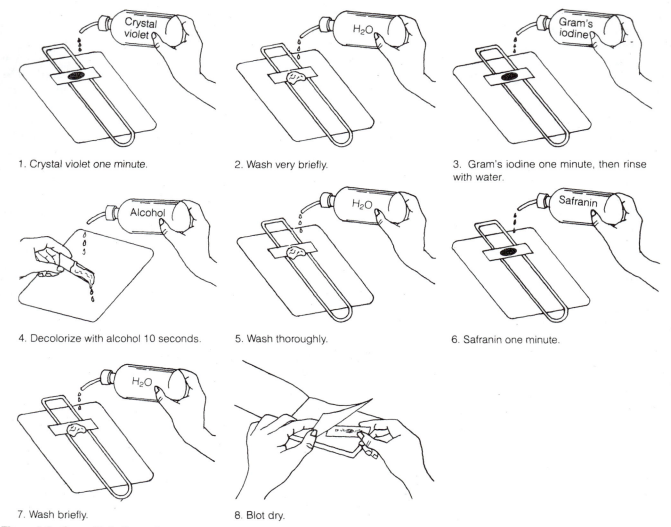

1. Crystal violet *one minute.*

2. Wash very briefly.

3. Gram's iodine one minute, then rinse with water.

4. Decolorize with alcohol 10 seconds.

5. Wash thoroughly.

6. Safranin one minute.

7. Wash briefly.

8. Blot dry.

Figure 9-2. Gram Stain Procedure

5. Cover the smear with the decolorizing agent, 95% ethanol, for 10 seconds. After 10 seconds, immediately wash all the ethanol from the slide with water to stop the decolorizing action of the ethanol.

• If the alcohol remains on the smear for more than 10 seconds, it will remove the crystal violet-Gram's iodine complexes from Gram positive as well as Gram negative bacteria and the staining procedure will not give accurate results. If the alcohol does not remain on the smear for at least 10 seconds or is diluted when added because of rinse water remaining on the smear, the crystal violet will not be leached out of either type of cell. Again the staining procedure will not give accurate results. If the decolorizing has been done properly, the Gram-positive cells will still be dark blue-violet while the Gram-negative cells will be clear and difficult to see.

6. Cover the smear with safranin. After 60 seconds wash the slide thoroughly with water.

7. Carefully blot the slide dry with a paper towel.

8. Locate the smear with the low power lens (10×). Then place a small drop of oil on the smear and rotate the oil immersion lens (100× objective) into place. The condenser diaphragm should be open so that as much light as possible is captured by the oil immersion objective.

• You have carried out the gram stain procedure correctly if the Gram-positive cells (*Staphylococcus* or *Bacillus*) are dark blue-violet and the Gram-negative cells (*Escherichia* and *Neisseria*) are light red.

9. Draw, label, and describe what you see.

Questions

1. Why must you use a young culture of organisms when doing the Gram stain?

2. If you leave the alcohol on too long, what would

you expect to see after Gram staining a mixture of *Escherichia* and *Staphylococcus?* If you do not leave the alcohol on long enough, what would you expect to see after Gram staining a mixture of *Neisseria* and *Bacillus?*

3. When you counterstain with safranin, why do the Gram positive bacteria not pick up safranin and stain red?

4. What would be the appearance of Gram positive (or Gram negative) bacteria if:

   a. you forgot to counterstain with safranin?

   b. the mordant were not used?

   c. the decolorizing step were omitted, too brief, or too long?

   d. crystal violet were replaced with methylene blue?

   e. 0.01% sulfuric acid were used instead of ethanol?

   f. your smear were too thick?

5. Why do bacterial species show characteristic arrangements of their cells?

B. ACID-FAST STAIN

The **acid-fast stain** is a useful procedure for distinguishing bacteria in the genera *Mycobacterium* and *Nocardia* from all other types of bacteria. These genera contain important human pathogens. *Mycobacterium leprae* is the cause of leprosy, while *Mycobacterium tuberculosis* is responsible for tuberculosis. *Nocardia asteroides* causes pulmonary nocardosis, a disease of the lungs that resembles tuberculosis. Bacteria in these genera are said to be **acid-fast bacteria** because they retain a primary stain (Ziehl-Neelsen carbolfuchsin) that is removed from other bacteria by a brief treatment with an acidified decolorizing agent (acid-alcohol). Acid-fast bacteria stain a bright red. Bacteria that lose the primary stain when treated with an acidified decolorizing agent are said to be **non-acid-fast bacteria.** Non-acid-fast bacteria are detected by use of a counterstain (methylene blue) that colors them a bright blue.

Mycobacterium and *Nocardia* are unusual because they have a high concentration of waxes in their cell wall. In some species the waxes may account for as much as 60% of the wall's weight. The waxes make the bacteria difficult to stain since charged dye ions do not readily penetrate the waxy layers of the wall. The primary stain is usually applied with the aid of heat to allow the dye to penetrate the waxy layer of the cell wall. Once stained, the acid-fast bacteria are difficult to decolorize. The acid-fast stain procedure (fig. 9-3) consists of the following steps.

STEPS		COMMENTS
1. Carbolfuchsin	5 min. (steamed)	Primary stain must be steamed for 5 min.
2. Water rinse	thoroughly	
3. Acid-alcohol	20 sec.	Decolorizing agent
4. Water rinse	thoroughly	
5. Methylene blue	1 min.	Counter stain
6. Water rinse	briefly	
7. Drain and blot dry		

Materials

Young cultures (less than 48 hours old) of the acid fast bacteria *Mycobacterium phlei* or *M. smegmatis*

Young cultures (less than 48 hours old) of the non-acid-fast bacteria *Bacillus subtilis, Escherichia coli,* and *Staphylococcus epidermidis.* (Tartar may substitute for cultures of non-acid-fast bacteria.)

Solutions of Ziehl-Neelsen carbolfuchsin and methylene blue

Solution of acid alcohol

Clean glass slides

Inoculating loops

Clothespins

Gas burner

Sink and tap water

Procedure

1. Prepare air dried, heat-fixed smears of each culture provided. You may wish to mix *S. epidermidis* with your acid-fast organisms so as to see the contrast between the two types of bacteria.

• Since the *Mycobacterium* is very waxy and does not mix with water easily, you must spend some time breaking up the clumps of bacteria with the inoculating loop. If you spread the bacteria out as much as possible, the bacteria will not be clumped and you will be able to see many individual mycobacteria.

2. Place the slide on a rack over a sink.

• Clothespins may be useful in stabilizing the slide on the rack.

3. Cover the fixed smear with carbolfuchsin and heat the stain so that it steams. The carbolfuchsin must steam for at least 5 minutes.

• In order to get the carbolfuchsin to steam, heat it from above with the flame from the gas burner. If the carbolfuchsin begins to evaporate, add fresh stain. DO NOT ALLOW THE DYE TO BOIL. Also, remove the flame

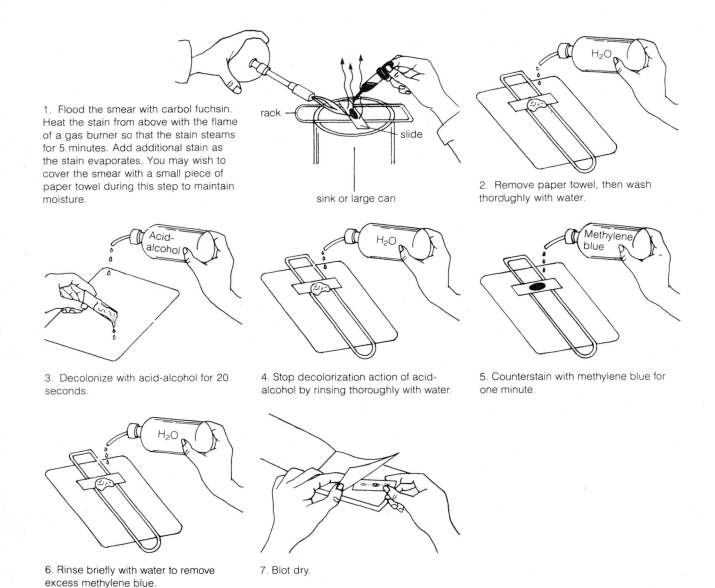

1. Flood the smear with carbol fuchsin. Heat the stain from above with the flame of a gas burner so that the stain steams for 5 minutes. Add additional stain as the stain evaporates. You may wish to cover the smear with a small piece of paper towel during this step to maintain moisture.

rack

slide

sink or large can

2. Remove paper towel, then wash thoroughly with water.

3. Decolonize with acid-alcohol for 20 seconds.

4. Stop decolorization action of acid-alcohol by rinsing thoroughly with water.

5. Counterstain with methylene blue for one minute.

6. Rinse briefly with water to remove excess methylene blue.

7. Blot dry.

Figure 9-3. Acid-Fast Stain Procedure

occasionally so that the slide does not become so hot that it breaks. Some laboratory instructors recommend that the student place a small piece of towel paper over the smear before adding the carbolfuchsin. This keeps the carbolfuchsin from evaporating and drying on the slide. The heating melts the waxes in the cells and allows the carbolfuchsin to enter. Once the waxes cool, the carbolfuchsin is trapped within the wax and within the cell.

4. If towel paper was used, remove it and discard in appropriate containers (not in the sink). Rinse the stained smear thoroughly with tap water and then drain the excess water from the slide.

5. Flood the slide with acid-alcohol. After decolorizing for 20 seconds, immediately and thoroughly wash with tap water to stop the decolorizing action of the acid-alcohol.

• The acid-alcohol removes the carbolfuchsin from cells that lack waxes in the cell wall but is unable to remove the stain from the vegetative cells with high concentrations of waxes. At this point in the procedure, the waxy cells are red while wax-free cells are clear and very difficult to see.

6. Counterstain the smear with methylene blue for 60 seconds.

• Methylene blue is a basic dye and binds to the negative charges on the wax-free (non-acid-fast organisms) vegetative cells. The counterstain does not readily penetrate or bind to the waxy cells. Thus, the waxy cells are red while the wax-free cells are blue.

7. Rinse the smear briefly with tap water. Methylene blue is a weak dye and it is easy to overrinse. Carefully blot dry with a paper towel.

8. Locate the smear with the low power lens (10×), then place a small drop of oil on the smear and rotate the oil immersion lens (100× objective) into place. The condenser diaphragm should be open so that as much light as possible is captured by the objective lens.

9. Draw, label, and describe what you see.

Questions

1. What would be the appearance of acid-fast and non-acid-fast bacteria if:

   a. plain ethanol were used instead of acid-alcohol?

   b. carbolfuchsin were not heated sufficiently?

   c. methylene blue were replaced with crystal violet?

   d. the counterstain were omitted?

   e. a thorough, long rinse occurred after the counterstain?

2. Of what practical importance is the acid-fast stain?

3. Which organism (genus and species) is responsible for the following diseases: pulmonary nocardosis, tuberculosis, leprosy.

4. *Mycobacterium tuberculosis* is not easily Gram stained. On the other hand *Mycobacterium phlei* or *Mycobacterium smegmatis* stain Gram-positive. Suggest a reason for the different staining characteristics of these mycobacteria.

5. What might be indicated by the presence of acid-fast organisms in a clinical specimen from the lungs? What might be indicated by finding these organisms in a skin lesion?

EXERCISE 10
STRUCTURAL STAINING

A. CAPSULE STAIN

A **capsule** is a thick layer of polysaccharide (occasionally polypeptide) material 1 to 2 μm thick that surrounds and adheres to some bacteria (fig. 10-1). Capsules have two major functions: they favor bacterial attachment to surfaces and protect bacterial populations from predators. Stringy polysaccharide material that binds bacteria to surfaces and to each other is called **glycocalyx.** When the glycocalyx becomes very thick, it is indistinguishable from a capsule.

Some bacteria are capable of causing disease because of their ability to synthesize capsules. For example, *Streptococcus mutans*, which produces a capsule, is one of the initiators of tooth decay because it is able to cling to teeth. The acids released by the bacteria degrade the enamel coating. The bacterium *Streptococcus pneumoniae* (pneumococcus), which causes pneumonia, is a very virulent pathogen for mice and immunologically compromised humans when it has a capsule. However, it does not cause disease in mice or humans when it is unable to produce a capsule. When pneumococci with capsules successfully enter the lungs, they are poorly phagocytized and inefficiently destroyed by the macrophages and monocytes in the lungs. Thus, pneumococci with capsules are able to reproduce, damage the lungs, and cause

pneumonia. On the other hand, pneumococci without capsules are efficiently phagocytized and destroyed

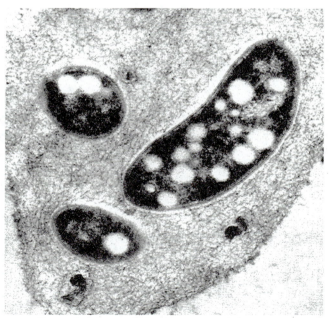

Figure 10-1. Bacterial Capsule. A transmission electron micrograph of a thin section of *Rhizobium trifolii* shows the capsule around these bacteria.

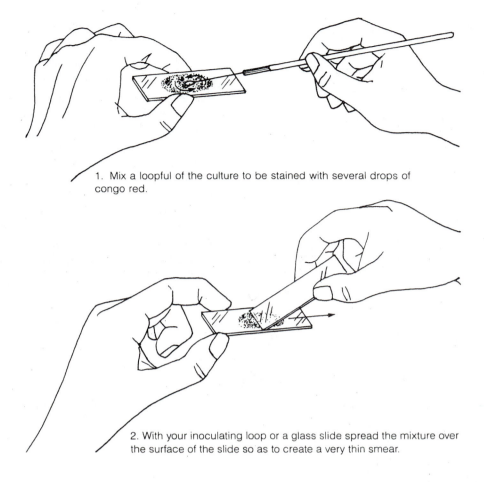

1. Mix a loopful of the culture to be stained with several drops of congo red.

2. With your inoculating loop or a glass slide spread the mixture over the surface of the slide so as to create a very thin smear.

3. Air dry. Do not heat fix.

4. Flood the slide with Maneval's stain for 1 minute. Drain excess dye. Do not rinse. Blot dry gently.

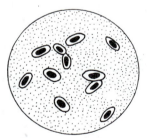

A good smear. Cells are well isolated and cellular detail is visible.

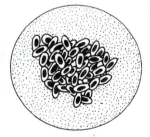

Too many cells. Cells are piled on top of each other and cellular detail is lost.

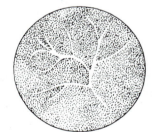

A thick smear. Stain covers the organisms and they are not visible.

Figure 10-2. Procedure for Capsule Staining

in the lung. Thus, the diseased state is averted. Approximately, 50% of normal humans carry encapsulated forms of the pneumococcus in their throats indicating that the ability to cause disease depends upon more than just the presence of a capsule.

Capsules are thick enough to be seen with the light microscope. However, since capsules are difficult to stain, they are generally observed in negative stains. The background is stained with an acidic stain while the cell is stained with a basic stain of another color.

The capsule remains unstained and is seen as a bright translucent area around the cell.

Materials

Young cultures (48 hours old) of the bacteria *Klebsiella pneumoniae* and *Micrococcus luteus* on TSA or nutrient agar slants

Solutions of congo red and modified Maneval's stain

Clean glass slides

Clothespins

Inoculating loops

Procedure

1. Place a loopful of congo red on a clean glass slide (fig. 10-2).

2. Mix a small sample or loopful of *Klebsiella* into the drop of congo red. Make sure the bacteria are well mixed.

- Obtain a very small amount of the culture with a presterilized inoculating loop and mix it in the congo red. Do not mix in too much of the specimen, if you do, the cells will be piled on top of each other and the capsule will be difficult to see around the organisms.

3. Mix a small sample or loopful of *Micrococcus* into the drop to create a mixture.

- You should have a mixture of bacteria. The rod-shaped bacterium has a capsule while the coccus does not.

4. Using another clean slide, spread the suspension of bacteria and congo red over the slide so that a very thin film results.

- The thinner the film is, the better. If the congo red film beads up, you have a greasy or dirty slide.

5. Air dry the thin film of congo red and cells on a warm hot-plate but DO NOT HEAT FIX.

- The congo red stains the background but does not stain the capsule or the cell. Congo red is an acidic stain (colored portion is negatively charged) and consequently is repelled by the many negative charges associated with the bacterial wall, membrane, and cytoplasm.

6. Cover the dried smear with modified Maneval's stain. After staining for 60 seconds drain off Maneval's stain (red) and carefully blot the remaining stain with paper towels or bibulous paper. DO NOT RINSE THE MANEVAL'S STAIN WITH WATER.

7. Locate the smear with the low power lens (10 ×), then place a small drop of oil on the smear and rotate the oil immersion lens (100 × objective) into position.

The condenser diaphragm should be open so that as much light as possible is captured by the objective lens.

- Maneval's stain (red) is a basic dye and stains the bacteria red because it associates with the many negative charges on the bacteria. Since neither the congo red nor the Maneval's stain associates with the capsule, it remains unstained and translucent. The reaction of the congo red with the Maneval's stain turns the congo red blue. Thus, the background may be blue rather than red.

8. Describe, draw, and label your findings.

Questions

1. Which of the bacteria that you examined possessed capsules?

2. Do all the *Klebsiella* in the smear have capsules? Do all the capsules have the same thickness?

3. Of what importance is the capsule to bacteria?

4. What role might a capsule play in the virulence of a bacterium?

5. What is dental plaque? What role does capsule (or glycocalyx) production by *Streptococcus mutans* play in its colonization of tooth surfaces and the development of dental caries?

6. Why are rocks in streams and ponds often slimy?

7. What is a glycocalyx?

B. ENDOSPORE STAIN

An **endospore** is a heat- and chemical-resistant resting form produced by members of certain bacterial genera in response to adverse environmental conditions. The only bacteria known to produce endospores belong to the following genera: *Bacillus, Clostridium, Desulfotomaculum, Sporolactobacillus, Thermoactinomyces,* and *Sporosarcina.* Endospores develop within the cell and only one endospore forms per cell (fig. 10-3). Consequently, this type of sporulation is not involved in reproduction.

Under favorable conditions, the endospore forming bacteria proliferate like other common bacteria. The growth and reproduction of the bacteria is sometimes referred to as **vegetative growth.** Under laboratory conditions many of the endospore formers may have generation times of less than an hour when they grow vegetatively.

When the environment becomes unfavorable for vegetative growth, the endospore former begins to **sporulate** or form an endospore. An unfavorable environment may be one in which the carbon, energy, or phosphate source is running low; toxic wastes are

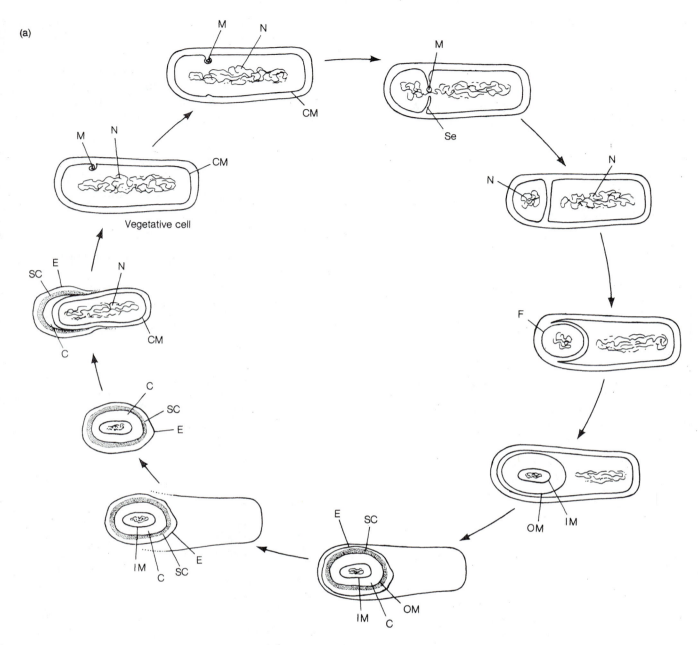

(a)

Figure 10-3. Life Cycle of Typical Sporeforming Bacterium.
(a) **Sporulation** begins when a committed cell divides into a mini-cell and a large cell. The sporulation process involves the engulfment of the mini-cell by the large cell, the loss of water by the mini-cell, and the development of various layers of material within and around the developing endospore. At 37°C, the sporulation process takes approximately 15 hours. The **germination** process involves the reconstitution of the endospore, the breakdown of the cortex and the rupture of the endospore coat. From the endospore emerges a vegetative cell. Germination takes approximately 1 hour. *(b)* Endospores may be spherical or oval. Each cell produces only one endospore. The endospore consists of a dehydrated core where the genome (G) is found, an inner membrane (IM) that surrounds the dehydrated core, a surrounding cortex (C), an outer membrane (OM), a spore coat (SC), and an exosporium (E).

(b)

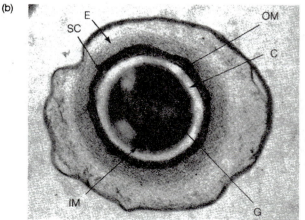

beginning to accumulate; the temperature is becoming unfavorable; or the environment is becoming hypertonic (as a result of desiccation). Under laboratory conditions, sporulation requires approximately 15 hours. The endospore is resistant to high temperatures and chemicals that are lethal to vegetative cells. Some endospores (such as those of *Clostridium botulinum*) can survive boiling in water for a number of hours. Desiccation or lack of nutrients has no effect on the viability of endospores. They can be kept dry and free of nutrients for many years without loss of viability. Endospores are highly resistant to ultraviolet light (UV) and not sensitive to antibiotics that kill vegetative cells.

A favorable environment stimulates endospores to **germinate** or develop into vegetative cells which resume active growth and reproduction. Under laboratory conditions, germination takes less than an hour. Germination is accompanied by the rapid loss of resistance to heat, toxic chemicals, desiccation, antibiotics, and UV light.

Because an endospore's outer coat is an effective barrier to chemicals, endospores generally stain poorly. Endospores can be stained, however, by using very hot dyes. The heat apparently causes the coat to expand so that the dye penetrates.

Materials

Old cultures (older than 48 hours) of *Bacillus subtilis*, *Clostridium sporogenes*, *Clostridium butyricum*

Solutions of malachite green (potential carcinogen) and safranin

Clean glass slides

Clothespins

Inoculating loops

Gas burner

Sink and tap water

Procedure

1. Prepare a thin smear of one of the assigned cultures. Air dry and heat fix.

2. Place the slide on a rack over a sink (fig. 10-4).

3. Cover the dried smear with malachite green and heat the stain so that it steams. The malachite green must steam for at least 5 minutes.

• One method of steaming is to heat the malachite green from above with the gas burner flame for 5 minutes. If the malachite green begins to evaporate, add fresh stain and reheat so that the stain shows rising vapors. After vapors are detected, remove the flame so that the slide does not become so hot that it breaks. Another method of steaming is to heat the

malachite green for 5 minutes from below with the steam from a boiling water bath. This method is very messy and is not recommended. Some laboratory instructors recommend that the student place a small piece of towel paper over the smear before adding the malachite green. This may keep the malachite green from evaporating unevenly and drying on the slide. The heating expands the spore coat and allows the malachite green to enter the endospore. Once the endospore cools, the malachite green is trapped within the endospore.

4. Cool the slide for about 1 minute before continuing.

5. If towel paper was used, remove it and discard it in an appropriate container. Wash the smear thoroughly with tap water.

• The tap water readily removes the malachite green from most bacterial vegetative cells but is unable to decolorize the endospores. At this point in the procedure, the endospores are green while the vegetative cells are clear and very difficult to see.

6. Counterstain the smear with safranin for 60 seconds. Do not heat the safranin.

• The counterstain is a basic stain and binds to the negative charges on the vegetative cell. The counterstain does not penetrate the endospore. Thus, the vegetative cells should stain red while the endospores, if present, should stain green.

7. Rinse the smear briefly with water. Carefully blot dry with a paper towel. Safranin is a weak dye and is easily removed from vegetative cells by overrinsing.

8. Locate the smear with the low power lens (10×), then place a drop of oil on the smear and switch to the oil immersion lens (100× objective) to examine the cells. The condenser diaphragm should be open so that as much light as possible is captured by the oil immersion lens.

9. Describe, draw, and label your findings.

Questions

1. What are the dimensions (length and diameter) of a typical vegetative cell and what are the dimensions of the endospore?

2. Where are most of the endospores found? Do you see any endospores within the vegetative cells? If you do, where are they located (terminal, central subterminal)?

3. What happened to the vegetative cells that gave rise to the free endospores?

4. Of what benefit is endospore formation to soil bacteria?

5. Which organism (genus and species) is responsible for each of the following diseases: anthrax, tetanus, botulism, and gas gangrene.

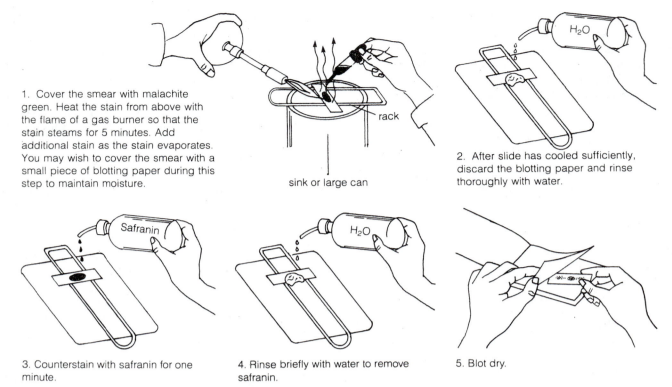

1. Cover the smear with malachite green. Heat the stain from above with the flame of a gas burner so that the stain steams for 5 minutes. Add additional stain as the stain evaporates. You may wish to cover the smear with a small piece of blotting paper during this step to maintain moisture.

rack

sink or large can

2. After slide has cooled sufficiently, discard the blotting paper and rinse thoroughly with water.

3. Counterstain with safranin for one minute.

4. Rinse briefly with water to remove safranin.

5. Blot dry.

Figure 10-4. Procedure for Endospore Stain

6. Do endospores from different organisms all have the same shape?

7. Some soil bacteria such as *Streptomyces* produce spores. What are some of the differences between endospores and the spores produced by *Streptomyces?*

8. Where do the endospore formers usually live?

9. What would be the results of the endospore stain if: you failed to use heat with the primary stain? you rinsed inadequately after malachite green? you rinsed excessively after safranin?

10. Do *Mycobacterium* species form endospores? What color would you expect a smear of a *Mycobacterium* species to take, if it was subjected to the endospore stain? Explain.

C. FLAGELLA STAIN

Bacterial flagella are long thin protein appendages that some bacteria use to propel themselves (fig. 10-5). Even though bacterial flagella may be 2 μm to 5 μm long, they cannot normally be viewed with the light microscope because they are generally less than 0.025 μm in diameter.

Flagella are found in arrangements characteristic of the species, with a single flagellum (**monotri-** **chous**), or flagella at one end (**polar**), or flagella at both ends (**amphitrichous**), or flagella that originate over the entire surface (**peritrichous**). Flagella at the poles of the cell may be single or multiple. The arrangement of flagella is sometimes useful in the identification of certain species of bacteria. Because of this, stains have been developed that allow flagella to be viewed with the light microscope. A chemical called a **mordant**, which precipitates on the flagella, is used to thicken them. The mordant is then stained and the thickened flagellar structures can be seen with the light microscope.

Motile species within the family Enterobacteriaceae (such as *Proteus*) have peritrichous flagella, while bacteria of the genus *Pseudomonas* typically show polar flagella.

Materials

Young slant cultures (24 to 48 hours) of *Proteus vulgaris* (peritrichously flagellated) and *Pseudomonas fluorescens* (polarly flagellated)

Solution of Gray's flagellar stain

Solution of acid ethanol

Clean glass slides

Small test tubes

Inoculating loops

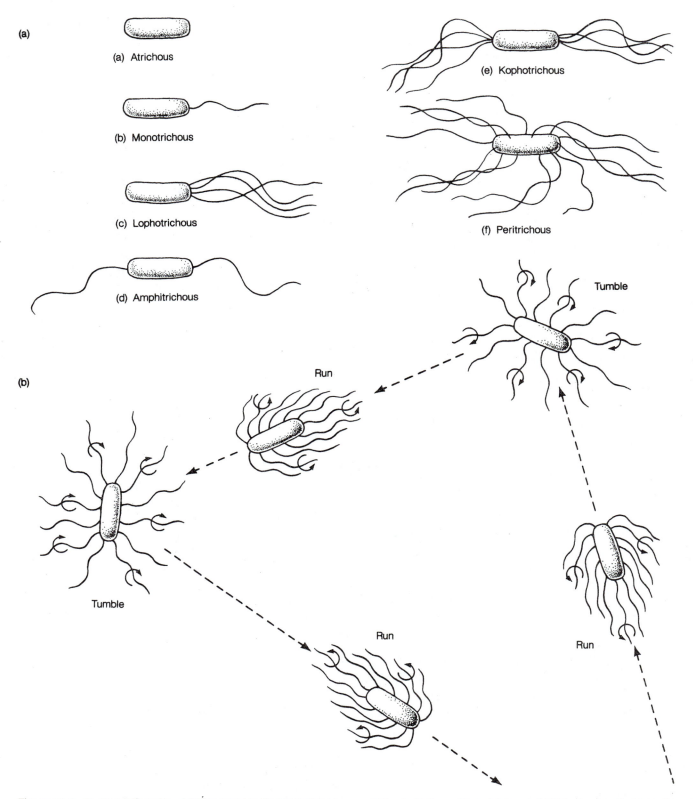

Figure 10-5. Bacterial Flagella. *(a)* This drawing illustrates the arrangement of bacterial flagella. Bacteria that have flagella only at their ends are polarly flagellated, while those that have flagella over much of their surface are peritrichously flagellated. The following nomenclature is sometimes used to describe the types of flagellation: A) atrichous; B) monotrichous; C) lophotrichous; D) amphitrichous; E) kophotrichous; F) peritrichous. *(b)* This drawing illustrates how a bacterium such as *Escherichia coli* might move in an aqueous environment. It moves in a straight line (run) when its flagella rotate in synchrony. When the flagella rotate asynchronously, the cell begins to tumble. When the flagella rotate synchronously again, the cell changes direction.

(a) Atrichous

(b) Monotrichous

(c) Lophotrichous

(d) Amphitrichous

(e) Kophotrichous

(f) Peritrichous

Run

Tumble

Tumble

Run

Run

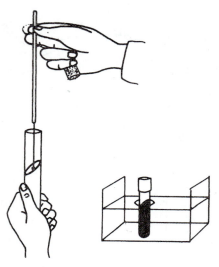

1. Transfer a small mass of cells to 0.5 ml of saline. Allow this bacterial suspension to sit undisturbed for 15 minutes.

2. Place a large drop of the slightly turbid bacterial suspension near the top of each of three alcohol cleaned slides and allow the drops to run down the slides.

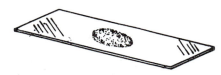

3. Air dry. Do not heat fix.

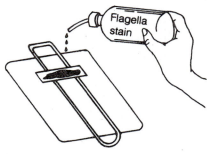

4. Stain each smear with freshly filtered flagella stain for a different time: 5 min., 10 min., and 15 minutes.

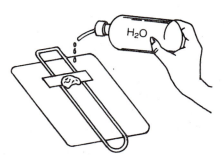

5. Gently rinse.

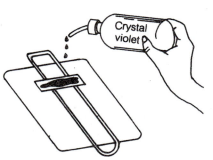

6. Counterstain with crystal violet for one minute.

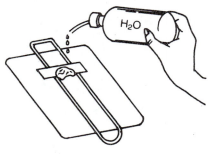

7. Gently rinse.

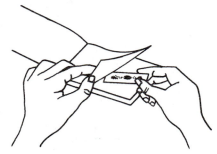

8. Blot dry.

Figure 10-6. Procedure for Flagella Stain

Pasteur (capillary) pipets

1 ml and 5 ml pipets

95% ethanol

Nitrocellulose filter (0.22 μm pores)

Clothespins

Gas burner

Sink and tap water

Procedure

1. Gently obtain a small mass of *Proteus* or *Pseudomonas* from a slant or plate and place the mass, without agitation, in 0.5 ml of saline in a small test tube. Allow the mass to sit undisturbed for 15 minutes so that bacteria diffuse from the mass (fig. 10-6).

2. While you are waiting for the bacteria to make the saline turbid, prepare the Gray's flagellar stain.

- Mix 9 ml of the mordant (5 ml of potassium alum + 2 ml of tannic acid + 2 ml mercuric chloride) with 0.8 ml of the stain (0.8 ml basic fuchsin). Then filter the mixture. The prepared stain may be provided by your instructor.

3. Rinse clean slides in 95% ethanol. Then flame the slides. Place the slides at a steep angle against a test-tube rack.

4. Place a large drop of the slightly turbid bacterial suspension near the top of a slide and allow it to run down the slide. Allow the slide to air dry but DO NOT HEAT FIX.
- Make three slides for each organism.

5. Place the slides on a staining rack over the sink.

6. Flood the smears with freshly prepared flagella stain. Stain one smear for 5 minutes, one for 10 minutes, and one for 15 minutes. Rinse GENTLY AND BRIEFLY with water (or dip into water).

7. Counterstain with crystal violet for 1 minute. Rinse GENTLY AND BRIEFLY with water and carefully blot dry with a paper towel or bibulous paper.

8. Locate the smears with the low power lens (10×), then switch to the oil immersion lens (100× objective).

9. Describe, draw, and label your findings.

Questions

1. What is the length and diameter of the cells? What does this indicate about the mordant?

2. What color did the cells stain? What color did the flagella stain?

3. Determine the length and diameter of the flagella.

4. How many flagella can you count on *Proteus?* How many can you count on *Pseudomonas?*

5. How are the flagella arranged on *Proteus* and on *Pseudomonas?*

6. Why does the procedure call for a young culture?

7. Why must the slides be free of grease and oil? Why must the bacteria be treated gently when making the smear and when staining and washing?

EXERCISE 11
DETECTION OF INCLUSIONS

Many different species of prokaryotic and eukaryotic microorganisms synthesize granules, which they store in their cytoplasm. These materials serve as nutrient reserves and provide readily available sources of carbon and energy which do not increase the water pressure within the cell or affect intracellular pH. In this exercise we will demonstrate the presence of metachromatic granules, poly-β-hydroxybutyric acid, starch, and glycogen granules as examples of storage granules.

A. METACHROMATIC (VOLUTIN) GRANULES

When some bacteria are stained with basic dyes such as toluidine blue, certain granules stain a dark blue-violet to red-violet, while the rest of the cytoplasm stains a light blue. Because the color of the granules is different from the color added, the granules are called **metachromatic granules.** Metachromatic granules, also known as **volutin,** are large clumps of polymerized phosphate that are found in the cytoplasm of some species of bacteria. The polymerized phosphate clumps may be an inorganic reserve material providing the cell with a readily available source of phosphate for biosynthesis. These granules are a way of saving excess phosphate ions without upsetting the ionic balance of the cell.

Metachromatic granules are quite common in certain Gram positive species of *Bacillus* and *Corynebacterium.* Thus, the detection of metachromatic granules sometimes helps in characterizing organisms that belong to these or several other related genera.

In this staining procedure you will stain granules with 1% toluidine blue, decolorize the cell with diluted H_2SO_4 (1:1000), treat with Gram's iodine, and counterstain with 1% eosin Y. This will yield dark black granules and a pink cytoplasm.

Materials

Tryptone phosphate agar slant cultures (24 to 48 hours old) of *Corynebacterium xerosis* and *Bacillus megaterium*

Solutions of 1% Toluidine blue or Loeffler's methylene blue

Diluted H_2SO_4 (1:1000)

Gram's iodine

1% eosin Y

Clean glass slides

Clothespins

Inoculating loops

Procedure

1. Prepare a smear of *Corynebacterium xerosis* or *Bacillus megaterium*.

2. Air dry and heat fix the smear.

3. Cover the smear with 1% Toluidine blue or with Loeffler's methylene blue for at least 1 minute. Rinse briefly with water.

4. Decolorize with 1:1000 H_2SO_4 for 1 to 2 seconds. Rinse briefly with water.

5. Treat with Gram's iodine for 1 minute. Rinse briefly with water.

6. Counterstain with 1% eosin Y for 30 seconds. Rinse briefly with water and carefully blot dry.

7. Locate the smears with the low power lens (10×), then switch to oil immersion (100×).

8. Describe, draw, and label your findings.

Questions

1. What color are the metachromatic granules and what are their dimensions? What color is the rest of the cytoplasm?

2. What are metachromatic granules and what importance may they have for bacteria?

3. What is volutin?

4. What is the purpose of the brief decolorization with dilute sulfuric acid?

5. What is the purpose of the Gram's iodine?

6. What is the purpose of the eosin Y?

B. POLY-β-HYDROXYBUTYRIC ACID (PHB)

One of the most common organic reserve materials in prokaryotes is poly-β-hydroxybutyric acid (PHB), a polymer of β-hydroxybutyric acid. β-hydroxybutyric acid is a ready source of carbon and energy, but it cannot accumulate in cells because in useful amounts it would lower the pH and increase the water pressure within the cell. β-hydroxybutyric acid can be stored, however, by polymerizing it. The polymer does not alter the pH or the water pressure within the cell.

Poly-β-hydroxybutyric acid forms a lipid-like granule which can be stained with lipid-soluble dyes such as Sudan black B. Sudan black B stains the poly-β-hydroxybutyric acid granules a dark blue. The staining procedure you will use stains lipid material dark blue and the cytoplasm red. If endospores are present, they remain unstained.

Lipid soluble stains are often referred to as "neutral" stains, since the dye is uncharged and colors the object by dissolving in the fat or lipid granule rather than binding to it because of attractive charges.

Materials

Nutrient agar slant cultures of *Bacillus cereus* and *Bacillus megaterium*

Solutions of Sudan Black B and safranin

Xylene (a potential carcinogen)

Clean glass slides

Inoculating loops

Clothespins

Gas burner

Sink with tap water

Procedure

1. Prepare a dried, heat-fixed smear of one of the cultures provided.

2. Flood the smear with Sudan Black B for 5 to 10 minutes.

3. Drain the excess stain and blot dry. Do NOT rinse.

4. Decolorize the smear with xylene for 15 to 20 seconds.

• Do not get the xylene on your skin and avoid breathing the vapors.

5. Drain the xylene and blot dry.

6. Counterstain the smear with safranin for 60 seconds.

7. Rinse the smear GENTLY AND BRIEFLY with water. Carefully blot dry with bibulous paper.

8. Locate the smear with the low power lens (10×), then switch to the oil immersion lens (100×).

9. Describe, draw, and label your findings.

Questions

1. What color did the PHB granules and cytoplasm stain? Were the endospores stained?

2. How large are the PHB granules?

3. Why do you need to use xylene?

4. Why are PHB granules stained by Sudan black B but not by safranin?

5. Why do cells store PHB? Of what value is PHB to the cell?

C. STARCH AND GLYCOGEN

Starch (amylose) is a polysaccharide made up of glucose units. The form of starch known as **alpha-amylose** is unbranched, while the form of starch called **amylopectin** is highly branched. **Glycogen** is similar to amylopectin, except that it is more highly branched.

Both starch and glycogen are examples of organic reserve materials produced by some microorganisms. Glucose is stored in these polymeric forms so that the osmotic balance of the cell is not upset. A high concentration of glucose units within a cell would increase the osmotic pressure within the cell and might result in the lysis of the cell. Polymers of glucose contribute to the osmotic pressure much less than the individual units of the polymers. In addition, starch and glycogen can be used as a source of energy and carbon. Because the bonds that hold together glycogen are identical to those found in amylopectin, glycogen is degraded by the same enzymes that degrade the starch.

Dextrans are branched polymers of glucose. However, the major backbone is held together by alpha 1,6.

When stained with iodine, amylose stains blue while amylopectin, and glycogen stain a dark red-brown to red-violet.

Materials

Slant cultures of the bacterium *Bacillus subtilis* and the yeast *Saccharomyces cerevisiae*

Solutions of Lugol's iodine or Gram's iodine

Clean glass slides and cover slips

Inoculating loops

Nitrocellulose

Clothespins

Gas burner

Sink and tap water

Procedure

1. Add a drop of Lugol's iodine to a clean slide.

2. Mix a small amount of the organism into the drop so that it becomes slightly turbid.

3. Place a cover slip over the preparation and find the organisms with the 10× objective. Then observe with the oil immersion lens.

4. Describe, draw, and label your findings.

Questions

1. What color are the granules in each of the organisms?

2. What does the color suggest about the chemical composition of the granules?

3. How large are the granules?

4. Explain how starch and glycogen differ.

5. Of what use are starch and glycogen granules to the cell?

EXERCISE 12
MORPHOLOGICAL UNKNOWN

The staining procedures introduced in this section are used by microbiologists in characterizing and identifying bacteria. Within a few minutes it is often possible to determine the group of organisms to which an unknown isolate belongs.

Gram staining and observing cell shape and arrangement divides bacteria up into at least four groups (fig. 12-1). An acid-fast stain of rods can determine whether or not you have an organism like *Mycobacterium*, while an endospore stain can indicate whether or not your unknown might be an endospore former such as *Bacillus*.

Gram-positive nonsporulating, non-acid-fast rods that have various shapes (**pleomorphic**), that line up parallel to each other like a picket fence (**palisade arrangement**), or that contain **metachromatic granules** are very likely **coryneforms** such as *Corynebacterium*.

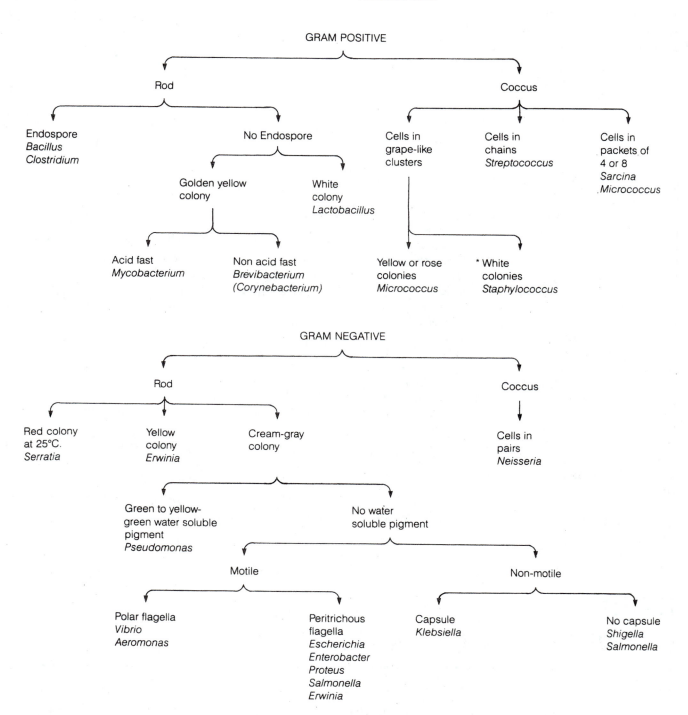

*Some species/strains have gold/yellow pigments.

Figure 12-1. Key for the Identification of Bacterial Genera Based on Morphological and Staining Characteristics.

Gram-positive cocci that form long chains are called **streptococci** and are probably related to *Streptococcus*, while Gram-positive cocci that form cuboidal packets of 4, 8, 16, or 32 cells are known as **sarcinae** and may be related to *Sarcina* or *Micrococcus*. Gram-positive cocci tending to occur in irregular "grape-like" clusters probably are related to *Staphylococcus* or *Micrococcus*. Cocci which are found in pairs are known as **diplococci**.

Gram-negative diplococci may be related to *Neisseria*, while Gram-positive diplococci may be related to *Streptococcus* or to *Staphylococcus*.

Materials

Unknown pure cultures that may include species of *Streptococcus, Staphylococcus, Micrococcus, Neisseria, Bacillus, Brevibacterium, Mycobacterium, Klebsiella, Escherichia, Proteus, Lactobacillus, Serratia, Erwinia Pseudomonas, Aeromonas*

Light microscope equipped with calibrated ocular micrometer

Lens paper

Immersion oil

Clean slides and cover slips

Depression slides

Vaseline

Stain reagents as used in earlier exercises

Trypticase soy agar (TSA)

Sterile Petri plates

Procedure

1. Carry out a Gram stain, acid-fast stain, capsule stain, endospore stain, and flagella stain on your unknown.

2. Determine the shape of your unknown and any distinctive arrangements of the cells. Measure your unknown.

3. Streak the unknown onto a TSA plate and incubate it 24 to 48 hours at 35°C. Characterize the colonial morphology and color of the unknown.

4. Describe, draw, and label your findings.

5. Use the key to determine which organism you have.

Questions

1. An endospore stain on *Mycobacterium* results in green vegetative cells. Explain why *Mycobacterium* stains this way. Does the green staining indicate that *Mycobacterium* has endospores?

2. Which bacteria produce water soluble pigments?

3. Which bacteria produce colored colonies?

4. Indicate the bacteria that typically grow in grape-like clusters, in long chains, and in packets of 4 or 8.

PART IV
ENUMERATION OF MICROORGANISMS

Much of what we know about microorganisms has come from studying microbial populations. Measuring the size of a microbial population provides valuable information that can be used to determine how microbes grow, to assess the microbiological quality of foods and beverages, and to monitor industrial processes. For example, one of the criteria used by public health agencies to establish the quality of milk is the total number of bacteria per ml of milk. High bacterial counts indicate that the milk was contaminated during milking, collection, or packaging and thus, is likely to spoil quickly and may carry disease-causing microorganisms. Population counts are also done in order to estimate the number of spoilage- or disease-causing microorganisms in foods and beverages. In later parts of this manual you will be using the techniques developed here to estimate the microbial populations of soils, foods, and milk.

Microbiologists measure the size of populations by counting the number of individual cells in the population or by measuring the weight of all the cells that make up the population. There are two common procedures used to measure the number of individuals in a population: the direct microscopic count, and the viable count. The **direct microscopic count** involves the enumeration of the total number of microbial cells in a sample, both living and dead, with the aid of a microscope. The **viable count** involves counting the number of living cells that can form colonies on agar plates or solid media.

Another commonly used procedure to determine population is the **turbidimetric method.** This technique is based on the principle that the greater the number of cells in the population, the greater the amount of light scattered by the sample. By measuring the amount of light transmitted through a population of microorganisms, you can estimate the numbers in the population. The most common unit used to measure turbidity is **optical density** (OD). The formula to calculate OD is:

$$OD = \log 100 - \log \%T$$

Where OD is the optical density and $\%T$ is the percent transmittance.

In practice, this calculation is seldom performed because most colorimeters have two scales, one in $\%T$ and the other in absorbance. All you have to do is read the absorbance value directly off the scale. When measuring the turbidity of a microbial population, absorbance can be equated to optical density. The term **absorbance** usually is employed when light is being absorbed by a solution, while OD is used when light is being scattered, as is the case with bacterial suspensions.

EXERCISE 13
DIRECT MICROSCOPIC COUNTS WITH A HEMOCYTOMETER

The direct microscopic count is a rapid method for enumerating microorganisms in suspension. This method determines the total number (both living and dead) of microorganisms in a suspension by counting the number of organisms in a known volume of fluid contained inside a counting chamber such as the hemocytometer.

The hemocytometer (fig. 13-1a) is a slide with a counting chamber 0.1 mm deep. On the bottom of the counting chamber there is an etched square divided into nine squares each 1 mm². The central square is divided into 25 smaller squares and each of these in turn are divided into 16 squares (fig. 13-1b). Therefore, within the central square millimeter there are a total of 400 small squares (25 × 16 = 400). The method for calculating the number of cells/ml of suspension is as follows:

(a) Count the number of cells in 5 of the (1/25)mm² squares (the central square and the 4 corner squares) in the central area of the hemocytometer. Count only those cells inside the square. Cells on the lines of the square may be counted as part of one square or as part of the other. You may count as part of the square only those cells that are on the top and left lines but not those on the bottom or right lines of the squares.

(b) Multiply the number of cells counted in the 5 squares by 5 to estimate the number of cells/mm² of surface. If the cells were diluted, multiply by the inverse of the dilution.

(c) Multiply the number of cells/mm² by 10 to obtain the number of cells/mm³.

(d) Multiply the number of cells/mm³ by 1,000 to convert cells/mm³ to cells/ml in the suspension.

For dilute cultures that require no dilutions, multiply the estimated number of cells in the central area (25 of the (1/25)mm² squares) by 10,000 (this number is derived from the logic described above) and express the results in cells/ml of suspension.

Materials

Overnight cultures of *Saccharomyces cerevisiae*
Hemocytometer slides (one for each 2 students)
Sterile 1 ml pipets
Sterile pasteur pipet with bulb
Screw-capped tubes with 9 ml of sterile saline
Compound microscope

Procedure

1. Mix the yeast suspension well using a vortex mixer or by swirling.
• It is important that the suspension is well-mixed so that the individual cells are evenly distributed throughout the fluid. Yeast cells are large enough to settle while the suspension is standing.

2. Make ¹⁄₁₀ and ¹⁄₁₀₀ dilutions of the yeast suspension.
• The ¹⁄₁₀ dilution is made by pipetting (see exercise 4D) 1 ml of the yeast suspension into a screw-capped tube containing 9 ml of sterile saline. The ¹⁄₁₀₀ dilution

(a)

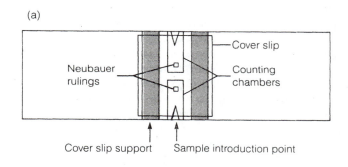

TOP VIEW

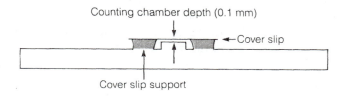

SIDE VIEW

Figure 13-1. The Hemocytometer and Its Use. *(a)* Top and side views of a hemocytometer illustrating the construction of a hemocytometer. *(b)* Enlarged view of the Neubauer ruling of the hemocytometer. Notice that the central counting area, a large square measuring 1 mm² and delimited by double lines, consists of 25 squares, each measuring ¹⁄₂₅ mm². Each of these squares, in turn, are subdivided into 16 squares, each measuring ¹⁄₄₀₀ mm².

(b)

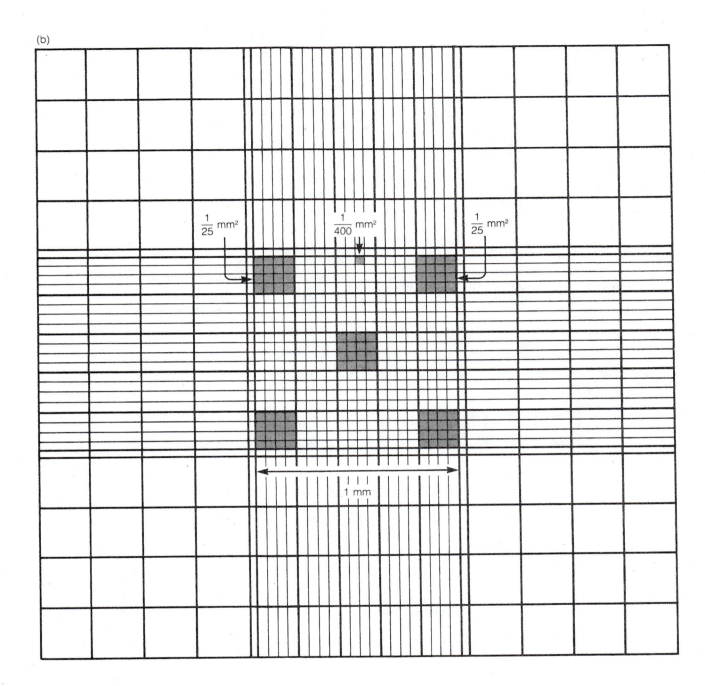

$\frac{1}{25}$ mm²

$\frac{1}{400}$ mm²

$\frac{1}{25}$ mm²

1 mm

is made by pipetting 1 ml of the well-mixed ¹⁄₁₀ dilution into another screw-capped tube containing 9 ml of sterile saline. The sample is diluted before mixing to obtain an optimal number of yeast cells in the suspension for counting.

3. Using a pasteur pipet, place a drop of the yeast suspension in each of the two counting chambers as illustrated in Figure 13-2. Use the highest dilution (¹⁄₁₀₀) first.

4. Place the hemocytometer on the microscope stage and focus on the central area using the low power (10×) lens. Count the number of cells in the central chamber using the high power (40×) lens. If there are too few yeast cells (less than 1 per (1/25)mm² square) repeat step 3 using a lower dilution of the yeast suspension.

• Count the number of yeast cells in five (1/25)mm² squares (central square and the four corner squares) and multiply the number by 5 to obtain the number of yeast cells in 1 mm². Multiply the results by 10,000 and express the number of yeast cells/ml of suspension.

5. Record your results and observations on the report sheet.

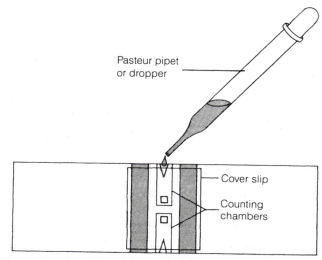

Pasteur pipet
or dropper

Cover slip

Counting
chambers

Figure 13-2. Procedure for Dispensing a Suspension of Micro-organisms into a Hemocytometer Chamber

6. Discard cultures and contaminated materials in a safe manner as directed by your instructor.

Questions

1. List two advantages of the direct microscopic method over the viable count method.

2. Cite one disadvantage of the direct microscopic method.

3. Cite two situations where the direct microscopic method would be desirable.

4. Calculate the total number of cells in a bacterial suspension, that shows 29 cells/mm² in the central square of the hemocytometer when a 1/10,000 dilution is counted.

5. What is the minimum number of cells that can be counted in each of five (1/25)mm² squares to obtain a statistically valid concentration?

EXERCISE 14
VIABLE COUNTS

The enumeration of microorganisms by **viable count** is based upon the assumption that each viable microorganism in a suspension will give rise to a single colony after incubation in suitable media under favorable conditions. After incubation, the number of colonies formed is counted to arrive at an estimation of the number of microorganisms in the original suspension.

The viable count is considered to be a **minimum count** because the number of colonies on the plate represents only those microorganisms that can multiply under the conditions that have been established. For example, a soil sample may contain both aerobic and anerobic bacteria, but if the plates are incubated in the presence of air, then only aerobic and facultative anaerobic bacteria will grow. Similarly, if a sample is plated onto a culture medium devoid of a growth factor like biotin, only microorganisms that do not require biotin in their diets will form colonies. In fact, the cultural and/or nutritional conditions can be manipulated so that only certain types of microorganisms will grow. This method sometimes is used to determine the presence of pathogens or spoilage microorganisms in water, foods, and dairy products.

Not all cells will give rise to a colony because certain microorganisms have a tendency to clump or aggregate. When plated onto suitable culture media, a clump will give rise to only one colony, regardless of how many cells are in the clump. For this reason microbiologists express the results obtained by viable plate counts as **colony forming units** (CFU)/ml rather than cells/ml. Ideally, only plates containing 30 to 300 colonies are counted because this will enhance the accuracy of the count. If the sample contains too many microorganisms (many more than 300 per plate) the microorganisms will be too crowded and will cover the entire plate. For this reason samples are sometimes diluted. If the sample is too dilute, however, the results obtained will not be statistically valid.

Viable counts have numerous applications in microbiology. When samples are tested for disease-causing or spoilage microorganisms, the viable count is used because only viable organisms (not dead ones) can cause infectious diseases or spoil foodstuffs. Industrial processes involving the synthesis of microbial-derived products commonly use viable plate counts to monitor the progress of the culture.

Materials for A and B

Sample of raw milk

Tubes containing 20 ml of plate count agar (PCA)

Petri dishes

Sterile screw-capped tubes with 9 ml sterile saline

1 ml pipets

Glass spreaders

Boiling water bath

Water bath set at 50°C

Incubator set at 35°C

Colony counter

A. PROCEDURE FOR THE POUR PLATE METHOD

First Period

1. Melt three tubes of sterile plate count agar (PCA) in a boiling water bath.

• Boiling for 10 minutes is usually sufficient to melt 10 to 20 ml of agar in tubes. It is important that the agar is completely melted; otherwise, there will be lumps in the poured plates.

2. Cool the medium in a water bath set at 50°C for 5 to 10 minutes.

• The medium must be cooled to 50°C because many microorganisms are killed when they are mixed with media at 55°C and above.

3. While the PCA is melting, make 10^{-1} through 10^{-4} dilutions of the milk sample (refer to figure 14-1 and appendix for proper procedure).

• Label three plates with your name, lab section, type of culture medium used, and the dilution (10^{-2}, 10^{-3}, or 10^{-4}).

4. Pipet 1 ml of the 10^{-2}, 10^{-3}, and 10^{-4} dilutions of the sample into appropriately labeled Petri dishes. While pipetting, lift the lid of the Petri dish just high enough to dispense the sample into the plate (fig. 14-1).

5. Carefully pour cool PCA (about 20 ml per plate) onto the sample within the dish and swirl the plate gently (5 times clockwise, 5 times counterclockwise, and 5 times back and forth) to mix the sample thoroughly with the agar.

• Avoid splashing the medium on the sides of the Petri dish while swirling. Also avoid creating foam or bubbles. Pour the agar gently and in a steady stream.

6. Allow the plates to solidify on a level surface (about 10 minutes). After the medium solidifies, place the plates in an inverted position in an incubator set at 35°C.

• Incubating the plates in an inverted position reduces the amount of condensation on the inside surface of the Petri dish lid and reduces the chances that

drops of water will fall onto colonies when the plate is turned right side up.

7. Incubate the plates for 24 to 48 hours.

Second Period

9. Count all colonies on plates showing 30 to 300 colonies.

• If available, use a colony counter to facilitate the counting procedure (your instructor will demonstrate how to use it). If there are no plates with 30 to 300 colonies, use the plate(s) having a count nearest 300 colonies.

10. Determine the viable plate count by multiplying the number of colonies appearing on the plates by the dilution factor (reciprocal of the dilution).

• Round off the results to two significant figures. Express the results as CFU/ml of milk.

11. Record the results on the report sheet.

Questions

1. Why was the milk sample diluted before plating?

2. Give three instances where viable plate counts would yield more meaningful data than the direct microscopic count.

3. What is the principle underlying the viable plate count method?

4. Why are the results of a plate count expressed as CFU/ml instead of bacteria/ml?

5. Modify the procedure for the pour plate method to count only anaerobic bacteria in the milk.

6. Compare and contrast the differences in morphology of surface and subsurface colonies.

B. PROCEDURE FOR THE SPREAD PLATE METHOD

1. Melt three tubes of PCA in a boiling water bath until completely melted.

2. Cool the melted PCA in a water bath set at 50°C.

3. Gently pour the PCA into sterile Petri dishes and allow the agar to gel on a flat surface.

• Allow 10 to 15 minutes for the agar to harden. Label the plates with your name, laboratory section, type of agar used, and the dilution (10^{-1}, 10^{-2}, 10^{-3}).

4. While the agar is hardening, prepare 10^{-1} through 10^{-3} dilutions of the milk sample.

• You may use the diluted milk samples that you prepared for the pour plate method.

5. Pipet 0.1 ml of the 10^{-1}, 10^{-2}, and 10^{-3} dilutions onto the surface of appropriately labeled PCA plates.

• The PCA must be completely solidified before the suspensions are pipeted.

6. Using a sterile glass spreader, spread the 0.1 ml of suspension over the entire surface of the agar.

• Your instructor will demonstrate how to sterilize the glass spreader and spread the suspension.

7. Incubate the plates in an inverted position in a 35°C incubator for 24 to 48 hours.

8. Count all colonies on plates exhibiting 30 to 300 colonies.

• Use a colony counter, if available, to facilitate the counting procedure.

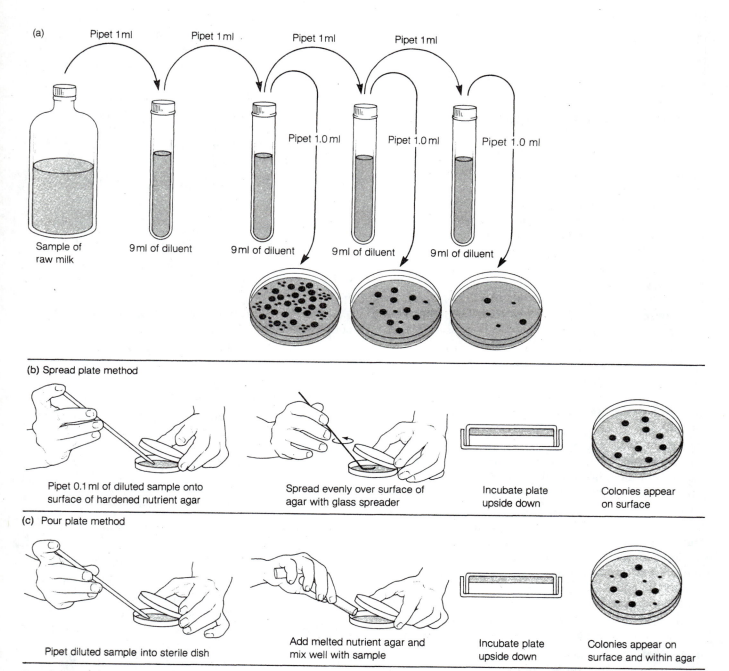

Figure 14-1. Viable Count Methods. *(a)* Viable counts are commonly carried out by dispersing a known volume of microbial suspension on an agar-solidified medium in a Petri dish. The Petri dish is then incubated for a specified amount of time and the colonies are counted. If the samples are expected to contain too many bacteria, they must be diluted before plating. *(b)* The **spread-plate** method involves spreading a known volume (usually 0.1 ml) of suspension over the surface of an agar-solidified medium. *(c)* The **pour-plate** method involves mixing a known volume (0.1–1.0 ml) of suspension in 15–20 ml of liquefied, cooled agar medium. After an incubation period, colonies that appear throughout the medium are counted.

9. Determine the viable count by multiplying the number of colonies counted by 10 (in order to obtain the number of CFU/ml, since only 0.1 ml was plated) and multiplying this product by the reciprocal of the dilution used.

10. Record the results on the report sheet.

11. Discard all materials and cultures in a safe manner as indicated by your instructor.

Questions

1. Would the viable counts obtained using the pour plate method differ from those obtained by the spread plate method?

2. Give two instances when the spread plate method would be more desirable than the pour plate method.

3. Give two instances when the pour plate method would be more desirable than the spread plate method.

4. Why is the viable plate count considered to be a minimal count?

EXERCISE 15
OPTICAL DENSITY MEASUREMENTS

The size of microbial populations in aqueous suspensions can be determined using optical density (OD) measurements. Optical density measurements are made using a colorimeter, an apparatus that emits a beam of light consisting of a narrow band of wavelengths. The beam of light is passed through a cell suspension and the amount of light transmitted by the suspension is measured. The amount of light transmitted is inversely proportional to the number of microorganisms in the suspension and the amount of light scattered and absorbed by the cells. The amount of light scattered and absorbed is determined by the size and shape of the cells.

The colorimeter can measure cell density in suspensions either in %T (percent transmittance) or OD (the amount of light absorbed and scattered). Microbiologists use OD as the unit of measurement because it is directly proportional to the cell density of the suspension.

In practice, the determination of population size using OD measurements is a two-step procedure (fig. 15-1). First, the colorimeter is calibrated to read 0 absorbance when no cells are present. This is done by filling a cuvette with the suspending dilutent (e.g., water or sterile nutrient broth), inserting it into the cuvette chamber of the colorimeter, and adjusting the colorimeter to read 0 absorbance. After the colorimeter is calibrated, a cuvette containing a sample of the bacterial population is inserted into the cuvette chamber and the optical density is measured.

(a)

Figure 15-1. Colorimeter. *(a)* The Bausch & Lomb Spectronic 20 colorimeter. *(b)* The turbidity of a cell-free medium is used to adjust the colorimeter to 0.0 absorbance. The turbidity of the culture is measured and the absorbance recorded. The absorbance is directly proportional to the number of bacteria in the sample and can be used to estimate the population size.

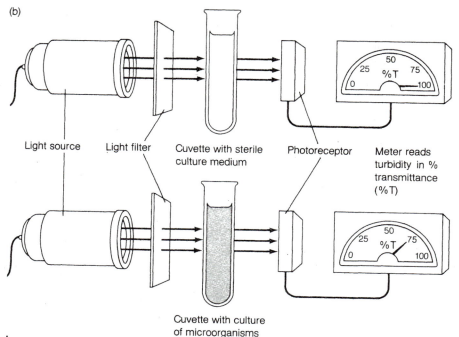

(b)

Light source   Light filter   Cuvette with sterile culture medium   Photoreceptor   Meter reads turbidity in % transmittance (%T)

Cuvette with culture of microorganisms

Figure 15-1. Continued

The OD of a suspension does not directly indicate the number of cells in the population, but only the light scattered by the population. To obtain an approximation of the population size, the OD values must be calibrated to express microorganisms as CFU/ml. To do this, the OD of various dilutions of a cell suspension is determined and viable counts of the corresponding dilution are made. A calibration curve is obtained by plotting viable counts on the X axis and OD values on the Y axis.

The Bausch & Lomb Spectronic 20 (fig. 15-1a) is one of the most commonly used colorimeters in microbiology. This exercise will use the Spectronic 20, although the same results can be obtained using any other colorimeter.

Materials for A and B

Overnight TSB culture of *Escherichia coli*

Tubes of plate count agar

Screw-capped tubes containing 9 ml of sterile saline

Screw-capped tubes containing 5 ml of sterile TSB

Petri dishes

Colorimeter cuvettes

Sterile 1 ml pipets

Sterile 10 ml pipets

Colorimeter

Boiling water bath

Water bath set at 50°C

Incubator set at 35°C

A. PROCEDURE FOR DETERMINING POPULATION SIZE USING THE POUR PLATE METHOD

First Period

1. Using the pour plate method, perform a viable count of an *Escherichia coli* culture. Plate, in duplicate, 1 ml of 10^{-6}, 10^{-7}, and 10^{-8} dilutions.

2. Incubate the plates in an inverted position at 35°C for 24 to 48 hours.

Second Period

3. Count the number of colonies appearing on plates with 30 to 300 colonies.

4. Record the results on the report form.

B. PROCEDURE FOR DETERMINING THE OD OF THE CULTURE

First Period

1. Prepare a series of five twofold dilutions (½, ¼, ⅛, ¹⁄₁₆, and ¹⁄₃₂) of the culture of *Escherichia coli*.
- The process is as follows (fig. 15-2):
  a. Label five tubes containing 5 ml each of TSB with the dilution value (i.e., ½, ¼, ⅛, ¹⁄₁₆, and ¹⁄₃₂).

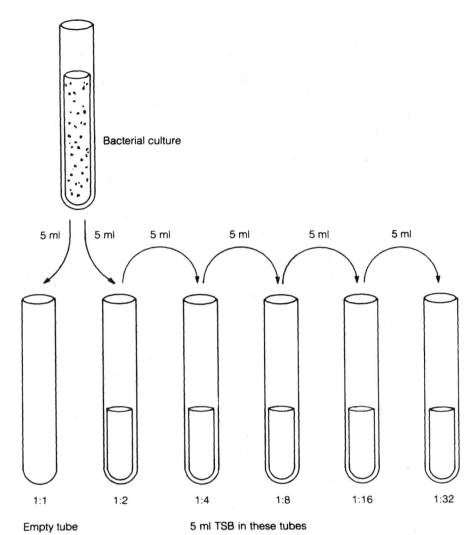

Bacterial culture

5 ml   5 ml   5 ml   5 ml   5 ml   5 ml

1:1   1:2   1:4   1:8   1:16   1:32

Empty tube   5 ml TSB in these tubes

Figure 15-2. Dilution Procedure

b. Mix the stock culture well by tapping the bottom of the tube several times with your index finger while holding the tube near the top with the other hand.

c. Transfer 5 ml of the culture suspension into the tube labeled ½. Mix well. This is a ½ dilution of the stock culture.

d. Transfer 5 ml of the ½ dilution into tube labeled ¼. Mix well. This is a ¼ dilution of the stock culture.

e. Repeat this procedure and make ⅛, ¹⁄₁₆, and ¹⁄₃₂ dilutions of the culture.

2. Calibrate the colorimeter to read 0.0 absorbance with sterile TSB medium as the blank.

• This is the suggested procedure using the Spectronic 20.

a. Set the wavelength at 686 nm.

b. Switch the Spectronic 20 on and allow the colorimeter to warm up for approximately 20 minutes.

c. Adjust the colorimeter to read 0 absorbance using the power/zero control knob.

d. Pipet approximately 5 ml of TSB into a cuvette. Make sure that there are no smudges or fingerprint marks on the walls of the cuvette. This may interfere with the path of the light and give erroneous readings.

e. Open the lid of sample holder, insert the cuvette with broth, then close the lid.

f. Adjust the meter to read 0.0 absorbance using the control knob.

g. Remove the cuvette containing the broth.

3. Read the absorbance of the undiluted stock culture and each of the 5 dilutions made.

- The procedure is as follows:

  a. Pipet approximately 5 ml of the sample into a clean cuvette. Use proper aseptic techniques since you are dealing with live microorganisms. Exercise care and do not contaminate the colorimeter.

  b. Insert the cuvette with the sample into the sample holder, close the lid, and read the absorbance displayed on the meter.

  c. Record the results in the report form provided.

  d. Remove the cuvette and dispose of the sample as indicated by your instructor.

  e. Repeat steps a through d for each of the culture dilutions.

4. Calculate the OD for the stock culture and each of the dilutions.

Second Period

1. Construct a calibration curve by plotting the OD on the Y axis and the viable counts on the X axis. Use the graph paper provided.

Questions

1. Explain the principle involved in OD determinations of microbial populations.

2. Give two instances when the turbidimetric method for counting microbial populations is indicated.

3. What is the purpose of constructing a calibration curve?

4. Why is OD used instead of %T?

5. Would the calibration curve obtained with *E. coli* in TSB be valid to measure an *Escherichia coli* culture growing in nutrient broth? Explain briefly.

6. Can the *E. coli* calibration curve be used to measure *Staphylococcus aureus* populations? Explain briefly.

PART V
ENVIRONMENTAL FACTORS AFFECTING GROWTH

Heat, hydrogen ion concentration (pH), water activity (available water), oxygen concentration, and light intensity are factors that dramatically affect the existence of microorganisms.

Heat is the kinetic energy that a group of atoms or molecules contains, while **temperature** is a measure of their average kinetic energy. The enzymes in living organisms that speed up chemical reactions have evolved to function best when they and their substrates have a given amount of energy. The temperature at which an enzyme functions most efficiently is known as its **optimum temperature.** When the temperature deviates from the optimum temperature, an enzyme's activity decreases. As the heat content of the enzyme and its substrate increases from the optimum heat content, the three-dimensional structure of the enzyme may become significantly and permanently altered, at which point it ceases to function. In the same manner, as the heat content decreases, the chemical reaction slows down until the three-dimensional structure of the enzyme is altered and it ceases to function.

The temperature sensitivity of enzymes determines the growth characteristics of microorganisms. The highest temperature at which an organism will grow is called the **maximum growth temperature,** while the lowest temperature at which an organism will grow is known as the **minimum growth temperature.** The minimum and maximum growth temperatures of an organism are not only dependent upon the temperatures at which a critical enzyme denatures or become nonfunctional, but also upon the temperatures at which the membrane becomes excessively fluid (melts) or excessively rigid (freezes). The flexibility and permeability characteristics of the membrane are maintained by the type of lipids in the lipid bilayer and the heat content of the membrane. As little as 5° to 10°C above the optimum temperature, a membrane may become excessively fluid. This leads to leakage of materials or cell lysis and death.

Hydrogen ions are protons that detach from various organic and inorganic molecules. A measure of the hydrogen ion concentration is

$$pH = -\log_{10}(H^+)$$

where H^+ is given as moles/liter. Pure water exists in two forms: HOH and $H^+ + HO^-$, in equilibrium with each other ($HOH \leftrightharpoons H^+ + HO^-$). Most of the water exists in the undissociated form. In fact, in pure water at 25°C, the concentration of hydrogen ions (as well as hydroxyl ions) is 10^{-7} moles/liter. Thus, the pH of pure water is 7. When the concentration of hydrogen ions and hydroxyl ions is equal, water is said to be neutral.

Acids are organic and inorganic molecules that release more hydrogen ions than hydroxyl ions. In an aqueous environment, the hydrogen ions associate with water molecules (H_3O^+). An acid such as lactic acid exists in two forms in equilibrium with each other: $CH_3-CHOH-COOH$ and $CH_3-CHOH-COO^- + H^+$. Most of the acid exists in the dissociated form.

Bases are organic and inorganic molecules that release more hydroxyl ions than hydrogen ions, or that bind hydrogen ions. A base such as an amine (CH_3-NH_2) picks up hydrogen ions and makes the environment more basic. On the other hand, a base such as sodium hydroxide (NaOH) releases hydroxyl groups (HO^-) that pick up hydrogen ions. This results in the formation of water.

The proteins (enzymes and transport systems) found in the plasma membrane are sensitive to the concentration of hydrogen ions in the medium. The three-dimensional structure of proteins and enzymes is altered when the concentration of hydrogen ions changes. Each organism has evolved enzymes that function optimally at a certain concentration of hydrogen ions. If the hydrogen ion concentration deviates from the optimum concentration, metabolism is disrupted. Although some bacteria can grow at pH values as low as 2 and others can grow at pH values as high as 10, most bacteria grow best at a pH close to 7 and do not grow well if the pH is below 6 or above 9. Fungi are able to grow over a wider range of pH values. Unlike bacteria, most fungi are able to grow well at pH values below 6 and consequently can be selected for on acidic media such as Sabouraud glucose agar (pH = 5.7), which inhibits the growth of most bacteria.

The **available water** in an environment is an important factor that affects the growth of microorganisms. A measurement of the available water in an environment is known as the **water activity** (a_w). If there is not a high enough concentration of water surrounding a cell, the water within the organism diffuses into the environment. This results in the shrinking of the protoplasm, a situation referred to as **plasmolysis.** Metabolic activity is inhibited when plasmolysis occurs, probably because materials within the cell have become so concentrated that they precipitate, disrupt regulation, and block enzyme activity.

Molecular oxygen (O_2) is usually thought of as being necessary for life. This is not always the case, however, since many organisms can grow and multiply in the absence of oxygen and some organisms are inhibited and killed in the presence of oxygen. Organisms that use oxygen require it for carrying out aerobic respiration; oxygen functions as an electron acceptor for waste electrons and hydrogen ions that were used to generate ATP. Microorganisms that can grow without oxygen obtain their energy in a number of ways. They may carry out fermentations, engage in anaerobic respiration, or carry out photophosphorylation (light-driven reactions of photosynthesis). In fermentations, waste electrons and hydrogen ions are accepted by organic compounds; while in anaerobic respiration, waste electrons and hydrogen ions are accepted by inorganic molecules such as sulfate and nitrate rather than molecular oxygen. Oxidized chlorophyll accepts spent electrons in microorganisms that carry out photophosphorylation.

Bacteria that are inhibited by oxygen lack enzymes that catalyze the conversion of superoxide (O_2^-) to hydrogen peroxide (H_2O_2), and the breakdown of hydrogen peroxide to nontoxic compounds. The enzyme **superoxide dismutase** rids the cell of toxic superoxide by converting it into H_2O_2, while two enzymes, **peroxidase** and **catalase,** break down toxic hydrogen peroxide. Superoxide and hydrogen peroxide constantly form in cells exposed to oxygen and must be rapidly converted to nontoxic compounds before they damage cellular components. Superoxide and hydrogen peroxide are very toxic because they react with enzymes and components of the plasma membrane and disrupt the cell. If an organism is missing superoxide dismutase or both enzymes involved in the elimination of hydrogen peroxide, it must avoid giving electrons to oxygen. It does this by existing in an anaerobic environment.

Visible light is electromagnetic radiation with a wavelength between 380nm (violet) and 760nm (red). A large number of microorganisms, prokaryotes and eukaryotes, use light as a source of energy to generate chemical energy in the form of adenosine triphosphate (ATP). The shorter the wavelength, the more energy associated with a quantum of light. Each quantum of light contains the energy given by the following formula: $E = ch/\lambda$, where E = energy in joules, h = Plank's constant = 6.63×10^{-35} joule-second, c = speed of light = 3×10^8 m/sec, and λ = wavelength of light. Bacteriocholorophyll, in the photosynthetic bacteria, absorbs light. This excites electrons in the bacteriochlorophyll that are transferred to an electron transfer system. The flow of electrons along an electron transfer system results in the creation of a hydrogen ion gradient and an electrical potential across internal membranes. The collapse of the hydrogen ion gradient and the electrical potential brought on by the flow of protons down the gradient provides the energy for making ATP from adenosine diphosphate (ADP) and inorganic phosphate (P_i).

Most photosynthetic microorganisms will not grow or will grow very slowly if they receive insufficient light with which to generate the electrical potentials and chemical energy they require. Thus, light is an important and sometimes essential environmental factor that affects the growth rate of most photosynthetic microorganisms.

In the following sections, you will be performing experiments that show the important effects that heat, hydrogen ion concentration, water availability, oxygen concentration, and light intensity have on the growth and survival of microorganisms.

EXERCISE 16
EFFECT OF TEMPERATURE
ON MICROBIAL GROWTH

Microorganisms are sometimes categorized according to the temperature at which they grow most rapidly (fig. 16-1). Organisms that have their optimum growth temperature between 0°C and 20°C are known as **psychrophiles.** Psychrophiles show significant growth within 2 weeks at 0°C. Those organisms that grow most rapidly between 20°C and 50°C are referred to as **mesophiles.** Many mesophiles grow slowly at temperatures as low as 5°C while a few grow slowly at temperatures as high as 55°C. Mesophiles that grow slowly at 5°C are often called psychrophiles by food microbiologists. **Psychrotrophic mesophile** (cold-nourished middle-temperature favoring organism) may be a better term for these organisms. Organisms that have their optimum growth temperature between 50°C and 100°C are called **thermophiles.** A thermophile grows very poorly or not at all at temperatures below 45°C.

Some organisms may be able to survive high temperatures for long periods of time even though they are unable to grow. These organisms are known as **thermoduric** organisms. Many of the endospore formers can withstand boiling for 5 to 15 minutes because their endospores are heat-resistant. The en-

dospore formers are thermoduric even though their vegetative cells are destroyed very rapidly by boiling, like other vegetative cells.

In this section, we will investigate the effect of temperature on a typical psychrophile, mesophile, and thermophile.

Materials

Marine broth culture of *Vibrio fischeri*

Trypticase soy broth (TSB) cultures of *Pseudomonas fluorescens, Escherichia coli, Mycobacterium phlei,* and *Bacillus stearothermophilus*

Tubes of marine broth

Tubes of trypticase soy broth

Incubators set at 4°C (a refrigerator will do), 18°C, 37°C, 45°C, 55°C, and 60°C

Procedure

First Period

1. Inoculate a set of six tubes of broth with one type of organism.

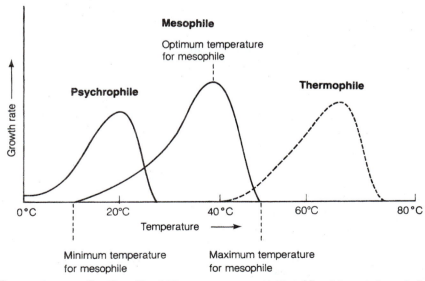

Figure 16-1. Effects of Temperature on the Growth of Microorganisms. *Thermophilic* microorganisms have their optimum growth temperatures between 50°C and 100°C. *Mesophilic* microorganisms grow most rapidly at temperatures between 20°C and 50°C while *psychrophilic* microorganisms have their optimum growth temperatures between 0°C and 20°C.

2. Clearly label the tubes with the name of the organism and the temperature at which they are to be incubated.

3. Incubate one tube at 4°C, 18°C, 37°C, 45°C, 55°C, and one tube at 60°C for 48 hours.

4. Follow the preceding instructions for each organism used. *Vibrio* should be grown in marine broth while the other organisms should be grown in TSB.

Second Period

1. Look for growth (turbidity, sediment, pellicle) in each of the tubes.

2. Record in tabular form the temperatures at which the organisms showed growth (+) or no growth (−).

Questions

1. Differentiate between heat and temperature.

2. Contrast minimum growth temperature, optimum growth temperature, and maximum growth temperature.

3. Why do organisms grow more slowly as the temperature deviates from the optimum growth temperature?

4. Define psychrophile, mesophile, and thermophile.

5. What is a thermoduric organism?

6. Does the experiment establish that *Escherichia* is a mesophile? Explain why.

7. Explain why *B. stearothermophilus* does not grow at 18°C, a temperature that humans consider comfortable.

EXERCISE 17
EFFECT OF pH ON MICROBIAL GROWTH

Microorganisms are often divided into three categories depending upon the pH at which they grow most rapidly. **Acidophiles** are organisms with a pH optimum between 2 and 5 that grow poorly or not at all at neutrality (pH = 7). Some acidophiles can grow at pH values as low as 1. Organisms with a pH optimum between 8.5 and 10 are known as **alkalophiles (alkalinophiles).** Alkalophiles grow at a pH of 10 or above, some to pH values as high as 12 but grow poorly or not at all at neutrality. Those organisms that cannot

grow if the pH deviates much from neutrality and have their optimum growth at pH values close to 7 may be called **neutralophiles.** Organisms that are tolerant of extremes in pH and able to grow slowly at low or high pH values are sometimes referred to as **acidotolerant (acidoduric)** and **alkalotolerant (alkaloduric),** respectively. Most fungi are acid-tolerant; in contrast to most bacteria, which are inhibited when the pH drops below 6 (fig. 17-1).

Microorganisms frequently alter the pH of their

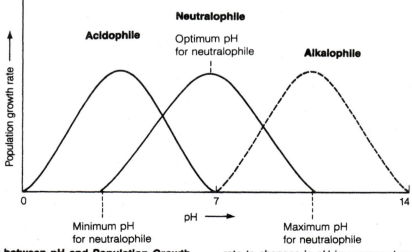

Figure 17-1. Relationship between pH and Population Growth Rate. Each microbial population responds to changes in environmental pH with changes in its growth rate. The response of the growth rate to changes in pH is expressed as a bell-shaped curve with a maximum, a minimum, and an optimum pH.

environment as they proliferate. Fermenting organisms may release so much acid that the pH drops as low as 3.5. Respiring and fermenting organisms metabolizing proteins and amino acids may release ammonium ions, which makes the environment alkaline.

Drastic changes of the pH occur rapidly in a closed environment such as a nutrient broth tube and can inhibit microbial growth and significantly limit the development of a culture. In order to prevent the pH from changing drastically, chemicals called **buffers** are added to media to inhibit rapid pH changes. A buffer that is frequently used in media consists of K_2HPO_4, a salt of a weak base; and KH_2PO_4, a salt of a weak acid. The weak base binds hydrogen ions when acids are produced, becomes a weak acid, and so buffers acidic environments. The weak acid donates hydrogen ions when ammonia is produced and becomes a weak base. The weak acid buffers strong bases (OH^- and NH_3) by supplying them with hydrogen ions.

In this exercise, you will see how the hydrogen ion concentration affects the growth of microorganisms.

Materials

Trypticase soy broth (TSB) cultures of *Lactobacillus bulgaricus*, *Escherichia coli*, and *Pseudomonas fluorescens*.

Sabouraud glucose agar (pH = 5.7) cultures of *Saccharomyces*, *Penicillium*, and *Rhizopus*

Tubes of TSB with pH values of 3, 4, 5, 6, 7, 8, 9, 10, 11, 12 in stoppered tubes

Procedure

First Period

1. Inoculate a set of ten TSB tubes having ten different pH values with a single type of bacterium or fungus.

2. Clearly label the tubes with the name of the organism and the pH of the tube.

3. Incubate the set of ten tubes at 28°C for 48 hours.

4. Follow the preceding instructions for each organism used.

Second Period

1. Look for evidence of growth (turbidity, sediment, pellicle) in each of the tubes.

2. Record in tabular form the pH at which the organisms showed growth (+) or no growth (−).

Questions

1. Differentiate between hydrogen ion concentration and pH.

2. What were the optimum pH values for the bacteria and fungi tested? Were there any differences between bacteria and fungi?

3. What were the minimum pH values for the bacteria and fungi tested? Were there any differences between bacteria and fungi?

4. What were the maximum pH values for the bacteria and fungi tested? Were there any differences between bacteria and fungi?

5. Why do organisms grow more slowly as the pH deviates from the optimum pH?

6. Does the experiment establish that *Escherichia* is a neutrophile? Explain why.

EXERCISE 18
EFFECT OF OSMOTIC PRESSURE ON MICROBIAL GROWTH

All cellular organisms require water to grow. Some must be continuously immersed in an aqueous environment, while others can grow on a surface if the humidity is sufficiently high. The concentration of a solute in water can affect the amount of water available to microorganisms and, because of this, can strongly affect the growth and survival of the organisms.

Liquid distilled water in a closed container is in equilibrium with water that evaporates into the enclosed atmosphere. The evaporated water creates a **vapor pressure,** P_w. A similar volume of salt water in a closed container produces a vapor pressure that is less than the vapor pressure produced by the distilled water, $P_w > P_s$. The amount of water available to microorganisms is measured as the **water activity (a_w),** a ratio of the vapor pressure of a solution (or material) to the vapor pressure of pure water ($a_w = P_s/P_w$). Pure water has a water activity of 1, while other materials have water activities that are less than one.

Most bacteria require a water activity above 0.90 for growth, although some common bacteria such as *Staphylococcus* can grow when the water activity is as low as 0.86. Some unusual bacteria such as *Halobacterium* grow when the water activity is as low as 0.65. Many fungi grow at water activities as low as 0.70, and a few can grow when the water activity falls to 0.60. In general, fungi are able to grow in drier environments than bacteria. Thus, fungi are usually responsible for the initial rotting of such things as woods, leathers, and textiles, as well as the primary spoilage of preserved or dried foods such as jams, jellies, jerky, fish, and fruit.

If there is not sufficient water available to a cell, water diffuses from the cell and the cell undergoes **plasmolysis** (fig. 18-1). Thus, in hypertonic environments where the concentration of solutes is high and the water activity is low, cells undergo plasmolysis. This type of environment is said to have a high **osmotic pressure.** When the concentration of solutes is very low and the water activity is very close to 1, water tends to diffuse from the environment into the cell. If the water pressure or osmotic pressure that builds up within the cell is sufficiently high it can rupture the cell. The lysis of the cell is known as **plasmoptysis.** In general, Gram-positive bacteria are more resistant to plasmoptysis than Gram-negative bacteria because Gram-positive bacteria have a thicker layer of peptidoglycan.

The a_w of an environment may decrease as water evaporates or as the solute concentration increases. Organisms capable of growth in dry environments are often referred to as **xerotolerant** microbes or as **xerophiles** while those that are capable of growth in environments with high solute concentrations are generally called **osmotolerant** microbes or **osmophiles.** Xerotolerant or osmotolerant organisms are defined as those capable of growth in dry environments or solutions (such as syrups) with an a_w as low as 0.90 but not as low as 0.85. Xerophiles or osmophiles, on the other hand, are defined as organisms capable of growth in dry environments or solutions with an a_w as low as 0.85. A 50% solution of glucose has an a_w of 0.90 while a 60% solution of glucose has an a_w of 0.85. A 15% solution of NaCl has an a_w of 0.895 while one that is 20% has an a_w of 0.845. Most of the commonly encountered xerotolerant and osmotolerant organisms as well as the xerophiles and osmophiles are filamentous fungi and yeasts. **Saccharophiles** are osmotolerant or osmophilic organisms that grow well in high concentrations of sugar.

Bacteria are osmophiles that have a requirement for high concentrations of sodium chloride (NaCl). Organisms that require high concentrations of NaCl can be divided into a number of groups. The **extreme halophiles** have a salt optimum between 2.5M and 5.2M of NaCl (15% to 30%) and grow poorly or not at all at lower concentrations. These concentrations of NaCl create solutions with a_w values of between 0.92 and 0.895. Extreme halophiles are also known as **obligate halophiles** and include such organisms as *Sarcina morrhuae*, *Halobacterium halobium* and *Halobacterium slinarum*. It is believed that *Halobacterium* requires very high concentrations of salt, positive sodium ions, to stabilize its membrane that contains many negative charges. *Halobacterium* and *Sarcina* species can grow in environments with as much as 30% NaCl (5M). Obligate halophiles are sometimes re-

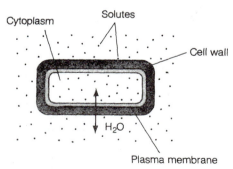

Cytoplasm · Solutes · Cell wall · H₂O · Plasma membrane

(a) Isotonic medium

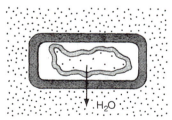

H₂O

(b) Hypertonic medium

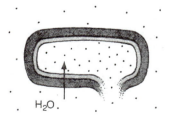

H₂O

(c) Hypotonic medium

Figure 18-1. Flow of Water in Isotonic, Hypertonic, and Hypotonic Environments. *(a)* Cell in an isotonic environment. In isotonic environments, the concentration of environmental solutes is equal to that of the cell. In this environment, cellular water shows no net movement in or out of the cell. *(b)* Cell in a hypertonic environment. In hypertonic environments, the concentration of environmental solutes is greater than those in the cell. Hence, there will be a tendency for cellular water to flow out into the environment. This results in the dehydration and shrinking (plasmolysis) of the cell. *(c)* Cell in hypotonic environment. In hypotonic environments, the concentration of environmental solutes is lower than those in the cell. Hence, environmental water will tend to flow into the cell. In the process, the cell swells from an excess of water. If the flow of water is too great, the cell may burst (plasmoptysis). Although cells with cell walls may simply swell and not burst, cells with weak cell walls (Gram negative bacteria) may burst due to excessive water intake.

sponsible for spoiling brines (12% NaCl) used to pickle olives. These organisms withstand the hypertonic environments and avoid plasmolysis because they accumulate potassium ions to concentrations between 2m and 5m.

Marine bacteria have their optimum growth with 3% to 15% NaCl (0.5M to 2.5M) and are referred to as **moderate halophiles.** Most bacteria that humans encounter grow best at very low concentrations of NaCl, below 3% (0.5M), and grow poorly or not at all at NaCl concentrations above 3%. A few common bacteria that have no requirements for NaCl, such as *Staphylococcus*, can grow well in the presence of 7.5% NaCl and are said to be **halotolerant** or **haloduric.**

This exercise is intended to acquaint you with compounds that can affect water activity and to show how some organisms are affected by high concentrations of NaCl and sucrose.

A. EFFECT OF NaCl ON MICROORGANISMS

Materials

Nutrient broth cultures of *Escherichia, Staphylococcus, Saccharomyces, Rhizopus,* and *Penicillium*

Nutrient agar plates with 0.5%, 1%, 3%, 5%, 10%, 15%, and 20% NaCl

Incubator set at 20°C

Procedure

First Period

1. Divide the 0.5% NaCl nutrient agar plate into five sectors. Inoculate each sector with a different organism.

2. Inoculate the remaining different NaCl plates (1, 3, 5, 10, 15, and 20%) in the same way.

3. Incubate the plates at room temperature for 2 to 4 days.

Questions

1. What is water activity? What is osmotic pressure?

2. Is there any relationship between water activity and osmotic pressure? Explain your answer.

3. At what water activities are most bacteria and fungi inhibited?

4. What are obligate halophiles? At what concentrations of NaCl do these organisms grow?

5. At what concentrations of NaCl do the halophilic marine bacteria grow?

6. What concentrations of NaCl do most bacteria prefer?

7. What concentration of NaCl is used to enrich for the haloduric *Staphylococcus?*

B. EFFECT OF SUCROSE ON MICROORGANISMS

Materials

Nutrient broth cultures of *Escherichia, Staphylococcus, Saccharomyces, Rhizopus,* and *Penicillium*

Nutrient agar plates with 1%, 5%, 10%, 20%, 40%, and 60% sucrose

Incubator set at 20°C

Procedure

First Period

1. Divide the 1% sucrose nutrient agar plate into five sectors. Inoculate each sector with a different organism.

2. Inoculate the remaining sucrose plates (5, 10, 20, 40, and 60%) in the same manner.

3. Incubate the plates at room temperature for 2 to 4 days.

Second Period

1. Look for growth on the plates.

2. Record in tabular form at which sucrose concentrations the organisms showed heavy growth (++), slight growth (+), or no growth (−).

Questions

1. Are the fungi you tested more or less resistant than the bacteria to high concentrations of salt and sugar?

2. At what sugar concentrations are the bacteria and fungi inhibited?

3. Differentiate acerophile, osmophile, saccharophile, and halophile.

EXERCISE 19
EFFECT OF OXYGEN
ON MICROBIAL GROWTH

Microorganisms are often divided into five categories depending upon their need for and response to molecular oxygen (O_2): obligate aerobes, obligate anaerobes, facultative anaerobes, aerotolerant anaerobes, and microaerophiles.

Those organisms that require oxygen for life are known as **obligate aerobes** or aerobes. These organisms generate their energy through aerobic respiration and require oxygen as the terminal electron and hydrogen ion acceptor. Examples of obligate aerobes are *Micrococcus luteus*, *Pseudomonas fluorescens*, and *Mycobacterium phlei*.

Organisms that are killed by oxygen, and consequently live only in oxygen-free environments are called **obligate anaerobes** or anaerobes. There are a number of reasons why organisms are killed by oxygen. They may be killed because they lack, or have very low levels of, superoxide dismutase, which converts the toxic superoxide (O_2^-) that forms under aerobic conditions into hydrogen peroxide (H_2O_2); they may lack both peroxidase and catalase, which convert hydrogen peroxide to nontoxic forms; or they may contain oxygen-sensitive enzymes. Obligate anaerobes generate their energy through fermentative processes or anaerobic respiration and thus do not require oxygen as the terminal electron and hydrogen ion acceptor. Examples of obligate anaerobes include *Bacteroides fragilis*, *Peptococcus anaerobius*, and *Clostridium tetani* among the chemoheterotrophs; and *Chromatium* and *Chlorobium* among the photoautotrophs. Anaerobic environments with less than 0.1% O_2 can be created in the laboratory by using the **GasPak anaerobic jar** (fig. 19-1). Anaerobic environments can also be achieved in **agar shake-cultures** (fig. 19-2) or the **pyrogallic acid-sodium hydroxide chemical system** (fig. 19-3). Anaerobes grow poorly or not at all at O_2 concentrations above 0.4%.

Some organisms are able to respire if oxygen is present but ferment if oxygen is limited or absent. These respiring and fermenting organisms that are capable of growth whether or not oxygen is present are called **facultative anaerobes.** Some examples of facultative anaerobes are *Escherichia coli* and *Staphylococcus aureus.*

One group of microorganisms ferments under both anaerobic and aerobic conditions. Molecular oxygen does not inhibit fermentation in these organisms; consequently they are fundamentally different from the other fermenting organisms. These organisms are sometimes referred to as **aerotolerant anaerobes.** Examples of these aerotolerant anaerobes include *Lactobacillus bulgaricus*, *Streptococcus lactis*, and *Clostridium histolyticum.* Notice that not all clostridia are obligate anaerobes.

Microaerophiles are organisms that require oxygen but only grow if the concentration of molecular oxygen is reduced below 15%. Most microaerophiles grow best at O_2 concentrations between 5 and 10%.

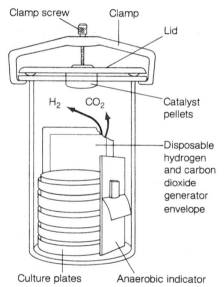

Figure 19-1. An Anaerobic System. The anaerobic system illustrated here is commonly used in the microbiology laboratory to cultivate anaerobic microorganisms. It consists of a jar with a screw-on lid, a hydrogen-carbon dioxide generator envelope, and a palladium catalyst. Water is added to the hydrogen-carbon dioxide generating envelope to activate it, and the envelope is placed inside the jar along with the cultures. An anaerobic indicator strip may also be included. The strip consists of a pad saturated with methylene blue solution, and it changes from blue to colorless when an anaerobic atmosphere inside the jar is developed. The hydrogen gas generated by the envelope reacts with any oxygen that may be present in the jar to form water. The reaction is catalyzed by the palladium catalyst, which permits the formation of water from hydrogen and oxygen to take place at room temperature. This reaction removes any free oxygen from inside the jar, creating an anaerobic environment. The CO_2 is needed to stimulate the growth of certain anaerobes.

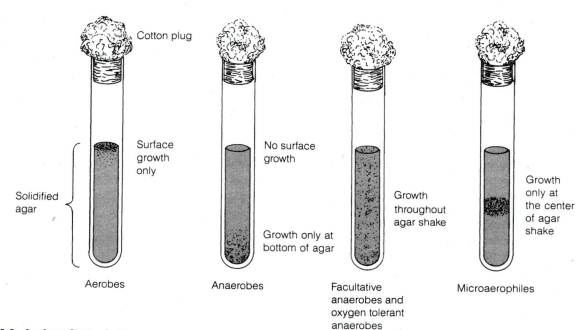

Figure 19-2. An Agar Shake-Culture

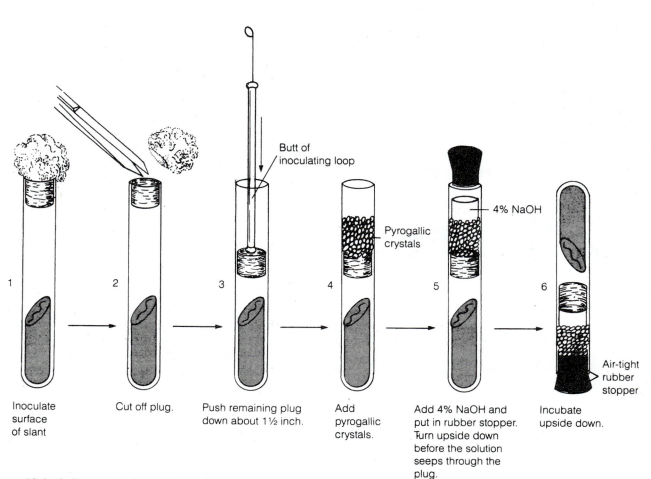

1 Inoculate surface of slant

2 Cut off plug.

3 Push remaining plug down about 1½ inch.

4 Add pyrogallic crystals.

5 Add 4% NaOH and put in rubber stopper. Turn upside down before the solution seeps through the plug.

6 Incubate upside down.

Figure 19-3. A Pyrogallic Acid-Sodium Hydroxide Chemical System

An example of a microaerophile is *Campylobacter fetus* ssp. *jejuni*, a common cause of diarrhea and acute gastroenteritis. Microaerophilic conditions are found in agar shake-cultures and can be created by using special GasPak envelopes in an "anaerobic jar." The special GasPak envelopes create an atmosphere of 5 to 12% CO_2 and 5 to 15% O_2. The normal atmosphere consists of 0.03% CO_2 and 20% O_2.

Some heterotrophic organisms require an atmosphere enriched in carbon dioxide in order to grow well. Organisms that grow well only when the carbon dioxide concentration is increased from 0.03% to 5% or more are referred to as **capnophiles.** Aerobic capnophiles require oxygen for their metabolism. *Neisseria sicca* is an example of a capnophile that reproduces well in an atmosphere of 4% CO_2, 16% O_2, and 80% N_2. In the laboratory, this type of atmosphere is easily created by using a **candle jar** (fig. 19-4).

This exercise shows how oxygen concentrations affect the growth of microorganisms, demonstrates the oxygen requirements of microorganisms, and acquaints you with various methods for cultivating anaerobic and microaerophilic organisms.

A. GROWING ORGANISMS IN AN ANAEROBIC GASPAK JAR

The GasPak system is generally used to grow anaerobes in the laboratory because it is simple to use and provides an atmosphere almost free of oxygen. The system creates an anaerobic atmosphere that is enriched with CO_2 and so promotes the growth of CO_2-requiring organisms. A disposable envelope containing chemicals that yield H_2 and CO_2 when water is added is also part of the system. A reaction catalyzed by palladium crystals converts the H_2 and O_2 to water (fig. 19-1).

Materials

TSB cultures of *Micrococcus luteus, Clostridium perfingens, Escherichia coli, Streptococcus lactis,* and *Neisseria sicca*

Petri plates

Tubes of plate count agar (PCA)

GasPak anaerobic jars with fresh palladium catalyst

oxygen indicators

H_2 and CO_2 generating envelopes (GasPak)

Distilled H_2O

10 ml pipet

Incubator set at 35°C

First Period

1. Pour ten plates of PCA.

2. Streak each of the cultures (*Micrococcus, Clostridium, Escherichia, Streptococcus,* and *Neisseria*) onto two plates each. One plate of each will be incubated anaerobically, the other aerobically.

3. Label each plate so that you know which organism and atmospheric conditions are involved.

4. Place one set of plates upside down in an incubator at 35°C for 48 hours. Place the second set of plates upside down in a GasPak anaerobic jar that has a fresh catalyst in the lid.

5. Open an oxygen indicator strip and place it inside the anaerobic jar so that the indicator is visible.

• White indicates anaerobic conditions while blue signifies aerobic conditions.

6. Tear off the corner of the hydrogen and carbon dioxide generating envelope and place the envelope inside the anaerobic jar. Add 10 ml of distilled water with a pipet to the envelope through the tear. Immediately place the top on the anaerobic jar and hand tighten it to seal the jar.

• The gas pack produces carbon dioxide and hydrogen, while the metal catalyst in the top rapidly converts the oxygen and hydrogen to water.

7. Incubate the sealed jar at 35°C for 48 hours.

Second Period

1. Before opening the anaerobic jar, check the oxygen indicator to make sure the jar became anaerobic.

• A white indicator indicates anaerobic conditions (all is well) but a blue indicator signifies aerobic conditions (not so good).

2. Open the anaerobic jar and check the plates for

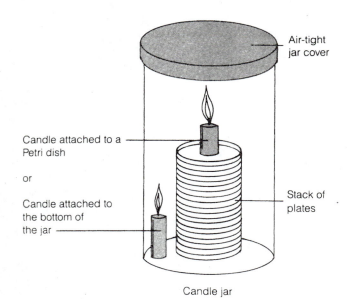

Candle jar

Figure 19-4. A Candle Jar

Air-tight jar cover

Candle attached to a Petri dish

or

Candle attached to the bottom of the jar

Stack of plates

growth. Check the set of plates that were incubated under normal atmospheric conditions.

3. Tabulate your results for each organism: heavy growth (+ +), slight growth (+), and no growth (−).

Questions

1. Explain how the atmosphere is created in an anaerobic jar. What are the concentrations of O_2 and CO_2?

2. Will an aerobic capnophile such as *Neisseria* grow in an anaerobic GasPak jar? Explain your answer.

B. GROWING ORGANISMS IN AGAR SHAKE-CULTURES

Agar shake-cultures provide a simple way to check an organism's response to molecular oxygen. They are easily prepared by melting a nutrient agar tube and then inoculating the melted nutrient after it has cooled to 47°C. Aerobes will grow only on the surface, anaerobes deep within the agar, and facultative anaerobes will grow throughout the agar medium (fig. 19-2).

Materials

TSB cultures of *Micrococcus luteus; Clostridium perfingens, Escherichia coli, Streptococcus lactis,* and *Neisseria sicca*

Tubes of plate count agar (PCA)

1 ml pipets

Incubator set at 35°C

Procedure

First Period

1. Melt five tubes of PCA and cool to 50°C.

2. Inoculate the tubes of PCA with *Micrococcus, Clostridium, Escherichia, Streptococcus,* and *Neisseria* so that there is a different organism in each tube.

• Use 0.1 ml of the culture to make the inoculations. Make sure that the temperature of the media is comfortable to the touch so that the organisms are not killed. Thoroughly mix the cultures and then allow them to solidify.

3. Incubate the organisms at 35°C for 48 hours.

Second Period

1. Look for evidence of growth in each of the tubes.

2. With drawings, indicate where in the tube each organism grew.

Questions

1. Differentiate between obligate aerobe, obligate anaerobe, facultative anaerobe, aerotolerant anaerobe, and microaerophile.

2. What are some of the reasons why an organism is an obligate anaerobe?

3. Where in a tube of solid medium would you expect the following organisms to grow?
   a. *Escherichia coli*
   b. *Clostridium perfringens*
   c. *Micrococcus luteus*
   d. *Pseudomonas fluorescens*
   e. *Streptococcus lactis*
   f. *Neisseria sicca*

4. In nature, where might you find obligate anaerobes?

C. GROWING ORGANISMS IN A PYROGALLIC ACID-NaOH SYSTEM

An inoculated slant can be made anaerobic by adding pyrogallic acid crystals and sodium hydroxide (NaOH) (fig. 19-3). When activated with NaOH, pyrogallic acid crystals reduce the oxygen in the tube to water.

Materials

TSB cultures of *Micrococcus luteus, Clostridium perfringens, Escherichia coli, Streptococcus lactis,* and *Neisseria sicca*

Slants of trypticase soy agar (TSA) with cotton plugs

Crystals of pyrogallic acid

4% NaOH solution

Scissors

Rubber stoppers

Procedure

First Period

1. Inoculate two sets of cotton-plugged TSA slants with *Micrococcus, Clostridium, Escherichia, Streptococcus,* or *Neisseria,* so that there is only one organism on each slant.

2. Label each slant so that you know which organism and atmospheric conditions are involved.

3. On one set of tubes (6), crop the cotton plugs using a pair of scissors so that the cotton is flush with the lip of the tube (fig. 19-3). Push the cotton into the tube with the butt of your inoculating loop until it almost

touches the edge of the agar slant. Seal the set of slants with rubber stoppers, invert the slants, and place them in an incubator at 35°C for 48 hours.

4. Crop and push the cotton plugs of the second set of tubes as you did in step 3 above.

5. Partially fill the space between the cotton plug and the top of the tube with crystals of pyrogallic acid.

6. Cover the crystals with 4% sodium hydroxide but make sure that there is enough room for a rubber stopper. Immediately stopper the slants with a rubber stopper and invert the slants.

7. Place the second set of inverted slants upside down in an incubator at 35°C for 48 hours.

Second Period

Check the two sets of plate count agar slants for growth.

Tabulate your results for each organism and each condition: heavy growth (+ +), slight growth (+), and no growth (−).

Questions

1. What is the atmosphere like in the pyrogallic acid-sodium hydroxide tubes?

2. Does the pyrogallic acid-sodium hydroxide system work as well as the anaerobic GasPak system for growing anaerobes?

D. GROWING ORGANISMS IN A CANDLE JAR

A candle jar (fig. 19-4) is frequently used to grow capnophilic organisms because it maintains a somewhat aerobic environment while increasing the CO_2 concentration. The burning of a candle reduces the oxygen from 20% to about 16% and increases the CO_2 from less than 0.4% to approximately 4%. The increased CO_2 stimulates the growth of capnophilic organisms.

Materials

TSB cultures of *Micrococcus luteus, Clostridium perfringens, Escherichia coli, Streptococcus lactis, Neisseria sicca*

Petri plates

Tubes of plate count agar (PCA)

Candle jars, candles, and matches

Incubator set at 35°C

Procedure

First Period

1. Make two sets of five plates of PCA.

2. Streak each of the cultures *(Micrococcus, Clostridium, Escherichia, Streptococcus, and Neisseria)* onto each of two plates.

3. Label each plate so that you know which organisms and atmospheric conditions are involved.

4. Place one set of plates upside down in an incubator at 35°C for 48 hours. Place the second set of plates upside down in a candle jar.

5. Light a candle (which is usually attached to the bottom of a Petri dish) and place the lit candle inside the candle jar. There must be enough room so that the flame does not burn the lid of the jar.

6. Screw the lid onto the candle jar so that it is airtight. The flame will go out in a few seconds.

7. Incubate the sealed jar at 35°C for 48 hours.

Second Period

1. Open the candle jar and check the plates for evidence of growth. Check the set of plates that were incubated under normal atmospheric conditions.

2. Tabulate your results for each organism: heavy growth (+ +), slight growth (+), and no growth (−).

Questions

1. The candle jar is used for growing what type of organisms?

2. Explain why you cannot use a candle jar to grow obligate anaerobes.

EXERCISE 20
EFFECT OF LIGHT
ON MICROBIAL GROWTH

Microorganisms that use light to generate energy in the form of electrical potentials or chemical energy are known as **phototrophs.** The prokaryotic phototrophs may be divided into four groups; the anaerobic **purple and green sulfur bacteria,** the facultatively aerobic **purple and green nonsulfur bacteria** (they grow best as anaerobic phototrophs), the aerotolerant **blue-green bacteria (cyanobacteria),** and the aerotolerant halophilic **red bacteria.**

The anaerobic purple and green sulfur bacteria use light to drive electrons from bacteriochlorophyll (photosystem-I) into electron transport systems. The flow of electrons through electron transport systems back to bacteriochlorophyll molecules pumps hydrogen ions out of the cell. This generates electrical potentials that are used to power the cell. When the purple and green sulfur bacteria divert electrons and hydrogen ions for the purpose of fixing carbon dioxide, the electrons and hydrogen ions lost are replaced by oxidizing hydrogen sulfide or molecular hydrogen. Examples of purple and green sulfur bacteria are *Chromatium* (purple), *Thiospirillum* (purple), *Chlorobium* (green), and *Pelodictyon* (green).

The facultative aerobic purple and green nonsulfur bacteria prefer to use light to generate energy. They generate energy in much the same way that the anaerobic purple and green sulfur bacteria do. These nonsulfur bacteria, however, may derive most of their carbon not by fixing carbon dioxide, but by assimilating simple organic compounds such as acetate and butyrate. These compounds supply the electrons and hydrogen ions for converting, for example, acetate into poly-β-hydroxybutyrate or butyrate and carbon dioxide into poly-β-hydroxybutyrate and glycogen. Examples of purple and green nonsulfur bacteria are *Rhodospirillium* (purple), *Rhodopseudomonas* (purple), *Rhodomicrobium* (purple), and *Chloroflexus* (green).

The aerotolerant cyanobacteria use light to drive electrons from chlorophyll (photosystem-II) into electron transport systems. The electrons are used to pump hydrogen ions out of the cytoplasm. This generates electrical potentials that are used to power the cell. The electrons ejected from photosystem-II never return to the system because they become part of photosystem-I and an electron transport system or they are continuously diverted to create reducing power for fixing carbon dioxide. The electrons from photosystem-II are replaced by oxidizing water. Examples of the cyanobacteria are *Anabaena*, *Spirulina*, *Oscillatoria*, and *Nostoc*.

The aerotolerant halophilic red bacteria use light to redistribute electrons along molecules of bacteriorhodopsin that span the plasma membrane. The redistribution of electrons causes hydrogen ions to be pumped out of the cell. The proton gradient created by pumping protons out of the cell represents an electrical potential that is used to power the cell. Simple organic compounds supply these organisms with the carbon compounds they require.

In this section you will investigate the effect light has on the growth of some of the photosynthetic microorganisms found in soil and pond water by constructing Winogradsky columns. This exercise will demonstrate that light is required for the growth of certain microorganisms.

Materials

Rich soil or lake shore mud, pond water

Calcium sulfate, cellulose (or shredded filter paper), 3M salt (NaCl), solution and potassium phosphate (dibasic and monobasic forms)

Four one-liter graduated columns, incandescent lights (fluorescent lights should not be used)

Pasteur pipets, long tipped

Procedure

First Period

1. Mix together enough rich soil (lake shore or estuary mud) and pond water to make enough mud to fill four one-liter graduated columns. Mix into the mud 2 or 3 tablespoons of cellulose (or a handful of shredded paper), 1 tablespoon of calcium sulfate, and ¼ teaspoon of each of the potassium phosphates.

2. Fill the 4 one-liter graduated columns with the enriched mud. Fill the final four inches of two of the columns with a 3M NaCl solution (17.5 g NaCl made up to 100 ml with water). Fill the remaining two col-

umns with lake or estuary water if possible. Cover the top of each graduated column with parafilm to reduce evaporation.

3. Incubate 2 of the columns (with and without the 3M salt solution) in the dark at room temperature for 2 to 3 weeks. The dark environment can be most easily achieved by wrapping the columns with aluminum foil. Incubate the remaining 2 columns in the light at room temperature for 2 to 3 weeks. The light environment can be most easily achieved by using incandescent lights.

Second Period (after 2 to 3 weeks)

1. Look for growth in the graduated columns kept in the light and in the dark.

2. Draw the graduated columns and indicate to scale exactly where you find the different colored organisms. Be sure to study the columns that were kept in the dark.

3. Sample some of the brightly colored areas with a long pasteur pipet and observe the organisms in a wet mount. Also carry out a Gram stain on the organisms you sample.

4. Draw the organisms you see associated with each colored area. Are the organisms mostly gram-positive or Gram-negatived?

Questions

1. Differentiate between the purple and green sulfur bacteria, the purple and green nonsulfur bacteria, the blue-green bacteria, and the halophilic red bacteria.

2. In the Winogradsky columns subjected to light, where are the different phototrophs located?

3. Where are the eukaryotic phototrophs located?

4. In the Winogradsky columns kept in the dark, are there any colored organisms? Explain your observation.

5. Where might you find the halophilic red bacteria in the Winogradsky column? Explain your answer.

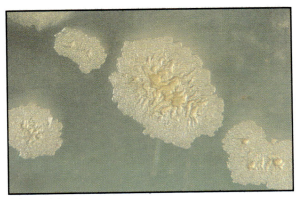

A bacterial colony.

A bacterial colony.

A fungal colony of *Microsporum cookeii*.

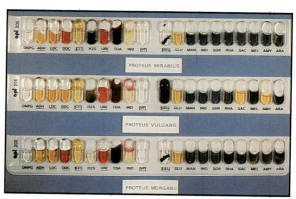

API 20E strips inoculated with *Proteus* sp.

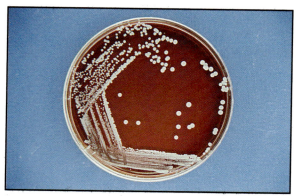

Blood agar plated with *Staphylococcus epidermidis*. No changes (gamma-reaction) are visible around the colonies.

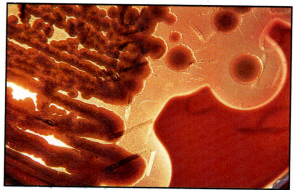

Blood agar plate with *Bacillus subtilis*. Extensive hydrolysis (beta-hymolysis) or clearing is visible around the colonies.

Casein agar plate with *Escherichia coli*. No changes are visible around the colonies.

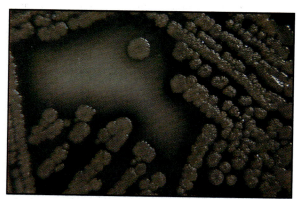

Casein agar plate with *Bacillus subtilis*. Extensive clearing of the casein is visible around the colonies.

Fat agar plate with *Escherichia coli*. No fat hydrolysis is indicated because the colonies are normal in appearance.

COLOR PLATE   6

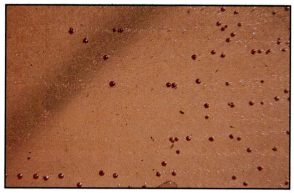

Fat agar plate with *Pseudomonas aeruginosa*. Fat hydrolysis is indicated by red colonies.

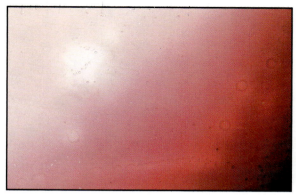

Starch agar plate with *Escherichia coli*. The dark purple color represents an iodine starch complex. A solid dark purple color around the colonies indicates that no starch hydrolysis has occurred.

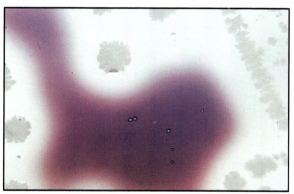

Starch agar plate with *Bacillus subtilis*. The dark purple color represents an iodine starch complex. Extensive clearing of the starch is indicated around the colonies because no dark purple develops when the iodine is added.

Oxidase test. A nutrient agar plate with *Pseudomonas fluorescens* is tested with a drop of oxidase reagent. Oxidase is indicated if the colonies turn black within an hour after a drop of reagent is placed on them.

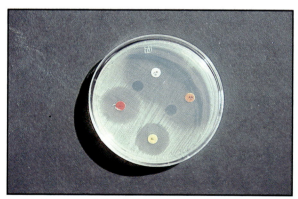

Antibiotic susceptibility. Disks impregnated with antibiotics are placed on a plate just spread with *Staphylococcus epidermidis*. P = penicillin, TE = tetracycline, AM = ampicillin, and E = erythromycin.

Eosin methylene blue (EMB) is a selective and differential medium. It contains lactose and the dyes eosin and methylene blue. The dyes inhibit most Gram positive bacteria. Dark purple colonies with a green sheen are an indication of lactose fermentation by *Escherichia coli*.

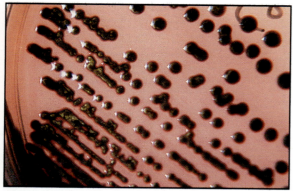

Colonies of *Klebsiella pneumoniae* on EMB. Since *Klebsiella* ferments lactose, colonies may also be dark purple with a green sheen.

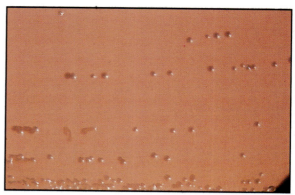

Colonies of a *Salmonella* sp. on EMB. The cream-colored to light red violet colonies indicate that lactose is not fermented.

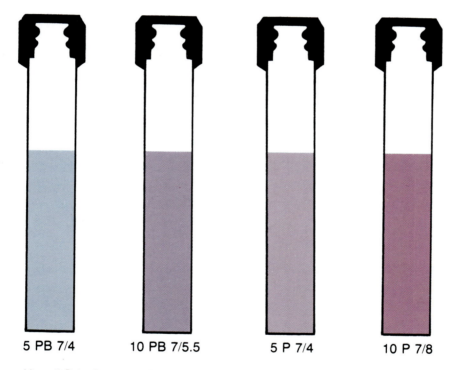

| 5 PB 7/4 | 10 PB 7/5.5 | 5 P 7/4 | 10 P 7/8 |

Munsell Color Standards. The Munsell color standards are required for the performance of the 1-hour and 3-hour resazurin reduction test of milk supplies. The 5P 7/4 color standard represents the end-point for resazurin milk mixtures, beyond which the milk sample is regarded as of unacceptable sanitary quality.

COLOR PLATE   8

PART VI
CONTROL OF MICROBIAL GROWTH

Microorganisms are responsible for the death and suffering of millions of people each year. In addition, they cause disease in farm animals and crops and spoil food and many manufactured goods. In order to reduce death and suffering and alleviate financial losses, a number of procedures and chemical compounds have been developed to control microorganisms. For ease of discussion, methods of control are often divided into **physical methods of control** and **chemical methods of control.**

Physical methods of control include hand washing and the use of surgical masks, sterile clothing, and gloves to restrict the transfer of microorganisms. Heat and ultraviolet light are important physical agents used to kill microorganisms.

Chemical methods of control include the use of gaseous sterilants, disinfectants, antiseptics, preservatives, antibiotics, and drugs that inhibit and kill microorganisms.

PHYSICAL METHODS OF CONTROL

Hand washing with soaps and mild disinfectants is an efficient way of removing microorganisms from the hands and thus reducing the chances of infecting animals and plants or contaminating food, drugs, or media. The use of surgical masks, sterile clothing, and gloves helps prevent infection of a patient during surgery.

Heating is the most efficient and cost-effective way of killing organisms and is used to sterilize glassware, heat-stable media, and equipment. These materials can be easily sterilized by raising them to a temperature of 121°C for 15 to 30 minutes in **autoclaves** (fig. VI-1). Cooling is a familiar physical method of control that inhibits the growth of organisms. Cooling and even freezing often does not kill microorganisms, and

consequently cannot be used when sterility is required.

Filtering is an important method of eliminating microorganisms from heat-labile liquid media (fig. VI-2) and from air in surgical rooms and cell culture rooms.

Because ultraviolet light is a potent mutagen, it is often used to kill microorganisms in the air and on the surface of working areas. Many cell culture rooms have ultraviolet lights installed in the ceilings. These UV lights are turned on during the night when no one is working in the rooms.

Gamma radiation from radioactive cobalt atoms is sometimes used to kill microorganisms in heat-labile medical supplies and some foods.

Aseptic techniques are the procedures used to prevent contamination of previously uncontaminated materials and to obtain and perpetuate pure cultures of microorganisms. A surgical team, performing even the most insignificant of procedures, employs aseptic techniques in order to prevent the contamination of the surgical incision or internal organs with microorganisms. Rudimentary aseptic techniques would include personnel washing their hands with antiseptics and mild disinfectants, treating bench tops and tables with disinfectants, sterilizing all materials and equipment used in an operation (including instruments, gloves, surgical drapes, masks and gowns), and maintaining their sterility by careful wrapping and handling.

CHEMICAL METHODS OF CONTROL

Antimicrobial agents are somewhat artificially divided into the following categories: antiseptics, disinfectants, sterilants, chemotherapeutic agents, pesticides, and preservatives. The distinction is sometimes blurred

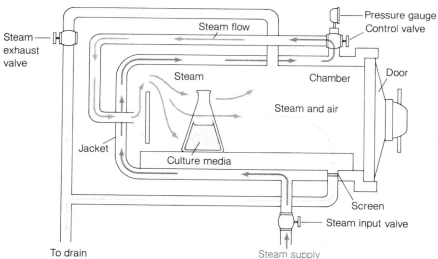

Figure VI-1. Autoclave. Autoclaves are used to sterilize heat-stable media and equipment. Steam initially enters the steam jacket around the autoclave chamber and heats up the autoclave. An object is placed in the chamber, the door is shut securely, and steam from the jacket is allowed to enter the chamber. Air is forced out of the chamber by the incoming steam until only pure steam is being forced out. Then the air exit is closed by the high temperature of the steam, and the steam pressure builds to 15 pounds per square inch above atmospheric pressure. Eventually, the temperature rises to 121°C. When the sterilization is completed, the steam exhaust valve opens and the steam flows out of the chamber.

because some of the chemicals can fall into a number of categories depending upon their concentration. For instance, formaldehyde at high concentrations (20%) is a sterilant, at intermediate concentrations (4%) it may be a disinfectant, and at very low concentrations (0.1%) it may be used as an antiseptic. Phenol at 5% is a toxic disinfectant but at 0.05% can be used as an antiseptic.

Antiseptics are chemicals that kill microorganisms but are relatively safe to use externally on humans and animals. Antiseptics include such chemicals as 70% isopropyl alcohol, 3% hydrogen peroxide,

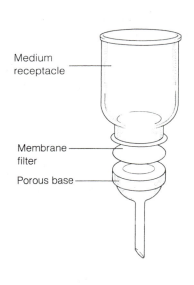

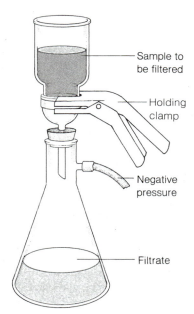

Figure VI-2. Filtration. Heat-labile (sensitive) media are generally "sterilized" by filtration through membrane filters. In fact, the medium is not really sterilized because viruses generally pass through the filters. Water samples are often checked for microorganisms by filtering. The microorganisms stuck on the filters can be detected by placing the filter on a selective and differential agar medium. The resulting colonies can then be counted and identified.

1% tincture of iodine, 1% silver nitrate, and 1% mercurochrome (a mercury-containing compound).

Disinfectants, chemicals that are usually more toxic than antiseptics, are used to kill microorganisms on floors, toilets, showers, bench tops, and equipment. Disinfectants include such chemicals as 0.5% chlorine bleach (Clorox), 3% phenol (carbolic acid), 5% quarternary ammonium compounds (Roccal and Zephrin), 80% ethyl alcohol, and 80% isopropyl alcohol. The alcohols are classified as disinfectants rather than antiseptics because at these concentrations they are not safe to use on mucous membranes. Many disinfectants are composed of a mixture of chemicals: Lysol disinfectant consists of 0.1% o-phenylphenol and 79% ethyl alcohol while Lysol cleaner is made up of 2.7% alkyldimethylbenzyl ammonium chlorides, 0.13% tetrasodium ethylenediamine tetraacetate, and 0.34% ethyl alcohol.

Chemical sterilants are compounds that are very toxic to life and must be used with great caution. Chemical sterilants such as the gases ethylene oxide (12%) and propiolactone are used to kill organisms in heat-labile materials such as plastic petri dishes, plastic pipets, plastic filtering systems, etc., and surgical implants. Liquid gluteraldehyde (20%) has been used to sterilize anesthesia tubing and surgical instruments since it rinses off easily with water.

Pesticides are chemicals that are used against insects, arachnids, nematodes, algae, fungi, and bacteria. For example, solutions of copper sulfate (Bordeaux mixtures) kill algae and fungi, 1% copper-8-hydroxy quinolate kills fungi and bacteria, and 400 mg/l ethylene bromide kills insects and arachnids in grains and on fruits as well as nematodes in soils.

Food preservatives are chemicals that are added to foods in order to inhibit the growth of microorganisms. Some of the preservatives are natural compounds such a vitamin C and propionic acid and probably are not dangerous to the consumer at the added concentrations. Many of the preservatives such as sodium nitrite, ethyl formate, sulfur dioxide, and polymyxin B may be dangerous if consumed in large amounts. Common food preservatives include sodium chloride, sugars, and acetic (vinegar), ascorbic (vitamin C), benzoic, lactic, propionic, sorbic, and sufurous acids. Fungi are inhibited by perservatives such as benzoic acid (benzoates), sorbic acid (sorbates), propionic acid (propionates), sulfur dioxide (sulfites), sodium diacetate, and ethyl formate while bacteria are commonly inhibited by vitamin C, sodium nitrite, sodium chloride, sugars, and polymyxin B.

Chemotherapeutic agents are chemical substances that possess a high degree of antimicrobial activity and that can be used internally either ingested or introduced intravenously. Chemotherapeutic agents that are produced by microorganisms are called **antibiotics.** The best-known antibiotics are penicillin (produced by a fungus) and chloramphenicol (produced by a bacterium). The chemotherapeutic agents that are synthesized by scientists are called **synthetic drugs.**

PRINCIPLES OF MICROBIAL KILLING

A variety of factors influence the amount of time required to sterilize or disinfect contaminated liquids or solid objects. The size of the microbial population is an important factor determining how long materials must be treated. The larger the contaminating population, the longer the treatment takes. Similarly, the larger the volume of the material, the longer sterilization or disinfection takes. The age of a microbial population also plays a role in treatments. Young actively-growing cultures are usually more susceptible to sterilizing agents than older ones in stationary phase. The amount of water present affects the time it takes to sterilize or disinfect a microbial population. The drier an environment, the longer it takes to kill contaminating microorganisms. In addition, the presence of organic matter may reduce the effectiveness of a disinfectant.

Various characteristics of the disinfecting or sterilizing agents determine their effectiveness. The higher the concentration of an antimicrobial or the higher the temperature, the more rapidly a population is killed. In addition, hot moist air kills more efficiently than dry air at the same temperature.

EXERCISE 21
PHYSICAL METHODS OF CONTROL

A. THE EFFECTIVENESS OF HAND WASHING

This exercise is intended to demonstrate how micro-organisms can be passed from one person to another by shaking hands and how hand washing can significantly reduce the number of organisms that are transferred.

Materials

Soil suspension (1 g fresh rich soil into 9 ml sterile saline)

Nutrient agar

Sterile Petri plates

Sterile cotton swabs

Sterile saline

Soap

Procedure

First Period

1. One student (out of a group of four) should pour a few drops of a soil suspension (1 g of fresh rich soil into 9 ml of sterile saline) onto the palm of the right hand and then smear the drops over the entire surface of the hand.

2. A second student should shake hands with the student who was inoculated with the soil suspension. A third student should then shake hands with the second, and a fourth student, in turn should shake hands with the third.

3. Sample the palm of the hands by using a sterile cotton swab dipped in sterile saline. The swab should be rubbed over the entire palm to insure picking up sufficient organisms.

4. Use the swab to spread any organisms picked up from the palm over the surface of a nutrient agar plate. Incubate the plate at room temperature (18°C) for 24 to 48 hours.

5. Each student in the group should then wash his or her hands thoroughly with soap and water.

6. Using a fresh sterile cotton swab dipped in sterile saline, swab the washed palms. The swab should be rubbed over the entire palm so that your procedure approximates that followed before.

7. Again, use the swab to spread any organisms picked up from the palm over the surface of a nutrient agar plate. Incubate the plate at room temperature for 24 to 48 hours.

Second Period

1. After growth appears on the plate, count the total number of colonies on the plates and determine whether any particular type of colony predominates.

Questions

1. How effective is hand washing in eliminating microorganisms?

B. THE EFFECTIVENESS OF HEATING

This exercise shows the relative effectiveness of boiling on populations of heat-sensitive and heat-resistant bacteria. This experiment's aim is to show that not all organisms die immediately and that many minutes of heating may be required to kill all the organisms in a culture. Some of the factors that influence the life of a population are size (volume), age, and the chemical and physical properties of the population.

Materials

TSB cultures of *Escherichia coli* and *Bacillus stearothermophilus*

Trypticase soy broth

Boiling water bath

Tubes of TSB

Incubators set at 35°C and 55°C

1 ml pipets

Procedure

First Period

1. Boil a sample of *Bacillus stearothermophilus* containing 10 ml of culture in TSB. After 1 minute of boiling, aseptically transfer 0.1 ml into a fresh tube of TSB using a sterile 1 ml pipet. Repeat the transferring procedure by sampling the culture after 5, 10, 15, and 30 minutes of boiling.

• Be sure to label clearly the broth tubes so that you know the amount of time you treated the samples. As a control, transfer 0.1 ml of untreated culture to fresh TSB.

2. Repeat the procedure above using *Escherichia coli*

3. Incubate *B. stearothermophilus* at 55°C and *E. coli* at 35°C for 24–48 hours.

Second Period

1. Check the broth tubes to see if growth has occurred. Record the results and determine the effectiveness of each treatment in killing microorganism.

Questions

1. When you boil *Bacillus stearothermophilus*, approximately how long does it take to kill all the organisms in the sample? Why does it take so long?

2. Why is *B. stearothermophilus* so resistant to heat?

3. How long does it take to kill a population of *E. coli* by boiling?

4. List some of the materials that are usually sterilized by autoclaving. List some materials that cannot be autoclaved.

C. THE EFFECTIVENESS OF FILTERING

This exercise demonstrates that liquid media can be cleared of microorganisms if an appropriate filter is used. Heat labile liquid media that cannot be heat sterilized are usually sterilized by filtration. Most bacteria can be filtered from a medium by filters with a pore size of 0.3 μm (fig. 21-1). Filters with a pore size of 0.1 μm are needed to eliminate mycoplasmas. Most viruses are too small to be filtered out by the filters commonly used to remove bacteria. Consequently, "filter sterilized media" may not be free of viruses.

Materials

Broth culture of *Escherichia coli*

Filtering apparatus presterilized by autoclaving

Filters (10 μm, 5 μm, 1 μm, 0.45 μm, 0.3 μm pore diameter)

Trypticase soy broth (TSB) tubes

One milliliter pipets

Incubator set at 35°C

Procedure

First Period

1. Filter a 10 ml sample of *Escherichia coli* through filters with various pore diameters. Begin with the filter with the smallest pores and progress to the largest.

2. Aseptically transfer 1 ml of the filtrates (filtered medium) that were originally contaminated with bacteria to sterile TSB tubes following each filtration. Incubate at 35°C for 24 to 48 hours.

• Be sure to include a control of uninoculated broth. Label all tubes with pore diameters and the organism filtered.

Second Period

1. Check the broth tubes for turbidity and determine which filters were most efficient in eliminating microorganisms from the filtrates.

Questions

1. Which pore sizes filter *Escherichia coli?*

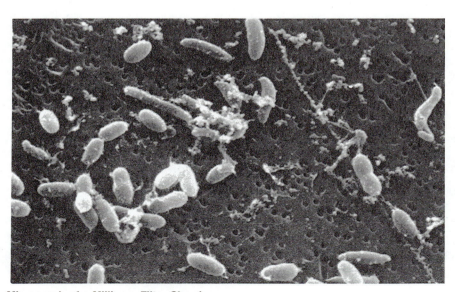

Figure 21-1. Electron Micrograph of a Millipore Filter Showing Bacteria that Have Been Captured on Its Surface

2. Name two things that are commonly "sterilized" using filters.

D. THE EFFECTIVENESS OF ULTRAVIOLET LIGHT

This exercise demonstrates that different types of organisms are killed at different rates by ultraviolet light. Bacterial chromosomes strongly absorb ultraviolet light at 260 nm. The absorbed energy usually causes the formation of **thymidine dimers** that alter the structure of the DNA. One repair system removes the thymidine dimers and replaces them with nucleotides that are complementary to the template. The extensive repair that may go on in highly damaged cells sometimes results in mutations when incorrect nucleotides replace the thymidine dimers. These mutations usually kill the cells.

Materials

Trypticase soy broth (TSB) cultures of *Serratia marcescens*, *Bacillus subtilis*, and *Escherichia coli*

Ultraviolet lights in protective boxes

Eye glasses that absorb ultraviolet light

Watch

Plate count agar (PCA)

Petri dishes

Glass spreaders

Incubators at 25°C and 55°C

Procedure

First Period

1. Make eight PCA spread plates using *Escherichia coli*, eight PCA spread plates using *Serratia marcescens*, and eight PCA plates using *Bacillus stearothermophilus*.

• Use 0.1 ml of the broth cultures as inocula for the plates.

2. Remove the lids from the plates and place them under ultraviolet lights that are shielded from sight (fig. 21-2).

• NEVER LOOK AT ULTRAVIOLET LIGHT UNLESS YOU ARE WEARING PROTECTIVE GLASSES THAT FILTER ALL THE UV WAVELENGTHS. ULTRAVIOLET LIGHT CAUSES RETINAL BURNS AND CAN LEAD TO BLINDNESS.

3. Expose the plates to ultraviolet light for 0, 10, 20, 30, 60, 120, 180, or 300 seconds.

• Be sure that you labeled the plates so that you know how long they were treated and which organism was treated.

4. Incubate the plates for 24 to 48 hours at 25°C.

Second Period

1. Count the colonies on each of the plates and plot the $\log_{10}$ number of organisms (colonies) against time of ultraviolet light treatment. Determine whether there is any difference in sensitivity to UV among the organisms.

Questions

1. How rapidly does UV destroy a bacterial culture? Why is *B. subtilis* more resistant to UV than *Serratia marcescens?* Why is *S. marcescens* more resistant to UV than *Escherichia coli?*

2. What two things are usually sterilized by using ultraviolet light?

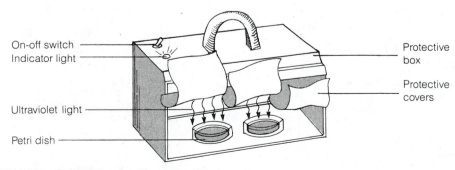

Figure 21-2. Ultraviolet Light on Petri Dishes in a Protective Box

On-off switch
Indicator light
Ultraviolet light
Petri dish
Protective box
Protective covers

EXERCISE 22
CHEMICAL METHODS OF CONTROL

Factors that influence the effectiveness of chemical control agents include the size of the microbial population, the organisms's susceptibility to the antimicrobial, the volume of material to be sterilized, the length of time the antimicrobial is used on the microorganisms, the concentration of the antimicrobial, the temperature, the water activity of the substance to be sterilized, and the amount of organic matter.

The microbiocidal efficiency of a chemical is often determined with respect to phenol, a commonly used disinfectant. The efficiency of a chemical relative to phenol is known as the **phenol coefficient.** The phenol coefficient is calculated by dividing the highest dilution of the antimicrobial of interest which kills all organisms after incubation for 10 minutes but not after incubation for 5 minutes, by the highest dilution of phenol that has the same characteristics. Chemicals that have a phenol coefficient greater than 1 are more effective than phenol.

The best chemotherapeutic agents are those that kill microorganisms rather than inhibit their growth. Agents that kill are called **microbiocidal,** while those that inhibit are **microbiostatic.** An antimicrobial agent may be microbiostatic at low concentrations but microbiocidal at high concentrations. Antimicrobials that are used to inhibit or kill bacteria are known as bacteriostatic and bactericidal, respectively.

The best chemotherapeutic agents are those that kill the microorganisms at risk but have little or no effect on the host. These chemotherapeutic agents are said to have a **selective toxicity.**

A. STUDYING THE FACTORS THAT MAKE AN ANTIMICROBIAL EFFECTIVE

This exercise is intended to show that there are a number of factors that affect the activity of antimicrobials. Antimicrobials do not kill a population immediately. The time of exposure required to kill a population depends not only on the size of the population, but also on the susceptibility of the microorganisms, the concentration of the antimicrobial, and the temperature.

Materials

Trypticase soy broth (TSB) cultures of *Staphylococcus aureus* and *Bacillus subtilis*

3% hydrogen peroxide

70% isopropyl alcohol

Bleach (5.25% hypochlorite)

Lysol Cleaner (2.7% alkyldimethylbenzyl ammonium chlorides)

Lysol Disinfectant (79% ethyl alcohol)

Sterile TSB for making dilutions (5 ml/tube)

Sterile TSB tubes for culturing microorganisms

Sterile Petri dishes

Sterile 5 ml and 1 ml pipets

Sterile test tubes

Ice bath

Incubator 35°C

Water baths set at 45°C and at room temperature

Procedure

First Period

1. For each antimicrobial, prepare six sets of tubes (each set containing five tubes) by making four two-fold dilutions (using sterile TSB as the diluent). The final volume in each tube should be 5 ml.

• See the appendix for an explanation of how to make dilutions and exercise 7D for the proper pipetting procedure.

2. Inoculate the undiluted and each of the diluted antimicrobials (three sets) with 0.5 ml of *Staphylococcus epidermidis.* Inoculate each of the tubes in the second three sets with *Bacillus subtilis.* Mix the contents by rotating the tubes between the palms of your hands.

• The inoculum should not be splashed onto the sides of the tube to insure that all microorganisms are subjected to the full strength of the antiseptic or disinfectant.

3. Incubate one *Staphylococcus* and one *Bacillus* set in an ice bath (0°C), one *Staphylococcus* and one *Bacillus* set at room temperature (18°C), and one *Staphylococcus* and one *Bacillus* set at 45°C.

4. Take a loopful of the mixture after 0, 5, 10, 15, and 20 minutes at the different temperatures and transfer it to a sterile tube of TSB, one for each incubation time.

• Be sure to label each tube with the organism used, the chemical tested, the temperature at which it was incubated, and the exposure time. Mix the culture tube well before sampling it.

5. Incubate the TSB tubes at 35°C for 24 to 48 hours.

Second Period

1. Determine which cultures show growth (+) and which do not show growth (−) by looking for signs of growth (turbidity, sediment, pellicle).

2. Plot the data for an antimicrobial so that you can see the effects of an organism's susceptibility, effects of temperature, effects of the exposure time, and effects of antimicrobial concentration.

3. Plot the data to determine which antimicrobials are most effective for one organism, one temperature, and one concentration.

Questions

1. Explain the difference between physical and chemical control methods. Give examples of each and when they are generally used.

2. Compare and contrast antiseptics, disinfectants, chemical sterilants, preservatives, and antibiotics. Give examples of each and the concentration at which they are generally used.

3. List the important factors (nine) that influence the effectiveness of a chemical control agent. Which of these factors also influences the effectiveness of a physical control agent such as ultraviolet light?

B. DETERMINING THE PHENOL COEFFICIENT

The effectiveness of an antimicrobial agent is frequently compared to that of phenol, an antimicrobial once widely used to kill microorganisms. The **phenol coefficient** is a measure of the effectiveness of an antimicrobial with respect to phenol. A phenol coefficient less than 1 indicates that the chemical is less effective than phenol while a phenol coefficient greater than 1 indicates that the chemical is more effective than phenol. The phenol coefficient (PC) is determined by dividing the highest dilution of phenol that

kills the organism after 10 minutes but not after 5 minutes (for example $1/90$) by the highest dilution of the test chemical that kills the organism after 10 minutes but not after 5 minutes (for example $1/350$). In this case, PC = $1/90 \div 1/350 = 3.89$

Materials

Broth culture of *Staphylococcus aureus* (24 hours)

Phenol (carbolic acid)

Sterile distilled H_2O

Lysol disinfectant (79% ethyl alcohol)

Sterile 5 ml and 1 ml pipets

Sterile TSB tubes

Incubator at 35°C

Procedure

First Period

1. Dilute phenol in sterile distilled water 1:80, 1:90, 1:100 and 1:110 so that the final volume in each tube is 5 ml. Inoculate each of these dilutions with 0.5 ml of a 24 hour broth culture of the test organism *(Staphylococcus aureus)*. Incubate at 20°C.

2. Dilute the chemical to be tested (Lysol Disinfectant) in sterile distilled water 1:100, 1:150, 1:200, 1:250, 1:300, 1:350, 1:400, 1:450, 1:500 so that the final volume in each tube is 5 ml. Inoculate each of these dilutions with 0.5 ml of a 24 hour broth culture of the test organism. Incubate 20°C.

3. At intervals of 5, 10, and 15 minutes, extract a loopful of the mixtures and inoculate sterile TSB with the loopfuls. Incubate the broths at 35°C for 24 to 48 hours.

Second Period

1. Shake the tubes. Determine which culture tubes show growth (+) and which do not show growth (−) by looking for signs of growth. Record the results in tabular forms.

2. Calculate the phenol coefficient.

Questions

1. Of what use is the phenol coefficient? Explain how it is calculated.

2. A chemical diluted $1/500$ with water kills an organism after 10 minutes but not after 5 minutes. Assume a $1/100$ dilution of phenol kills the organism after 10 minutes but not after 5 minutes. What is the phenol coefficient of the chemical?

C. TESTING THE BACTERICIDAL OR BACTERIOSTATIC EFFECT OF ANTIBIOTICS AND DRUGS

Some antimicrobials do not kill organisms but merely inhibit growth and reproduction. Other antimicrobials are inhibitory if used at low concentrations but kill if used at high concentrations. Because of this, it is important to know the **minimum inhibitory concentration** (MIC) and the **minimum killing concentration** (MKC) for an antimicrobial with each organism that might be present.

In this exercise, the MIC is defined as the lowest concentration of the antimicrobial that inhibits growth while the MKC is defined as the lowest concentration of antimicrobial that kills.

Materials

Trypticase soy broth culture of *Escherichia coli* (24 hours)

TSB for dilutions (5 ml/tube)

TSB tubes for culturing

TSB with 0.08 M sulfanilamide (10 ml/tube)

TSB with 400 μg/ml streptomycin (10 ml/tube)

TSB with 200 units/ml penicillin (10 ml/tube)

Sterile test tubes

Sterile 5 ml and 1 ml pipets

Incubator set at 35°C

Procedure

First Period

1. Using sterile TSB as the diluent, dilute a TSB that contains 0.08 M of sulfanilamide to obtain a series of TSB tubes with 0.08 M, 0.04 M, 0.02 M, 0.01 M, and 0.005 M sulfanilamide. Include a control tube with no sulfanilamide.

2. Using TSB as the diluent, dilute a TSB that has 400 μg/ml of streptomycin so as to make a series of TSB tubes with 400 μg/ml, 200 μg/ml, 100 μ/ml, 50 μg/ml, and 25 μg/ml of streptomycin. Include a control with no antibiotic.

3. Using TSB as the diluent, dilute a TSB that has 200 units/ml, of penicillin so that you have a series of TSB tubes with 200 units/ml, 100 units/ml, 50 units/ml, 25 units/ml, and 10 units/ml of penicillin. Include a control tube with no antibiotic.

THE EXPERIMENT CAN BE MADE LESS COMPLEX BY USING ONLY ONE ANTIMICROBIAL.

4. Inoculate each of the tubes with 0.1 ml of a 24 hour culture of *Escherichia coli* growing in TSB. Incubate at 35°C for 48 hours.

Second Period

1. Shake the tubes vigorously and then determine which tubes show growth (+) and which do not show growth (−). Record the results in tabular form.

2. Transfer a loopful of culture from those tubes that SHOW NO GROWTH to fresh TSB free of antibiotics or drugs. Incubate the tubes at 35°C for 24 to 48 hours.

• Be sure to label the tubes as to the organisms used and the concentration of antimicrobial in the original tube.

Third Period

1. Shake the tubes vigorously and determine which tubes show growth (+) and which do not show growth (−). Record the results in the table provided.

2. Determine at which concentration of each antimicrobial you find a bacteriostatic effect and/or a bacteriocidal effect.

• In those tubes where growth occurred, indicate the concentrations at which the antimicrobials are bacteriostatic; in those tubes where no growth occurred, indicate the concentrations at which the antimicrobials are bacteriocidal.

Questions

1. What is the difference between an antimicrobial that is bacteriostatic and one that is bactericidal?

2. Explain how each of these chemotherapeutic drugs works:

    a. sulfanilamide

    b. penicillin

    c. streptomycin

D. DETERMINING THE SELECTIVE TOXICITY OF ANTIBIOTICS AND DRUGS

Some antimicrobials affect a large number of different organisms and are known as **broad spectrum** antimicrobials. Others are effective against a small group of organisms and consequently they are known as **narrow spectrum** antimicrobials.

The best antimicrobials are those that kill undesirable microorganisms at low concentrations but have no toxic effects toward the host (plant, animal or human).

In this exercise you will discover that antimicrobials have a **selective toxicity.** Some antimicrobials affect one organism but not another.

Materials

TSB cultures of *Staphylococcus epidermidis* and *Saccharomyces cerevisiae (24 hours)*

TSB tubes with 0.02 M sulfanilamide

TSB tubes with 100 units/ml penicillin

TSB tubes with 10 units/ml mycostatin

1 ml pipets

Incubator set at 35°C

Procedure

First Period

1. Inoculate each of the three different antimicrobial tubes with 0.1 ml of a 24 hour culture of *Staphylococcus epidermidis* growing in TSB. Inoculate a second set of antimicrobials with 0.1 ml of a 24 hour culture of *Saccharomyces cerevisiae.* Incubate at 35°C for 48 hours.

- Three tubes of TSB constitute a set. Each tube has one of the following antibiotics or drugs: 0.02 M sulfanilamide, 100 units/ml penicillin, or 10 units/ml mycostatin.

Second Period

1. Shake the tubes vigorously and then determine which tubes show growth (+) and which do not show growth (−). Record the results.

Questions

1. Why is selective toxicity important in choosing a chemotherapeutic drug?

2. Which of the chemotherapeutic drugs you tested show a selective toxicity?

PART VII
METABOLIC ACTIVITIES
OF MICROORGANISMS

Microorganisms grow and multiply by using raw materials found in their environment. The available nutrients in the environment may consist of simple molecules such as H_2S and NH_4^+, or complex organic molecules such as proteins and polysaccharides. Microbes oxidize these nutrients to obtain energy and modify them to build precursors for the synthesis of necessary cellular components such as cell walls, plasma membranes, and flagella.

Microbes utilize nutrients in a variety of ways. The metabolism of nutrients often produces byproducts that can be used to aid in the identification of microorganisms.

In this section you will study and demonstrate experimentally some of the diverse metabolic activities of microorganisms. You will accomplish this by observing the ability of microorganisms to use and degrade complex molecules such as starches, fats, proteins, and nucleic acids; and simpler molecules such as amino acids and sugars. You also will use the results of the exercises to characterize and identify an unknown bacterium.

EXERCISE 23
HYDROLYSIS OF LARGE EXTRACELLULAR MOLECULES

Many nutrients found in nature are large, complex organic molecules (macromolecules) such as proteins, polysaccharides, lipids, and nucleic acids. These chemical compounds serve as sources of building blocks (precursors) and energy to many microorganisms. Macromolecules must be **hydrolyzed** before they can be utilized by the cell. Hydrolysis is the breakdown of a molecule by the addition of water. Even microorganisms like *Paramecium* or *Amoeba*, which can engulf whole particles of food, must hydrolyze the particles inside vacuoles before nutrients can enter the cytoplasm of the cell.

Certain microorganisms have adapted to living in environments rich in complex molecules by secreting into the environment enzymes, called **exoenzymes,** that can catalyze the hydrolysis of macromolecules into simpler ones (e.g., proteins to amino acids or polysaccharides to sugars). The small molecules are then transported into the cytoplasm of the cell, where they can be used as sources of energy or as precursors for the synthesis of cellular components (fig. 23-1).

A. HYDROLYSIS OF POLYSACCHARIDES

Starch is a polysaccharide composed of repeating units of the sugar glucose (fig. 23-2). It is found in many plants and thus is commonly present in many environments colonized by microorganisms. When starch is hydrolyzed in the presence of exoenzymes called **amylases** it is ultimately degraded to maltose (a disaccharide consisting of two glucose units) and glucose. The sugars are then transported into the cytoplasm of the cell and used as carbon and energy sources.

Starch reacts chemically with iodine to produce a blue-black color. The blue-black color develops when iodine molecules insert themselves in the hollow portion of the spiral-shaped starch (amylose) molecule. The iodination of the starch results in a molecule that absorbs most visible light except blue. If the starch has been broken down into maltose or glucose, however, no such color develops because there is no spiral to which iodine can associate. The absence of color when iodine is added to a medium can be used as a simple method for demonstrating starch hydrolysis.

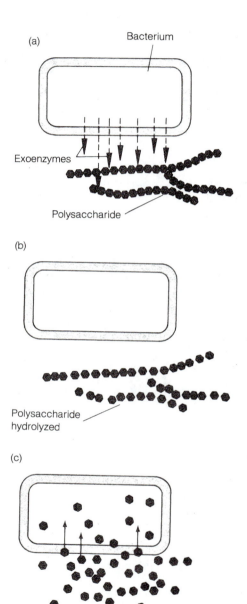

Figure 23-1. Function of Exoenzymes in Nutrient Digestion. Exoenzymes are secreted by the cell to act on large, insoluble food molecules (polysaccharides, proteins, lipids, etc.) present in the immediate environment. The exoenzymes break these large molecules into constituent molecules (simple sugars, amino acids, etc.), which are then transported into the cell to be used as nutrients.

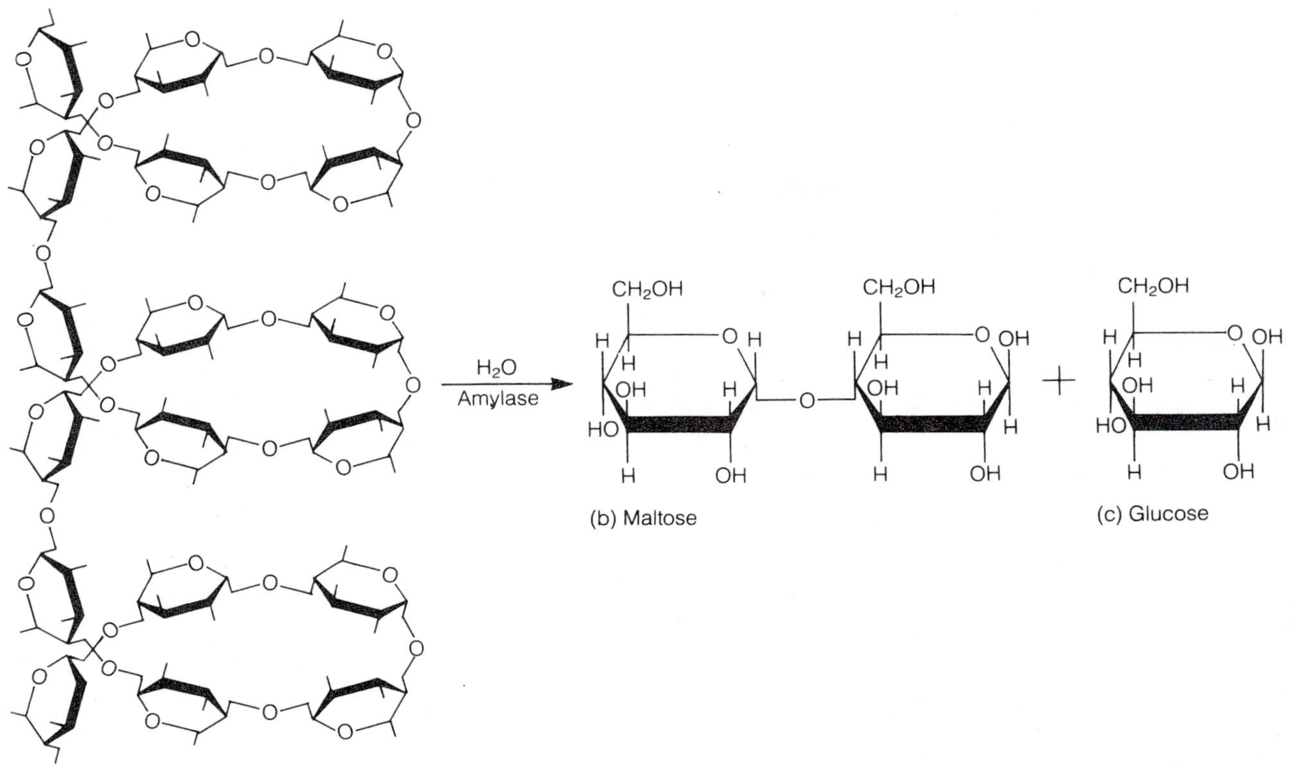

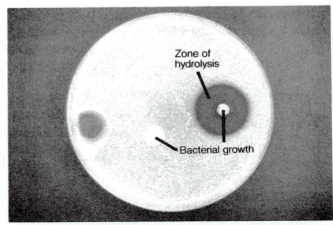

(b) Maltose

(c) Glucose

H_2O
Amylase

(a) Starch (amylose) molecule

Figure 23-2. A Starch Molecule (Amylose)

When bacteria produce amylase on an agar plate containing starch, the starch near the bacterial growth is hydrolyzed. If a few drops of an iodine solution (e.g., Gram's iodine) are added to the agar plate, a colorless area is seen surrounding the growth (fig. 23-3). If no starch has been hydrolyzed, however, the entire surface will show the blue-black color.

Materials

Broth (or slant) cultures of *Escherichia coli* and *Bacillus subtilis*

Tubes of starch agar

Sterile Petri dishes

Inoculating loop

Boiling water bath

Water bath set at 50°C

Incubator set at 35°C

Procedure

First Period

1. Melt 2 tubes of starch agar by placing them in a boiling water bath for approximately 10 minutes.

2. When the starch agar is completely melted, cool the agar by placing the tubes in a water bath set at 50°C for approximately 10 minutes.

3. While the agar is cooling label the bottom of 2 Petri plates with your name, type of culture medium, date, and the name of the organism used.

4. Pour one tube of starch agar into each of the two Petri dishes. Spread the melted agar over the entire

Zone of hydrolysis

Bacterial growth

Figure 23-3. Hydrolyzed Starch Surrounding a Bacterial Growth

surface of the plate by swirling the dish carefully so as not to splash on the sides or top of the dish.

5. Allow the starch agar to solidify by placing the pour plates on a cool, level surface.

6. Spot-inoculate the starch agar plates with the corresponding culture as illustrated in figure 23-4.

• The starch agar plate should be inoculated by placing a loopful of the culture on the center of the plate and spreading it over an area about 0.5 cm in diameter (approximately half the size of a dime).

7. Incubate the plates in an inverted position at 35°C for 24 to 48 hours.

Second Period

1. Examine the plates for starch hydrolysis by adding several drops of Gram's iodine to the plates (fig. 23-5).

• Use only the amount of iodine solution needed to barely cover the surface of the agar. Hydrolysis of starch is evidenced by a clear zone around the bacterial growth. Areas where no starch hydrolysis has taken place will acquire a blue-black color following the addition of the iodine solution (fig. 23-3).

2. Measure the diameter of the hydrolysis zone and record the results on the report form.

3. Discard all cultures as indicated by your instructor.

Questions

1. What determines the size of the zone of hydrolysis?

2. Name some foods that would be spoiled by starch hydrolyzing bacteria.

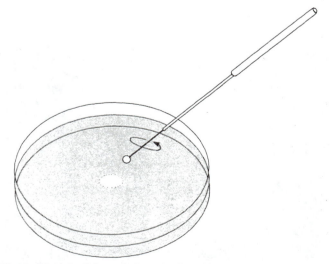

Figure 23-4. Spot-inoculation of a Starch Agar Plate

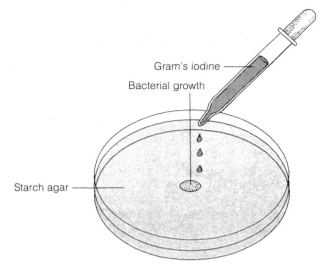

Figure 23-5. Testing for Starch Hydrolysis

B. HYDROLYSIS OF PROTEINS

Casein is a protein found in milk. Like all proteins, it is composed of **amino acids** (fig. 23-6). The amino acids, when available to the cell, can serve as carbon and nitrogen sources as well as energy sources for certain microorganisms.

When milk is mixed with culture media such as plate count agar (PCA), the casein in the milk makes the medium cloudy. The cloudiness is due to the fact that casein, as present in the milk, is complexed with calcium ions, forming calcium caseinate. This substance does not dissolve in the agar but forms a colloid instead, making the PCA turbid. When microorganisms produce the exoenzyme **caseinase,** which catalyzes the hydrolysis of casein, the area immediately surrounding the colony will appear clear because the casein molecules are digested. The resulting amino acids will then dissolve in the aqueous environment and the turbidity in the area of casein hydrolysis disappears. This phenomenon provides a simple and effective means of detecting the degradation of casein by microorganisms.

Protease is a generalized term designating an enzyme that hydrolyzes proteins.

Materials

Broth (or slant) cultures of *Escherichia coli* and *Bacillus subtilis*

Tubes of plate count agar (PCA)

Tubes containing 2 ml of sterile skim milk (heated to 116° for 20 minutes)

Sterile Petri dishes

Inoculating loop

Boiling water bath

(a) Generalized formula for an amino acid

Amino group

Carboxyl group

(b) Formation of a peptide bond

Peptide bond

(c)

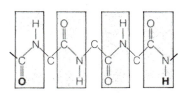

Figure 23-6. Structure and Composition of Proteins. *(a)* Generalized formula of an amino acid. Amino acids consist of a carboxyl group, an amino group, and an R group. The R group may be an atom or a group of atoms. *(b)* A peptide bond. Amino acids are joined together by covalent bonds, called peptide bonds, to form polypeptides. Peptide bonds form between the carboxyl group of one amino acid and the amino group of the other. *(c)* The primary structure of proteins consists of the linear arrangement of amino acids.

Water bath set at 50°C

Incubator set at 35°C

Procedure

First Period

1. Melt 2 tubes of plate count agar by placing them in a boiling water bath for approximately 10 minutes.

2. When the PCA is completely melted, cool the agar by placing the tubes in a water bath set at 50°C for approximately 10 minutes.
• While the agar is cooling, label 2 Petri plates with your name, date, type of culture medium, and the name of the organism used.

3. Pour 2 ml of sterile skim milk into each of the 2 labeled Petri dishes.
• If the milk is cold when mixed with the agar (step 4), it will cause jelling. To avoid this, warm the sterile milk in a water bath (50°C) for a few minutes and then pour into the Petri dishes.

4. Add the cooled PCA to the Petri dishes containing the skim milk and mix thoroughly by swirling the dish carefully so as not to splash on the sides or lid of the dish.

5. Allow the skim milk agar to jell on a cool, level surface.

6. Spot-inoculate the skim milk agar plates with the corresponding culture.
• The plates should be inoculated by placing a loopful of the culture on the center of the plate and spreading it on an area about 0.5 cm in diameter (approximately half the size of a dime).

7. Incubate the plates in an inverted position at 35°C for 24 to 48 hours.

Second Period

1. Examine the plates for casein hydrolysis.
• Hydrolysis of casein is evidenced by a zone of clearing around the bacterial growth. Areas where no casein hydrolysis has taken place will remain cloudy.

2. Measure the zone of hydrolysis and record the results on the report form.

3. Discard all cultures as indicated by your instructor.

Questions

1. Based on the results you obtained in this and the previous exercise, which types of environments might *Escherichia coli* and *Bacillus subtilis* inhabit? Would it be likely that *E. coli* grow on soils rich in plant materials? Explain. How about *B. subtilis?*

2. Which types of foods would casein hydrolyzing bacteria spoil?

3. How might protease- and DNase-producing microorganisms be involved in causing a disease?

C. HYDROLYSIS OF NUCLEIC ACIDS (DNA)

Certain microorganisms are capable of producing exoenzymes that catalyze the breakdown of nucleic acids. One such exoenzyme, **DNase,** catalyzes the hydrolysis of DNA into nucleotide monophosphates. Nucleotides in DNA consist of purines and pyrimidines (nitrogen bases) covalently bonded to the 1′ carbon of a deoxyribose and a phosphate group attached to the 5′ carbon of the same sugar (fig. 23-7). Nucleotides are digested further by microorganisms into phosphates, pentoses, and nitrogen bases before they are transported into the cell.

Extracellular DNase activity is an important criterion in the identification of certain species of microorganisms and can also be used as a measure of the pathogenic potential of certain medically-important bacteria. For example, DNase activity is frequently associated with pathogenic strains of *Staphylococcus aureus*.

Extracellular DNase can be detected by culturing microorganisms in a DNA-containing agar that also contains the dye methyl green. Methyl green combines with DNA to form a green-colored compound. When DNA is hydrolyzed, the color fades. Colonies releasing DNase can be detected by a colorless zone around them.

Materials

Broth (or slant) cultures of *Escherichia coli* and *Serratia marscescens*

Plates of DNase test agar with methyl green

Inoculating loop

Incubator set at 35°C

Procedure

First Period

1. Label the bottom of 2 Petri plates with your name, date, type of culture medium, and the name of the microorganism used.

2. Spot-inoculate the DNA agar plates with the corresponding culture.
• The plate should be inoculated by placing a loopful of the culture on the center of the plate and spreading it on an area about 0.5 cm in diameter.

3. Incubate the DNA agar plates in an inverted position at 35°C for 24 to 48 hours.

Second Period

1. Examine the plates for DNA hydrolysis.
• Hydrolysis of DNA is evidenced by clearing of the methyl green around the colony.

2. Measure the diameter of the zone of DNA hydrolysis and record the results on the report form.

3. Discard all cultures as indicated by your instructor.

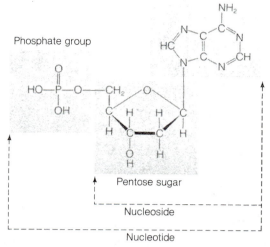

(a)

Heterocyclic nitrogen base

Phosphate group

Pentose sugar

Nucleoside

Nucleotide

Figure 23-7. Components of Nucleotides. Nucleotides are the building units of nucleic acids. Each is composed of a pentose, a heterocyclic nitrogen base, and a phosphate group. *(a)* A heterocyclic nitrogen base covalently bonded to the pentose forms a nucleoside. A phosphate group covalently bonded to a nucleoside forms a nucelotide.

Questions

1. The disease-causing potential of *Staphylococcus aureus* is sometimes measured by the bacterium's ability to hydrolyze DNA. Why would such a test be of use in establishing the disease-causing potential of this organism? Discuss a possible mechanism by which *S. aureus* causes disease.

D. HYDROLYSIS OF FATS

Microorganisms often inhabit environments or colonize foods that are rich in animal or vegetable fats. Such materials as butter, tallow, corn oil, and coconut oil are largely made of fats known as triglycerides. When triglycerides are hydrolyzed by microorganisms that produce fat-digesting exoenzymes called **lipases,** long-chain fatty acids and glycerol are produced (fig. 23-8). Lipase-producing microorganisms use these molecules as sources of carbon and energy. **Rancidity** is a type of spoilage caused when lipolytic bacteria hydrolyze fats in foods. The disagreeable flavors and odors of rancid foods are due to the various fatty acids released by the hydrolysis of fats.

When fats in an agar-solidified culture medium are hydrolyzed by lipases released by cells, the surrounding medium will be acidified due to the release of fatty acids. By adding a pH indicator to the culture medium it is possible to detect the hydrolysis of fats as a color change in the medium. The color change will depend

STEROID

Progesterone

TRIGLYCERIDE

Glycerol group

PHOSPHOLIPID

Polar group (hydrophilic)

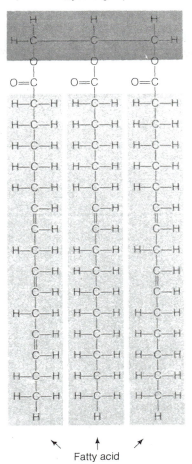

Fatty acid

Figure 23-8. Chemical Structures of Representative Lipids

upon the pH indicator used. Spirit blue agar with Bacto lipase reagent (Difco) has a pale lavender color when freshly prepared and it turns to a royal blue color around lipolytic colonies due to acidification.

Materials

Broth (or slant) cultures of *Proteus mirabilis* and *Staphylococcus epidermidis*

Plates of spirit blue agar with 3% Bacto lipase reagent (Difco). An emulsion of 100 ml of olive oil in 400 ml of warm water with 1 ml Tween 80 can be used instead of the Bacto lipase reagent. Add 30 ml of the emulsion to 1000 ml of sterile spirit blue agar still in liquid form at about 55°C

Inoculating loop

Incubator set at 35°C

Procedure

First Period

1. Label the bottom of 2 Petri plates with your name, type of culture medium, date, and the name of the organism used.

2. Spot-inoculate the spirit blue agar plates with the corresponding cultures.
• A plate should be inoculated by placing a loopful of the culture on the center of the plate and spreading it over an area about 0.5 cm in diameter.

3. Incubate the spirit blue agar plates in an inverted position at 35°C for 24 to 48 hours.

Second Period

1. Examine the plates for fat hydrolysis.
• Hydrolysis of fats is evidenced by the development of a blue zone around the bacterial growth. Areas where

no fat hydrolysis has taken place will retain the original color of the medium (pale lavender).

2. Measure the diameter of the zone of hydrolysis and record the results on the report sheet.

3. Discard all cultures as indicated by your instructor.

Questions

1. Which foods are likely to be spoiled by lipolytic bacteria?

2. Is there any relationship between production of exoenzymes and Gram staining properties? Explain.

EXERCISE 24
FERMENTATION OF CARBOHYDRATES

The ability of microorganisms to ferment different sugars, and the type of products formed when the sugars are fermented, are very useful characteristics for identification. **Fermentations** are energy-producing metabolic reactions in which organic molecules serve both as electron donors and as electron acceptors. For example, when the sugar glucose is fermented by lactic acid bacteria it is first oxidized to pyruvate, producing energy. The pyruvate then serves as an electron acceptor and is reduced to form lactic acid. The outcome of this series of reactions is the net production of 2 ATP molecules per molecule of glucose oxidized. A given carbohydrate may be fermented to a number of different end products depending upon the microorganisms involved (fig. 24-1). These end products, which can be acids, alcohols, or other organic molecules or gases, are characteristic of particular microbes and can be used to identify them.

Not all sugars can be fermented by bacteria that are capable of carrying out fermentations. Before the sugars can be used, they must be transported into the cell and modified. For example, before the sugar lactose, a disaccharide composed of a glucose and a galactose, can be fermented, at least two enzymes must act upon it: lactose permease and β-galactosidase. Lactose permease transports the sugar into the cell, while β-galactosidase catalyzes the hydrolysis of lactose into glucose and galactose, which can then be fermented. If a microorganism lacks the ability to synthesize either of these two enzymes, it will not be able to use lactose as an energy source.

If fermenting bacteria are grown in a liquid culture medium containing a sugar such as glucose, they may produce acids as byproducts of the fermentation. The acids released into the medium lower its pH. If a pH indicator such as brom-cresol purple (BCP) or phenol red (PR) is included in the medium, the production of acids can be detected by a color change to yellow (fig. 24-2).

Gases produced during fermentation can be detected by using a small, inverted tube (a **Durham** tube) within the culture medium. If gas is produced, some of the fluid inside the Durham tube will be displaced, thus entrapping a gas bubble as an indication of gas production (fig. 24-2).

A. FERMENTATION OF GLUCOSE, LACTOSE, SUCROSE, MALTOSE, AND MANNITOL

Materials

Broth cultures of *Escherichia coli*, *Staphylococcus aureus*, *Micrococcus luteus*, and *Saccharomyces cerevisiae*

Tubes of BCP-glucose, BCP-sucrose, BCP-lactose, BCP-maltose, and BCP-mannitol with Durham tubes (PR may be used instead of BCP as a pH indicator)

Inoculating loop

Test tube racks

Incubator set at 35°C

Procedure

First period

1. Label 5 sets of tubes each composed of a tube of BCP-glucose, BCP-lactose, BCP-sucrose, BCP-maltose, and BCP-mannitol broths with your name, date, and type of culture medium. Label the first set *Escherichia coli*, the second set *Staphylococcus aureus*, the third set *Micrococcus luteus*, the fourth set *Saccharomyces cerevisiae*, and the last set *Control*.

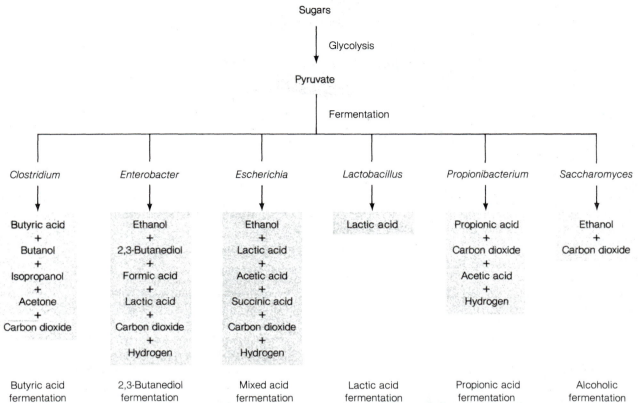

Sugars

↓ Glycolysis

Pyruvate

↓ Fermentation

Clostridium	*Enterobacter*	*Escherichia*	*Lactobacillus*	*Propionibacterium*	*Saccharomyces*
Butyric acid + Butanol + Isopropanol + Acetone + Carbon dioxide	Ethanol + 2,3-Butanediol + Formic acid + Lactic acid + Carbon dioxide + Hydrogen	Ethanol + Lactic acid + Acetic acid + Succinic acid + Carbon dioxide + Hydrogen	Lactic acid	Propionic acid + Carbon dioxide + Acetic acid + Hydrogen	Ethanol + Carbon dioxide
Butyric acid fermentation	2,3-Butanediol fermentation	Mixed acid fermentation	Lactic acid fermentation	Propionic acid fermentation	Alcoholic fermentation

Figure 24-1. Outline of Major Fermentation Pathways. Microorganisms produce various waste products when they ferment glucose. The byproducts released are often characteristic of the microorganism and can be used as an aid in their identification.

2. Inoculate the first 4 sets of carbohydrate fermentation broths with the corresponding bacterial cultures. Leave the last set uninoculated.

• Exercise care not to introduce air bubbles into the inverted tube.

3. Incubate all 5 sets of tubes in an incubator set at 35°C.

Second Period

1. Examine the carbohydrate fermentation tubes for evidence of acid and/or gas production. Use figure 24-2 as a reference.

• ACID production is detected by the development of a yellow color in the medium. GAS production is detected by the appearance of gas bubbles within the

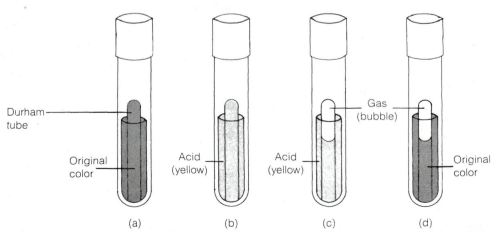

Figure 24-2. Carbohydrate Fermentation Patterns of Some Microorganisms. (a) Uninoculated control (or respiration), (b) Lactic acide fermentation, (c) A type of mixed acid fermentation, (d) Alcoholic fermentation.

Durham tube. Check uninoculated controls to make sure that bubbles have not developed within the Durham tube. This sometimes occurs during incubation because oxygen dissolved in the medium during cold storage becomes less soluble in water at the elevated temperature of incubation and may become entrapped in the tube.

2. Record the results on the report sheet.

3. Discard all cultures as indicated by your instructor.

Questions

1. Some bacteria ferment glucose but not sucrose although sucrose is a disaccharide made up of glucose and fructose. Explain why these bacteria use glucose but not sucrose.

2. What type of metabolism is carried out by *Micrococcus luteus?*

3. Can *Staphylococcus* multiply in an anaerobic environment? Can *Micrococcus?* Explain.

B. METHYL RED TEST

The methyl red test is used to detect a specific type of fermentation called **mixed acid fermentation** (fig. 24-1). Some bacteria ferment glucose and produce large quantities of acidic end products that lower the pH of the medium below 5.0. The addition of the pH indicator methyl red is used to detect this acidity. Methyl red is red at a pH of 4.4 but yellow at a pH of 6.2.

A mixed acid fermentation is detected by growing bacteria in a glucose-containing medium, and, after an incubation period, adding some methyl red reagent to the culture medium. If a mixed acid fermentation has taken place, the methyl red will remain red. If no mixed acid fermentation has occurred, however, a yellow color will develop. This procedure is useful in identifying certain fermenting bacteria, principally those that colonize the human gastrointestinal tract, commonly called **enterics.** Some of these bacteria are human pathogens and must be identified quickly to allow prompt and efficacious treatment. The methyl red test, as well as other carbohydrate fermentation tests, is routinely used to identify these microorganisms.

Materials

Cultures of *Escherichia coli* and *Enterobacter aerogenes*

Tubes of MR-VP (methyl red-Voges Proskauer) broth

Methyl red reagent

Clean test tubes

Inoculating loop

Test tube rack

Incubator set at 35°C

Procedure

First Period

1. Label 2 tubes of MR-VP broth with your name, lab section, type of culture medium, and the name of the microorganism used.

2. Inoculate the tubes of MR-VP broth with the corresponding bacterial culture.

3. Incubate the tubes of MR-VP broth at 35°C for 24 to 48 hours.

Second Period

1. Pour (aseptically) approximately 2 ml of the culture to be tested into a clean test tube. Repeat the procedure with the other culture.

• AVOID SPILLING THE CULTURES BECAUSE THESE BACTERIA ARE POTENTIAL HUMAN PATHOGENS AND MAY REPRESENT A HEALTH HAZARD.

2. Add 5 drops of the methyl red reagent to each of the 2 tubes of MR-VP cultures.

3. Read the results as follows:

POSITIVE: A red (not orange) color is evident immediately after the addition of the methyl red reagent.

NEGATIVE: The culture medium turns yellow or orange.

4. Record the results on the report sheet.

5. Discard all cultures and materials as indicated by your instructor.

Questions

1. Name five different organic compounds produced as byproducts in the mixed acid fermentation.

2. If you were to develop a new culture medium to test for mixed acid fermentation, which organic molecule must you include in order to obtain the desired results? (HINT: What is being converted into mixed acid products?)

C. VOGES-PROSKAUER TEST

The Voges-Proskauer test identifies organisms that carry out a **2,3-butanediol fermentation.** When bacteria ferment sugars producing 2,3-butanediol as a major product (fig. 24-1), they accumulate this compound in the medium. The addition of 40% KOH and a 5% solution of alpha naphthol in absolute ethanol will reveal the presence of **acetoin (acetylmethylcarbinol)**, a precursor in the synthesis of 2,3-butanediol. The acetoin, in the presence of KOH, will develop a pink color. This reaction is expedited by the addition of α-naphthol. The development of color is more noticeable on the portion of the culture exposed to air because some of the 2,3-butanediol is oxidized back to acetoin, hence increasing the intensity of the color reaction. So, in order to obtain more clear-cut results, the MR-VP cultures with added reagents should be shaken by tapping gently at the bottom of the tube so that a vortex is created along with some foaming. Aeration increases the rate of oxidation of 2,3-butanediol to acetoin.

Materials

Cultures of *Escherichia coli* and *Enterobacter aerogenes*

Tubes of MR-VP broth

Clean test tubes

40% aqueous KOH

5% α-naphthol in absolute ethanol

Test tube rack

Incubator set at 35°C

Procedure

First Period

1. Label 2 tubes of MR-VP broth with your name, lab section, type of culture medium, and the name of the organism used.

2. Inoculate the tubes of MR-VP broth with the corresponding bacterial culture.

3. Incubate the tubes of MR-VP broth at 35°C for 24 to 48 hours.

Second Period

1. Pour (aseptically) approximately 5 ml of the culture to be tested into a clean test tube.

• AVOID SPILLING THE CULTURE BECAUSE YOU ARE USING LIVING CULTURES OF POTENTIALLY PATHOGENIC MICROORGANISMS.

2. Add 10 drops of the 40% KOH reagent and 15 drops of α-naphthol to each of the 2 tubes of MR-VP cultures then mix well.

• Shake the mixture well so that the culture is aerated. This will increase oxidation of 2,3-butanediol to acetoin and give more clear-cut results. Results should be evident after 30 minutes. Be careful not to splash the cultures while mixing.

3. Read the results as follows:

POSITIVE: A red color develops within 30 minutes after the addition of the reagents.

NEGATIVE: No color change takes place.

4. Record the results on the report sheet.

5. Discard all contaminated cultures and materials as indicated by your instructor.

Questions

1. Can a bacterium that ferments via the 2,3-butanediol pathway ferment also via the mixed acid route?

2. Why does shaking the MR-VP culture with KOH and α-naphthol increase the rate of development of the red color?

EXERCISE 25
RESPIRATION OF CARBOHYDRATES

Carbohydrates (e.g., sugars) are commonly used as a source of energy by microorganisms. Many microorganisms use sugars (e.g., glucose) for energy by oxidizing them to pyruvate during **glycolysis** (fig. 25-1). Glycolysis yields a net 2 molecules of ATP per glucose molecule oxidized. In the absence of molecular oxygen, many microorganisms resort to fermentation pathways following pyruvate formation, to eliminate electrons and protons (H). In the presence of an external electron acceptor (e.g., O_2), pyruvate is oxidized

to acetate (acetyl-CoA), which then enters the Krebs' cycle (fig. 25-1), where it is further oxidized. In the Krebs' cycle, some ATP is synthesized once more by substrate level phosphorylation, but the main function of the cycle is to oxidize (remove electrons and protons) various molecules. The electrons and protons obtained from the Krebs' cycle enter electron transport systems, where more energy in the form of ATP is generated by oxidative phosphorylation. The electrons and protons are ultimately donated to an electron

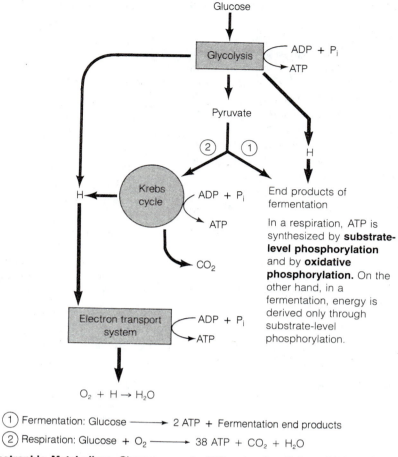

Glucose

Glycolysis $\quad$ ADP + P$_i$ / ATP

Pyruvate

② ①

Krebs cycle $\quad$ ADP + P$_i$ / ATP

CO$_2$

H

End products of fermentation

In a respiration, ATP is synthesized by **substrate-level phosphorylation** and by **oxidative phosphorylation**. On the other hand, in a fermentation, energy is derived only through substrate-level phosphorylation.

H

Electron transport system $\quad$ ADP + P$_i$ / ATP

$$O_2 + H \rightarrow H_2O$$

① Fermentation: Glucose $\longrightarrow$ 2 ATP + Fermentation end products

② Respiration: Glucose + O$_2$ $\longrightarrow$ 38 ATP + CO$_2$ + H$_2$O

Figure 25-1. Overview of Chemotrophic Metabolism. Chemotrophic organisms oxidize fuel molecules, such as glucose, in a series of enzyme-catalyzed reactions such as glycolysis, the Krebs cycle, and the electron transport system. The end result of these oxidations is the extraction of energy from fuel molecules plus its conservation in ATP molecules. If the oxidation of fuel molecules includes the Krebs cycle and the electron transport system, the process is called respiration. If the oxidation of fuel molecules is to end products of fermentation, the process is called a fermentation.

acceptor such as O_2 or nitrate (NO_3^-) that is found in the cell's environment. If O_2 is the electron acceptor, this process is known as **aerobic respiration.** If NO_3^- or other inorganic molecules other than O_2 serve as electron acceptors, the process is known as **anaerobic respiration.**

In this exercise you will demonstrate experimentally some cell functions that are associated with life in the presence of air by performing an oxidase test and a catalase test. The oxidase test demonstrates the presence of cytochrome oxidase, an enzyme found in the electron transport system of bacteria. The catalase test detects the presence of the enzyme catalase, which is involved in the detoxification of H_2O_2. Hydrogen peroxide is a common byproduct of aerobic respiration which must be removed by the cell because it is cytotoxic. You will also demonstrate the phenomenon of anaerobic respiration, in which bacteria use an inorganic molecule other than molecular oxygen (in this

case, nitrate) as the terminal electron acceptor during respiration.

A. OXIDASE TEST

The oxidase test detects the presence of **cytochrome oxidase,** the terminal cytochrome system AA_3 found in some bacteria. Cytochrome oxidase transfers electrons from cytochrome c to oxygen.

This test is a useful procedure in the clinical laboratory because some pathogenic species of bacteria, such as *Neisseria gonorrhoeae* and *Pseudomonas aeruginosa*, are oxidase-positive, while many other bacteria are negative. When colonies of oxidase-positive bacteria are treated with the **oxidase reagent** (dimethyl-p-phenylenediamine oxalate), they turn black within 30 minutes. The color development occurs because cytochrome oxidase oxidizes the reagent. The oxidized reagent is black and the reduced form is colorless. The intensity of the color increases as more and more molecules of the oxidase reagent are converted by the enzyme to the colored compound.

Materials

Cultures of *Escherichia coli* and *Pseudomonas fluorescens* on nutrient agar plates

Oxidase reagent: a 1:1 (vol/vol) mixture of 1% α-naphthol in 95% ethanol and 1% aqueous dimethyl-p-phenylenediamine oxalate

Pasteur pipets with rubber bulbs

Procedure

1. Add 1 to 2 drops of the oxidase reagent to an isolated colony of *Escherichia coli* on a nutrient agar plate. Repeat the procedure using *Pseudomonas fluorescens*.

• Oxidase-positive colonies may turn black within a few minutes after the addition of the oxidase reagent.

   In some cases the reaction is slow. Consequently, the colonies may take up to 30 minutes to turn black.

2. Record the results on the report sheet.

3. Discard all materials as indicated by your instructor.

Questions

1. What is the significance of the oxidase reaction?

2. Outline a simple procedure for the presumptive identification of *Neisseria gonorrhoeae* in clinical specimens using the oxidase reagent.

B. CATALASE TEST

Catalase is an enzyme that catalyzes the breakdown of hydrogen peroxide (H_2O_2) to water and O_2. This enzyme is important to aerobic microorganisms because it detoxifies H_2O_2. Hydrogen peroxide is toxic to cells because it inactivates essential cellular enzymes. Hydrogen peroxide forms during aerobic metabolism when components of the respiratory chain donate electrons to molecular oxygen. Since hydrogen peroxide is a common byproduct of aerobic metabolism, microorganisms that grow in aerobic environments must remove this toxic substance. Catalase is one of two enzymes commonly synthesized by microorganisms to detoxify hydrogen peroxide; the other one is called **peroxidase.** No oxygen is evolved from the breakdown of hydrogen peroxide by peroxidase.

$$H_2O_2 \xrightarrow{\text{peroxidase}} 2H_2O$$
$$NADH + H^+ \quad NAD^+$$

The catalase test is very useful in differentiating between groups of microorganisms. For example, the morphologically-similar *Streptococcus* and *Staphylococcus* are differentiated, in part, using the catalase test. Species of *Streptococcus* are catalase-negative, while species of *Staphylococcus* are catalase-positive.

The catalase activity of microorganisms is detected by adding a few drops of 3% H_2O_2 to a colony and looking for the formation of oxygen bubbles all around the colony. The chemical reaction catalized by catalase is as follows:

$$H_2O_2 \xrightarrow{\text{catalase}} H_2O + \tfrac{1}{2}O_2$$

Materials

Cultures of *Streptococcus faecalis* and *Staphylococcus epidermidis* on nutrient agar plates

Catalase reagent: 3% aqueous H_2O_2

Pasteur pipets with rubber bulbs

Procedure

1. Add a few drops of the catalase reagent to an isolated colony of *Staphylococcus epidermidis* appearing on the nutrient agar plate. Repeat the procedure using *Streptococcus faecalis*.

• Catalase activity is detected by the formation of oxygen bubbles, no matter how small, all over the colony.

2. Record the results on the report sheet.

3. Discard all materials and cultures as indicated by your instructor.

Questions

1. What is the importance of catalase to microorganisms? To microbiologists? Explain your answer.

2. How would you employ the catalase test in the identification of the Gram-positive coccus *Streptococcus pyogenes* isolated from a throat culture in view of the fact that *Staphylococcus aureus*, another gram-positive coccus, can also be isolated from the throat?

3. Do anaerobic bacteria require catalase? Explain your answer briefly.

C. NITRATE REDUCTION

Some microorganisms are able to use molecules other than oxygen as terminal electron acceptors during respiration. When this occurs, it is called **anaerobic respiration.** Nitrate is used by some organisms as a terminal electron acceptor. When nitrate (NO_3^-) serves as an electron acceptor, it is reduced to nitrite (NO_2^-). Some microorganisms can also reduce nitrite to molecular nitrogen (N_2).

The ability of microorganisms to reduce nitrate is an important criterion in identification. For example, *Escherichia coli* and *Pseudomonas aeruginosa* can both use nitrate as a terminal electron acceptor. However, *E. coli* reduces it only to nitrite, while *P. aeruginosa* reduces it all the way to N_2. By contrast, *Staphylococcus epidermidis* is unable to utilize nitrate as a terminal electron acceptor.

The nitrate test is usually performed by culturing microorganisms in nutrient broth containing 0.5% KNO_3 and a Durham tube. After a suitable period of incubation, the cultures are examined for evidence of gas in the Durham tube and for the presence of nitrite ions in the medium. The gas trapped in the Durham tube is probably a mixture of N_2 and CO_2. The N_2 is from the complete reduction of nitrate and the CO_2 is from the citric acid cycle, which produces this gas as a byproduct of respiration (anaerobic respiration). The nitrite ions are detected by the addition of sulfanilic acid and α-naphthylamine to the nitrate broth culture. Any nitrite in the medium will react with the reagents to produce a red or pink compound.

Since α-naphthylamine is listed as a carcinogen, **sulfamic acid** can be used instead of sulfanilic acid and α-naphthylamine to detect the presence of nitrite ions. If sulfamic acid is used, addition of a few drops of a sulfamic acid solution (4g sulfamic acid in 100 ml of aqueous, 20% H_2SO_4) to the broth culture containing nitrite ions will cause the production of gas bubbles (N_2) from the reduction of NO_2^- to N_2. In this exercise we will use the sulfanilic acid and α-naphthylamine reagents because these are commonly used in clinical laboratories. If sulfamic acid is to be used, follow the procedures as outlined in the manual, replacing the reagents with sulfamic acid. Detect the presence of nitrite ions by noting the bubbles emanating from the culture.

Materials

Cultures of *Escherichia coli*, *Pseudomonas fluorescens* and *Staphylococcus epidermidis*

Nitrate broth tubes (nutrient broth + 0.5% KNO_3) with Durham tubes

Inoculating loop

Sulfanilic acid solution

Alpha-naphthylamine solution

Zinc dust

Test tube rack

Pasteur pipets and rubber bulbs

Incubator set at 25° to 35°C

Fresh soil

Procedure

First Period

1. Label 4 tubes of nitrate broth with your name, date, type of culture medium, and the microorganism or soil sample to be used. Label a 5th tube "Control."

• The control tube serves to determine if the medium is sterile and to determine if any bubbles develop in the Durham tube as a result of oxygen coming out of solution.

2. Inoculate the 4 tubes with the appropriate material. Leave the control tube uninoculated.

3. Incubate all 4 tubes at 35°C.

Second Period

1. Observe the cultures for evidence of growth. Also examine the control tube to ascertain that the medium was sterile.

2. Examine the Durham tubes for evidence of gas. Compare the amount of gas in the Durham tube with that of the control tube (if any).

3. Add approximately 1 ml of sulfanilic acid and 1 ml of alpha-naphthylamine to each of the tubes and look for the development of a pink or red color in the medium. The color development is an indication that nitrate has been reduced to nitrite and that anaerobic respiration has taken place.

In tubes where no reactions have occurred, add a few grains of zinc dust to determine whether or not nitrate has been reduced to ammonium.

• If nitrate is present in the medium, the medium will turn pink or red because the zinc will reduce nitrate in the medium to nitrite, which will then react with the reagents to form the colored compound.

4. Record the results on the report sheet.

5. Discard all cultures as indicated by your instructor.

Questions

1. Which of the three test organisms is negative for nitrate reduction? What does this tell you about the metabolic capabilities of organism?

2. Explain what the reagents do when they are added to the culture. What would positive results indicate? Negative?

3. A nitrate broth tube shows growth but no evidence of gas or nitrate production. Upon addition of Zn dust, gas is produced. What information can you obtain from this?

EXERCISE 26
UTILIZATION OF AMINO ACIDS

Amino acids are organic molecules containing an acid group (COOH), an amino group (NH_2), and an R-group (R), as illustrated in figure 26-1. The amino acids used by cells differ from each other because of the R-group.

Microorganisms use amino acids to build proteins, as precursors for the synthesis of other cellular components, and sometimes as a source of energy. Amino acids are modified in various ways when they are metabolized. This exercise demonstrates some of the ways in which microorganisms modify amino acids. Since some of these modifications are not carried out by all species of microorganisms, the byproducts of amino acid modifications can be used as aids in the identification of unknown specimens.

A. INDOLE PRODUCTION

The amino acid tryptophan is a component of nearly all proteins, and is therefore available to microorganisms as a result of protein breakdown. Some bacteria, like *Escherichia coli*, are able to use tryptophan to satisfy their carbon needs (fig. 26-2).

Escherichia coli produces an enzyme called **tryptophanase** that catalyzes the removal of the indole residue from the tryptophan. Indole accumulates in the culture medium as a waste product, while the rest of the tryptophan molecule (pyruvate and NH_4^+) is used to satisfy nutritional needs.

The production of indole from tryptophan by microorganisms can be detected by culturing them in a medium rich in tryptophan. The tryptophan is generally provided in the form of tryptones, polypeptides that contain many tryptophan residues. The accumulation of indole in the medium can be detected by adding Kovac's reagent, which reacts with indole, giving a water-insoluble, bright red compound on the surface of the medium (fig. 26-3).

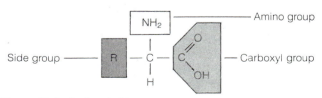

Figure 26-1. Amino Acids

Materials

Cultures of *Escherichia coli*, *Enterobacter aerogenes*, and *Proteus vulgaris*

Tubes of SIM (sulfide-indole-motility) agar deeps

Kovac's reagent

Pasteur pipets with bulbs

Inoculating loop

Incubator set at 35°C

Procedure

NOTE: SIM medium can be used to test for both indole and H_2S production in the same tube. Hence, if you perform the indole test, you are also performing the test for production of H_2S. Observe H_2S production first, before the addition of Kovac's reagent.

First Period

1. Label 3 tubes of SIM with your name, date, type of culture medium, and the name of the bacterium tested.

2. Inoculate the tubes of SIM with their corresponding organisms by stabbing the agar deep (fig. 26-4) about ¾ of the way to the bottom.

3. Incubate the cultures at 35°C for 48 hours.

Second Period

1. Add 10 drops of Kovac's reagent to each of the 2 SIM cultures.
• The presence of indole is indicated by the development of a bright red layer on the top portion of the culture. The color develops within a few seconds after the addition of Kovac's reagent.

2. Record the results on the report sheet.

3. Discard all cultures as indicated by your instructor.

Questions

1. What component in SIM deeps makes this medium suitable for testing indole production by bacteria?

B. HYDROGEN SULFIDE PRODUCTION

Many proteins, including those found in eggs, are rich in the amino acids cysteine and methionine. These amino acids are released when proteins are hydro-

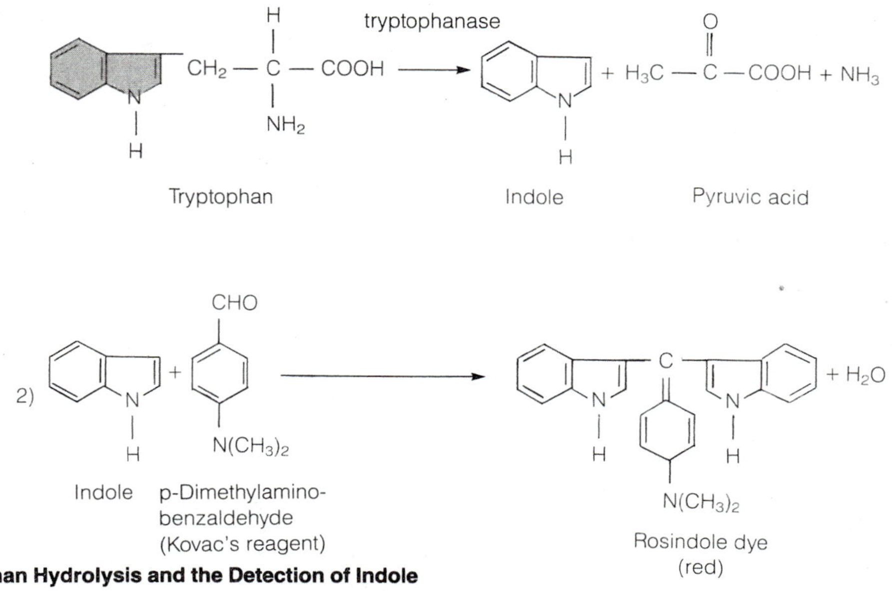

Figure 26-2. Tryptophan Hydrolysis and the Detection of Indole

lyzed and can be taken up by microorganisms to satisfy their nutritional needs. The production of hydrogen sulfide (H_2S) by the microbial alteration of sulfur-containing amino acids is an indication that microorganisms modify sulfur-containing amino acids to satisfy their nutritional requirements. The smell typically associated with rotten eggs may be due to the release of H_2S from sulfur-containing amino acids.

Bacteria that produce a **desulfurase** when cultivated in a medium rich in sulfur-containing amino acids produce H_2S. Ferrous ions incorporated in the culture medium react with the H_2S, producing a black, insoluble compound, ferrous sulfide (FeS). Thus, in this kind of culture medium, the production of H_2S by bacteria is demonstrated by the blackening of the culture medium (fig. 26-3).

There are several commercially prepared culture media designed for the detection of H_2S produced by microorganisms. All of these media contain polypeptides rich in sulfur-containing amino acids and ferrous ions. SIM (sulfide-indole-motility medium) is one such medium. This medium is rich in sulfur-containing amino acids as well as in tryptophan. Therefore, if desired, you can conduct the H_2S production test concurrently with the indole test.

Materials

Tubes of SIM (sulfide-indole-motility agar)

Cultures of *Escherichia coli, Enterobacter aerogenes,* and *Proteus vulgaris*

Inoculating loop

Incubator set at 35°C

Procedure

First Period

1. Label 3 tubes of SIM with your name, date, type of culture medium, and the name of the bacterium tested.

2. Inoculate the tubes of SIM with their corresponding organism by stabbing the agar deep (fig. 26-4) about ¾ of the way to the bottom.

3. Incubate the cultures at 35°C for 48 hours.

Second Period

1. Examine the tubes for the development of a blackening of the medium.

• Any blackening of the medium is an indication of H_2S production.

2. Record the results on the report form.

3. Discard all cultures as indicated by your instructor.

Questions

1. What components in the SIM medium are necessary to test for hydrogen sulfide production?

2. Why does the SIM turn black when H_2S is produced?

C. LYSINE DECARBOXYLASE TEST

Decarboxylation is the removal of a carboxyl group (fig. 26-5) from organic molecules. Decarboxylation of amino acids generally results in the release of CO_2. This process is carried out by many microorganisms.

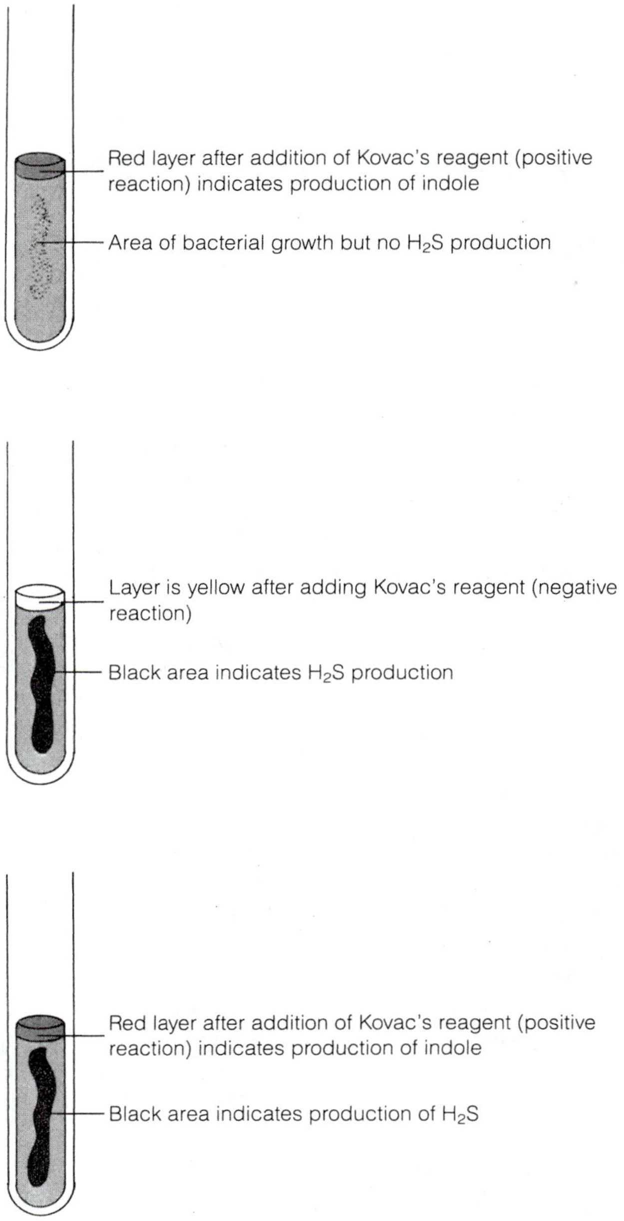

Red layer after addition of Kovac's reagent (positive reaction) indicates production of indole
Area of bacterial growth but no H_2S production
Layer is yellow after adding Kovac's reagent (negative reaction)
Black area indicates H_2S production
Red layer after addition of Kovac's reagent (positive reaction) indicates production of indole
Black area indicates production of H_2S

Figure 26-3. Patterns of Reaction that Can be Seen in SIM Deeps

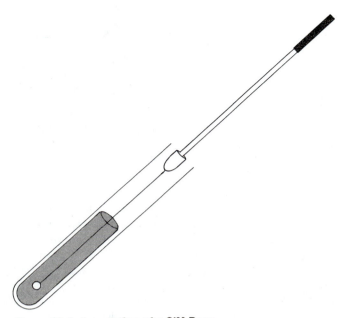

Figure 26-4. Inoculation of a SIM Deep

The decarboxylated molecule is used as a precursor for the synthesis of other molecules that are needed by the cell.

Decarboxylation of amino acids also serves as a means of neutralizing acid environments. When microorganisms ferment, they often produce acidic waste products that make the medium inhospitable. **Decarboxylases** (enzymes), many of which are acid-activated, remove acid groups from amino acids, producing **amines** that make the medium more alkaline.

Decarboxylation of lysine can be detected by cultivating microorganisms in a medium containing lysine, a fermentable sugar such as glucose, and a pH indicator (usually brom-cresol purple) for visual detection of pH changes. The acids from the fermentation of glucose will lower the pH of the medium and

1)
Cysteine
cysteine desulfurase
$H_2S + H_3C - C(=O) - COOH + NH_3$
Pyruvic acid
H_2S + $FeSO_4$ → FeS + H_2SO_4
2) Hydrogen sulfide Ferrous sulfate Ferrous sulfide Sulfuric acid

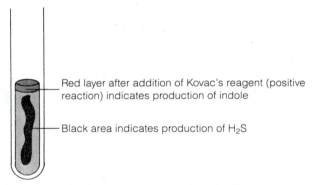

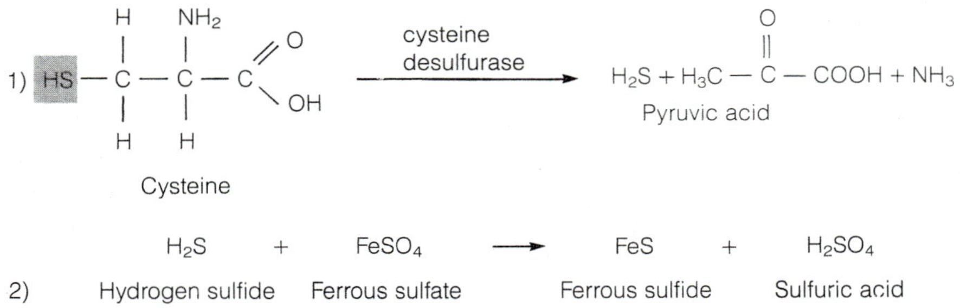

Figure 26-5. Release of Hydrogen Sulfide from the S-containing Amino Acid Cysteine

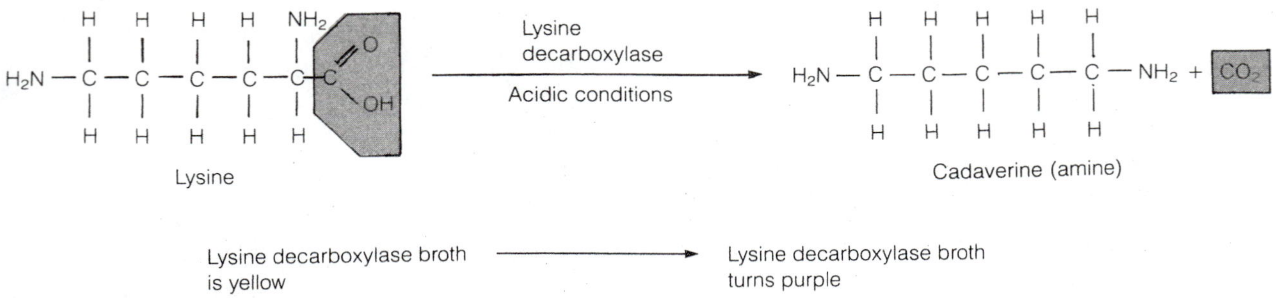

Figure 26-6. Decarboxylation of Lysine

cause a color change in the brom-cresol purple from purple to yellow. At this point, decarboxylation of lysine (which is acid-activated) will occur, resulting in the production of amines and the subsequent neutralization of the medium. This will result in a color change from yellow back to purple, indicating that lysine has been decarboxylated (fig. 26-6).

The lysine decarboxylase test is a useful test in differentiating among bacteria, principally those that are commonly found in the human gastrointestinal tract.

Materials

Cultures of *Enterobacter aerogenes* and *Citrobacter freundii*

Moeller's lysine decarboxylase broth with lysine (LDC)

Moeller's decarboxylase control broth without lysine (DC)

Sterile mineral oil

Sterile Pasteur pipets

Test tube rack

Incubator set at 35°C

Procedure

First Period

1. Label 2 sets of tubes (a tube of LDC and one of DC broth) with your name, date, type of culture medium, and the name of the bacterium used.

2. Inoculate the tubes of LDC and DC broth with their corresponding organisms.

3. Layer about 1 ml of sterile mineral oil on top of the inoculated medium with a sterile Pasteur pipet.

• Be careful not to touch the inoculated medium or the inside of the culture tube with the pipet because this will contaminate the mineral oil if the pipet is reused.

4. Incubate the cultures at 35°C for 48 hours.

Second Period

1. Examine the cultures for color changes in the medium.

• Lysine decarboxylation is indicated by the presence of a purple color in the LDC broth AND a yellow color in the DC broth. If the LDC broth is yellow instead of purple, lysine may not have been decarboxylated. If both tubes (LDC and DC) are purple, lysine was not decarboxylated because the bacterium does not produce acidic products from the fermentation of glucose, and hence will not activate the lysine decarboxylase enzyme. Some organisms may reduce the BCP and render it colorless. If this occurs, the pH of the medium will have to be measured with paper pH indicators.

3. Record the results on the report form.

4. Discard all cultures as indicated by your instructor.

Questions

1. Why is the lysine decarboxylation test considered

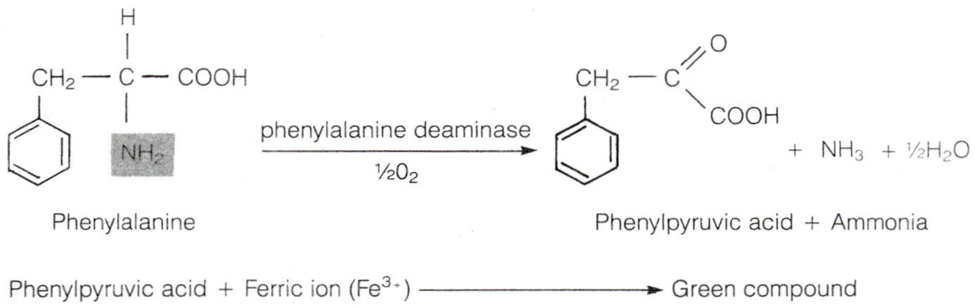

Figure 26-7. Phenylalanine Deamination

negative if both the LDC and the DC broth cultures are purple?

2. Why does the medium always turn yellow, regardless of the ability of the microorganism to decarboxylate?

3. Why does the medium turn purple when lysine is decarboxylated?

D. PHENYLALANINE DEAMINATION

Deaminases catalyze the removal of amino groups (NH_2) from amino acids and other NH_2-containing molecules (fig. 26-7). The resulting organic acids can be used by microorganisms in biosynthesis. In addition, the deamination detoxifies inhibitory amines.

The phenylalanine deaminase test can be used to differentiate among bacteria, including those that inhabit the gastrointestinal tract. The deamination of phenylalanine produces phenylpyruvic acid, which reacts with ferric chloride ($FeCl_3$) to form a green-colored compound. Hence, phenylalanine deaminase activity can be detected by growing bacteria on a phenylalanine-containing agar medium and adding a few drops of 10% solution of $FeCl_3$. The development of a green color indicates that phenylalanine has been deaminated.

Materials

Cultures of *Escherichia coli* and *Proteus vulgaris*

Slants of phenylalanine agar

10% aqueous solution of ferric chloride

Pasteur pipets with bulbs

Test tube rack

Incubator set at 35°C

Procedure

First Period

1. Label 2 tubes of phenylalanine agar with your name, date, type of culture medium, and the name of the bacterium to be tested.

2. Inoculate the slants of phenylalanine agar with their corresponding organisms.

3. Incubate the cultures at 35°C for 48 hours.

Second Period

1. Add 5 to 10 drops of 10% $FeCl_3$ to each of the 2 cultures and tap the side of the culture tube gently to mix the reagent.

• The presence of phenylpyruvic acid (product of phenylalanine deamination) is indicated by the development of a green color. The color develops within a few minutes after the addition of reagent to the culture. If a green color does not develop within 5 minutes, no deamination has occurred.

3. Record the results on the report form.

4. Discard all cultures as indicated by your instructor.

Questions

1. Cite three ways in which phenylalanine can be used by *E. coli*.

2. Give three ways in which amino acids can be used by *E. coli*.

3. Outline a simple scheme for the differentiation of *Escherichia coli* and *Enterobacter aerogenes* isolated from a sample, using the tests that you performed in this exercise and in exercise 25.

EXERCISE 27
UTILIZATION OF
CITRATE, GELATIN, AND UREA

Tests for citrate utilization, urea hydrolysis, and gelatin liquefaction are very useful in the differentiation of microorganisms. The citrate utilization test is one of the IMViC (Indole, **M**ethyl red, **V**oges-Proskauer and

Citrate) test series used in differentiating among coliform bacteria. The urea hydrolysis test is often used to differentiate *Salmonella*, a common enteric pathogen, from *Proteus*, a urinary tract pathogen, which can also be found in stool specimens as a nonpathogen. Gelatin hydrolysis is a trait used to differentiate among certain species of *Pseudomonas* and among enteric bacteria.

A. CITRATE UTILIZATION TEST

The citrate test determines the ability of microorganisms to use citrate as the sole source of carbon and energy. Simmons' citrate agar, a chemically defined medium with sodium citrate as the sole carbon source, NH_4^+ as a nitrogen source, and the pH indicator bromthymol blue, is commonly used for this test. When microorganisms utilize citrate, they remove the acid from the medium, which raises the pH and turns the pH indicator from green to blue. A color change in the medium from green to blue indicates that the microorganism tested can utilize citrate as its only carbon source.

Materials

Cultures of *Escherichia coli* and *Enterobacter aerogenes*

Slants of Simmons' citrate agar

Inoculating loop

Test tube rack

Incubator set at 25° to 35°C

Procedure

First Period

1. Label 2 slants of Simmons' citrate agar with your name, date, type of culture medium, and the name of the organism tested.

2. Inoculate the slants of Simmons' citrate agar with a very light inoculum of the indicated organism.

• If the agar slants are inoculated heavily, growth often occurs even when the organism is unable to use citrate as its only carbon and energy source. Consequently, ambiguous results will be obtained.

3. Incubate the cultures at 35°C for 48 hours.

Second Period

1. Examine the cultures for evidence of citrate utilization by noting growth and a color change from green to blue.

2. Record the results on the report sheet.

3. Discard all cultures as indicated by your instructor.

Questions

1. Why is a chemically defined medium necessary to test for citrate utilization?

2. Outline a simple scheme for the differentiation of *E. coli* and *E. aerogenes* isolated from a sample using the IMViC reactions.

3. What problem will you encounter when testing for citrate utilization by microorganisms that require growth factors?

B. GELATIN HYDROLYSIS

Gelatin is a fibrous protein prepared by boiling bone, cartilage, and other animal connective tissue. This protein, when cooled, forms a gel. Certain microorganisms are able to break down gelatin molecules and use the resulting amino acids as nutrients. The hydrolysis of gelatin by these microorganisms is catalyzed by a proteolytic exoenzyme sometimes called **gelatinase.** Digested gelatin is no longer able to gel and it becomes liquefied. The ability of certain microorganisms to digest gelatin is an important characteristic in their differentiation. For example, *Serratia marscescens* can be differentiated from *Klebsiella pneumoniae* or *Escherichia coli* by its ability to digest gelatin. Gelatin hydrolysis is used sometimes to assess the pathogenic potential of certain strains of microorganisms. This is because the production of gelatinase can often be correlated with the production of enzymes that allow a microbe to break down tissue and spread in a lesion.

In the laboratory, gelatin liquefaction is tested by stabbing a semisolid medium containing nutrient broth and gelatin with microorganisms. The inoculated medium is then incubated for a suitable period of time and subsequently examined for liquefaction. Since gelatin can liquefy when incubated at 35°C, whether or not the microorganisms hydrolyzed the gelatin, it is necessary to refrigerate the nutrient gelatin cultures for about 30 minutes before the results of the gelatin hydrolysis test are observed. If gelatin has been hydrolyzed, the medium will remain liquid after refrigeration. If gelatin has not been hydrolyzed, however, the medium will resolidify during the period of refrigeration.

Materials

Cultures of *Escherichia coli* and *Bacillus subtilis*

Deeps of nutrient gelatin

Inoculating loop

Test tube rack

Incubator set at 35°C

Ice water bath or refrigerator

Procedure

First Period

1. Label 2 nutrient gelatin deeps with your name, date, type of culture medium, and name of organism tested.

2. Inoculate each of the nutrient gelatin deeps with the appropriate bacterium by stabbing the medium ¾ of the way to the bottom.

3. Incubate the inoculated tubes at 35° until the next laboratory period.

Second Period

1. Examine the nutrient gelatin deeps for evidence of growth and then place them in an ice water bath or in a refrigerator for approximately 30 minutes.
• This is necessary because nutrient gelatin cultures are usually liquid if they were incubated at 35°C, whether or not gelatin was hydrolyzed.

2. Examine the cooled tubes of nutrient gelatin for liquefaction.
• This can be done by gently slanting the tubes and noting whether the surface of the medium is fluid or solid. If the nutrient gelatin remains in liquid form after refrigeration it is an indication that gelatin has been hydrolyzed by the microorganism. Some organisms hydrolyze gelatin very slowly. Consequently, all negative cultures should be reincubated for a minimum of one week before they are discarded.

3. Record the results on the report form and draw the approximate shape of the area hydrolyzed by the culture.

Questions

1. Explain why nutrient gelatin is not solidified by using agar as most other solid media are.

2. Can gelatin hydrolysis be important to microorganisms, even though they are not pathogenic? Explain your answer. Where in nature might you find such microorganisms?

C. UREA HYDROLYSIS

Some microorganisms are able to produce an enzyme called **urease,** which splits urea into ammonium and carbon dioxide (fig. 27-1). Urease activity is detected by growing microorganisms in a medium containing

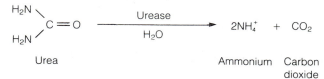

Figure 27-1. Urea Hydrolysis

urea and a pH indicator (usually phenol red). When urea is hydrolyzed, NH_4^+ accumulates in the medium and makes it alkaline. A color change from orange-red to violet-red is an indication of urea hydrolysis.

Materials

Cultures of *Escherichia coli* and *Proteus vulgaris*

Slants (with a 1″ butt) of Christensen's urea agar

Inoculating loop

Test tube rack

Incubator set at 35°C

Procedure

First Period

1. Label 2 slants of Christensen's urea agar with your name, date, type of culture medium, and the name of the organism used.

2. Inoculate the slants of urea agar with the indicated organism.

3. Incubate the cultures at 35°C for 48 hours.

Second Period

1. Examine the cultures for evidence of a color change from orange-red to violet-red.
• If the color change is difficult to detect because the entire surface of the slant has been covered with growth, comparing the color of the slant surface with that of the bottom of the agar butt will help in making color comparisons. Urea hydrolysis is indicated when the agar has been changed from a orange-red color to a violet-red color.

2. Record the results on the report sheet.

3. Discard all cultures as indicated by your instructor.

Questions

1. What components in the medium are necessary to test for urea hydrolysis?

2. Why is a pH indicator included in the formulation of Christensen's urea agar?

EXERCISE 28
IDENTIFICATION OF
AN UNKNOWN BACTERIUM

The identification of unknown bacterial isolates is a common task of microbiologists. In the clinical laboratory, it is essential that microorganisms isolated from a patient be identified promptly and accurately so that the proper treatment can be initiated without delay. It is also important that microorganisms isolated from foods and beverages implicated in food poisoning be identified so that the appropriate control measures can be initiated and the outbreak stopped.

Bacteria may be identified on the basis of morphological, cultural, and biochemical characteristics. Morphological criteria such as the shape, size, and arrangement of cells are usually not sufficient to identify a bacterium unequivocally. Other characteristics, such as Gram stain and acid-fast staining properties, colonial growth patterns, fermentation reactions, and assimilation of amino acids must be used. Table 28-1 summarizes the characteristics that you will use in this exercise to identify an unknown.

An unknown may exist as a pure culture or as a mixed culture. If the unknown is not pure, it must first be purified before proceeding with its identification. Purification usually is accomplished by streaking the mixture onto agar media. The purity of a culture (even if it was pure at one time) should be verified by performing the Gram stain and determining if all the bacterial cells seen in the stained smear have the same general shape and arrangement.

Once the purity of the isolate has been established, the culture can be subjected to a variety of tests to obtain a morphological and biochemical profile that will aid in identification. Each test should be controlled to make sure that the culture medium is sterile and suitable for the test performed, the reagents are able to detect the desired products, and that your techniques are adequate. The suitability of the culture medium can be ascertained by inoculating it with both a culture known to give a positive reaction and one that gives a known negative. Known positive and negative controls for each test are listed in table 28-1. These organisms are the same ones suggested in previous exercises.

Once a biochemical and morphological profile is obtained for each of the isolates, they are identified by using a **dichotomous key** (fig. 28-1). The key will provide you with a series of mutually exclusive choices.

By choosing the characteristics that fit the unknown, eventually you are led to a possible genus.

It is always a good idea to verify your results by reading the description of the identified isolate in a reference book. *Bergey's Manual of Systematic Bacteriology* is the major reference. Use the latest edition available in your library or classroom and compare the description of your isolate with the results that you obtained. Try to explain any discrepancies between your results and those published in *Bergey's Manual*. Ask your instructor for assistance only after you have exhausted all other recourses. Remember, this is YOUR unknown, not your instructor's.

Materials

Cultures of unknowns

Cultures of bacteria used as controls for the tests listed in table 28-1

Plates of trypticase soy agar (TSA)

Slants of TSA

The various culture media used in exercises 23 through 27

Incubator set at 35°C

Procedure

First Period

1. Label the bottom of two TSA plates with your name, date, and unknown identification number. Label one plate for incubation at room temperature and the other at 35°C.

2. Streak both plates with your unknown sample in order to obtain well-isolated colonies of the unknown culture.

3. Incubate one of the plates at room temperature and the other at 35°C.

Second Period

1. Examine the plates for the presence of well-isolated colonies.

TABLE 28-1
CULTURAL AND BIOCHEMICAL TESTS USED TO IDENTIFY UNKNOWNS

	CONTROL MICROORGANISMS	
TEST*	POSITIVE	NEGATIVE
Growth at 20°C (16)	*Pseudomonas fluorescens*	*Bacillus stearothermophilus*
Growth at 35°C (16)	*Escherichia coli*	*Pseudomonas fluorescens*
Growth at 55°C (16)	*Bacillus stearothermophilus*	*Escherichia coli*
Colony characteristics (6)		
form		
elevation		
margin		
consistency		
pigmentation		
Gram stain (9)	*Staphylococcus epidermidis*	*Escherichia coli*
Acid-fast stain (9)	*Mycobacterium phlei*	*Escherichia coli*
Cell morphology (3)		
Capsule (10)	*Klebsiella pneumoniae*	*Staphylococcus aureus*
Endospores (10)	*Bacillus subtilis*	*Escherichia coli*
Motility (3)	*Proteus vulgaris*	*Staphylococcus aureus*
Glucose fermentation (24)	*Escherichia coli*	*Micrococcus luteus*
Lactose fermentation (24)	*Escherichia coli*	*Micrococcus luteus*
Maltose fermentation (24)	*Escherichia coli*	*Micrococcus luteus*
Sucrose fermentation (24)	*Proteus vulgaris*	*Micrococcus luteus*
Mannitol fermentation (24)	*Staphylococcus aureus*	*Micrococcus luteus*
Methyl red test (24)	*Escherichia coli*	*Enterobacter aerogenes*
Voges-Proskauer test (24)	*Enterobacter aerogenes*	*Escherichia coli*
Catalase test (25)	*Staphylococcus epidermidis*	*Streptococcus faecalis*
Oxidase test (25)	*Pseudomonas fluorescens*	*Escherichia coli*
Starch hydrolysis (23)	*Bacillus subtilis*	*Escherichia coli*
Casein hydrolysis (23)	*Bacillus subtilis*	*Escherichia coli*
DNA hydrolysis (23)	*Serratia marcescens*	*Escherichia coli*
Fat hydrolysis (23)	*Proteus mirabilis*	*Staphylococcus epidermidis*
Indole production (26)	*Escherichia coli*	*Enterobacter aerogenes*
H_2S production (26)	*Proteus vulgaris*	*Escherichia coli*
Nitrate reduction (25)	*Pseudomonas fluorescens*	*Staphylococcus epidermidis*
Gelatin hydrolysis (27)	*Bacillus subtilis*	*Escherichia coli*
Lysine decarboxylation (26)	*Enterobacter aerogenes*	*Citrobacter freundii*
Phenylalanine deamination (26)	*Proteus vulgaris*	*Escherichia coli*
Urea hydrolysis (27)	*Proteus vulgaris*	*Escherichia coli*
Citrate utilization (27)	*Enterobacter aerogenes*	*Escherichia coli*

* Numbers in parenthesis indicate the exercise in which the characteristic was discussed.

• Since your unknown may be contaminated with a mixture of bacterial species, carefully examine the colonies on each plate to determine the extent of contamination and the characteristics of the unknown.

2. Mark well-isolated colonies of each of the types appearing on the plate by circling their location with a grease pencil on the bottom of the Petri plate.

• Make sure that the colonies you select are on the streak pattern. Colonies off the streak lines may be contaminants.

3. Characterize selected colonies as follows:

    a. determine the colony's shape, margin, elevation, consistency, and color.

    b. determine the temperatures at which the organism grows by inoculating trypticase soy broths. Incubate the broths at 4°C, 20°C, 35°C, 45°C, and 55°C.

    c. determine the Gram reaction for each of the isolated organisms.

4. Subculture the purified organism onto slants of TSA.

5. Record your observations on the unknown report form.

Third Period

1. Verify the purity of your cultures by doing the following:

    a. examine the plates for the presence of one predominant colony. Make sure that you examine only those colonies appearing along the path of streaking.

    b. perform a Gram stain on several representative colonies in each of the plates.

- If the culture is pure, all the colonies will have cells of the same morphology and Gram staining characteristics.

2. a. If the cultures are pure, prepare fresh subcultures of each isolate on TSA slants. These two subcultures represent a stock that will serve as a source of bacteria for the tests to be performed in the fourth (and any subsequent) period of this exercise.

   b. If your cultures are not pure, repeat the procedures outlined the second period of this exercise.

Fourth Period

NOTE: This portion of exercise 28 can be performed in one period or it may be stretched out over a few laboratory periods and integrated with each of the procedures performed in exercises 23 through 27.

1. Using the plates or slants prepared in the previous period, perform each of the tests indicated on the unknown report form.

- The procedures for these tests are described in exercises 23 through 28. The exercise number appears in parenthesis next to the name of the test in table 28-1.

Fifth Period

1. Using the dichotomous key in figure 28-1, identify your unknown bacteria.

2. Compare the results you obtained with those published in *Bergey's Manual of Systematic Bacteriology*. Using the unknown report form, write a paragraph or two explaining your results, how you arrived at the identity of your unknowns, and any discrepancies that may be evident between your results and the descriptions in *Bergey's Manual*.

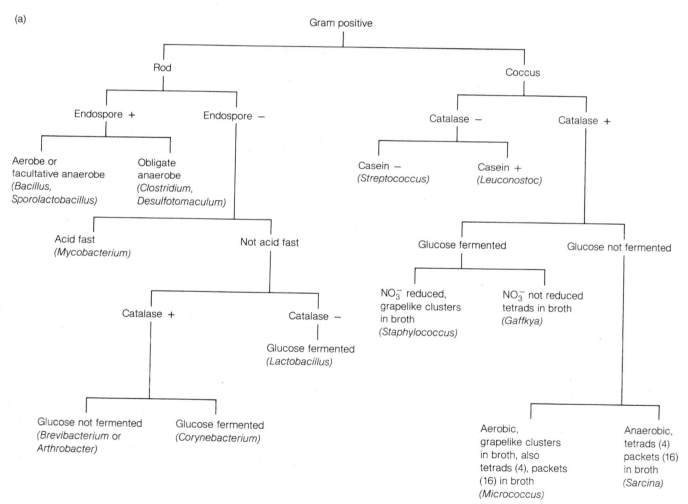

Figure 28-1. A Key for the Identification of Common Bacteria. *(a)* Key to selected Gram positive bacteria. *(b)* Key to selected Gram negative bacteria.

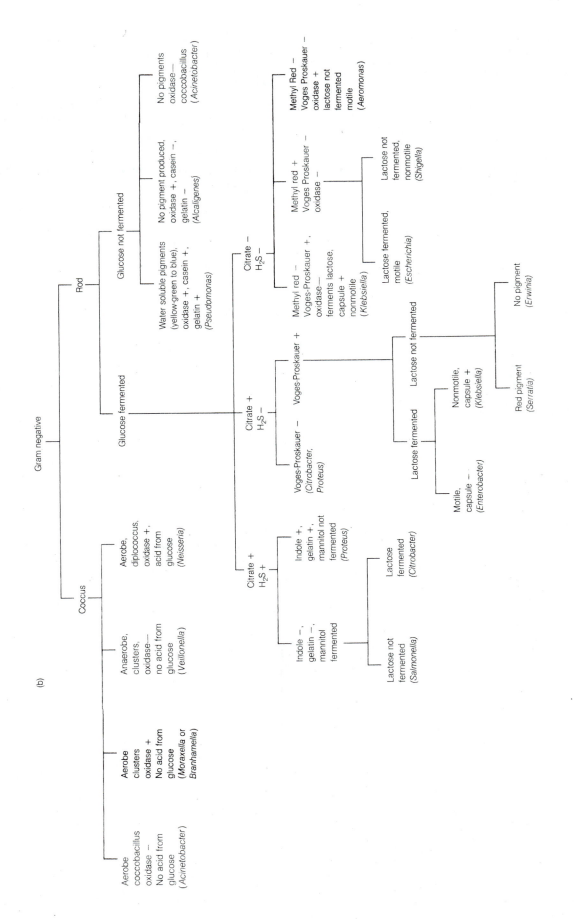

(b)

Questions

1. Would you expect that discrepancies between your results and those published in *Bergey's Manual* occur even though you have the correct results? Explain your answer.

2. Why is it important to test control microorganisms along with your unknown when you perform the biochemical tests? Would you need control organisms when you perform the Gram stain or the acid-fast stain?

3. Why shouldn't you perform biochemical tests using the colonies isolated in the second period of this exercise?

4. Modify the dichotomous key in figure 28-1 so that the glucose fermentation test is the first differentiating criterion instead of the gram stain. Did you experience any difficulties identifying your unknown using your modified key?

5. Is using dichotomous keys the only way to identify bacteria? Explain your answer.

PART VIII
SURVEY OF
EUKARYOTES AND CYANOBACTERIA

Eukaryotic microorganisms are generally single-celled organisms (or simple aggregations of cells) that have a nucleus and intracellular organelles. Eukaryotic microorganisms include the fungi, the protozoa, and the algae. They are widely distributed in our environments and comprise an essential component of all ecosystems. These microorganisms are found colonizing soil and water, in the air, and some are capable of reproducing in and on living organisms. Many of these microorganisms perform useful functions in our environment, such as detoxification of pollutants, recycling of nutrients, and protecting animals and plants from harmful microbes. Some, however, can cause disease if they reproduce in living tissue.

This section involves the study of the major groups of eukaryotic microorganisms; the fungi, the unicellular algae, and the protozoa. You will be working with both prepared specimens and living organisms in order to become familiar with their salient characteristics, mode of reproduction, and species diversity.

EXERCISE 29
THE YEASTS AND MOLDS

The fungi are a widely distributed group of eukaryotic microorganisms. They usually obtain their nutrients by decomposing dead organic matter. For this reason most fungi are called **saphrophytes.**

The fungi characteristically form filaments called **hyphae**. These are elongated, tubular cells or chains of tubular cells whose cytoplasms are generally shared. The hyphae may be completely devoid of cross walls **(septa)** or may have septations that are evenly distributed along their length. Aseptate hyphae, also called **coenocytic** hyphae, are characteristic of a number of fungal classes such as the Oomycetes and the Zygomycetes, while septate hyphae are commonly found in the Ascomycetes, Basidiomycetes, and Deuteromycetes. Fungi usually have cell walls that contain chitin as the major structural component. Hyphae ab-

sorb nutrients from the environment and increase in length by cell division. Hyphae may give rise to specialized reproductive structures called **spores** (fig. 29-1). **Conidia** are asexual spores.

The fungal colony, or **thallus,** consists of a mass of hyphae called the **mycelium.** The thallus is derived from a bit of hypha or from a germinating spore. Those filamentous fungi that form colonies primarily composed of hyphae are referred to as **molds.**

Yeasts are fungi that do not form hyphae. They are unicellular organisms that reproduce by budding or fission and form colonies resembling those of bacteria. Some yeasts are used to ferment grape juice to make wine, to make bread, and to produce certain types of cheese. Others, especially those in the genera *Candida* and *Cryptococcus*, can cause serious, even fatal human infections.

The fungi, as a group, exhibit a remarkable variability in morphology which is used to classify them (table 29-1). The majority of the fungi are placed into groups based on the development and morphology of their asexual and sexual spores. The development of sexual spores can be determined by culturing two compatible strains of a species of fungus on an agar plate containing suitable nutrients. When the fungi grow they come in contact with each other and mate. The zygotes resulting from the mating process develop and give rise to characteristic sexual structures (figs. 29-2 to 29-4). In this exercise you will study a number of different fungi and demonstrate the mating of two compatible strains of a common fungus, *Rhizopus*, and the subsequent development of a zygospore.

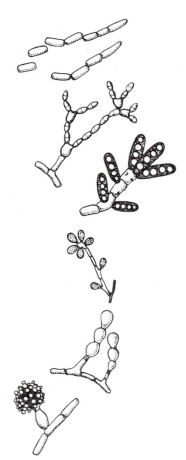

Figure 29-1. Spores

A. EXAMINATION OF CULTURES

Materials

Cultures of *Aspergillus, Penicillium, Alternaria, Geotrichum, Saccharomyces,* and *Rhizopus* on Sabouraud glucose agar

Prepared slides of *Morchella* and *Puccinia*

Sabouraud agar slants

Sterile glass coverslips

Sterile glass slides

Lactophenol cotton blue stain (LPB)

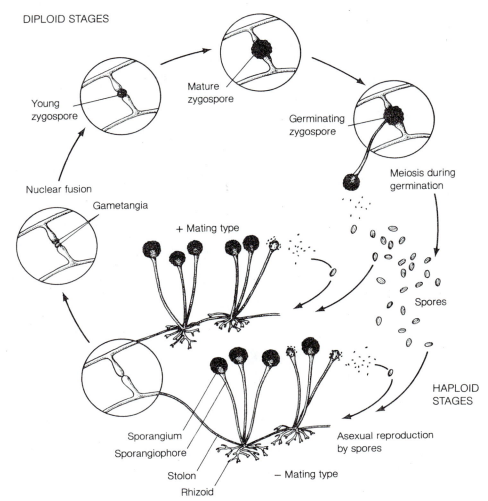

DIPLOID STAGES

Young zygospore

Mature zygospore

Germinating zygospore

Nuclear fusion

Meiosis during germination

Gametangia

+ Mating type

Spores

HAPLOID STAGES

Asexual reproduction by spores

Sporangium

Sporangiophore

Stolon

Rhizoid

− Mating type

Figure 29-2. Life Cycle of *Rhizopus*, a Typical Zygomycete. Compatible hyphae (+ and −) fuse to form a zygote called the zygospore. The mature zygospore develops a thick coat and can remain dormant for many months. After the dormancy period, the zygospore cracks open, producing a sporangium filled with sporangiospores. The spores can germinate, producing new thalli. Vegetative hyphae can also give rise to sporangia by asexual means. When compatible hyphae fuse, the cycle begins again.

Inoculating needles with tips bent at 90° angles

Dissecting (teasing) needles

Sand jar with disinfectant

Procedure

First Period

1. Examine the fungal colonies. Characterize them using the following criteria: texture of colony, color of upper surface, color of lower surface, and appearance of the colony.

2. Prepare tease mounts in LPB (fig. 29-5) of the fungi provided. Tease mounts are prepared as follows:

(a) place a small drop of LPB on the center of an appropriately labeled glass slide.

(b) Using a bent dissecting needle (fig. 29-5), asep-tically transfer a small mass of mycelium (about the size of a pinhead) to the drop of LPB.

(c) Using a pair of disinfected dissecting needles, pick apart the hyphal mass to separate individual strands of hyphae. Clean the needles by swirling them in a sand jar, then flame the tips to sterilize.

(d) Using disinfected forceps, gently place a clean coverslip over the teased preparation (avoid the formation of air bubbles).

3. Examine each of the preparations first using the 10× objective lens and then the 40× objective lens.

• Do not use the oil immersion lens while examining tease mounts of fungi. Use the following characteristics to describe each of the fungi: morphology of hyphae, pigmentation (color) of hyphae, presence or absence of cross walls (septa), width of hyphae, length

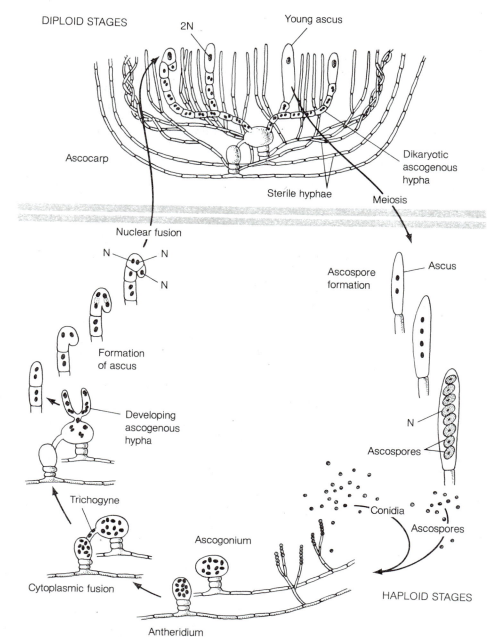

DIPLOID STAGES

2N

Young ascus

Ascocarp

Dikaryotic ascogenous hypha

Sterile hyphae

Meiosis

Nuclear fusion

N — N

N

Ascospore formation

Ascus

Formation of ascus

Developing ascogenous hypha

N

Ascospores

Trichogyne

Ascogonium

Conidia

Ascospores

Cytoplasmic fusion

HAPLOID STAGES

Antheridium

Figure 29-3. Life Cycle of a Typical Ascomycete. Asexual reproduction takes place when conidia, which develop from hyphae, germinate to produce new fungal thalli. During sexual reproduction, two compatible fungal elements come together. The fusion of the fungal elements results in the formation of asci containing four to eight ascospores.

of distance between septa (if present), size, shape, and color of spores, arrangement (if detectable) of spores on hyphae.

4. Study the microscopic characteristics of the prepared slides. Record your observations in the report form.

5. Discard and put away cultures and materials as indicated by your instructor.

B. COVERSLIP PREPARATION

Materials

Cultures of *Aspergillus, Penicillium, and Alternaria* on Sabouraud glucose agar

Cornmeal agar deeps

Petri plates

Sterile glass coverslips

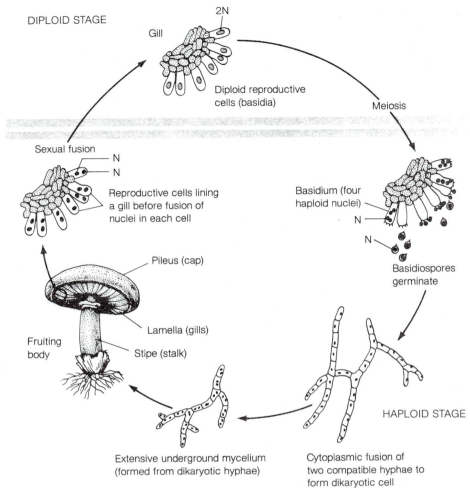

DIPLOID STAGE

Gill

2N

Diploid reproductive cells (basidia)

Meiosis

Sexual fusion

N
N

Reproductive cells lining a gill before fusion of nuclei in each cell

Basidium (four haploid nuclei)

N

N

Basidiospores germinate

Pileus (cap)

Fruiting body

Lamella (gills)

Stipe (stalk)

HAPLOID STAGE

Extensive underground mycelium (formed from dikaryotic hyphae)

Cytoplasmic fusion of two compatible hyphae to form dikaryotic cell

Figure 29-4. Structure and Life Cycle of a Basidiomycete

Glass slides

Lactophenol cotton blue stain (LPB)

Inoculating needles with tips bent at 90° angles

Sand jar with disinfectant

Scalpel

Forceps

Procedure

First Period

1. Pour 3 plates of cornmeal agar (CMA) with about 20 ml each. Allow the plates to harden on a level, cool surface.

2. With a presterilized (alcohol-flamed) spatula or scalpel cut a block of agar (with a surface area of 1 cm²) from the center of each of the 3 plates. Place the blocks beside the holes (fig. 29-6a).

3. Using aseptic techniques, inoculate each of the 4 sides of the agar block with a small portion of the *Aspergillus* culture (fig. 29-6a).

4. Using presterilized forceps (alcohol-flamed), place a sterile coverslip over the inoculated block of agar and then press gently on the top of the coverslip with the forceps to insure good contact between the culture and the coverslip.

• If you are sterilizing the coverslips by dipping them in alcohol and then flaming them, shake off the excess alcohol before flaming. Otherwise the heat will crack the coverslip.

5. Repeat steps 3 and 4 using *Alternaria* and *Penicillium*.

6. Incubate all three plates at room temperature in an upright position for approximately a week.

Second Period

1. Examine the cultures for evidence of growth.

• The agar blocks should have fungal growth on all

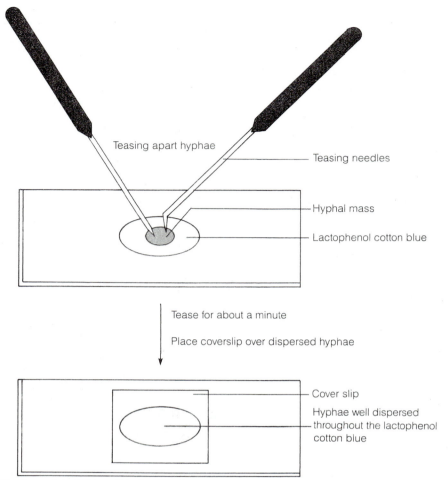

Teasing apart hyphae

Teasing needles

Hyphal mass

Lactophenol cotton blue

Tease for about a minute

Place coverslip over dispersed hyphae

Cover slip

Hyphae well dispersed throughout the lactophenol cotton blue

Figure 29-5. Procedure for Preparing a Tease Mount of Hyphae

4 sides and the hyphae should be growing on the coverslip.

2. Place a drop of LPB on each of 3 slides that have been labeled with the name of one of the fungi cultivated.

• The name of the fungus should be placed at the edge of the slide so as to not interfere with the preparation.

3. Using a pair of forceps, gently lift the coverslip from the *Aspergillus* culture and place it, *fungal growth side down*, directly on the LPB drop (fig. 29-6b).

• You may make a semi-permanent preparation of this slide culture by sealing all 4 edges of the coverslip with fingernail polish. Apply 2 to 3 coats, 30 minutes apart.

4. Repeat steps 2 and 3 using the other two cultures.

5. Using the 10x objective lens first and then the 40x lens, characterize the cultures as you did previously. Record your observations on the report sheet provided.

• Coverslip preparations are very useful in examining fungi because they allow you to view the specimens in a natural state. If the coverslips are handled carefully, the arrangement of hyphae and reproductive structures can be seen in an undisturbed state.

Questions

1. List five characteristics that can be used to differentiate fungi.

2. To which class (see table 29-1) do the following fungi belong? Explain your answer.

    a. *Aspergillus*

    b. *Penicillium*

    c. *Geotrichum*

    d. *Morchella*

    e. *Agaricus* (common mushroom)

    f. *Puccinia*

    g. *Saccharomyces*

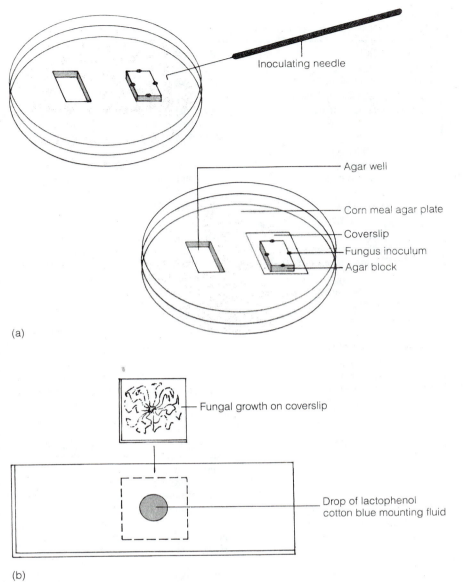

(a)

Inoculating needle

Agar well

Corn meal agar plate

Coverslip

Fungus inoculum

Agar block

Fungal growth on coverslip

Drop of lactophenol cotton blue mounting fluid

(b)

Figure 29-6. A Coverslip. *(a)* Preparation. *(b)* Procedure for making a coverslip preparation.

3. From the observations that you have made, define a fungus.

4. Can one species of fungi produce more than one type of spore?

5. Differentiate between a Zygomycete, a Deuteromycete, a Basidiomycete, and an Ascomycete.

6. How could you differentiate between the microscopic appearance of *Aspergillus* and *Penicillium?*

7. Which technique is more useful in determining the differences and similarities between fungi, the LPB tease mount or the coverslip preparation? Explain briefly.

C. SEXUAL REPRODUCTION IN *RHIZOPUS*

Materials

Cultures of + and − strains of *Rhizopus* on Sabouraud glucose agar

Plates of Sabouraud glucose agar

Inoculating needles with a bent tip

Glass slides and coverslips

Lactophenol cotton blue stain (LPB)

TABLE 29-1
SUMMARY OF THE MAJOR GROUPS OF FUNGI

CLASSIFICATION	CHARACTERISTICS
Kingdom: Fungi Division: Gymnomycota	Naked (wall-less) cells.
Subdivision: Acrasiogymnomycotina Class: Acrasiomycetes	Myxamoeba→pseudoplasmodium (made up of individual amoeba)→sorocarp. Example: *Dictyostelium*

Subdivision: Plasmodiogymnomycotina
   Class: Protosteliomycetes
   Class: Myxomycetes
	Myxamoeba→plasmodium (a single multi-nucleated macroscopic cell)→sporangium. Examples: *Physarum, Fuligo*

Division: Mastigomycota	Flagellated cells.

Subdivision: Haploidmastigomycotina
   Class: Chytridiomycetes
   Class: Hyphochytridiomycetes
   Class: Plasmodiophoromycetes.
	Haploid hyphae, flagellated zoospores, hyphae coenocytic, many aquatic forms. Examples: *Synchytrium, Allomyces; Rhizidiomyces; Plasmodiophora.*

Subdivision: Diplomastigomycotina Class: Oomycetes	Diploid hyphae, flagellated zoospores. Hyphae coenocytic. Many aquatic forms. Examples: *Phytophthora* and *Plasmopara*
Division: Amastigomycota	Nonflagellated cells.

Subdivision: Zygomycotina
   Class: Zygomycetes
   Class: Trichomycetes
	Haploid, coenocytic hyphae. Sexual spore is a zygospore. Asexual spore is the sporangiospore borne inside sporangia. Examples: *Rhizopus* and *Mucor.*

Subdivision: Ascomycotina Class: Ascomycetes	Haploid, septated hyphae with septal pore. Sexual spore is the ascospore borne inside ascus. Asci frequently develop inside ascocarp. Asexual spore is usually the conidium. Examples: *Neurospora, Morchella,* and *Saccharomyces* (yeast).
Subdivision: Basidiomycotina Class: Basidiomycetes	Hyphae with dolipore septum. Clamp connections seen in certain hyphae. Sexual spores are basidiospores borne on basidia. Examples: *Puccinia* and *Amanita*

Subdivision: Deuteromycotina
   Class: Deuteromycetes
    (Fungi Imperfecti)
	Yeast-like or filaments. Hyphae resemble those of the Ascomycetes. Sexual stage unknown. Parasexual cycle in some species. Conidia of various types produced. Examples: *Trichosporon, Penicillium, Aspergillus, Candida, Alternaria,* and *Geotrichum.*

Procedure

First Period

1. Pour a plate of Sabouraud glucose agar and allow it to harden on a cool, level surface.

2. Label one end of the plate with "*Rhizopus* (+)" and the opposite end with "*Rhizopus* (−)".

3. Inoculate both ends of the plate with a small amount of mycelium of the appropriate strain of *Rhizopus* and then incubate the plate, in an inverted position, for 2 days to 1 week.

Second Period

1. Examine the plate for evidence of growth.
- Growth is detected by the presence of fluffy, aerial, grayish hyphae. Handle these cultures with care because some species of *Rhizopus* may cause disease in humans. In addition, when the lid of the plate is opened,

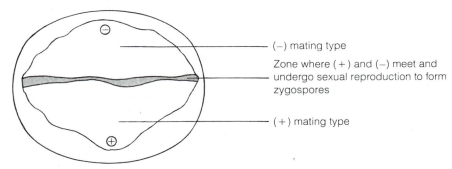

(−) mating type

Zone where (+) and (−) meet and undergo sexual reproduction to form zygospores

(+) mating type

Figure 29-7. Demonstration of the Sexual Stage of *Rhizopus*

millions of sporangiospores may be released into the air. The air-borne spores may contaminate other cultures for some time to come.

2. Note the dark line near the middle of the plate, where the + and the − strains meet (fig. 29-7). The dark line reflects the presence of zygospores.

3. Using the inoculating needle, prepare a wet mount of the fungal material in the dark area of the plate, using LPB as the mounting fluid.

• If you use a teasing needle to tease apart the stringy mycelium, sterilize the needle to avoid contaminating the lab with spores. Sterilize the teasing needle by swirling it in a sand jar and then flaming its tip.

4. Record the results on the report sheet. Make a detailed illustration of your observations.

• Make sure that you characterize the following structures: zygospores, suspensors, hyphae, sporangia, and sporangiospores.

5. Discard all cultures as indicated by your instructor.

Questions

1. What does the presence of zygospores indicate?

2. What events lead to the development of zygospores?

3. Explain two ways in which Zygomycetes such as *Rhizopus* reproduce asexually.

4. What sexual structures and spores are produced by the Ascomycetes? The Basidiomycetes? The Deuteromycetes?

EXERCISE 30
THE PROTOZOA

The protozoa constitute a diverse group of single-celled eukaryotic chemoheterothrophs found in many different environments. There are four major groups of protozoa: the **amebas,** the **flagellates,** the **ciliates,** and the **sporozoans.** These groups are differentiated from each other largely based on their mode of locomotion, mode of cell division, and life cycle stages. The characteristics of each of these groups are summarized in table 30-1.

Many of the protozoa are structurally and physiologically complex. For example, some have multiple nuclei and different types of nuclei (macronuclei and micronuclei). Some of them grow in several hosts and have distinct developmental stages. On the other hand, many protozoa gather nutrients from their environment by phagocytosis or by absorbing them in a dissolved form.

Some protozoa cause severe diseases that afflict millions of humans throughout the world. For example, malaria, African sleeping sickness, leishmaniasis, and toxoplasmosis collectively afflict more than 1 billion humans. Coccidians are protozoans that are common pathogens of many animals. Some of these, particularly those in the genus *Eimeria*, can cause fatal diseases in wild and domestic animals. For example, *Eimeria tenella* can spread very rapidly among chickens and kill most of a flock within a few days.

Protozoa also perform useful functions in nature. Ciliated protozoa in the rumen of cattle and sheep help these animals digest and assimilate foods that otherwise could not be used.

Protozoa exhibit a variety of morphological characteristics that can be used in their identification. Human protozoan pathogens seldom need to be cultured in the clinical laboratory in order to identify them. Instead, microbiologists rely on morphological fea-

TABLE 30-1
CHARACTERISTICS OF THE MAJOR GROUPS OF PROTOZOA

TAXONOMIC GROUP	LOCOMOTION	ASEXUAL REPRODUCTION	SEXUAL REPRODUCTION	GENERA OF IMPORTANCE
Phylum: Ciliophora Subphylum: Ciliata (ciliates)	Ciliary movement	Transverse fission	Conjugation	*Balantidium* *Paramecium* *Vorticella*
Phylum: Sarcomastigophora* Subphylum: Sarcodina (amebas)	Pseudopodia	Binary fission	Gamete fusion	*Entamoeba* *Amoeba* *Naegleria*
Subphylum: Mastigophora (flagellates)	Flagellar movement	Binary fission	Not seen	*Giardia* *Trypanosoma* *Trichomonas*
Phylum: Apicomplexa Class: Sporozoa (sporozoans)	Usually not motile	Schizogony (multiple fission)	Gamete fusion	*Plasmodium* *Toxoplasma* *Elmeria* *Cryptosporidium*

*fleshy, flagellated organisms.

tures of the protozoa seen with the light microscope in clinical specimens.

In this exercise you will be introduced to the major groups of protozoa, using both prepared slides of major human pathogens and living specimens obtained from cultures or from natural environments. You should observe these organisms with the light microscope, measure their size, note their mode of locomotion, and make drawings of them. You should use figure 30-1 to recognize by sight some of the protozoa in the samples of pond water that you will be examining.

A. EXAMINATION OF CULTURES

Materials

Cultures of *Paramecium, Amoeba, Chilomonas, Didinium,* and *Vorticella*

Glass slides and coverslips

Pasteur pipets and bulbs

Procedure

1. Prepare wet mounts of *Paramecium, Amoeba, Chilomonas, Didinium,* and *Vorticella*.

2. Examine each of the preparations in detail, using the compound microscope at 400× magnification. Make detailed drawings of each of the protozoa, paying special attention to their size, shape, intracellular structures, mode of locomotion, and the way they gather nutrients.
• Use PROTOSLO.

3. Record your observations in the report form.

4. Put away the samples and materials as indicated by your instructor.

B. EXAMINATION OF POND WATER FOR PROTOZOA

Materials

Sample of pond water

Glass slides and coverslips

Pasteur pipets and bulbs

Procedure

1. Prepare several wet mounts of the sample of pond water that you collected or that was provided by your instructor.
• Use PROTOSLO, a viscous fluid that is mixed in equal volumes with your pond water sample, to slow down the rapid movements of the protozoa.

2. Using a compound microscope, make detailed observations of at least five different types of protozoa.

3. Record your observations on the report form and attempt to identify the organism by group (ciliate, flagellate, ameba, sporozoa) and by name.

4. Put away the pond water samples and other materials as indicated by your instructor.

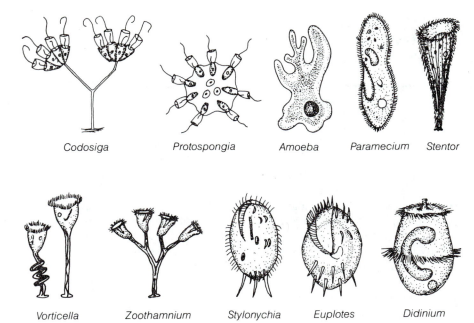

Codosiga   Protospongia   Amoeba   Paramecium   Stentor

Vorticella   Zoothamnium   Stylonychia   Euplotes   Didinium

Figure 30-1. Protozoans

C. EXAMINATION OF PREPARED SLIDES

Materials

Prepared slides of *Giardia*, *Entamoeba*, *Plasmodium*, *Trypanosoma*, and *Balantidium*

Procedure

1. Using a compound microscope at 1000× magnification, carefully examine the prepared slides of human pathogens provided by your instructor.

• These slides are prepared from actual clinical specimens and therefore you may have to scan the slide for a few minutes before you find a suitable microscopic field to examine. You should scan the slide at 100× first and then examine in detail the desired field at 400× and 1000× magnification.

2. Record your observations on the report form.

3. Blot the oil off the slides with a soft cloth or tissue paper and put them away as indicated by your instructor.

Questions

1. Define "protozoa" based on your observations of the specimens you examined.

2. How does *Chilomonas* differ from *Amoeba*? From *Paramecium*?

3. State four of the criteria you used to identify the various protozoa in the pond water. Why did you choose these criteria? How did you know that the organisms you were examining were protozoa and not bacteria, algae, or metazoans?

4. What types of specimens were used to view each of the following:

    a. *Giardia*

    b. *Entamoeba*

    c. *Plasmodium*

    d. *Trypanosoma*

    e. *Balantidium*

5. Were you able to see all the stages of the life cycle of *Plasmodium* in the slides that you examined? Which of the stages were most prominent?

6. Using a textbook and the observations that you made as a basis, diagram the life cycle of each of the parasites listed in question 4.

7. Why is it important to be able to recognize the various stages in the life cycles of protozoan pathogens?

8. Of what use is the morphology of a parasite to a clinical microbiologist?

EXERCISE 31
THE ALGAE

The algae (fig. 31-1) are photosynthetic prokaryotes and eukaryotes that are found primarily in aquatic environments. The algae are very diverse in size and shape. This variability is reflected by the fact that the algae are placed in three kingdoms: Monera, Protista, and Plantae. The **cyanobacteria,** or blue-green algae, are prokaryotic organisms in contrast to all the other algae that are eukaryotic. The cyanobacteria bear a superficial resemblance to the eukaryotic algae and carry out a type of photosynthesis that is similar to that of eukaryotes.

Algae are classified on the basis of their motility, photosynthetic pigments, mode of reproduction, and the chemical composition of their cell wall. A common scheme for the classification of the algae, along with their salient characteristics, is presented in table 31-1.

The algae are important members of the biosphere. They produce much of the oxygen in the atmosphere and are an important source of food for aquatic communities. In aquatic environments, the phytoplankton, the algae found floating on the surface of the water, are the primary source of nutrients for fish and waterfowl.

Algae are an importnt source of food, food additives, and agar. The large brown algae (kelp) are cultivated and harvested in large quantities to be used as food or food additives. Some of the red algae (e.g.,

TABLE 31-1
SOME CHARACTERISTICS OF THE CYANOBACTERIA AND THE EUKARYOTIC ALGAE

DIVISION	PIGMENTS FOR PHOTOSYNTHESIS*	NUMBER OF THYLAKOIDS PER STACK	FLAGELLAR CHARACTERISTICS	CELL WALL COMPOSITION	NUTRIENT RESERVE
Chlorophyta (Green Algae)	Chl a, Chl b α, β, γ, -carotenes xanthophylls	2–5	1 to many apical, equal in length	Cellulose	Starch
Phaeophyta (Brown algae)	Chl a, Chl c α, β -carotenes xanthophylls fucoxanthine	2–6	2, lateral, unequal in length	Cellulose, Alginic acid	Laminarin, Mannitol
Rhodophyta (Red algae)	Chl a α, β-carotenes phycocyanin allophycocyanin phycoerythrin xanthophylls	1	Not present	Cellulose, Xylans	Floridean, Starch
Chrysophyta (Diatoms)	Chl a, Chl c β, ε -carotenes xanthines fucoxanthines	3	2, apical	Cellulose, Silica, $CaCO_3$	Chrysolaminarin, Oils, and fats
Pyrrophyta (Dinoflagellates)	Chl a, Chl c β-carotene xanthophylls	3	2, one apical and 1 lateral	Cellulose, mucilagenous substances	Starch
Euglenophyta (Euglenoids)	Chl a, Chl b β-carotene xanthophylls	2–6	1 to 3 (to 7) apical or subapical	None	Paramylon
Cyanophyta (Cyanobacteria)	Chl a β-carotene phycobiliproteins	0	Not present (gliding molility)	Peptidoglycan	Lipids, Starch

* Chl = chlorophyll

Gelidium and *Gracilaria*) are harvested and processed to extract agar-agar, the jelling agent of laboratory culture media.

Certain species of algae, primarily the dinoflagellates, sometimes produce toxins that, when ingested by humans, can cause severe disease and sometimes even death.

In this exercise you will study the single-celled algae in the kingdoms Monera and Protista by using pure cultures and samples of pond water.

A. EXAMINATION OF ALGAL CULTURES

Materials

Cultures of *Anabaena*, *Oscillatoria*, *Nostoc*, *Euglena*, *Chlamydomonas*, *Spirogyra*, *Chlorella*, *Ceratium*, *Cyclotella*, and *Scenedesmus*

Glass slides and coverslips

Pasteur pipets and bulbs

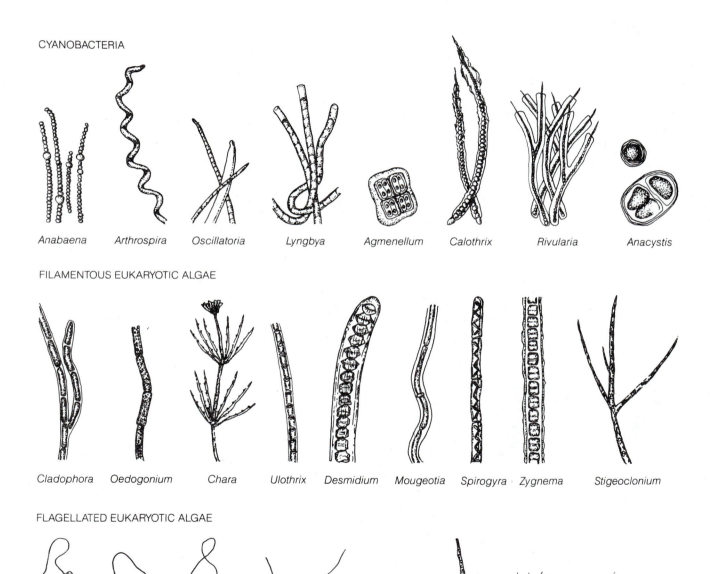

CYANOBACTERIA

Anabaena   *Arthrospira*   *Oscillatoria*   *Lyngbya*   *Agmenellum*   *Calothrix*   *Rivularia*   *Anacystis*

FILAMENTOUS EUKARYOTIC ALGAE

Cladophora   *Oedogonium*   *Chara*   *Ulothrix*   *Desmidium*   *Mougeotia*   *Spirogyra*   *Zygnema*   *Stigeoclonium*

FLAGELLATED EUKARYOTIC ALGAE

Euglena   *Phacus*   *Lepocinclis*   *Chlamydomonas*   *Peridinium*   *Ceratium*   *Gonium*   *Volvox*

Figure 31-1. Eukaryotic Algae

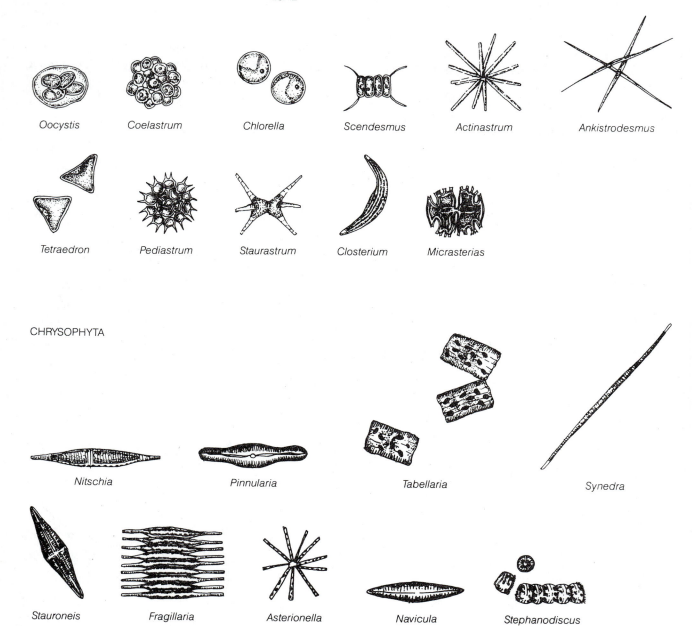

Oocystis   *Coelastrum*   *Chlorella*   *Scendesmus*   *Actinastrum*   *Ankistrodesmus*

Tetraedron   *Pediastrum*   *Staurastrum*   *Closterium*   *Micrasterias*

CHRYSOPHYTA

Nitschia   *Pinnularia*   *Tabellaria*   *Synedra*

Stauroneis   *Fragillaria*   *Asterionella*   *Navicula*   *Stephanodiscus*

Figure 31-1. Continued

Procedure

1. Prepare wet mounts of the various algae provided by your instructor and examine them with a compound microscope at $100\times$ and $400\times$ magnification.

2. Make careful observations and drawings of these organisms on the report form.

• Pay special attention to cell size, shape, mode of locomotion, and intracellular structures. Use the report table as a guide to the expected characteristics.

B. EXAMINATION OF POND WATER FOR ALGAE

Materials

Sample of pond water

Glass slides and coverslips

Pasteur pipets and bulbs

Procedure

1. Prepare several wet mounts of the pond water sample you collected or that was provided by your instructor.

2. Examine in detail the various organisms in the pond water with a compound microscope.

• When examining the pond water, note the characteristics of the algae that differentiate them from protozoans or metazoans.

3. Make detailed observations and drawings of five different algae in the pond water. Identify these species by using a textbook and figure 31-1 as a guide.

4. Record your observations and drawings on the report form.

5. Put away the cultures and pond water as indicated by your instructor.

Questions

1. Based on your observations of pure cultures of algae and those found in pond water, write a definition of an alga.

2. In which way(s) do *Anabaena* and *Oscillatoria* differ from *Spirogyra* or *Chlorella*?

3. Name five criteria that you used to identify the various algae in the pond water.

4. How were you able to differentiate the algae from the protozoa in the pond water?

5. How would you differentiate between:

    a. a cyanobacterium and a green alga

    b. a diatom and a dinoflagellate

    c. a green alga and a diatom

PART IX
THE VIRUSES

Viruses are noncellular infectious agents that consist of a nucleic acid, a covering protein coat, and sometimes a membranous envelope. The protein coat is often referred to as the virus **capsid.** Viruses only propagate when they are within host cells. In their extracellular forms they are metabolically inactive. Thus, viruses are referred to as obligate intracellular parasites. Viruses that infect bacteria, plants, and animals have been studied most extensively (fig. IX-1).

Bacterial viruses, called **bacteriophages** or **phages,** occur in a wide assortment of shapes and sizes and may contain double-stranded DNA (ds-DNA), single-stranded DNA (ss-DNA), double-stranded RNA (ds-RNA), or single-stranded RNA (ss-RNA) **genomes.** A complete set of virus's hereditary material is referred to as its genome. Phages are usually classified on the basis of their nucleic acid and their structure, which may be **complex, icosahedral** (20-faced solid geometric body), **cylindrical,** or **enveloped.** Most bacteriophages are complex or icosahedral. The complex, ds-DNA T-phages such as T4, the icosahedral ss-DNA phages such as ØX174, the icosahedral ss-RNA phages like QB, and the cylindrical ss-DNA phages such as fd and M13 are some of the most-studied phages.

Most plant viruses contain ss-RNA and are cylindrical or icosahedral. Many plant viruses have segmented genomes that are packaged into separate capsids. All segments are usually required for infectivity and propagation. The cylindrical ss-RNA plant virus that causes tobacco mosaic disease has been studied extensively for over a century.

Nearly all animal viruses are either icosahedral or enveloped and have ss-RNA or ds-DNA genomes. The best-known animal viruses are those that cause diseases. The icosahedral ss-RNA picornaviruses are responsible for diseases such as hepatitis-A, polio, and hoof and mouth disease; enveloped ds-DNA viruses are responsible for such diseases as hepatitis-B, herpes, smallpox, and warts; while the enveloped ss-RNA retroviruses are the cause of hepatitis non-A non-B, AIDS and some types of cancers.

DETECTION AND ENUMERATION OF VIRUSES

Generally, a virus infection causes observable changes in the host cells. For example, some animal viruses (such as those that cause smallpox, herpes, and chicken pox) cause the **lysis** of the host cells. This lysis is seen as sores. A great many plant viruses also lyse plant cells and this is seen as brown spots or mottled areas on the plant. Many bacterial viruses lyse their host cells as they proliferate. The lysis of bacteria growing

(a)

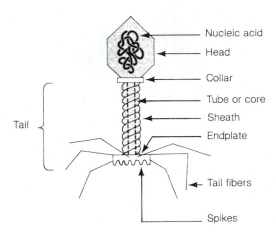

Nucleic acid

Head

Collar

Tube or core

Sheath

Endplate

Tail

Tail fibers

Spikes

Figure IX-1. Viruses. (a) Bacteriophage T4. (b) Tobacco mosaic virus. (c) Polio virus.

139

(b)

(c)

ISOLATION OF VIRUSES

Viruses are ubiquitous. Almost any population of bacteria, plants, or animals is infected with one or more bacteriophages, plant viruses, or animal viruses, respectively. Bacteriophages are so plentiful that they can be isolated easily from waters rich in bacteria and organic nutrients. Phages can be detected because they produce plaques on lawns of susceptible bacteria. The bacteriophages, for the most part, have very specific hosts. Some phages are so specific that they will infect only one bacterial strain within a species.

The **host range** of the plant viruses is not as restricted as the host range of the bacteriophages. A particular plant virus may infect various species and genera of plants. Many plant viruses that are transmitted by insects may infect the insect as well as the plant. Some plant viruses, such as tobacco mosaic virus, can be easily isolated from nature by rubbing material from plant lesions onto young plants of the same type.

The host range of animal viruses is also broader than that of bacteriophages. It is common for animal viruses, such as the herpes viruses, to be able to infect a variety of mammals. Most mammalian viruses are difficult and dangerous to isolate. Consequently, their isolation and study requires appropriate containment facilities. Exercise 33 is concerned with the isolation of a bacteriophage, while exercise 34 is concerned with the isolation of a plant virus.

VIRUS REPLICATION

The mechanisms of replication for a number of bacterial viruses have been studied in detail. For ease of

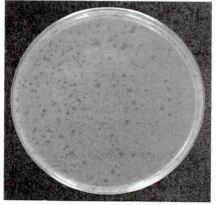

Figure IX-2. Viral Lesions. A lawn of bacteria contains clear areas that represent bacteriophage plaques or virus colonies that have developed from a single bacteriophage.

confluently on a nutrient agar plate produces clearings in the **lawn of bacteria.** These clearings, called **plaques,** each represent a colony of viruses that have arisen from a single infectious agent (fig. IX-2). Consequently, the number of plaques in a bacterial lawn can be used to determine the number of viruses that are present. Exercise 32 will demonstrate how the number of viruses in a sample can be determined.

140   PART IX

discussion, the replication process is generally divided into a number of steps: phage attachment or adsorption; virus genome penetration into the host cell; virus RNA synthesis, protein synthesis, and genome replication; phage assembly; and release (fig. IX-3).

Depending upon the type of phages, **adsorption** may occur on the host's wall, flagella, or pili. Some of the phages that attach to the wall degrade part of the wall and create a passage through which their genomes can pass. **Genome penetration** occurs when the phage's hereditary material passes from the capsid into the cytoplasm. The phages that attach to the flagella are thought to inject their genomes into the hollow flagella that penetrate the wall and plasma membrane. Similarly, the phages that bind to pili may inject their genomes into the hollow pili that breach the wall

and plasma membrane. When the pili are absorbed, the virus genomes gain entrance.

Once in the cell, the phage genomes direct the **synthesis** of virus mRNA, proteins, and new genomes. The host's polymerases, ribosomes, and nutrients are used to synthesize virus components, although the larger viruses direct the synthesis of some of their own enzymes.

After the appropriate components for a part are present, the components undergo **self-assembly.** In the case of bacteriophage T4, proteins self-assemble into a capsid, tail, and tail fibers. Then a headful of DNA associates with the capsid and the capsid, tail, and tail fibers self-assemble into a complete virus.

Phage **release** varies depending upon the virus. Some viruses, such as T4, produce lysozyme that degrades the peptidoglycan of the host's cell wall. This

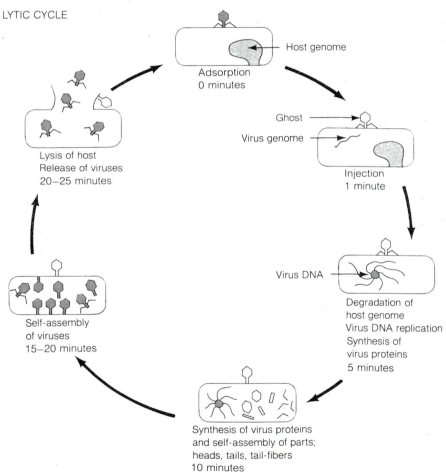

LYTIC CYCLE

Host genome

Adsorption
0 minutes

Ghost

Virus genome

Injection
1 minute

Lysis of host
Release of viruses
20–25 minutes

Virus DNA

Degradation of
host genome
Virus DNA replication
Synthesis of
virus proteins
5 minutes

Self-assembly
of viruses
15–20 minutes

Synthesis of virus proteins
and self-assembly of parts;
heads, tails, tail-fibers
10 minutes

Figure IX-3. Virus Replication. The lytic cycle of T4 bacteriophage begins with the adsorption of the virus to a host cell. Next, the virus genetic information is injected into the host's cytoplasm. Early T4 mRNA and protein syntheses begin approximately 2–3 minutes after infection, while host DNA begins to be degraded and virus DNA begins to be synthesized approximately 5 minutes after infection.

Viral components begin to self-assemble about 10 minutes after infection, and viral components assemble themselves into viruses 15–20 minutes after infection. The viruses escape from their host cells approximately 20–25 minutes after infection by breaking open the cells.

breach in the wall allows the viruses to escape from the host. The cylindrical phages such as M13 appear to grow like pili from the cell and do not usually lyse the host cell.

Exercise 35 is devoted to illustrating the kinetics of virus replication.

PHAGE TYPING

Since some bacteriophages have a very narrow **host range,** they can be used to characterize unknown strains of bacteria. Exercise 36 is concerned with phage typing of bacteria.

EXERCISE 32
PHAGE ASSAY

In the course of their replication, many bacterial viruses cause the lysis of their host cells. These viruses are said to undergo a **lytic cycle of infection.** When lytic viruses grow on a lawn of cells, they spread to neighboring cells, lysing an area in the lawn (called a **plaque**) that may be a millimeter in diameter or larger. Because a single virus can give rise to a plaque, the number of plaques in a lawn can be related to the number of functional viruses. Each plaque is said to be due to a **plaque-forming unit (PFU).** Thus, the number of functional viruses or plaque-forming units in a sample can be determined by making appropriate dilutions and mixing these dilutions with host cells.

When determining the **titer** (concentration) of phages, plates that contain between 30 and 300 plaques should be used. Plates with fewer or greater numbers become statistically unreliable. Plates with many plaques underestimate the number of phages because many phages in the sample land together and are counted as one. Plates with few plaques will give titers that vary drastically because of the inaccuracies of the pipeting.

The plaque-forming units are determined by counting the number of plaques on a plate and multiplying this number by the inverse of the dilution. For example, a plate that received 1 ml of a 10^{-9} dilution and had 25 PFUs would indicate a total of 25 $\times$ 10^9 PFUs/ml in the culture.

Some bacteria *(Bdellovibrio)* produce plaque-like lesions in bacterial lawns. Cultures of these types of bacteria and phages can be distinguished from each other because bacteria can be seen with the light microscope, they can be filtered, they (generally) can be grown on artificial media, and they are killed by chloroform.

Materials

Culture of *Escherichia coli* B, sample of bacteriophage T2 or T4 coliphage with a titer of $10^9 - 10^{10}$ phage/ml

Trypticase soy agar (TSA) plates

5 ml soft agar in tubes

Dilution tubes containing 9 ml of saline (0.8% NaCl solution)

Sterile 0.1 ml pipets

Incubator set at 37°C

Procedure

First Period

1. Make 10^{-9}, 10^{-8}, and 10^{-7} dilutions of the bacteriophage sample.
Be sure to use a separate pipet for each transfer. If the titer of the coliphage is unknown, also check the 10^{-6}, 10^{-5}, and 10^{-4} dilutions.

2. Transfer 0.1 ml of the 10^{-9} dilution to 5 ml of melted soft agar at 50°C. Also transfer 0.3 ml of *E. coli* to the tube of melted soft agar. Mix by rapidly rotating the tube between the palms of your hands. Then pour the mixture onto a prepoured solidified plate of TSA. Slightly tilt and then rotate the plate to further mix and spread the soft agar evenly over the TSA. Repeat the same procedure for the 10^{-8} and 10^{-7} dilutions.

3. Incubate the plates right side up at 37°C for 24 hours. Refrigerate the plates after 24 hours until they can be studied.

Second Period

1. Record the numbers of plaques at each dilution and determine the titer of the phage sample.

Questions

1. If dilutions of 10^{-9}, 10^{-8}, 10^{-7}, and 10^{-6} yield an average of 8, 30, 350, and 1700 plaques per plate, respectively, what is the best estimate for the number of plaque-forming units per ml of the undiluted sample? Do not forget that you used 0.1 ml of the dilutions.

2. Did you observe any bacterial growth within the plaques? What could this growth be due to? Why were these bacteria not killed?

3. Some bacteria consume other bacteria, producing plaque-like lesions. How might you distinguish between cultures of plaque-producing bacteria and bacteriophages?

4. Is the number of plaque-forming units the same as the number of virus particles? Explain.

5. What factors might account for a lysate having 10^9 plaque-forming units/ml but having 10^{11} virus particles/ml as determined by using the transmission electron microscope?

6. What kind of problems might there be in determining the PFUs in a culture of lysogenic phage? How would these problems be overcome?

EXERCISE 33
BACTERIOPHAGE ISOLATION

Bacteriophages generally can be isolated from any source where there are large numbers of bacteria: soil, human and animal feces, and raw sewage. Usually the number of bacteriophages is so low in a sample that a direct isolation will not be successful; thus, bacteriophages in a sample must be enriched before attempting an isolation.

In the procedure you will use, the number of coliphages in a sample of fresh sewage or manure is increased by mixing the sewage or manure with a broth culture of *Escherichia coli*. Next, the bacteria and debris are removed by centrifugation and then by filtration. The filtrate should contain sufficient coliphages so that an isolation will be successful.

Raw sewage should be collected and handled aseptically so that pathogenic microorganisms that might be in the sewage do not cause disease. Sewage may contain the following pathogens: *Salmonella* (typhoid fever), *Shigella* (dysentery), *Vibrio* (cholera), *Campylobacter* (gastroenteritis), picornaviruses (polio and hepatitis-A), hepatitis-B virus (hepatitis-B), retrovirus (hepatitis non-A non-B), *Entamoeba* (amebic dysentary), and *Giardia* (giardiasis).

Materials

24 hour culture of *Escherichia coli* B, 50 ml sample of fresh manure collected in screw-capped jars

Trypticase soy agar (TSA) plates

Trypticase soy broth (TSB)

5 ml soft agar in tubes

250 ml Erlenmeyer flask

Dilution tubes containing 9 ml of saline (0.8% NaCl solution)

Centrifuge and sterile centrifuge cups

Sterile filter apparatus

Sterile filters with 0.5 μm pores

Sterile 1 ml pipets

Incubator set at 37°C

Procedure

First Period

1. Aseptically mix together 5 ml of TSB, 5 ml of the *E.*

coli culture, and 50 ml of a fresh manure sample in a 250 ml Erlenmeyer flask.

2. Incubate at 37°C for 24 hours.

Second Period

1. After incubation, harvest the mixture by centrifuging at 2500 rpm for 20 minutes. Use 100 ml centrifuge bottles.

2. Carefully decant the supernatant fluid into a sterile 250 ml flask.

• Autoclave the bottle with the pellet of manure and bacteria.

3. Pour the supernatant fluid through a sterile filter aparatus. Use a filter with a pore diameter of 0.5 μm.

4. To a melted soft agar tube at 50°C add 0.3 ml of fresh *E. coli* and 0.3 ml of the filtered phage solution. Mix together by rapidly rotating between the palms of the hand. Pour the mixture onto a prepoured solidified plate of TSA. Tilt the plate and swirl to mix and distribute the mixture evenly.

5. Prepare a 10^{-1} dilution of the phage solution. Use 0.2 and 0.1 ml of the phage solution as well as 0.5 ml of the 10^{-1} dilution to make plates to detect phage in concentrated lysates.

6. Incubate the plates right side up at 37°C for 24 hours. Refrigerate after 24 hours until the plates can be studied.

Third Period

1. Count the plaques on each plate and determine the number of PFU/ml in the enriched filtrate.

Questions

1. Explain why it is necessary to enrich for bacteriophage in manure or sewage and how this enrichment is accomplished.

2. Explain how you would isolate a bacteriophage that proliferates specifically on *Pseudomonas*.

3. Sometimes you will observe different types of plaques on the *E. coli* lawns. What might this indicate? How might you test your hypotheses?

EXERCISE 34
PLANT VIRUS ISOLATION

One of the easiest viruses to isolate is tobacco mosaic virus (TMV), a cylindrical (helical) single-stranded RNA virus that infects tobacco and other plants. In some strains of tobacco, such as *Nicotiana tabacum xanthi*, TMV produces distinct yellow-green to brown lesions on leaves, while in other strains, such as *Nicotiana tabacum burley*, the virus spreads over large areas of the leaves and creates white to yellow-green mottling on the dark green leaves ("mosaic" leaves). Because of the mottling that is produced in *Nicotiana tabacum burley*, the disease is known as **tobacco mosaic disease.**

Determining the number of **lesion forming units (LFU)** in plant extracts is not quite as easy as it is in bacterial extracts. When tobacco plants of different ages are quantitatively inoculated with the same solution of TMV, different values for the LFU are obtained. Younger plants develop more lesions than older plants. In addition, it has been found that younger leaves on a plant develop more lesions than older leaves on the same plant. Since the age of the plant tissue drastically affects the LFU values, it is important that all experiments carefully monitor and consider the age of the plant and the leaves.

In this exercise, viruses will be obtained from tobacco leaves of *Nicotiana tabacum burley* that show extensive mottling. A number of mottled leaves are ground using a mortar and pestle. A little saline is added to make a slurry. Measured samples of the slurry are placed on young leaves of a tobacco plant that produce distinct lesions (*Nicotiana tabacum xanthi* or *Nicotiana glutinosa*), and then rubbed over the leaf with a sterile glass spreader. The plant should show lesions in approximately three weeks.

Materials

Infected tobacco plants (*Nicotiana tabacum burley*). Young uninfected *Nicotiana tabacum xanthi* or *Nicotiana glutinosa*

Dilution tubes containing 9 ml of saline (0.8% NaCl solution)

Mortar and pestle

Sterile 1 ml pipets

Glass spreaders and jar with 95% ethanol

Procedure

First Period

1. Strip a tobacco leaf showing mosaic disease from an infected plant and grind it in 2 ml of saline using a sterile mortar and pestle.

2. Dilute the extract 1/10, 1/100, 1/1000, and 1/10,000 with sterile saline.

3. Spread 0.1 ml of the undiluted extract with a sterile glass spreader over a leaf on a disease-free plant. With a piece of tape, mark the leaf and the dilution.

4. Spread 0.1 ml of each of the leaf extract dilutions onto separate disease-free tobacco leaves and mark the leaves with the dilution.

5. Inoculate a leaf with sterile saline as a control.

6. Incubate the tobacco plants at room temperature or at 18°C for 3 to 4 weeks.

Second Period

1. After incubating for 3 to 4 weeks, check the leaves for the development of lesions or mosaic disease. Follow the experiment for 8 weeks.

2. Record the number of lesions at each dilution.

Questions

1. Explain why some infected tobacco plants yield distinct lesions while others give a mottled appearance.

2. How does TMV differ from the T-even coliphages?

3. What problem is there with plant viruses in determining the LFUs in an extract?

4. Were the LFUs you determined from the different dilutions consistent?

EXERCISE 35
ONE-STEP GROWTH CURVE

Most of the commonly encountered bacteria proliferate by binary fission. This means that after a generation one cell will become two and then after another generation the two will become four and so on until nutrients become limiting or wastes inhibit growth. Thus, bacterial populations grow exponentially to the base 2 and the population can be described by the following equation: $N_f = N_o2^n$ where n = number of generations and N_f and N_o are the final and initial numbers in the population, respectively.

Bacteriophages proliferate by infecting cells and by causing the cells' biosynthetic machinery to synthesize virus parts. When the virus parts self-assemble, a large number of viruses are formed very rapidly within the host cells. The hereditary information from a single virus can direct the synthesis of hundreds of new viruses in one generation. This type of proliferation can be visualized by constructing a **one-step growth curve.** The one-step growth curve is made by plotting the increase in viruses over time. Even though bacteriophages may increase by a hundredfold or more every generation, they are growing exponentially but with a base much larger than 2. The bacteriophage population that expands by a hundredfold each generation might be described by the following equation: $N_f = N_o100^n$.

In an experiment to obtain data for a one-step growth curve, a culture of *Escherichia coli* is infected with T4 phage. Samples are removed every 5 minutes in order to determine how many phages are present. For approximately 20 to 25 minutes, depending upon the temperature and other conditions, there is no increase in the number of phage in the mixture. Then there is a dramatic hundredfold rise in the titer within 5 or 10 minutes. The period of time from infection to the rise in titer is known as the **latent period** (fig. 35-1). If the host cells are artificially lysed during the development of the bacteriophages, no infectious viruses are present for most of the latent period. The period of time from infection to the appearance of infectious bacteriophages within the host cells is known as the **eclipse period.**

Materials

Bacteriophage suspension of T4 with a titer of 5×10^9/ml

Exponentially growing culture of *E. coli* at about 5×10^8/ml

3 ml of phage top agar

Trypticase soy agar plates

Dilution tubes containing 9 ml of saline (0.8% NaCl solution)

Sterile KCl broth: 9 ml/tube and 0.9 ml/tube

Sterile 10 ml, 5 ml, 1 ml, and 0.1 ml pipets

Water bath at 37°C

Ice bath

Clean empty tubes

Procedure

First Period

1. The instructor will mix 0.1 ml of the phage (5×10^9/ml) with 9.9 ml of an exponentially growing culture of *E. coli* (5×10^8/ml) and allow the phage to adsorb for 2 minutes in a 37°C water bath (fig. 35-2).
- This represents a 10^{-2} dilution (the number of PFUs is 5×10^7/ml).

2. After 2 minutes, the instructor will dilute the 10 ml mixture by transferring 1 ml into 99 ml of fresh TSB in the 37°C water bath.
- This 1/100 dilution reduces the chances of adsorptions and represents a 10^{-4} dilution of the original lysate (the number of PFUs is 5×10^5/ml).

3. Each student group should immediately take 1 ml of the diluted adsorption mixture and dilute it again in 9 ml of KCl broth in the 37°C water bath.
- This transfer should not take more than 5 minutes (total time elapsed since mixing is 7 minutes). This represents a 10^{-5} dilution (the number of PFUs is 5×10^4/ml).

4. Sample 1 ml of the KCl broth mixture at 10, 15, 20, 25, 30, 35, 40, and 45 minutes. Place the 1 ml samples in an ice bath to inhibit further development.

5. Each of the samples in the ice bath should be serially diluted in ice cold KCl broth through 10^{-9} as shown in step 5 of figure 35-2.
- At this stage you should have 5 tubes for each sampled time (a total of 40 tubes).

6. For each of the 10^{-6} to 10^{-9} dilutions made from each sampled time add 0.1 ml of the dilution and 0.3 ml of *E. coli* to 3 ml of melted top agar at 50°C (see step 6 of fig. 35-2).
- 0.1 ml of the 10^{-9}, 10^{-8}, 10^{-7}, and 10^{-6} dilutions

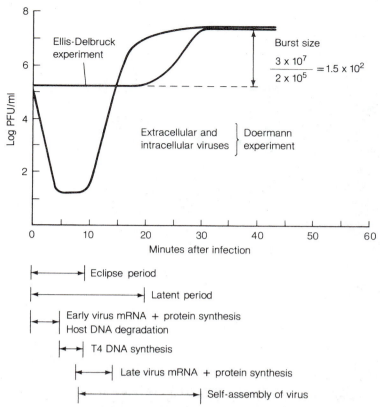

Figure 35-1. One-Step Growth Curve. Within minutes after T4 DNA enters *E. coli*, early virus mRNA and proteins are synthesized. In addition, the host's DNA is degraded. Approximately 5 minutes after infection, T4 DNA begins to be replicated. Viral proteins self-assemble into mature viruses 10 to 15 minutes after infection. During this same period, late viral m RNA and proteins begin to be synthesized. Some of these late proteins will promote the lysis of the host cell and the escape of the viruses. Cells begin to break open approximately 20 minutes after infection and extracellular phage begin to appear. **Ellis-Delbruck experiment.** Bacteriophage (2×10^5/ml) are mixed with bacteria (5×10^8/ml) so that only one phage adsorbs to a bacterium.. The diluted culture of infected bacteria is subsequently sampled every 5 minutes. For the first 20–25 minutes after T4 infects *E. coli*, the number of plaques remains the same. The time during which no "new" phage appear is called the **latent period.** After the latent period, the number of phage in the culture suddenly increases approximately 70-fold and then levels off. The rise in phage number represents the release of virus progeny. **Doermann experiment.** If infected bacteria are broken open at various times after infection, the number of plaques decreases more than 100-fold during the first 10 minutes of the infection. Phage begin to form within the cell approximately 10 minutes after infection. The time during which no phage progeny can be detected within the cells is called the **eclipse period.**

will give around 0.5 (a plaque every two plates), 5, 50, and 500 plaques, respectively, if the original titer is 5×10^9/ml. If the original titer is less than 5×10^9/ml, adjust the dilutions so that there will be a plate with between 30 and 300 plaques. The plaques should increase approximately a hundredfold between 20 and 30 minutes.

7. Pour the inoculated top agar onto solidified TSA plates.

• If you follow this procedure correctly, you should have a total of 32 plates.

8. Incubate the plates right side up for 24 hours. At the end of 24 hours the plates should be refrigerated.

Second Period

1. After incubation for 24 hours, count the number of plaques produced for each dilution and multiply by the inverse of the dilution factor to determine the number of phages that would have appeared in the original adsorption mixture.

2. Construct a one-step growth curve. Plot PFU/ml against time.

3. Determine the latent period and the burst size.

Questions

1. What was the latent period and the burst size for the T-even coliphage studied?

2. How would you alter the one-step growth experiment to determine the eclipse period for the T-even coliphage?

3. You carry out a one-step growth experiment on a T-even bacteriophage and obtain the data given below.

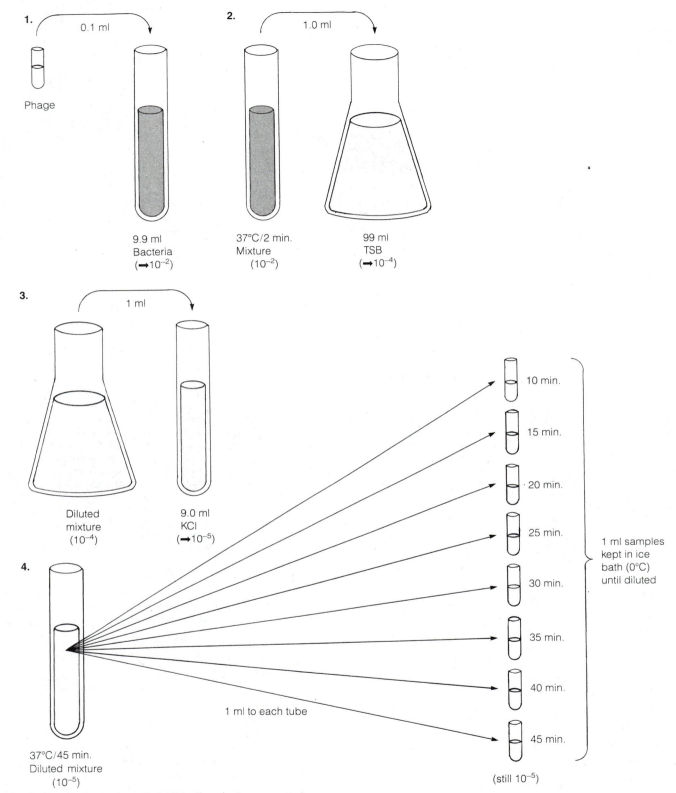

1. 0.1 ml

Phage

9.9 ml
Bacteria
($\rightarrow 10^{-2}$)

2. 1.0 ml

37°C/2 min.
Mixture
(10^{-2})

99 ml
TSB
($\rightarrow 10^{-4}$)

3. 1 ml

Diluted
mixture
(10^{-4})

9.0 ml
KCl
($\rightarrow 10^{-5}$)

4.

37°C/45 min.
Diluted mixture
(10^{-5})

1 ml to each tube

10 min.

15 min.

20 min.

25 min.

30 min.

35 min.

40 min.

45 min.

1 ml samples
kept in ice
bath (0°C)
until diluted

(still 10^{-5})

Figure 35-2. Procedure for Obtaining Data to Construct a One-Step Growth Curve

5. Procedure followed for each timed sample. Keep all small dilution tubes at 0°C.

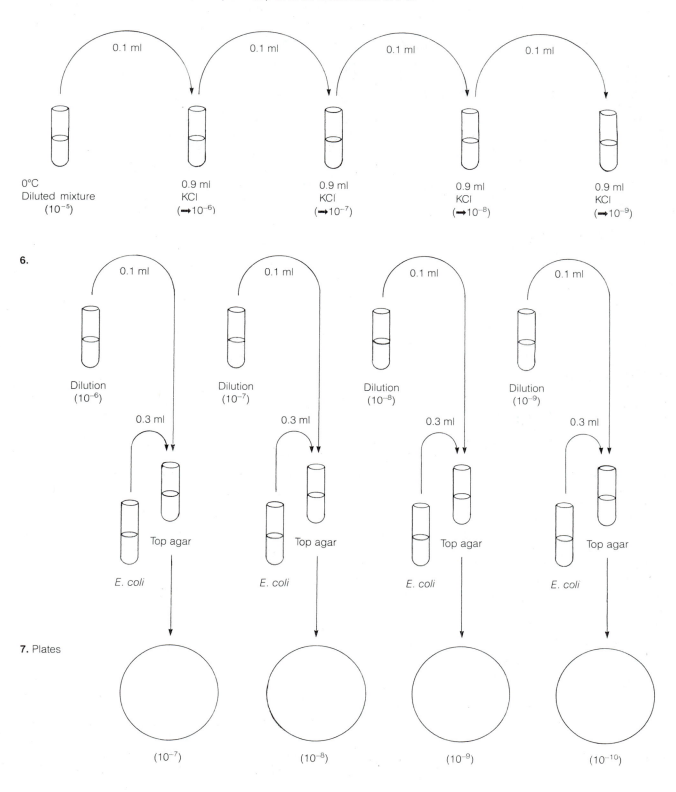

0°C
Diluted mixture
(10^{-5})

0.9 ml
KCl
($\rightarrow 10^{-6}$)

0.9 ml
KCl
($\rightarrow 10^{-7}$)

0.9 ml
KCl
($\rightarrow 10^{-8}$)

0.9 ml
KCl
($\rightarrow 10^{-9}$)

6.

0.1 ml

0.1 ml

0.1 ml

0.1 ml

Dilution
(10^{-6})

Dilution
(10^{-7})

Dilution
(10^{-8})

Dilution
(10^{-9})

0.3 ml

0.3 ml

0.3 ml

0.3 ml

Top agar

Top agar

Top agar

Top agar

E. coli

E. coli

E. coli

E. coli

7. Plates

(10^{-7})

(10^{-8})

(10^{-9})

(10^{-10})

(a) Plot the data on graph paper. (b) What is the burst size for the virus? (c) What is the latent period? (d) Indicate what the data might look like if you altered the experiment to determine the eclipse period.

TIME (MINUTES) AFTER ADSORPTION	DILUTION			
	10^{-7}	10^{-8}	10^{-9}	10^{-10}
	PFU/ML			
5	50, 40	6, 3	0, 0	0, 0
10	45, 55	2, 4	0, 0	0, 0
15	40, 45	3, 8	1, 0	0, 0
20	100,120	15, 10	0, 2	0, 0
25	CLEAR	CLEAR	100,120	15, 10
30	CLEAR	CLEAR	130,150	20, 15
35	CLEAR	CLEAR	125,120	10, 20
40	CLEAR	CLEAR	130,140	15, 20

EXERCISE 36
PHAGE TYPING OF
STAPHYLOCOCCUS AUREUS

Certain bacterial strains can be differentiated from each other by the use of bacteriophages that have a narrow host range. For example, pathogenic strains of *Staphylococcus aureus* are differentiated and identified by observing which bacteriophages are able to grow on them. Strains of bacteria are identified by the bacteriophages that grow on them (fig. 36-1). For example, if phages 6, 7, 71, 77, 80, and 83 grow on a particular strain, the strain's identification number is (6, 7, 71, 77, 80, 83). **Phage typing,** as this differentiation is called, is used when strains are difficult to tell apart or cannot be distinguished with the usual biochemical tests, stains, and structural features.

Materials

Cultures of *Staphylococcus aureus* strains ATCC 27691, 27693, and 27697. Unknown *Staphylococcus aureus* strain. Samples of bacteriophage types 47, 52A, 71, and 81

Trypticase soy TSA agar plates

Top agar (5 ml/tube)

Sterile glass capillary tubes

0.1 ml pipet

Incubator set at 37°C

Water bath set at 50°C

Procedure

First Period

1. Transfer 0.3 ml of a *Staphylococcus aureus* strain to 5 ml of melted top agar held at 50°C. Mix the bacteria and top agar by rapidly rotating the tube between the palms. Pour the mixture over the surface of a solidified TSA plate and evenly distribute it over the surface by tilting and swirling the plate and mixture, respectively (fig. 36-1). Similarly, spread the other strains onto separate plates.

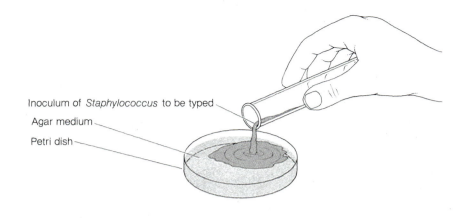

Inoculum of *Staphylococcus* to be typed
Agar medium
Petri dish

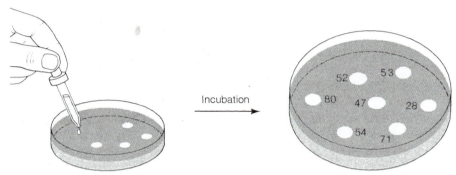

Incubation

52   53
80   47   28
54   71

Figure 36-1. Procedure for Phage Typing of *Staphylococcus aureus.* A suspension of an unknown strain of *Staphylococcus aureus* is spread over the surface of a nutrient agar. Drops of different bacteriophages are applied to the plate. After incubation, the pattern of phage susceptibility and the phage group number are determined by noting which phages formed plaques on the bacterial lawn.

• Before adding the bacteria to the agar, make sure that the agar is at a comfortable temperaure (50°C or less).

2. Using sterile glass capillary tubes, spot the plates with each of the bacteriophages.

• Use as small a drop as possible so that the drops will not run into each other. Be sure to label each phage spot and its location. Use a separate capillary tube for each phage.

3. Incubate the plates at 37°C

Second Period

1. Record which phages produce plaques.

PART X
GENETICS OF MICROORGANISMS

Although all the microorganisms in a pure culture belong to the same species (or strain) and are very similar, they are not all identical. Some of the microorganisms in the culture carry **mutations** that make them genetically different. Mutations are changes in the hereditary material. Variations due to such changes are referred to as **genotypic variations.**

Mutations are generally due to the replacement of a base pair $(A=T)$ with another base pair $(T=A, G\equiv C, C\equiv G)$, the addition of one or more base pairs, the deletion of one or more base pairs, or the inversion or translocation of a region. The organisms that differ genetically from the majority are known as **mutants.** For a particular gene, the number of mutants is about one out of every million (10^{-6}) organisms in a population. Consequently, if there are a billion organisms in a bacterial population, there are approximately 1,000 organisms that carry a mutation in a particular gene $(10^{-6} \times 10^{9} = 1,000)$. Since an organism such as *Escherichia coli* has about 4000 genes, calculations indicate that there are approximately four million mutants in a bacterial population of a billion organisms (1,000 mutants/gene $\times$ 4,000 genes).

Some of the most commonly encountered mutants are organisms that cannot synthesize essential building blocks, or **growth factors,** such as amino acids, nucleosides, and vitamins. These mutants will grow only if supplied the needed growth factor. Organisms that have these types of growth requirements are called **auxotrophs,** while organisms that do not need growth factors are referred to as **prototrophs.**

Mutations may make organisms incapable of using a particular carbohydrate as their source of carbon and energy, or they may make organisms resistant to antibiotics, heavy metals, or virus infections. Mutations may also result in the loss of structural components such as capsules, pili, and flagella.

Genetic variation in bacteria may also be due to conjugation, transduction, and transformation. Some bacteria, known as **donor** or male bacteria, are able to transfer DNA to **recipient** or female bacteria when they contact each other. The unidirectional transfer of DNA from a donor to a recipient upon contact is called **conjugation.** The transfer of DNA from one bacterium to another by a virus, and the subsequent alteration of the host cell's hereditary material is known as **transduction.** DNA from lysed cells, called free or naked DNA, is sometimes picked up by bacteria. When this absorbed DNA results in genetic variation, the process is called **transformation.**

Microorganisms often change their physiology and appearance temporarily because of external factors. Changes in physiology and appearance that are not due to permanent changes in the hereditary material but instead are due to environmental factors are referred to as **inducible phenotypic variations.** The most extensively studied examples of inducible phenotypic variation involve those gene systems that degrade carbohydrates and those that synthesize amino acids.

When an organism such as *E. coli* is growing in a glucose minimal medium, it contains very low levels of the enzymes that catabolize or break down the sugar lactose. In contrast, the same organism growing in a lactose minimal medium contains high levels of the enzymes that catabolize lactose. The sugar lactose is responsible for inducing the synthesis of enzymes involved in its metabolism. If the organism is transferred from lactose to glucose minimal medium, the synthesis of the enzymes involved in lactose metabolism is immediately repressed.

In the following sections, you will conduct experiments that demonstrate the difference between inducible phenotypic variation and genotypic variation; illustrate how the physiology of a cell can be modified by external factors; and demonstrate the difference between conjugation, transduction, and transformation.

EXERCISE 37
PHENOTYPIC AND GENOTYPIC VARIATION

A. ILLUSTRATING PHENOTYPIC VARIATION

Environmental factors can alter the physiology and appearance of organisms while not altering their hereditary material. For example, organisms such as *Pseudomonas fluorescens*, *Pseudomonas aeruginosa*, and *Serratia marcescens* synthesize pigments at low temperatures but not at high temperatures (fig. 37-1). *Alcaligenes viscolactis* produces a capsule at low temperatures but not at high temperatures, and *Lactobacillus casei* requires pyridoxine as a growth factor at pH 7 but not at pH 5.

If the bacterium *Serratia marcescens* is grown at 25°C, the colonies that develop are bright red because the cells produce the pigment prodigiosin. If this same bacterium is grown at 37°C, the colonies lack red pig-

ment and are cream-colored. It is believed that one of the enzymes involved in the synthesis of the red pigment is denatured, or its synthesis is blocked, at 37°C. In this exercise, you will demonstrate that the *Serratia marcescens* pigment is always produced at 25°C but never at 37°C. The temperature-dependent pigment production is an example of inducible phenotypic variation.

Materials

Trypticase soy broth culture of *Serratia marcescens*

Trypticase soy agar

Sterile Petri plates

Incubators set at 25°C and at 37°C

Procedure

First Period

1. Streak *Serratia marcescens* onto two trypticase soy agar plates.

2. Incubate one group of plates upside down at 25°C and a second group upside down at 37°C. Plates should be incubated for 48 hours.

Second Period

1. After growth appears on the plates, confirm that pigment production occurs at 25°C but not at 37°C.

2. Streak two plates with material from a single red colony and incubate one plate at 25°C and the second plate at 37°C.

3. Similarly, streak two plates with material from a single colorless colony and incubate one plate at 25°C and the second plate at 37°C.

Third Period

1. After growth appears on the plates, confirm that pigment production occurs at the lower temperature but not at 37°C.

2. Record the results.

Questions

1. What is a mutation?

2. What is a mutant?

3. If the mutation rate for a particular gene is one in ten thousand (10^{-4}) and the population consists of

Figure 37-1. Phenotypic Variation. *Serratia marcescens* grown at 25°C produces a red pigment but produces no pigment at all if grown at 37°C.

one million organisms, how many mutants are there in the population for this particular gene?

4. What is a prototroph and what is an auxotroph?

5. Compare and contrast genotypic variation with inducible phenotypic variation.

6. Explain what transformation is.

7. What role does conjugation play in genetic variability?

8. What is transduction?

9. How might transformation, conjugation, and transduction be distinguished experimentally?

10. How can you experimentally demonstrate that the change in colony color at 25°C and at 37°C in *Serratia* is due to inducible phenotypic variation rather than genotypic variation?

11. Does your answer to question 10 exclude the possibility that the higher temperature selects for white mutants (in a red population) and the lower temperature selects for red mutants (in a white population)?

B. ISOLATING A MUTANT AND ILLUSTRATING GENOTYPIC VARIATION

Environmental factors may alter the physiology and appearance of organisms by changing their hereditary material. For example, if the bacterium *Serratia marcescens* is exposed to ultraviolet light for a few minutes, some of the organisms will permanently lose the ability to synthesize their red pigment. In this case, one of the genes responsible for production of the pigment has become defective due to the irradiation. The *Serratia* mutant will no longer produce red colonies.

In this exercise, you will show that mutants are unable to produce the pigment at any temperature. This permanent loss of pigment production is an example of genotypic variation.

Materials

Trypticase soy broth culture of *Serratia marcescens*

Trypticase soy agar

Sterile Petri plates

Inoculating loop

Ultraviolet light, protective box and eyeglasses

Incubator set at 25°C and 37°C

Procedure

First Period

1. Make six nutrient agar spread plates using *Serratia marcescens*. Use 0.1 ml of the broth culture.

2. Remove the lids from the plates and place the plates under ultraviolet lights.

• Never look at ultraviolet light unless you are wearing protective glasses that filter all the UV wavelengths. Ultraviolet light causes retinal burns and can lead to blindness.

3. Treat the plates with ultraviolet light for 0, 6, 30, 60, 90, 120, and 150 seconds, respectively.

• Label the plates so that you know how long they were treated.

4. Incubate the *Serratia* plates for 24 to 48 hours at 25°C.

Second Period

1. Study the plates that have well isolated colonies. Most of the colonies should be red. Look for colonies or sectors within colonies that are colorless. Pick a colorless colony or sector and streak it onto two TSA plates. Pick four other colorless colonies or sectors and streak each of them onto their own pair of plates.

• Be careful not to pick up any red material if you pick from a sectored colony.

2. Incubate one of each pair of plates at 25°C and the second of each pair at 37°C for 48 hours.

Third Period

1. After growth has appeared, check the mutant colonies to see if they have remained colorless at the two temperatures.

2. Record the results.

Questions

1. How many of the red colonies show sectoring after the population is subjected to ultraviolet light?

2. Can you see any relationship between time of ultraviolet treatment and the number of colorless and/or sectored colonies?

3. How many of the colorless colonies picked from the plate incubated at 25°C turned out to produce the red pigment when they were restreaked and incubated at 25°C again?

C. ISOLATING STREPTOMYCIN-RESISTANT MUTANTS (GRADIENT PLATE TECHNIQUE)

Many bacterial populations susceptible to antibiotics become resistant to specific antibiotics when subjected to them. This is because an antibiotic selects for the rare (10^{-6} to 10^{-9}) mutants that have undergone **spontaneous mutations** to antibiotic resistance. Resistance to an antibiotic may be due to a spontaneous mutation that causes a certain protein to be altered or not to be made. For example, an enzyme normally inhibited by an antibiotic may be altered so that the drug no longer has an effect. On the other hand, the transport system responsible for bringing an antibiotic into the cell may be changed. As a result, the antibiotic cannot reach its site of action within the cell. These types of alterations are known to be responsible for streptomycin resistance. Once a mutation has made a cell resistant to an antibiotic, the presence of the antibiotic in the environment selects against the susceptible organisms and for the resistant one and its daughters. In the case of bacteria like *E. coli* that can have a generation time as short as 20 minutes, a sensitive population can become mostly resistant in 24 hours.

There are several problems associated with the isolation of antibiotic resistant mutants. First, one must use a large population of cells because the frequency of streptomycin resistant mutants in a population is usually less than 10^{-8}. This necessitates a population of at least 10^8 cells. However, identifying a single antibiotic resistant cell in a population of 10^8 cells is not always easy. One of the most commonly used methods of identifying this type of mutant is to put the population into an environment in which only the mutant can grow.

In exercise 37°C, you will find streptomycin-resistant bacteria by plating a population of approximately 5×10^9 bacteria onto a medium that contains a concentration gradient of streptomycin (fig. 37-2). This procedure is known as the **gradient plate technique.**

Materials

Trypticase soy broth culture of *Escherichia coli*

Trypticase soy agar

TSA + 0.1 mg/ml of streptomycin

Sterile Petri plates

Glass spreaders

Large jar with 95% ethanol

Incubator set at 30°C

Procedure

First Period

1. Prepare a trypticase soy agar plate with a gradient of streptomycin.
• In the bottom of a tilted Petri dish, pour 15 ml of trypticase soy agar and allow it to harden so that a slant is created. On the bottom of the plate, clearly mark the side of the plate where the agar is the thickest (−) to indicate low concentrations of streptomycin and where the agar is the thinnest (+) to indicate

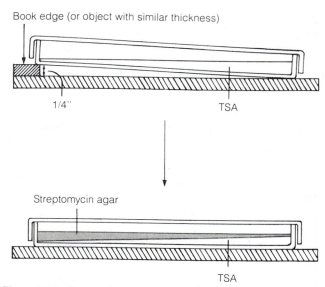

Figure 37-2. Preparation of a Streptomycin Gradient Agar Plate

high concentrations of streptomycin. After the agar has solidified, return the plate to a level surface and add 15 ml of the TSA streptomycin agar.

2. Place 0.1 ml of an overnight (24 hour) culture of *E. coli* onto the surface of the gradient plate and spread the culture evenly over the surface.

3. Incubate at 30°C for 48 hours.

Second Period

1. Record the pattern of growth on the gradient plate.

2. Locate three or four colonies that have grown in the area where there is a high concentration of streptomycin. With a sterile inoculating loop, streak the colonies toward the highest concentrations of streptomycin.

3. Reincubate at 30°C for 48 hours.

Third Period

1. Prepare trypticase soy agar plates with a uniform concentration of streptomycin.
• Place the Petri dishes on a level table and pour trypticase soy agar-streptomycin and allow it to harden.

2. Pick the organisms from the streptomycin gradient plate that grew in the area where there is a high concentration of streptomycin. Streak them onto the plates with a uniform concentration of streptomycin.

3. Streak out wild-type organisms onto a plate with a uniform concentration of streptomycin as a control.

4. Incubate at 30°C for 48 hours.

Fourth Period

1. Record the results.

Questions

1. Discuss the types of mutations that can make an organism resistant to an antibiotic.

2. Why did we use a streptomycin gradient plate to isolate streptomycin-resistant mutants rather than a plate with a uniform concentration of streptomycin?

3. Where might you find antibiotic-resistant organisms in nature?

D. REPLICA PLATING

In the isolation of the streptomycin-resistant *Escherichia* mutants, streptomycin functions as a selective agent. It inhibits the growth of wild-type bacteria but does not induce mutations. The mutants are present in the population before it is subjected to streptomycin. Thus, streptomycin only selects for mutants already present. How can it be shown that streptomycin selects for but does not cause the mutation? The **replica plating technique** can be used to demonstrate that mutants were present in a population previous to its exposure to a selective agent (fig. 37-3).

In this exercise, we wish to demonstrate that mutations making an organism resistant to an antibiotic are already present in a population before it comes into contact with the antibiotic. It is technically difficult, however, to isolate a mutant out of a population of 10^9 cells without using a selective agent. Consequently, the isolation of a mutant that has never been subjected to a selective agent requires a number of steps to pinpoint the mutant.

Materials

Trypticase soy broth culture of *Escherichia coli*

Trypticase soy agar, 30 ml/plate

TSA + 0.1 mg/ml of streptomycin

Sterile Petri plates

Replica plating block

Sterile squares of velveteen

Heavy rubber bands

Procedure

(The first two steps may be done for the student in order to shorten the procedure.)

First Period

1. Spread 0.1 ml of an overnight wild-type culture of *Escherichia* onto plates containing 30 ml of TSA.

• These plates are known as the **master plates.** Each plate may contain 0 to 10 streptomycin-resistant bac-

teria, depending upon the frequency of these mutants in the population.

2. Incubate the plates at room temperature for 12 hours (overnight).

Second Period

1. Place a sterile velveteen patch over a replica plating block and secure it with a rubber band.

• Make sure that the velveteen is stretched tightly and is flat. Mark the edge of the velveteen with a marking pencil so that you will be able to orient your plates.

2. Mark the bottom of the master plate with a marking pencil so that you will be able to orient your plate with the mark on the velveteen.

3. Press the master plate gently onto the velveteen surface so that organisms will come off onto the velveteen. Remove the master plate and reincubate it at room temperature.

4. Mark the bottom of a fresh TSA-streptomycin plate for orientation. Press the agar surface against the velveteen surface so that organisms come off the velveteen onto the agar surface. Make sure the marks on the plate and on the velveteen coincide.

• This TSA-streptomycin plate is known as the **replica plate** since it contains a replica of the organisms on the master plate.

5. Incubate the replica plate at room temperature for 48 hours.

Third Period

1. If you find discrete patches of growth on the TSA-streptomycin replica plate, determine as precisely as possible the corresponding sites on the master plate. With a swab, sample the area on the master plate where the streptomycin-resistant organism is growing and spread the organisms onto a fresh TSA plate (lacking streptomycin). This is a second master plate.

2. Incubate the second master plate for 12 hours at room temperature.

Fourth Period

1. Replica plate the second master plate onto a TSA-streptomycin plate.

2. Incubate the second replica plate and the second master plate at room temperature for 48 hours.

Fifth Period

1. You should find many more colonies on the second TSA-streptomycin replica plate than you did on the first TSA-streptomycin replica plate. If you do, this indicates that you picked a streptomycin-resistant mutant on the first master plate.

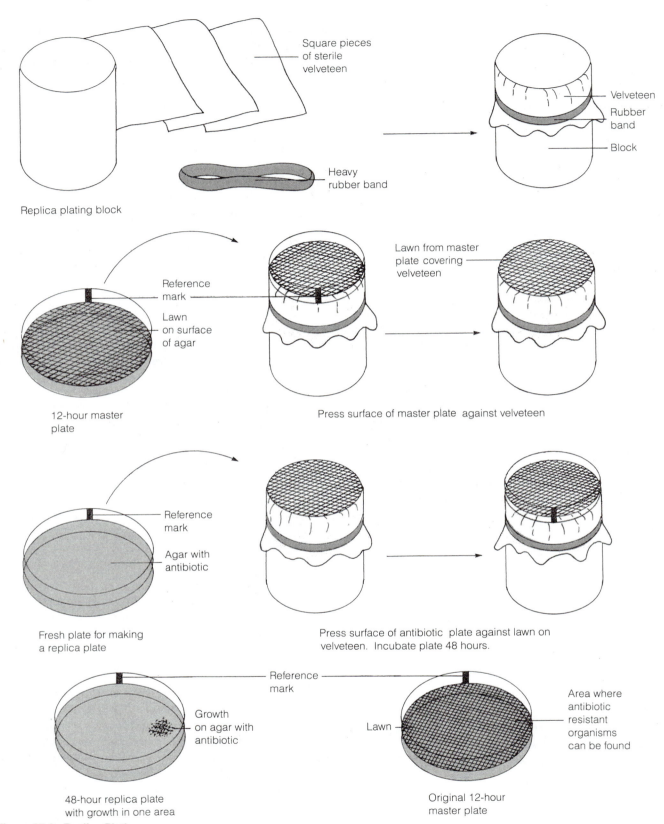

Square pieces of sterile velveteen

Heavy rubber band

Velveteen

Rubber band

Block

Replica plating block

Reference mark

Lawn on surface of agar

12-hour master plate

Lawn from master plate covering velveteen

Press surface of master plate against velveteen

Reference mark

Agar with antibiotic

Fresh plate for making a replica plate

Press surface of antibiotic plate against lawn on velveteen. Incubate plate 48 hours.

Reference mark

Growth on agar with antibiotic

48-hour replica plate with growth in one area

Lawn

Area where antibiotic resistant organisms can be found

Original 12-hour master plate

Figure 37-3. Replica Plating

Questions

1. Does inducible phenotypic variation occur in higher organisms? List some examples.

2. How can you prove that a selective agent is not inducing mutations?

3. Why are there more drug-resistant genes in plasmids today than in the 1940s?

EXERCISE 38
GENE REGULATION

Environmental factors may alter an organism's physiology temporarily by turning gene systems on or off. For example, the disaccharide lactose induces the synthesis of the enzymes involved in its metabolism. In the absence of lactose, these enzymes are not synthesized. In another case, the amino acid tryptophan in the medium represses the synthesis of the enzymes involved in the synthesis of tryptophan. In the absence of this amino acid in the medium, the enzymes involved in its biosynthesis are made.

The enzymes involved in the initial metabolism of lactose are coded for by genes in what is called the **lactose operon** (fig. 38-1). In the absence of lactose, the expression of the genes in the lactose operon is prevented by a **repressor protein.** The repressor acts as a **negative control** element. When lactose is present, a small amount of the sugar is converted to allolactose that combines with and inactivates the repressor. Under these conditions, the genes of the lactose operon are expressed.

Enzyme systems such as the lactose operon that are turned on by environmental factors are known as **inducible operons.** Operons like the lactose operon that are turned off by a protein repressor are under

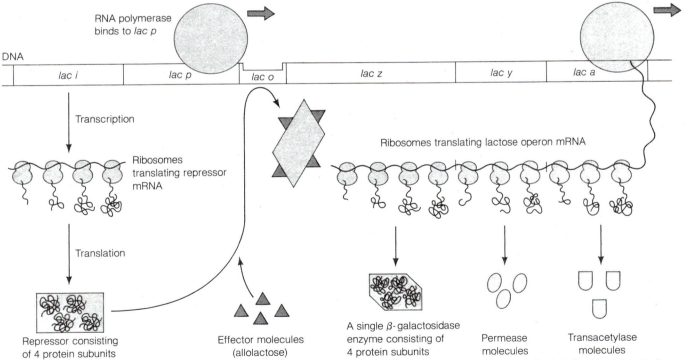

Figure 38-1. An Induced Lactose Operon. When lactose is present, the lactose operon is induced and the structural genes are transcribed. The effector (allolactose) interacts with the repressor and inactivates it so that it can no longer bind to the operator site (*lac o*). This allows RNA polymerase to attach to the promoter site *lac p* and initiate transcription of the structural genes. The resulting mRNA (polycistronic mRNA) contains the information for the synthesis of the enzymes necessary for the catabolism of lactose.

negative control. Some inducible operons are under **positive control;** that is, they require the presence of a protein for their expression. Examples of inducible operons that are under positive control are the arabinose operon and the maltose operon.

The repressor of the lactose operon is coded for by the *lac i* gene. The repressor binds to the operator site, *lac o,* and blocks the attachment of the RNA polymerase to the promoter site, *lac p.* The RNA polymerase must attach to the promoter site and move past the operator site if the lactose operon is to be expressed. Lactose is the natural **inducer** of the lactose operon. It is believed that it is converted to allolactose that acts as the **effector** and actually binds to the repressor, inactivating it. Other chemicals, such as thiomethyl-beta-D-galactoside (TMG), isopropyl-thio-beta-D-galactoside (IPTG), and ortho-nitro-phenyl-beta-D-galactoside (ONPG), can act as inducers of the lactose operon.

The regulation of the lactose operon is not only under the negative control of the repressor but also under positive control of another protein called the **catabolite repressor protein (CRP)** and an effector molecule, **cyclic adenosine monophosphate (cAMP)** (fig. 38-2). A complex of CRP and cAMP, which binds to the promoter site, is required for the expression of the lactose operon. If cAMP is not present in sufficiently high concentrations within the cell, CRP-cAMP complexes do not form and do not help the RNA polymerase bind to the promoter site. CRP by itself is unable to bind to *lac p.* The concentration of cAMP varies drastically depending upon the metabolic state of the cell.

If the cell is transporting nutrients and metabolizing rapidly, cAMP is excreted from the cell and cellular levels of cAMP are very low. The cAMP level is also thought to be controlled by an inhibition of the enzyme, adenylcyclase, that synthesizes cAMP. On the other hand, if the cell is transporting nutrients and metabolizing slowly, cAMP concentrates within the cell. Thus, a cell that is metabolizing rapidly will have little or no CRP-cAMP, while a cell that is metabolizing slowly will have effective concentrations of CRP-cAMP. Consequently, the lactose operon in the presence of lactose will be regulated (turned off or on) depending upon the cell's metabolic state. This mechanism is similar to a thermostat, in that it maintains an optimum concentration of lactose enzymes within the cell.

In the experiment below, you will discover that cells growing in a mineral glycerol medium do not produce the enzymes for lactose metabolism while organisms induced by lactose or TMG do. When glucose and lactose are added together, no lactose enzymes are produced because glucose drastically reduces the level of cAMP and this inhibits the expression of the lactose operon. This inhibition of the lactose operon is known as **catabolite repression.** Each carbohydrate produces some level of catabolite repression, that can be alleviated by the addition of cAMP.

One of the enzymes involved in the metabolism of lactose is beta-galactosidase (lactase), which is coded for by the *lac z* gene. Beta-galactosidase cleaves lactose into galactose and glucose. It also cleaves other beta-galactosides. ONPG, for example, is cleaved by beta-galactosidase into galactose and the yellow nitrophenolate ion. This is a very useful reaction since it can be used to detect beta-galactosidase. Cells disrupted by sonication or toluene can be tested for beta-galactosidase by adding ONPG. If beta-galactosidase is present, a yellow color generally develops within 30 to 60 minutes at 37°C. The intensity of color development within a certain time period is an indication of the amount of beta-galactosidase.

Materials

Culture of *Escherichia coli* K12 (*lac*$^+$ ATCC e23725) grown in mineral glycerol broth +0.05% yeast extract

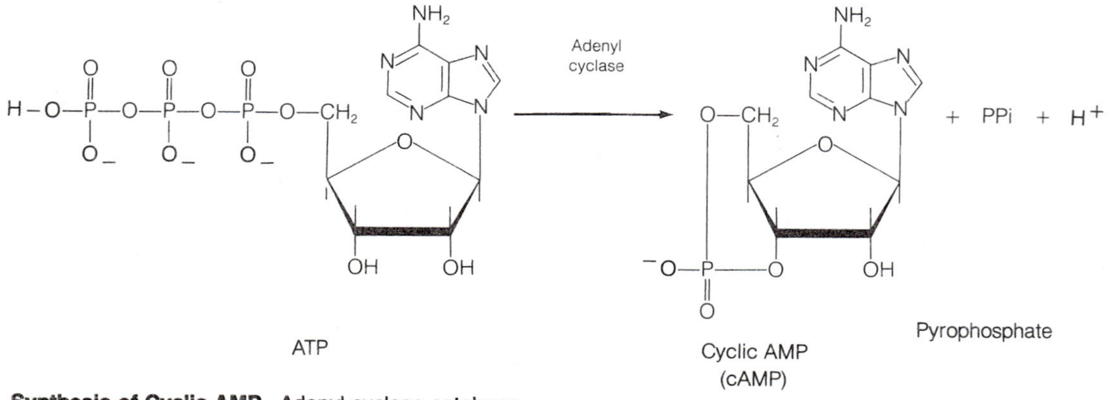

Figure 38-2. Synthesis of Cyclic AMP. Adenyl cyclase catalyzes the conversion of ATP into cyclic AMP.

Sterile minimal 0.2% glycerol broth (supplemented with 0.05% yeast extract)

Sterile 10% solution of glucose

Sterile 10% solution of lactose

1% ONPG (ortho-nitrophenyl-beta-D-galactoside)

10% TMG (thiomethylgalactoside)

2% potassium carbonate (K_2CO_3)

10^{-1}M cAMP = 100 mM cAMP (DO NOT AUTOCLAVE)

Sterile 500 ml flasks

Klett colorimeter or Bausch and Lomb Spectronic 20 spectrometer

Shaking water bath at 35°C

Vortex mixer

Ice bath

Toluene

Parafilm

Clean, large test tubes

Procedure For
Activating an Inducible Enzyme System

The first two steps may be done for the student in order to shorten the procedure.

First Period

1. Inoculate 50 ml of minimal 0.2% glycerol broth with a loopful of *E. coli* (*lac*$^+$) grown on TSA.

2. Incubate overnight at 35°C.

Second Period

1. Prepare six flasks each with 45 ml of minimal 0.2% glycerol medium. Label the flasks CONTROL, LAC, GLU, GLU + cAMP, LAC + GLU, and TMG respectively.
- If you wish to simplify the experiment, use only four flasks: CONTROL, LAC, GLU, and LAC + GLU.

2. Transfer 5 ml of the overnight culture to each of the six flasks. Add nothing more to the CONTROL flask. To LAC add 1 ml of the 10% lactose solution, to GLU add 1 ml of the 10% glucose solution, to GLU + cAMP add 1 ml of the 10% glucose solution and 1 ml of the 10^{-1}M cAMP solution, to LAC + GLU add 1 ml of the 10% glucose solution and 1 ml of the 10% lactose solution, and to TMG add 1 ml of the 10% TMG solution.

3. Incubate the flasks at 35°C in a shaking water bath for 1 hour.

4. Transfer 5 ml of the 1-hour cultures to fresh test tubes in ice and adjust the density of the cultures with ice cold sterile minimal glycerol and yeast extract medium so that the cultures all have the same turbidity.

Dilute the denser cultures so that they have the same turbidity as the least turbid culture.

5. Transfer 1 ml of the density adjusted cultures to fresh test tubes and add 0.2 ml (5 drops) of toluene. Cover the tubes with parafilm and mix. Then add 1 ml of 1% ONPG to each test tube, cover the tubes again with parafilm, and mix.

6. Incubate all tubes at 35°C for 15 to 35 minutes or until a yellow color develops in one of the tubes.

7. Quickly add 2 ml of 2% potassium carbonate to each of the tubes to stop the reaction. Be sure to mix.

8. Determine the intensity of the color for each of the tubes using a Klett colorimeter (blue filter) or Spectronic 20 spectrophotometer (wavelength = 425 nm).

9. Record your results on the report sheet provided.

Questions

1. What is an operon?

2. Using the lactose and arabinose operons as examples, explain the difference between negative control and positive control.

3. Distinguish between an inducer and an effector (actual inducer) molecule. Are the following molecules inducers or effectors: lactose, allolactose, TMG, cAMP, and ONPG? Are inducers and effectors ever the same?

4. Explain the difference between inducible operons and repressible operons.

5. To which proteins do cAMP and allolactose bind?

6. Are the levels of cAMP within a cell high or low when the cell is metabolizing rapidly? What happens to the cAMP within a cell when a cell is metabolizing rapidly? Will the level of lactose enzymes be high or low in a rapidly metabolizing cell? Explain your answer.

7. Distinguish between negative control repression and catabolite repression in the lactose operon.

8. Of what advantage to a cell is negative control repression?

9. Of what advantage to a cell is catabolite repression?

10. Why were the cells in the experiment treated with toluene?

11. Why was ONPG added to the cells after toluene treatment?

12. Why did some of the tubes turn yellow and others remain colorless?

13. Why does the intensity of the yellow color indicate the level of lactose enzymes?

14. Why should the GLU + LAC tube develop less yellow color than the LAC or TMG tubes?

15. Did the GLU + cAMP tube develop more yellow color than the GLU tube? Explain the result.

EXERCISE 39
TRANSFORMATION

Some bacteria are able to take up pieces of DNA from their environment. These pieces of DNA may recombine with homologous regions of the bacterial genome. If genetically different, the new DNA may change or **transform** the genotype of the organism. The alteration of an organism's genotype by the uptake and incorporation of cell-free DNA (naked DNA) is known as cell **transformation.**

In order for a cell to efficiently take up pieces of chromosomal DNA, the cell must enter into a **competent state.** Competence in some cells occurs during early stationary phase or during starvation in iron-deficient minimal media. The change in physiology and membrane structure allows bacteria to take up pieces of chromosomal DNA and carry out recombination between this DNA and homologous regions of the genome.

In the Gram-negative bacterium *Haemophilus parainfluenzae*, pieces of chromosomal DNA are taken up by outer membrane vesicles called **transformasomes.** The transformasomes carry the DNA into the cytoplasm where single-stranded regions are created by nucleases. Special enzymes guide recombination between the pieces of DNA and homologous regions of the bacterial genome.

In the Gram-negative bacterium *Azotobacter vinelandii*, transformation by naked plasmids does not require the bacterium to be in any specific phase of the growth curve. Thus, the competent state required for transformation by pieces of chromosomal DNA may reflect the induction of a special enzyme system needed for processing DNA and recombining it with the host's genome.

The uptake of DNA by cells is an important step in gene cloning. When a gene is cloned by modern genetic engineering techniques, it is first inserted into a self-replicating **plasmid.** Plasmids are small circular pieces of DNA that exist independently of the main genome. This plasmid, containing the gene to be cloned, is mixed with cells that can host the plasmid. The host cells (usually bacteria or yeast) may then take up this plasmid DNA from the environment. The plasmid generally does not recombine with the host's genome, but does alter the host's genotype. Plasmids may contain genes that confer antibiotic resistance and are selected by adding the antibiotic to the growth medium.

In the exercise that follows, you will demonstrate the process of transformation in the bacterium *Acinetobacter calcoaceticus*. You will release the DNA from a wild-type or prototrophic strain of the bacterium, and use this DNA to transform an auxotrophic strain which requires the amino acid tryptophan.

Materials

24 hour TSB culture of *A. calcoaceticus* prototroph (trp^+) and tryptophan auxotroph (trp^-)

Sterile lysis mixture: 0.1 M NaCl, 0.015 M sodium citrate, and 0.05% sodium dodecylsulfate

Minimal 0.2% sodium acetate agar plates

Sterile 0.1 ml, 1 ml, and 5 ml pipets

Sterile test tubes

Sterile distilled water

60°C water bath

Ice bath

Centrifuge and centrifuge tubes

Sterile capillary tubes

37°C incubator

Procedure

First Period

1. Extract DNA from the *A. calcoaceticus* prototroph (trp^+).

   a. Centrifuge 2 ml of the overnight prototroph for 10 minutes at 5000 rpm.

   b. Pour off the supernatant and resuspend the cell pellet at the bottom of the centrifuge tube in 2 ml of lysis mixture.

   c. Incubate in the 60°C water bath for about 30 minutes. At the end of 30 minutes place the extract in an ice bath to keep it cold.

2. While waiting for the lysis of the prototroph, spread 0.1 ml of the overnight *A. calcoaceticus* auxotroph (trp^-) onto a minimal 0.2% sodium acetate plate. Using a marking pencil, divide the bottom of the plate into three sections.

3. At the end of the 30-minute incubation of the prototroph, make a 10^{-1} and a 10^{-2} dilution in cold lysis mixture.

4. With a sterile capillary tube, carefully place a drop of the undiluted lysate onto one section of the plate

spread with the auxotroph. Place a drop of the 10^{-1} lysate into a second section, and a drop of the 10^{-2} lysate into the third section of the plate, using separate sterile capillary tubes.

5. Repeat step 4, using a fresh plate with no bacteria (control plate).

6. Incubate the plates at 37°C for 48 hours.

Second Period

1. Count the number of colonies that appear within each spot.

• Make sure you record the number of colonies within the spots on the control plate and on the spread plate. Also make sure you record the number of colonies outside the spots on the spread plate.

2. Streak at least five of the colonies from within the spots onto fresh minimal 0.2% sodium acetate agar to make sure that they really are prototrophs.

3. Incubate the plates at 37°C for 48 hours.

Third Period

1. Check the minimal 0.2% sodium acetate plates for growth. Record your results and determine the percent phototrophs.

Questions

1. What is transformation?

2. Why does the prototroph (trp^+) but not the auxotroph (trp^-) grow on minimal 0.2% acetate?

3. Are the number of colonies on the plates related to the dilution of the lysate? Explain why.

4. How do you know that the colonies on the spread plate are not due to prototrophs that were not killed in the lysate?

5. How do you know that the colonies on the spread plate are not due to revertants of the auxotroph? The spontaneous reversion rate is about 10^{-8}.

6. How do you know that the colonies on the spread plate are not growing because enough tryptophan was carried over from the overnight culture to provide growth?

7. What is competence and what does it have to do with transformation?

8. How was competence achieved in the transformation experiment?

9. Is it possible to transform other bacteria such as *Bacillus* and *Escherichia*?

EXERCISE 40
TRANSDUCTION

Some bacterial viruses (**bacteriophages**) package host cell DNA within their protein coat (**capsid**) as they assemble. If these bacteriophages infect new bacterial hosts, they may transmit the bacterial genes to the new host and alter its genotype. This process of genetic conversion mediated by a virus is known as a **transduction.**

Bacteriophages that transduce bacteria with acquired bacterial DNA are divided into two classes, **generalized transducing phages** (fig. 40-1) and **specialized transducing phages.** The generalized transducing phages are able to pick up any piece of host DNA while the specialized transducing phages are only able to incorporate a specific region of a bacterial genome.

In the exercise that follows, you will demonstrate the process of transduction using the generalized transducing phage P22 and its host bacterium *Salmonella typhimurium*. The phage P22 capsid is large

enough to hold a piece of bacterial DNA that codes for 10 average size proteins.

You will begin this exercise by growing phage P22 on a *Salmonella* prototroph (a wild-type that has no growth requirement). Most of the phage population will be normal phages but a few will contain various pieces of bacterial DNA. If only one capsid of every 10,000 that develop in the bacterial population picks up any bacterial genes and the titer is 10^{11} capsids per ml, the result would be 10 million transducing capsids per ml. Only four hundred capsids, picking up different regions of the bacterial genome, are theoretically enough to carry all the bacterial genes.

Using the phage P22 lysate prepared on a *Salmonella* prototroph, you will transduce a *Salmonella* that is auxotrophic for two amino acids, tryptophan and cysteine ($trpD^- cysB^-$). The phage lysate and the auxotroph are mixed together before plating on a minimal glucose medium with tryptophan. The colonies that

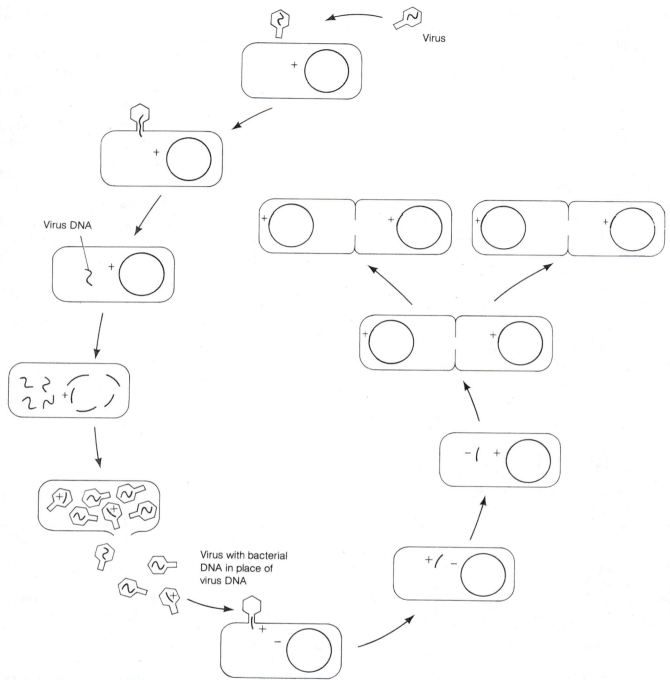

Virus

Virus DNA

Virus with bacterial
DNA in place of
virus DNA

Figure 40-1. Generalized Transduction. Generalized transducing bacterial viruses invade bacteria by injecting their hereditary material into the bacteria. The virus DNA directs the synthesis of new virus DNA and proteins. As the viruses develop within the bacterial cytoplasm, some of the fragmented bacterial DNA is incorporated within virus coats instead of virus DNA. The "pseudoviruses" with the bacterial DNA are able to inject the bacterial DNA into other bacteria. The bacterial DNA injected by the pseudoviruses may recombine with the bacterial genome giving the bacterium a new trait.

appear on the plate are composed of cells that can synthesize cysteine. The number of tryptophan and cysteine transductants can be determined by picking the colonies to minimal glucose medium.

When the DNA from **transducing particles** enters a cell, it may recombine with the host genome and become part of it. This can result in a recombinant that is genetically altered and that gives rise to daughter cells that will be genetically identical. Under these circumstances, a normal size colony develops. Frequently, however, the DNA does not recombine with the host genome, yet is expressed in the infected cell.

Since this DNA is not replicated unless it becomes part of the host's genome, there will be only one cell giving rise to a colony that contains the foreign DNA. The **abortive transduction** results in a minute colony compared to that which forms after a **completed transduction.**

This exercise demonstrates how viruses can be used to transduce bacterial genes into bacteria and transform the genotype of a cell.

Materials

24 hour TSB culture of *Salmonella typhimurium* prototroph (ATCC e23564 LT2 wild) and tryptophan and cysteine auxotroph (ATCC e23595 LT2 *trpD⁻cysB⁻*)

Wild-type phage P22 lysate, or phage P22 lysate grown on a *Salmonella* prototroph

Sterile KCl broth tubes, 9 ml/tube to be used for diluting phage lysate

Minimal 0.4% glucose agar

Minimal 0.4% glucose 0.1% tryptophan agar

Top agar

Chloroform

Sterile 0.1 ml, 1 ml, and 10 ml pipets

Screw-capped tube or bottle to store phage lysate

Centrifuge and centrifuge tubes

Incubator set at 37°C

Water bath set at 50°C

Sterile toothpicks

Vortex mixer

Procedure

The steps in periods 1 and 2 may be carried out by the instructor with the students beginning at period 3.

First Period

1. Prepare a lysate of P22 using wild-type (prototroph) of *Salmonella typhimurium.*

   a. Add 0.1 ml of phage P22 (5×10^9 PFU/ml) to 10 ml of a fresh broth culture of the *Salmonella* prototroph (5×10^8/ml). This gives a multiplicity of infection of 0.1. The multiplicity of infection should be less than 1. Incubate for 3 hours at 37°C or until the broth culture becomes clear.

   b. Add 0.2 ml of chloroform and vortex the contents to lyse any remaining cells. Centrifuge the phage suspension to pellet the debris at the bottom of the centrifuge tube. Save the phage in the supernatant fluid in a bottle to which you have added a small amount of chloroform. The titer expected is between $10^{11} - 10^{13}$ PFU/ml.

• Chloroform should be used with care because it has been shown to be a carcinogen.

Second Period

1. Make a 10^{-10} dilution of the phage lysate using sterile KCl broth.

2. Determine the titer of the phage lysate by using 0.1 ml of 10^{-7}, 10^{-8}, 10^{-9}, and 10^{-10} dilutions of the phage lysate. Add 0.1 ml of the phage dilution to 0.2 ml of the wild-type *Salmonella* in 5 ml of cooled melted top-agar. Pour the contents of the mixture onto a solid TSA plate and incubate the plates at 37°C for 24 hours.

Third Period

1. Spread 0.1 ml of the *trpD⁻cysB⁻* auxotroph on each of 6 minimal glucose plates supplemented with tryptophan.

2. Spread 0.1 ml of the phage lysate (with a titer of 10^{11} PFU/ml) grown on the prototroph onto 2 of the plates containing the auxotroph. Also spread 0.1 ml of a 10^{-1} dilution of the phage lysate on 2 other plates containing the auxotroph. The remaining 2 plates of the auxotroph will serve as bacterial control plates. Spread 0.1 ml of the phage lysate onto 2 sterile minimal glucose tryptophan plates as a control for bacteria in the phage lysate (phage control).

3. Incubate the transduction plates and control plates at 37°C for 48 hours.

Fourth Period

1. Check the minimal glucose plates supplemented with tryptophan for cysteine prototrophs (*cys⁺*). Count the number of cysteine prototrophs on the control plates on which only the auxotroph or the phage lysate was spread.

2. Determine whether tryptophan has been cotransduced with cysteine by using sterile toothpicks to transfer the cysteine prototrophs to minimal glucose and minimal glucose tryptophan plates. Stick the end of the toothpick into the transductant and then stab the minimal glucose and minimal glucose tryptophan plates in identical positions. Pick at least 50 transductants.

3. Incubate the plates for 48 hours at 37°C.

Fifth Period

1. Calculate the percentage of colonies that are able to grow on the minimal glucose medium lacking tryptophan.

Questions

1. What is transduction?

2. How do you know that the colonies that developed in the transduction experiment are not due to prototrophs that were not killed in the lysate?

3. How do you know that the colonies that developed

in the transduction experiment were not due to re-vertants of the auxotroph? *The spontaneous reversion rate is about* 10^{-8}.

4. How do you know that the colonies that developed in the transduction experiment did not grow because there was enough cysteine carried over from the over-night culture or the phage lysate?

5. What is the back mutation frequency of the auxotroph?

6. What is the transducing frequency of the phage?

7. Explain what abortive transduction is and how it can be distinguised from completed transduction.

8. What is the frequency of abortive transduction and completed transduction?

9. What is the frequency of cotransduction of the cys-teine (*cys*) and tryptophan (*trp*) genes? From a genetic map, determine how far apart in minutes these gene systems are.

10. How frequently is a piece of bacterial DNA incor-porated into the P22 capsid?

11. How can you use generalizing transducing phages to order (map) the genes of a bacterium?

12. Explain the difference between generalized trans-ducing phages and specialized transducing phages.

EXERCISE 41 CONJUGATION

Bacteria that contain special plasmids called **fertility factors (F-factors)** are able to transfer hereditary in-formation to other bacteria of the same species (fig. 41-1). The transfer of hereditary material from a **donor** bacterium to a **recipient** bacterium requires contact between the bacteria. The unidirectional transfer of hereditary material that requires physical contact be-tween the participants is known as **conjugation.**

There are three types of donor bacteria: F^+ donors that transfer plasmids containing genes not normally found on the main genome, F' donors that transfer plasmids having genes found on the main genome, and **Hfr** donors that transfer part or all of the main genome. The recipient bacterium lacks a fertility factor and consequently is known as the F^- recipient (fig. 41-2).

In gram-negative bacteria, the fertility factor con-tains the genes that code for the pilus proteins, the genes that are required for plasmid replication, and the genes that are required for plasmid transfer. Fer-tility factors may also carry genes that inhibit the growth of viruses, that code for proteins that kill bacteria (bac-teriocins), that confer resistance to heavy metals, or that make a cell resistant to antibiotics and chemicals.

Since most plasmids are less than 100,000 base pairs long, complete copies generally can be trans-mitted to F^- recipients. Because the bacterial genome is so long (most are greater than 3,500,000 base pairs long), however, the entire genome is seldom trans-ferred.

A. MAPPING MUTATIONS AND GENES WITH F' AND F⁻ MUTANTS

When an F' and an F^- mutant with mutations in dif-ferent genes are mated with each other, the plasmid and chromosome in the recipient cell **complement** each other. That is, the donor and recipient genes add up to a complete set of functional genes (fig. 41-3). Consequently, the recipient's deficiency is rectified. If the plasmid is not lost upon cell division, all the daughter cells of the recipient are F'. When genes of the F' plasmid and the main genome complement, growth occurs because all the required genes are pres-ent and functional. If, for example, F' leucine auxo-trophic bacteria are spotted onto a lawn of F^- thre-onine auxotrophic bacteria, all the recipients will grow in the spotted area and a region of confluent growth will occur after incubation (fig. 41-3).

On the other hand, when F' and F^- mutants car-rying mutations in the same gene are mated with each other, the plasmid and the chromosome in the recip-ient cell do not complement each other. Thus, the recipient's deficiency is not usually rectified. If the mutations affect different nucleotides in the same gene, however, the genes on the plasmid and on the chro-mosome may undergo recombination and a func-tional gene will be created out of two nonfunctional genes. This occurs very rarely, but when it does, the recipient will grow (fig. 41-3). If F' auxotrophic bacteria

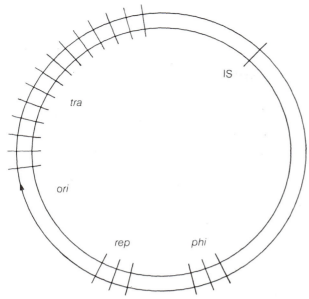

Figure 41-1. Fertility Factor. A fertility factor is a plasmid that can transfer itself or the main genome during conjugation to a recipient bacterium. The *tra* genes are required for the transfer of genetic material. The *ori* region is the region of DNA that is initially transferred to the recipient. The *rep* genes are required for the replication of the plasmid. The *phi* genes are involved in the inhibition of certain phages. The *IS* sites are regions of DNA that allow the plasmid to integrate into the main genome.

are spotted onto a lawn of F⁻ auxotrophic recipients, which have a mutation in the same gene, a very few of the recipients will undergo recombination and will grow in the spotted area. A few colonies will be seen within the spotted area when recombination occurs. If the mutations overlap, no recombinants can form and so no colonies will appear in the spotted areas.

When a number of mutants are mated with each other, the pattern of complementation, recombina-

tion, or no growth can often be used to determine how many genes are involved and to order (map) the genes and mutations relative to each other. A large set of F′ and F⁻ point and deletion mutations from a multigene system such as the L-arabinose operon and its regulatory operon is necessary if the genes and mutations are to be mapped. Since it is difficult for most laboratories to obtain and maintain a large set of mutants, this exercise will be limited to only a few crosses that demonstrate complementation, recombination, and the absence of these two phenomena.

Materials

24 hour minimal glucose (supplemented with vitamin B1, threonine and leucine) cultures of F′*lac* and F⁻*lac Escherichia coli* point and deletion mutants in the lactose operon: ATCC e25251 K12 F′*lac y⁻* B1⁻, ATCC e25255 K12 F′*lac z⁻* B1⁻, ATCC e23724 K12 F⁻ *lac y⁻* B1⁻ *thr⁻leu⁻*, ATCC e23722 K12 F⁻*lac z⁻* B1⁻, ATCC e25253 K12 F⁻ *lac z⁻* B1⁻, ATCC e25254 K12 F⁻ △*lac*

Minimal 0.4% lactose agar (supplemented with vitamin B1, threonine, and leucine. Vitamin B1 (thiamine) = 0.1 μg/ml, DL−amino acids = 0.1 mg/ml each.)

0.1 ml pipets,

Sterile glass capillaries

Incubator set at 37°C

Procedure

First Period

1. Spread 0.1 ml of the F⁻ mutants onto separate minimal lactose agar plates.

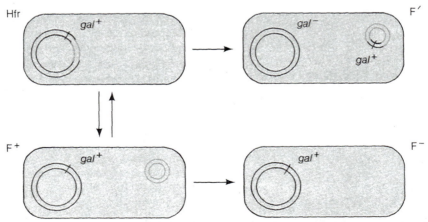

Figure 41-2. The Relationships Among Different Mating Types. The F⁺ bacterium is haploid and contains one or more plasmids. The F′ bacterium contains one or more plasmids with genes from the main genome. The Hfr bacterium is haploid and its plasmid is integrated into the main genome. The F⁻ bacterium is haploid and lacks a plasmid or an integrated plasmid. The bacteria may change from one form to another, as illustrated.

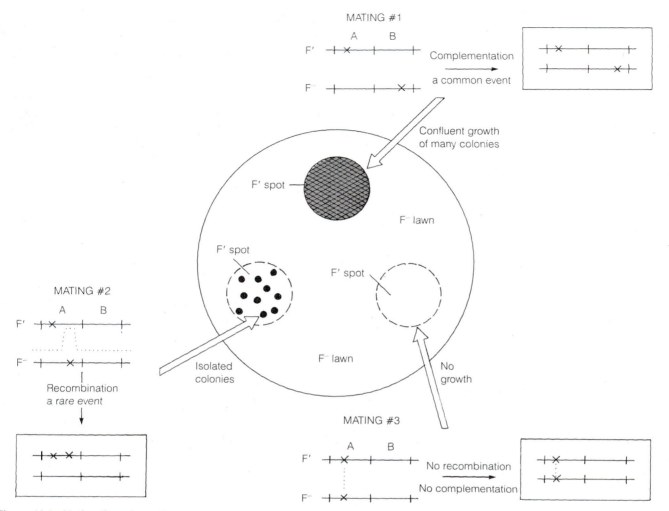

Figure 41-3. Mating Experiment Using F′ and F⁻ Mutants

- Use at least two different *lac y⁻* or two different *lac z⁻* mutations in the mating so that you will be able to observe recombination.

2. With glass capillaries, sample the F′ mutants. Spot the F′ mutants onto each of the spread plates.

- Make sure that the spot for a particular F′ mutant is in the same location on each of the spread plates. As a control, also spot a sterile minimal lactose agar plate with the F′ mutants.

3. Incubate the plates at 37°C for 24 hours.

Second Period

1. After 24 hours check the plates to look for complementation within the spots.

2. Reincubate the plates after you have recorded your results. Complementation (confluent growth or large colonies) = C, and no growth = −.

Third Period

1. After a total of 48 hours, check the plates to look for recombination within the spots that showed no growth and record your results. Recombination (colonies present after 48 hours but not after 24 hours) = R, and no growth = −.

Questions

1. What role does conjugation play in genetic variability?

2. In the F′ × F⁻ matings, how do you know that the colonies on the plates are not due to revertants of the mutants?

3. If a donor and a recipient have mutations in the same gene, can the donor's gene complement the recipient's gene? Explain your answer.

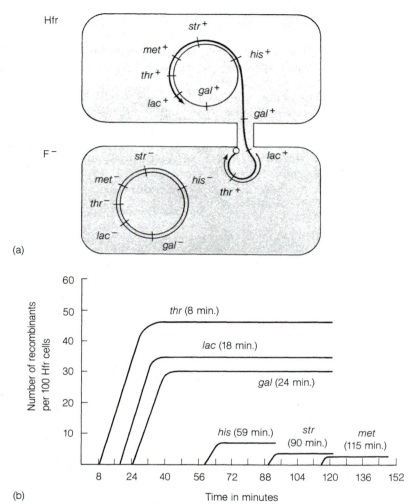

(a)

(b)

Figure 41-4. The Use of Conjugating Bacteria to Map Genes. *(a)* A prototrophic Hfr and an auxotrophic F⁻ are mixed together and allowed to mate for various periods of time. Samples of the bacteria are agitated to discontinue conjugation and are then plated on selective media that do not allow the parents to grow. *(b)* The order of the genes entering the female bacterium is *thr, lac, gal, his, str,* and *met.* Consequently, if the mating is disrupted soon after it is initiated, only the first gene system (*thr*) has time to enter the female. Thus, bacteria grow initially only on the plates that have all the nutrients except threonine. The time at which recombinants appear on each type of plate indicates the time the gene systems enter the female bacterium. The order of the gene systems can be determined from this data.

4. In F′ × F⁻ matings, how can you distinguish between complementation and recombination? Explain why the results are different.

5. When two mutations complement each other, what does this indicate about the mutations?

6. If a bacterium has a mutation in the lactose operon and this mutation does not complement or recombine with a number of different mutations in the lactose operon, what might this indicate about the organism's mutation?

7. Distinguish between an F⁺, F′, Hfr, and F⁻.

8. Draw a circular plasmid and indicate the origin of transfer with an arrow head, the location of the genes required for plasmid transfer with the symbol *tra,* and the genes required for plasmid replication with the symbol *rep.*

B. MAPPING GENES BY INTERRUPTING MATINGS BETWEEN Hfr AND F⁻ MUTANTS

Hfr prototrophs and F⁻ point mutants can be used to map genes. When Hfr and F⁻ mutants carrying mutations in different genes are mated, the Hfr's chromosome and the recipient cell's chromosome can recombine with each other. Thus, the recipient's deficiency is rectified and it will grow where neither the Hfr nor the F⁻ would.

Matings between Hfr prototrophs and F⁻ point mutants are carried out in broth. It takes approximately 100 minutes at 37°C for the Hfr to transfer one strand of its main genome to the F⁻. The genes enter the F⁻ in the same order in which they are linked. The genes that enter the recipient first are those that are closely linked to the **origin of transfer locus** on the F-factor (fig. 41-4). Genes close to the origin of transfer locus are transferred early while genes further away are transferred late. Usually, a mating pair does not remain together long enough for the entire genome to be transferred from the Hfr to the F⁻. Because the mating is easily interrupted, the further the genes are from the origin of transfer locus, the less likely they are to be transferred and the number of recombinants decreases.

In a typical mapping experiment, an Hfr prototroph that is streptomycin-sensitive (*str*ˢ) is mated with an antibiotic resistant F⁻ that cannot utilize a number of carbohydrates or synthesize various amino acids. For example, a streptomycin resistant (*str*ᴿ) F⁻ might be unable to utilize lactose (*lac*⁻) or galactose (*gal*⁻), or unable to make leucine (*leu*⁻). At various times after mixing the Hfr and the F⁻ together, samples are taken and shaken vigorously (vortexed) to disrupt the mating and interrupt the transfer of DNA. The sample is then plated onto a medium that does not permit the growth of the donor or the recipient, but allows the growth of recombinants. For example, the mixture might be plated onto minimal glucose streptomycin medium. This would select against the Hfr prototroph (which is streptomycin-sensitive) and the recipient (which is unable to synthesize luecine). Only recombinants that develop from recipients that receive the wild-type genes for leucine synthesis will grow. In the mating experiment in part B, you will use a selective medium to select against the donor and differentiate between the recipient and the recombinant.

This exercise will demonstrate how bacteria transfer their genes in a unidirectional manner when they conjugate and how Hfr and F⁻ conjugation can be used to map bacterial genes.

Materials

24 hour TSB culture of Hfr *Escherichia coli* prototroph (CGSC strain 5461) that is sensitive to streptomycin and that has an origin of transfer at 0 minutes and that transfers clockwise

24 hour TSB culture of F⁻ *Escherichia coli* that is *leu*⁻*lac*⁻*gal*⁻ and *str*ᴿ (CGSC strain 5581)

Minimal agar medium with lactose, leucine, and streptomycin (MLac + Str + Leu); minimal agar medium with galactose, leucine, and streptomycin (MGal + Str + Leu); and minimal agar medium with glucose and streptomycin (no leucine) (MGlu + Str)

Sterile 0.1 ml, 1 ml, and 10 ml pipets

Vortex mixer

Minimal broth (minimal medium, with no carbon or energy source)

Incubator set at 37°C

Glass spreaders and jar with 95% ethanol

Procedure

First Period

1. Mix 1 ml of the Hfr and 20 ml of the F⁻*leu*⁻*lac*⁻*gal*⁻*str*ᴿ in a 250 ml flask and incubate at 37°C in a slowly shaking water bath for 5 minutes to allow mating pairs to form.

2. Prevent the bacteria from proliferating by transferring 0.1 ml of the mating mixture to 100 ml of warm (37°C) minimal broth.

3. Sample the minimal broth mating mixture 10, 15, 20, 25, 30, and 40 minutes after the mating is initiated by transferring 1 ml of the mixture to 4 ml of fresh minimal broth. Vortex the mating mixture for 15 seconds immediately after sampling so that the mating is interrupted.

4. Plate 0.1 ml of the interrupted cells onto each of the following agar plates: MGlu + Str, MLac + Str + Leu, and MGal + Str + Leu at each sampled time.

5. Incubate the plates 48 hours at 37°C.

Second Period

1. Count the colonies that appear on the various agar media that were plated with samples from the different interruption times.

• Colonies on the MGlu + Str are *leu*⁺ recombinants, those on the MLac + Str + Leu are *lac*⁺ recombinants, while those on the MGal + Str + Leu are *gal*⁺ recombinants.

2. Graph the number of each type of recombinant against time of interruption. Extrapolate the curves to zero colonies and estimate the time at which each marker entered the recipients.

Questions

1. In the Hfr × F⁻ mating, how do you know that the colonies on the spread plates are not due to the Hfr becoming resistant to streptomycin or to the recipient reverting?

2. Why is the Hfr × F⁻ mating interrupted? How would the results change if the mating were not interrupted?

3. Why is the Hfr × F⁻ mating diluted into minimal medium after mating pairs form? How would the re-

sults change if the mating were not diluted into minimal salts medium?

4. Draw a plasmid integrated into a circular bacterial chromosome. Indicate the important regions on the plasmid. Which part of the plasmid enters the recipient first and which part enters last?

5. Of what importance are the streptomycin markers on the Hfr and the F⁻?

6. How might transformation, conjugation, and transduction be distinguished experimentally?

EXERCISE 42
AMES TEST

Each year in the United States more than 800,000 persons will develop cancer, and more than 400,000 will die from the disease. Ninety percent of the cancers are caused by ultraviolet light from the sun and from chemicals and radioactive atoms in our food, air, and water. Chemicals on clothing, in medicines, in tobacco smoke, and in hair dyes can cause cancer. Tobacco smoke alone has been estimated to be responsible for about 100,000 cancer deaths in the United States each year. Cancers of the lung, tongue, mouth, lip, larynx, bladder, and pancreas have been linked to tobacco smoke.

At this point in our understanding of cancer, it appears that two or more alterations in the hereditary material are required to transform a normal cell into a cancerous cell. The first mutation or translocation in the genome might result in the development of a transformed cell that might proliferate (divide) continuously. These proliferating cells would die out after a small number of generations, much like normal cells do. Thus, they would not be cancerous. The second mutation or translocation might lead to the development of a cancerous population of cells if the alteration allowed the cells to have a large number of generations.

Ultraviolet light, radioactive atoms, and various chemicals that induce mutations are known as **mutagens,** and those that also cause cancer are called **carcinogens.** Physical and chemical factors that induce mutations are sometimes referred to as **initiators** of cancer. While *all* mutagens must be considered potential carcinogens, there are a few chemicals that do not appear to cause mutations yet induce cancer. Thus, not all carcinogens are mutagens. Asbestos is an example of a chemical that does not cause mutations but does cause cancer. Asbestos is a potent **promoter** of cancer because it apparently stimulates the proliferation of mutated cells increasing the chances that a second mutation will occur that converts the cell into a cancerous one.

Some chemicals are not mutagenic or carcinogenic as they occur in the environment, but only become mutagenic and carcinogenic when they are altered chemically within an organism. Benzopyrene, found in tobacco smoke, charcoal-broiled foods, car exhausts, and smoke from coal and wood, only becomes mutagenic in bacteria and carcinogenic in animals when it is altered by mammalian cells. Similarly, arsenic only becomes mutagenic in bacteria and carcinogenic in animals when it is altered by mammalian enzymes.

Carcinogens generally act synergistically; that is, small amounts of carcinogens that individually would cause very few cancers, together cause many more than expected from the sum of their individual effects. Thus, it is important to detect even low levels of carcinogens in the environment to assess their impact. Some carcinogens are more potent if taken in low concentrations for long periods of time than if given in one large dose. For example, aflatoxins, produced by certain species of fungi growing on peanuts, wheat, rice, corn, and other grains, are mutagens and carcinogens if low doses are consumed for a number of years. A few large doses that exceed the amounts taken over a long period generally do not induce cancers.

The Ames test is a simple, relatively inexpensive procedure developed by Dr. Bruce Ames at the University of California at Berkeley that can be used to detect mutagens, and thus, potential carcinogens. Histidine auxotrophs of the bacterium *Salmonella typhimurium* are used to detect back-mutations to histidine prototrophy. The bacterium also is unable to produce a normal cell wall or the enzymes required for DNA repair. The defective cell wall allows chemicals to penetrate the cells more easily, while the defective DNA repair system increases the number of mutants. This increases the sensitivity of the Ames test.

In the Ames test, *Salmonella* histidine auxotrophs in top agar, supplemented with minute amounts of biotin and histidine, are spread over the surface of a minimal glucose medium. The minute amounts of biotin and histidine allow the auxotroph to divide a few times but not enough to develop visible colonies. These divisions are necessary so that mutations can occur in the cells affected by the mutagen. Filter paper disks saturated with the chemical to be tested are placed on the surface of the agar in the center of the plate. The chemical diffuses out from the disk and creates a concentration gradient. Chemically induced and spontaneous mutations result in back-mutations to histidine prototrophy. Colonies of revertant *Salmonella* are visible in the soft agar 48 hours after incubation.

Since some compounds (such as benzopyrene and the fungal aflatoxins) are not mutagenic unless they are chemically altered by enzymes found in mammalian liver cells, a liver homogenate (S-9) is generally added to the soft agar. In order to simplify the Ames test, we will not add the liver homogenate.

This exercise will demonstrate how chemical mutagens and carcinogens can be detected and show that Petri plates sterilized with ethylene oxide can induce mutations.

Materials

24 hour TSB culture of *Salmonella typhimurium* histidine auxotroph (*his*$^-$) Ames strain TA-1538

Minimal glucose agar plates

Tubes of top agar (3 ml)

Sterile biotin-histidine solution (12.2 mg biotin + 10.5 mg L-histidine HCl in 100 ml of distilled water)

Hair dye containing 4-nitro-o-phenylenediamine

Thick slurry of old motor oil, cigarette ashes, or chimney ashes

Sterile 0.1 ml, 1 ml, and 5 ml pipets

Plastic Petri plates sterilized with ethylene oxide

Glass plates sterilized by autoclaving (or plastic Petri plates sterilized by gamma radiation)

Propipet

Filter paper disks

Sterile test tubes

Sterile distilled water

48°C water bath

Centrifuge and centrifuge tubes

Procedure

First Period

1. Using a propipet (do not pipet by mouth), add 0.2 ml of the *Salmonella* histidine auxotroph and 0.2 ml of the biotin-histidine solution to the melted top agar in a 48°C water bath.

- The *Salmonella*-contaminated pipets should be autoclaved before washing.

2. Mix the top agar by rapidly rolling the tube between the palms of the hands, and then pour the contents over the surface of a minimal glucose agar plate. Make sure that the top agar is uniformly spread over the surface of the plate by gently swirling the top agar. Allow the top agar to harden for a few minutes before continuing.

3. Holding a filter paper disk with forceps, dip it into the chemical to be tested. Blot the excess chemical on the disk onto a paper towel and then place the disk in the center of the plate.

- The forceps and disks should be sterile.

4. Make a control plate by dipping a disk in sterile distilled water and placing the disk in the center of a plate.

5. Use glass Petri dishes sterilized by autoclaving and plastic Petri dishes sterilized with ethylene oxide to test for the effect residual ethylene oxide may have on causing back-mutations. Use water-soaked disks as controls.

- Each student should use duplicate plastic and glass dishes.

6. Incubate at 37°C for 48 hours.

Second Period

1. Count the number of colonies that appear on each plate. Make sure you record the number of colonies that develop on the control plates and where most of the colonies are located with reference to the disks.

2. Streak some of the colonies onto minimal glucose agar to make sure that they really are prototrophs.

3. Incubate the new plates at 37°C for 48 hours.

Third Period

1. Check the minimal glucose plates to make sure that the organisms are prototrophs.

Questions

1. What is the purpose of the minute amounts of biotin and histidine in the top agar?

2. Why is the greatest number of revertants not seen along the edge of the filter paper disk?

3. How do you know that the colonies on the plate are not due to spontaneous revertants?

4. How do you know that the colonies on the plate are not growing because too much biotin and histidine were added to the top agar?

5. What does the Ames test demonstrate?

6. What is the relationship between mutagens and carcinogens?

7. Are all mutagens potential carcinogens? Are all carcinogens potential mutagens? Explain your answer.

8. What kind of carcinogens will not be detected by the Ames test (two types)?

9. What is the advantage of using a *Salmonella* strain that has a defective cell wall and a defective DNA repair system?

10. How can compounds that must be modified by mammalian enzymes before they are mutagenic be discovered using the Ames test?

PART XI
MICROBIAL ECOLOGY

Microbial ecology is the study of the interrelationships among the various bacteria, fungi, algae, protozoa, and viruses and the diverse environments in which they are found. These interrelationships are extremely complex, and are governed by the physico-chemical parameters of the specific environment as well as the various biochemical activities of the microbes themselves.

A microbial **ecosystem** is defined as the microorganisms in a given environment and the abiotic (nonliving) surroundings in which those organisms are found. The organisms inhabiting a specific habitat are considered to be a **community.** The community contains **populations** of individual species. The community may contain populations considered to be true inhabitants (indigenous or autochthonous) of that habitat as well as those organisms which have been transported to the habitat (allochthonous) and will in a relatively short period of time probably be eliminated.

Microorganisms are believed to be ubiquitous; that is, they are found throughout the biosphere. Some habitats are heavily populated, while others are sparsely populated. Some communities contain a wide variety of populations, while others may be restricted to a very few. The extent of the variety is referred to as **species diversity.**

Microbial ecologists are interested in the habitat and niche of a particular microorganism. The **habitat** is a given area that has a degree of uniformity in its physico-chemical characteristics. The **niche** is the role of the organism in that habitat. Some organisms, considered to be **specialists,** have very narrow niches; while others, or **generalists,** perform a wider variety of processes. The latter are not affected as greatly when changes occur in their habitats.

EXERCISE 43
POPULATION DYNAMICS (GROWTH CURVE)

Microbial growth is an increase in cell numbers (or an increase in microbial mass). In unicellular organisms, growth usually implies an increase in number. Some bacteria are capable of rapid growth, with doubling (**generation**) times of less than 20 minutes during peak growth periods. Theoretically, a single cell that could double every 20 minutes in an ideal environment would achieve a mass 4,000 times that of the earth in 48 hours!

There are four distinct stages in a bacterial population growth curve. They are the **lag** phase, in which the cells become adjusted to their new environment; the **logarithmic** or **exponential** phase, in which the cells grow as rapidly as possible and at a constant rate; the **stationary** phase, in which growth ceases; and the **death** phase, in which the numbers of viable cells decrease. A typical bacterial growth curve is shown in fig. 43-1.

There are a variety of ways to measure microbial growth. These usually involve either following the changes in numbers or weight, the increase in some component of the cells (e.g., DNA or total protein), or the turbidity of the population.

Materials

Overnight culture of *E. coli* in nutrient broth + 1% glucose

250 ml side-arm flasks containing 25 ml nutrient broth + 1% glucose

Klett-Summerson colorimeter or Spectronic 20

9 ml dilution blanks

1 ml pipets

Shaking water bath, 37°C

Plate count agar

Incubator set at 37°C

Procedure

First Period

1. Aseptically transfer 1 ml of *E. coli* from the overnight culture to a side-arm flask containing 25 ml NB + 1% glucose. Measure the turbidity on the Klett-Sumerson colorimeter.

2. Remove a 1 ml sample from the flask. Prepare dilutions through 10^{-7}. Plate in duplicate to yield plates with final dilutions of 10^{-5} to 10^{-7}. Add melted and cooled PCA. Swirl gently. Allow to solidify. Incubate at 37°C for 24 to 48 hours.

3. Repeat the colorimetric measurements and plate counts at 30 minute intervals for the desired length of the exercise. As the turbidity increases, plate dilutions of 10^{-7} to 10^{-9}.

4. Discard the flasks as indicated by your instructor.

Second Period

1. Arrange the plates from the various sample times in order of lowest to highest dilution. Determine which dilution provided the duplicate set of plates showing between 30 and 300 colonies. Enumerate the colonies on that set of plates and determine the number/ml at the various sample times.

2. Plot a growth curve like the one shown in fig. 43-1 on semilogarithmic paper, using viable cells/ml versus

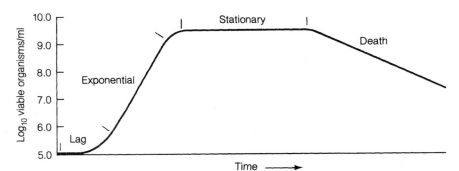

Figure 43-1. The Bacterial Growth Curve. A population of bacteria growing in a broth medium will undergo a series of changes in its rate of growth that reflect changes in its environment and phys-iology. These phases are the lag phase, the exponential phase, the stationary phase, and the death phase.

time on the x and y axis, respectively.

3. Plot a second curve on standard graph paper using turbidity (Klett units) versus time. How can the plots drawn in steps 2 and 3 be correlated?

4. Discard the plates as indicated by your instructor.

Questions

1. If 1 ml aliquots of a culture grown overnight in nutrient broth were inoculated into fresh nutrient broth and into minimal salts + glucose, in which medium would you expect the longest lag phase? Why?

2. Name four factors which might decrease the duration of the logarithmic (log) phase in a given medium.

3. If an organism is inoculated into several flasks of similar broth, then shows a generation time of 45 minutes at 15°C, 42 minutes at 20°C, 38 minutes at 25°C, 21 minutes at 30°C, 32 minutes at 35°C, and no growth at 40°C, what is the optimum growth temperature? Why?

4. Why might a growth curve using a turbidometer give inconsistent results with pigment-producing organisms or with an endospore former?

EXERCISE 44
MICROBIAL ANTAGONISM

Populations of organisms are rarely found as pure cultures in nature. Instead, they are usually part of a community. Despite the many biological and non-biological stresses placed on a community, the populations within that community are usually very adept at adjusting to changes in their habitat. This ability to adjust to environmental change and maintain stability is termed **homeostasis.**

Both beneficial and harmful associations between populations contribute to the homeostasis of a community. The term **antagonism** refers to interactions in which one species is detrimentally affected by the actions of another species. Examples of antagonism are competition, amensalism (i.e., one population is supressed due to the presence of a toxin produced by another), parasitism, and predation.

This series of exercises is designed to illustrate some of these interrelationships that may occur in natural habitats.

A. COMPETITION

When organisms are allowed to grow in the laboratory in pure culture, they are freed from the pressures of having to compete with other species for nutrients or other limiting factors. The pure culture thus can achieve a growth rate dependent on its own genetic abilities and the given growth conditions. When, however, this same species is placed in an environment with the same nutrients, and another species is introduced

that also utilizes the nutrients, the two species may compete for resources. As a result, the growth rates of both species may be retarded.

Materials

Overnight broth cultures of *S. aureus* and of *E. coli* grown in nutrient broth + 1% glucose

250 ml Erlenmeyer flasks with 25 ml NB + 1% glucose

Shaking water bath at 37°C

1 ml pipets

9 ml and 9.9 ml dilution blanks

Eosin methylene blue (EMB) agar

Phenol red mannitol salt (PRMS) agar

Incubator set at 37°C

Procedure

First Period

1. Inoculate one 250 ml flask with 1 ml of the overnight culture of *E. coli* culture. Inoculate a second flask with 1 ml of the overnight culture of *S. aureus*. Inoculate a third flask with 1 ml of each of the cultures. (The class may be divided into groups, with each group being responsible for one flask.)

2. Remove 1 ml from each of the three flasks and make appropriate dilutions of each through 10^{-7} (fig. 44-1). Place the flasks in the 37°C shaking water bath.

3. From the dilution series containing the *E. coli*, plate

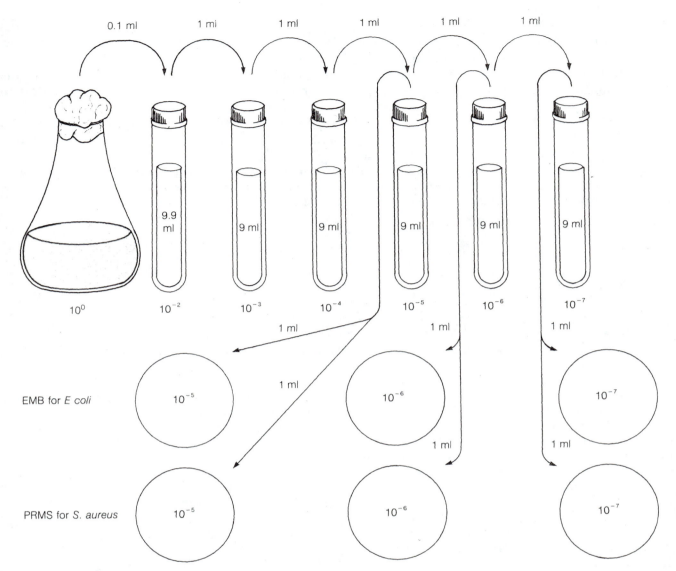

Figure 44-1. Serial Dilutions and Selective Plating for *E. coli* and *S. aureus*

to give final dilutions of 10^{-5} through 10^{-7}, using EMB agar.

4. Repeat with the dilution series containing *S. aureus*, but use PRMS agar.

5. For the dilution blanks containing the mixed culture, repeat as in steps 3 and 4 above, plating to each of the media.

6. Repeat steps 2 through 5 every 30 minutes for the duration of the period.

7. Incubate all plates at 37°C. Discard the flasks as indicated by your instructor.

Second Period

1. Arrange all the EMB plates from the *E. coli* flask according to time and dilution. For each of the times sampled, determine the number of *E. coli*/ml.

2. Arrange all the PRMS agar plates from the *S. aureus* flask as in step 1 above. Determine the number of these organisms/ml for the time periods tested.

3. Repeat steps 1 and 2 for the organisms growing in the mixed culture flask.

• The staphylococci are inhibited on the EMB due to the presence of the dyes eosin and methylene blue, while the *E. coli* are inhibited on the PRMS agar due to the high (7.5%) salt content of that medium.

4. Record the data. Plot graphs for each of the organisms as pure cultures and when growing competitively.

5. Discard all plates as indicated by your instructor.

Questions

1. What is the difference between a population and a community?

2. Given the same nutrients, what factors might be involved in a microbial habitat that may provide one population with a competitive advantage over another population if both are capable of using the same nutrients?

3. Why are eosin methylene blue and phenol red mannitol salt agars suitable media for the isolation and detection of *E. coli* and *S. aureus*?

B. AMENSALISM

In nature, many organisms inhabiting a specific habitat produce compounds that inhibit the growth of other organisms. In some cases a substance produced by one population of microorganisms may kill another population. For example, some antibiotics are waste products of microorganisms that may kill other microbes or impede their growth. A relationship where one organism inhibits or kills another (usually due to the release of a toxic material) is known as **amensalism.**

Materials

Broth cultures of *E. coli, Pseudomonas fluorescens, Staphylococcus epidermidis, Bacillus cereus,* and *Penicillium* sp.

Nutrient agar

1 ml pipets

Incubator set at 30°C

Procedure

First Period

1. Inoculate each of three tubes of melted and cooled nutrient agar with 1 ml of *E. coli*. Pour into three plates. Swirl gently. Allow to solidify.

2. Inoculate each of three additional tubes with 1 ml of *S. epidermidis*. Pour into three plates, swirl gently, and allow to solidify.

3. With an inoculating loop, place a streak of the pseudomonad down the middle of one plate containing the *E. coli* and a similar streak on one plate containing the *Staphylococcus*, as shown in fig. 44-2.

4. Repeat with a second pair of plates, streaking this time with the *Bacillus*.

5. Repeat with the third set, using the *Penicillium*.

6. Incubate the plates at 30°C for 2 to 7 days.

Second Period

1. Observe the plates for inhibition of the *E. coli* and *S. epidermidis* by the pseudomonad, bacillus, and *Penicillium*.

• Remember while making your observations that one of the control organisms is Gram-positive, the other Gram-negative.

2. Record your observations. Discard the plates as indicated by your instructor.

Questions

1. Anaerobic metabolism of molecules, such as amino acids and sugars, by microorganisms can yield end products that are toxic. What other end products of sugar or amino acid metabolism may be toxic to other microbes in the habitat?

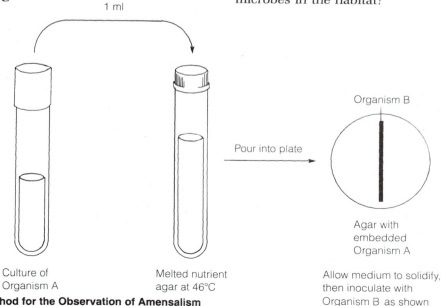

1 ml

Culture of Organism A

Melted nutrient agar at 46°C

Pour into plate

Organism B

Agar with embedded Organism A

Allow medium to solidify, then inoculate with Organism B as shown

Figure 44-2. Streak Method for the Observation of Amensalism

2. Why might inhibitory compounds produced by microorganisms affect other microorganisms that have a specific Gram reaction?

3. Would you consider oxygen evolution by microorganisms to be amensalistic? Explain.

4. Would you think that a compound produced by a population that exhibits an amensalistic reaction to a second population might also be stimulating to yet another population in the community? Explain.

EXERCISE 45
MICROBIAL SYNERGISM

Synergism is a relationship in which organisms benefit each other. In addition, the term describes changes that occur in a habitat when two or more species grow together, changes that none of the organisms could perform when growing alone in pure culture. In many habitats, natural selection favors synergistic organisms just as it may favor generalists over specialists with respect to ecological niches.

Examples of synergism are widespread in nature, and include more rapid cellulose digestion in mixed cultures, gas production by joint participation, and the combined synthesis of complex products.

Materials

Lactose fermentation broth with Durham tubes

Sucrose fermentation broth with Durham tubes

Broth cultures of *Staphylococcus aureus*, *Proteus vulgaris*, and *Escherichia coli*

Incubator set at 37°C

Procedure

First Period

1. Inoculate *E. coli* into one lactose and one sucrose fermentation tube. Inoculate *P. vulgaris* into a second set, and *S. aureus* into a third. These are the control tubes.

2. Inoculate both *E. coli* and *S. aureus* into a fourth set of tubes, *E. coli* and *P. vulgaris* into a fifth set, and *S. aureus* and *P. vulgaris* into the final set.

3. Incubate all tubes at 37°C for 24 hours.

Second Period

1. Examine each tube for acid and gas production, comparing the pure culture controls with the mixed culture tubes.

2. Determine which pair of organisms exhibited synergism. Discard all tubes as indicated by your instructor.

Questions

1. How might you explain the gas produced synergistically, when the organisms growing as pure cultures were incapable of gas production?

2. How might you explain accelerated cellulose degradation in a synergistic environment?

3. Why would you expect that organisms capable of synergism might be favored over their nonsynergistic counterparts in some habitats?

EXERCISE 46
MICROBIAL POPULATIONS IN SOIL

The concentrations and types of microorganisms vary greatly from one habitat to another. The predisposing abiotic factors that determine numbers and diversity include soil composition, pH, moisture content, and soil depth. Some soils, such as those with an acidic pH, are expected to have a higher ratio of fungi; rich garden soil may contain numerous actinomycetes; water logged soils may have a disproportionate number of anaerobes.

A. ISOLATING MICROBIAL POPULATIONS

It should be apparent that no single bacteriological medium can support the growth of all the diverse populations found in nature. A standard plate count is representative only of those organisms present in the sample and capable of growth on the medium used under the incubation conditions imposed. Hence, plates incubated aerobically would not include obligate anaerobes. Specific populations (e.g., nitrogen fixers, endospore formers, anaerobes) require special nutrients or conditions for their isolation.

Materials

Plate count agar

Sabouraud agar

Starch-casein agar (see Exercise 54)

Cyclohexamide stock solution (see Exercise 54)

80°C water bath

50°C water bath

1 ml pipets

9 ml dilution blanks

99 ml dilution blanks

Soil samples

Incubator set at 30°C

Procedure

First Period

(The instructor may wish to divide the class into groups, with each group responsible for one of the procedures.)

1. Add 11 grams of soil to 99 ml of sterile saline in a bottle. This is a 10^{-1} dilution. Make dilutions through 10^{-6} as indicated in fig. 46-1. Proceed as follows for the desired organisms.

    a. for total bacteria, place 1 ml on the plates of the 10^{-4} through 10^{-6} dilutions into duplicate plates. Add melted and cooled PCA. Swirl gently. Allow to solidify and incubate at 30°C for 48 hours.

    b. for fungi, place 1 ml of the 10^{-3} through 10^{-5} dilutions into duplicate plates. Add melted and cooled Sabouraud agar. Swirl gently. Allow to solidify and incubate at 30°C for 5 to 7 days.

    c. for actinomycetes, pour three plates of starch-casein agar. Add 2 ml aliquots from the 10^{-4} through 10^{-6} ml dilution blanks to three tubes respectively of melted and cooled starch-casein agar. Add 1 ml cyclohexamide solution to each tube. Roll the tubes between your palms to mix the contents. Pour onto the appropriately labeled plates of solidified starch-casein agar, swirl,

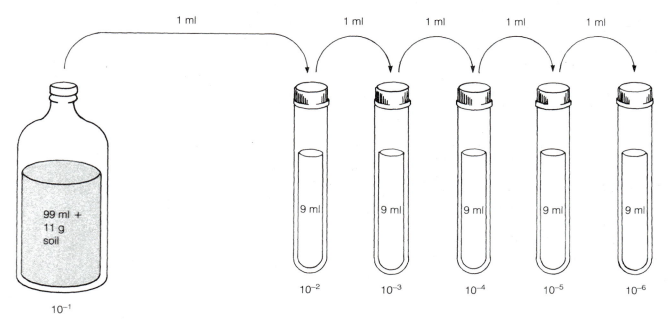

Figure 46-1. Dilution Series for the Enumeration of Microorganisms from Soil

and allow to solidify. Incubate at 30°C for 5 to 7 days.

    d. for endospores, heat the original 10^{-1} soil suspension at 80°C for 15 minutes (make sure the sample attains a temperature of 80°C before you begin timing). Make dilutions through 10^{-5} and place 1 ml of the 10^{-3} through 10^{-5} dilutions into labeled duplicate plates. Add melted and cooled PCA. Swirl gently. Allow to solidify and incubate at 30°C for 48 hours.

2. Discard all dilution blanks as indicated by your instructor.

Second Period

1. Arrange the plates labeled *total bacteria* in order of lowest to highest dilution. Determine the dilution that yielded between 30 and 300 colonies. Count the colonies on that plate and determine the aerobic plate count (APC)/g soil.

2. Repeat with those plates labeled *endospores*.

- As this sample was heated to 80°C for 15 minutes, the only growth on the plates should be due to those endospores *(not endospore formers)* that were present in the sample.

3. Determine the ratio of endospores to total bacteria in the soil sample. Discard these plates as indicated by your instructor.

Third Period

1. Arrange the starch-casein agar plates in order of lowest to highest dilution to count the actinomycetes.

- These organisms appear dull or chalky when covered with aerial hyphae. The colonies may also adhere tightly to the agar and have a leather-like texture. A magnification of $100 \times$ will verify hyphal growth. Fungi should be inhibited by the cyclohexamide, but if such fungal growth is suspected, it can be differentiated by the larger hyphal diameter of the fungi. Determine the number of actinomycetes/g soil.

- YOU MAY WISH TO KEEP THESE PLATES, IF ACTINOMYCETES ARE GROWING, FOR EXERCISE 49B.

2. Arrange the Sabourad agar plates in order of lowest to highest dilution. This medium, due to its high sugar and acidity, tends to inhibit bacterial growth. Determine the number of molds and yeasts/g soil.

- Yeast colonies, which appear similar to bacterial colonies, may be differentiated by simple stains.

3. Discard all plates (if desired).

Questions

1. Why is Sabouraud agar the preferred medium to use when isolating fungi?

2. What value does cyclohexamide have in the isolation of actinomycetes?

3. How can you differentiate between actinomycetes and filamentous fungi?

4. Does heating of the soil sample to 80°C for 15 minutes isolate endospores, or endospore formers, or both? Would fungal spores survive this temperature?

B. ISOLATION OF AN ENDOSPORE FORMER

Some genera of bacteria, most notably species of the genera *Bacillus* and *Clostridium*, undergo a life cycle that involves the formation of a heat-resistant, dormant structure called an **endospore.** This structure is produced when environmental conditions in the habitat have become detrimental for normal vegetative growth. Essentially, all the biochemical activities of the vegetative cell become centered on spore formation until synthesis of the endospore is completed. The cell may then rupture and liberate the endospore that contains the species' genome.

    The outer walls of the endospore are tightly bonded partially due to a high content of the amino acid cysteine that binds proteins together. The impervious outer wall, the dry core, and the presence of dipicolinic acid in the cortex all protect the species from heat, dessication, ultraviolet light, and toxic chemicals.

    Endospores are not very permeable to stains; hence, they cannot be stained by conventional methods. In order to stain an endospore, it must be heated. The heating alters the spore so that the stain can penetrate. The first stain used, containing malachite green, also stains vegetative cells. Since this dye does not form a strong bond with cellular material, it is easily washed from vegetative cells with water. Water will not, however, penetrate the walls of the endospore. Consequently, they retain the color of the malachite green.

    Endospore-forming bacteria are simple to isolate from any habitat. Samples need merely to be heated to a temperature sufficient to kill all vegetative organisms but not to destroy existing endospores.

Materials

Gram stain reagents and dyes

Malachite green

Plate count agar

Sporulating agar

Soil samples

9 ml dilution blanks

99 ml dilution blanks

80°C water bath

Incubator set at 30°C

Procedure

First Period

1. Proceed with the isolation and enumeration of endospores as described in exercise 46A.

Second Period

1. Enumerate the number of endospores/gram of soil sample.

2. Perform Gram stains on the predominating colonial types. If you observe Gram-positive rods, streak onto sporulating agar. Incubate at 30°C for 48 hours.

3. Discard the plates as indicated by your instructor.

Third Period

1. Perform endospore stains of the colonies on the sporulating agar.
- Note the position of the endospores within the cells that have not yet lysed.

2. Discard all plates as indicated by your instructor.

Questions

1. What Gram reaction do all known endospore formers share?

2. What, in one word, is the function of an endospore? How does this differ from the function of fungal spores?

3. Name three diseases of humans that are caused by endospore-formers.

4. What in the outer walls of the endospore is responsible for its impermeability?

C. ISOLATION OF AN ANTIBIOTIC PRODUCER

The soil contains a number of microorganisms that produce compounds that inhibit the growth of or kill other microorganisms (antibiotics). The mode of action of these antibiotics vary; some inhibit cell wall synthesis, others may affect either DNA, RNA, or protein synthesis or function. Some affect only prokary-

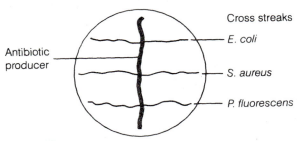

Figure 46-2. Cross-streak Method for the Isolation of Antibiotic Producers

otic cells, others only eukaryotes, and still other antibiotics affect both cell types. Some antibiotics may be used therapeutically or prophylactically in humans and other animals to treat disease; others are toxic and cannot be used.

Among the most common antibiotic producers in the soil are the actinomycetes such as *Streptomyces griseus* (streptomycin), the endospore-formers such as *Bacillus subtilis* (bacitracin), and the molds such as *Penicillium chrysogenum* (penicillin).

Materials

Streptomyces stock culture (or an isolate from exercise 46 or 54)

Stock cultures of *Escherichia coli, Staphylococcus aureus,* and *Pseudomonas fluorescens*

Plate count agar

Incubator set at 30°C

Procedure

First Period

1. Streak a single line of the actinomycete down the center of a plate of PCA. Incubate at 30°C until growth becomes evident (5 days or more). The slow-growing actinomycete must grow and have an opportunity to produce antibiotics before the plate is inoculated with the faster growing bacteria.

Second Period

1. After growth of the actinomycete has occurred, cross-streak with the test organisms as shown in figure 46-2.

2. Incubate the plate at 30°C for 24 to 48 hours.

Third Period

1. Observe the plate for inhibition of the test organisms by the actinomycete.
- Remember which of the organisms are Gram-positive and which are Gram-negative.

2. Discard all plates as indicated by your instructor.

Questions

1. What microorganisms are the major producers of antibiotics in the soil?

2. Name an antiobiotic that affects cell wall synthesis. Name one that affects protein synthesis. Name one that affects eukaryotes.

3. Are all antibiotics produced by soil microorganisms suitable for treating human diseases? Explain.

4. What is the difference between antibiotics and chemotherapeutic drugs?

EXERCISE 47
THE NITROGEN CYCLE

All organisms need nitrogen for the synthesis of proteins, nucleic acids, and other nitrogen-containing compounds. However, the biosphere we live in has a limited supply of nitrogen in the forms required by most living organisms. Furthermore, the various populations in the biosphere require their nitrogen in different forms.

It is apparent that, for the varied populations making up an ecological community, the available nitrogen must be recycled through its various usable forms.

There must be sufficient nitrogen present in the organic form (e.g., amino acids) to nourish those organisms requiring organic nitrogen. Similarly, there must also be sufficient nitrogen in forms such as ammonium, nitrite, or nitrate, to satisfy those organisms with inorganic needs. Given the limited amount of usable, available nitrogen, the organisms involved in cycling nitrogen through its various forms are critical to the survival of a particular ecological community. An illustration of the nitrogen cycle is given in figure 47-1.

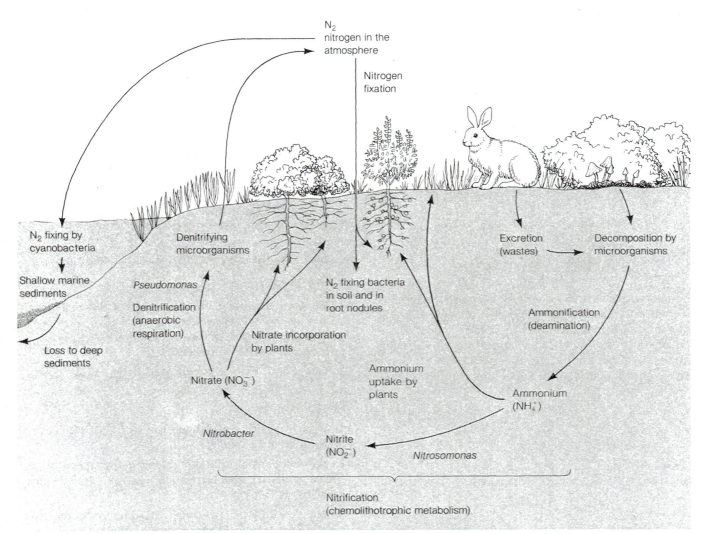

Figure 47-1. The Nitrogen Cycle. The nitrogen cycle is an important process that takes place in nature and insures an adequate supply of nitrogenous compounds to living organisms. The cycle involves a sequence of transformations of organic and inorganic nitrogen compounds. The transformations are due to the metabolic activities of microorganisms, plants, and animals.

A. AMMONIFICATION

Nitrogen in living organisms is primarily organic and is found in proteins and nucleic acids. When plants or animals die, these organic compounds become subject to microbial decomposition. Similarly, waste materials of living plants and animals may also contain nitrogenous organic material (e.g., urea) that also is subject to microbial activity.

In a given ecosystem in which organic material is present (e.g., hamburger, a compost pile), microbial decomposition can take place. The large organic macromolecules such as proteins and nucleic acids may be hydrolyzed by the decomposers in the ecosystem to amino acids, nucleosides, etc. These compounds may then be assimilated by decomposers and ammonium (NH_4^+) released into the habitat. The release of NH_4^+, performed by a wide variety of chemoheterotrophs, is termed **ammonification.** As plants may absorb NH_4^+ into the cells of growing tissues, but cannot take up nitrogen-containing organic molecules, the importance of these ammonifying microbes becomes apparent.

Materials

4% peptone broth

Garden soil

Stock cultures of *Escherichia coli* and *Pseudomonas fluorescens*

Nessler's reagent

Spot plates

Inoculating loops

Pasteur pipets and rubber bulbs

Incubator set at 30°C

Procedure

First Period

1. Inoculate three tubes of 4% peptone broth, one with a loop of *E. coli*, one with a loop of *P. fluorescens*, and the third with a pinch of soil.

2. Incubate the tubes at 30°C for 3 to 4 days.

Second Period

1. Place a few drops of Nessler's reagent into each of 4 depressions on a spot plate.

2. Using Pasteur pipets, add a few drops of culture medium from each of the three tubes of peptone broth into respective depressions containing the Nessler's reagent. Add a few drops of sterile 4% peptone to the fourth depression containing Nessler's reagent. This mixture serves as a control.

3. Observe the mixtures for a yellow color more intense than the control. A faint yellow color indicates a slight amount of ammonium (NH_4^+) as observed in the control. A deeper yellow is indicative of more NH_4^+, and a brownish precipitate of a larger amount.

4. Record the results for the soil and test organisms and discard the tubes as indicated by your instructor.

Questions

1. Would you expect ammonification to be performed by autotrophs or heterotrophs? Explain.

2. Why is 4% peptone a suitable medium to detect ammonification?

3. If an organism were capable of liberating ammonium in this medium, would it similarly be capable of liberating ammonium from a specific compound such as urea?

4. What organisms might use ammonium as a nitrogen source?

B. NITRIFICATION

Some of the NH_4^+ liberated into the ecosystem by ammonifiers is assimilated by plants and microbes as a source of their nitrogen for biosynthesis. A good proportion, however, of the NH_4^+ is used by specific chemoautotrophic bacteria as their energy source. This oxidation of NH_4^+ is a two-step process and is performed by two different groups of chemoautotrophic (chemolithotrophic) nitrifiers, but for the same purpose: to obtain energy.

The first step in nitrification is the oxidation of the NH_4^+ to nitrite (NO_2^-) by nitrifiers such as *Nitrosomonas*. The NO_2^-, liberated into the ecosystem as a waste product is then used by the second group of nitrifiers and is oxidized further to nitrate (NO_3^-). This reaction is performed by organisms such as *Nitrobacter*. The respective reactions can be depicted as follows:

$$\text{(a)} \quad NH_4^+ + 1\tfrac{1}{2}O_2 \rightarrow NO_2^- + 2H^+ + H_2O$$

$$\text{(b)} \quad NO_2^- + \tfrac{1}{2}O_2 \rightarrow NO_3^-$$

The NO_3^- can then be used by plants and some microbes as their source of nitrogen for assimilation into amino acids and other nitrogen-containing organic compounds.

Materials

Ammonium broth, 20 ml/6 oz prescription bottle

Nitrite broth, 20 ml/6 oz prescription bottle

Soil samples

Nessler's reagent

Spot plates

Trommsdorf's reagent

Diphenylamine

Concentrated sulfuric acid

Procedure

First Period

1. Inoculate a bottle of each medium with about 1 gram of soil.

2. Incubate the bottles on their sides for 1 week at room temperature.

Second Period

1. Place a drop of 1:3 sulfuric acid and three drops of Trommsdorf's reagent into a depression on a spot plate.

2. Pipet a drop of culture medium from the NH_4^+ broth into the depression. An intense blue-black color indicates the presence of NO_2^-.

3. Perform the Nessler's test as described in exercise 50A from the same bottle to test for residual NH_4^+.

4. To test for NO_3^-, check the NO_2^- broth for residual NO_2^- as described in steps 1 and 2. If no blue-black color appears, test for NO_3^-.

5. To test for NO_3^-, mix one drop of diphenylamine, 2 drops of concentrated sulfuric acid, and one drop of NO_2^- broth in a spot plate depression. A blue-black color is positive for NO_3^-.

6. Perform Gram stains of smears prepared from the culture medium.

7. Repeat steps 1 through 6 weekly for up to three weeks, then discard all materials as indicated by your instructor.

Questions

1. Name the principle genera of bacteria involved in nitrification.

2. Are these organisms autotrophs or heterotrophs?

3. Why are these organisms oxidizing the nitrogen compounds in question?

4. Could these reactions occur in soils made anaerobic from water saturation?

C. DENITRIFICATION

Nitrate (NO_3^-) is the preferred form of nitrogen for most plant life. In well-aerated and nourished soils, this form is available due to the activities of the ammonifiers and nitrifiers. However, if the ecosystem becomes deprived of available oxygen (due, for example, to heavy rainfall that saturates the soil), some bacteria will use NO_3^- molecules as a terminal electron acceptor. The net result of this activity, called **anaerobic respiration,** is the reduction of NO_3^- to atmospheric nitrogen (N_2). As this form of nitrogen is not usable by most living organisms, and not at all by plants and animals, this step represents a loss of available nitrogen to much of the ecosystem.

Some denitrifiers (e.g., some pseudomonads) respire aerobically in the presence of oxygen, and anaerobically (using NO_3^-) in the absence of oxygen. Other denitrifiers (e.g., some clostridia) are obligate anaerobes and use nitrate reduction as their sole means of eliminating electrons.

Not all denitrifiers reduce NO_3^- to N_2; some reduce NO_3^- only to NO_2^-, some to NH_4^+, and others to both of these latter forms.

Materials

Soil

Pseudomonas denitrificans stock culture

Escherichia coli stock culture

Nessler's reagent

Sulfamic acid

Zinc dust

Nitrate broth tubes with inverted Durham tubes

Incubator set at 30°C

Procedure

First Period

1. Inoculate three tubes of nitrate broth, one with *P. denitrificans*, one with *E. coli*, and the third with a pinch of soil.

2. Incubate the tubes at 30°C for 1 week.

Second Period

1. Test the tubes for nitrate reduction.
- DO NOT SHAKE THE TUBES. PROCEED IN THIS ORDER:

   a. *nitrate reduction to nitrogen gas.* Observe the Durham tube for appreciable gas formation. This would be indicative of nitrate reduction to atmospheric nitrogen. Discard the tube if positive. If negative, proceed to (b).

   b. *nitrate reduction to ammonium.* Remove a few drops of medium from the culture tube and perform the Nessler's test on a spot plate as described in exercise 50A. Continue with step (c).

c. *nitrate reduction to nitrite.* Add a few drops of sulfamic acid to the nitrate broth tube and shake the tube. The presence of intense, prolonged fizzing (due to the reduction of nitrite gas by the sulfamic acid) is considered a positive test. If (b) and (c) are negative, proceed to (d).

d. *no nitrate reduction.* Add a pinch of zinc dust to the nitrate broth and shake the tube. An intense, prolonged fizzing (due to the reduction of nitrate to nitrite by the zinc and the subsequent reduction of the nitrite to nitrogen gas by the previously added sulfamic acid) indicates the presence of nitrate.

2. Discard the tubes as indicated by your instructor.

Questions

1. What is the difference between aerobic and anaerobic respiration?

2. What is the difference between anaerobic respiration and fermentation?

3. Would you consider denitrification to be advantageous or detrimental to soil fertility? Why?

4. Are all denitrifiers obligate anaerobes?

5. Must all denitrifiers be chemoheterotrophs?

D. NITROGEN FIXATION

Almost 80% of the air we breathe is atmospheric nitrogen gas (N_2). While this provides our biosphere with abundant nitrogen, it is in a form that most living organisms cannot use. This is because most organisms lack the enzyme system—and hence the genetic ability—to reduce atmospheric nitrogen to the amino form ($-NH_2$) required for incorporation into organic molecules.

There are, however, a few species of prokaryotic organisms that can carry out this reaction. These organisms, called nitrogen fixers, can bring nitrogen from the atmosphere into the biological nitrogen cycle. The most common of these prokaryotes are the free-living bacterium *Azotobacter*, which fixes nitrogen in the soil; the free-living photosynthetic bacterium *Nostoc*, which performs similar reactions in fresh waters; and the symbiotic bacterium *Rhizobium*, which fixes nitrogen only when growing in the roots of leguminous plants such as sweet peas, alfalfa, clover, and lupines. *Rhizobium* cannot fix nitrogen when it is free-living in the soil, but must rely on fixed nitrogen.

Materials

Gram stain reagents

Legumes with root nodules

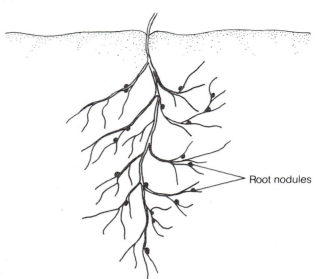

Figure 47-2. Root Nodules on a Legume

Azotobacter nitrogen free broth, 50 ml/250 ml Erlenmeyer flask

Forceps

Procedure

First Period

1. Locate the nodules on the roots of the legumes (see fig. 47-2). Using clean forceps, remove a nodule and place it between two microscope slides. Crush the nodule.

2. Allow the suspension on the slide to air dry. Heat fix and Gram stain. Observe for large, swollen, misshapen Gram-negative rods.

3. Inoculate the nitrogen-free broth with 1 gram of soil. Incubate the flask at room temperature for 1 week. DO NOT SHAKE THE FLASK.

Second Period

1. Without shaking it, examine the flask for growth and pellicle formation.

• *Azotobacter* is aerobic and can form a pellicle in order to remain on the surface of the broth.

2. Perform Gram stains from the surface and from the bottom of the broth.

• You may observe endospores in stains from the bottom of the sample (some clostridia can fix nitrogen). You may also notice encysted *Azotobacter* in the stained smears from surface material. As no fixed nitrogen was placed in the medium except for the slight amount added with the soil, the sole nitrogen source is atmospheric nitrogen. Hence growth, at least initially, is due to the presence of nitrogen fixers.

Questions

1. What is the difference between free-living nitrogen fixers and symbiotic nitrogen fixers?

2. How might you construct a medium to isolate a free-living nitrogen fixer such as *Azotobacter?*

3. Why might crop rotation designed to utilize legumes be advantageous to soil fertility?

4. Are there any organisms, other than a few prokaryotes, that are known to be nitrogen fixers?

5. Draw the life cycle of *Azotobacter.*

6. What is the function of *Azotobacter* cysts?

EXERCISE 48
THE SULFUR CYCLE

The sulfur cycle consists of sulfur transformations carried out by microorganisms. The steps of the sulfur cycle are found in figure 48-1. These steps should be compared to those of the nitrogen cycle.

When proteins are decomposed, the sulfur-containing amino acids methionine and cysteine are broken down and hydrogen sulfide (H_2S) may be released. The hydrogen sulfide may then be oxidized to sulfur (S) and sulfate (SO_4^{-2}) by chemoautotrophic bacteria, or used as a source of reducing power (hydrogens) by some types of photosynthetic bacteria. The liberated sulfate can then be assimilated by organisms requiring this form of sulfur, or reduced under anaerobic conditions by organisms using the SO_4^{-2} as a terminal electron acceptor.

A. THE WINOGRADSKY COLUMN

The Winogradsky column is a simple and economical means of growing a variety of soil and aquatic microorganisms. The column provides a gradient of oxygen, allowing for aerobic and anaerobic life. It was initially devised by Sergei Winogradsky to study a variety of soil organisms.

The Winogradsky column is, in essence, a small pond. Many photosynthetic bacteria inhabit the mud and anaerobic regions of the water where organic matter and sulfide are present and where infrared light, which is absorbed by the photosynthetic pigments of the organisms, can penetrate.

The bottom of the column provides microorganisms with a source of carbon (both organic, from cellulose, and inorganic, from carbonate) and sulfur. Cellulose-decomposing bacteria hydrolyze the cellulose to simple sugars that are fermented to release acids. The acids react with the carbonate to release carbon dioxide for the photosynthetic organisms. Other decomposition products such as citrate, pyruvate, and acetate are used by other microorganisms in the minipond. This column will be used in succeeding exercises to isolate and examine a variety of microorganisms involved in the sulfur cycle.

Materials

Graduated cylinder

Calcium sulfate

Bottom ooze (mud)

Garden soil

Decomposed plant material (e.g., leaf litter, grass clipping, paper toweling)

Pond, river, or stream water

Plastic wrap

Aluminum foil

Light with 60-watt bulb

Procedure

First Period

1. Mix 50 g calcium sulfate, 40 g of bottom ooze, and 10 g of decomposed plant material in a bucket.

2. Pack moderately well into the bottom ⅔ of a glass cylinder.

3. Cover with pond water to a depth of about 1 inch from the top of the cylinder. Cover the top with plastic wrap and a rubber band. Cover with aluminum foil so light cannot penetrate the cylinder (see fig. 48-2).

4. Place on a shelf at room temperature for 2 to 4 days.

Second Period

1. Remove the aluminum foil.

2. Expose the column continuously to light from a 60-watt bulb at a distance of 12 to 20 inches.

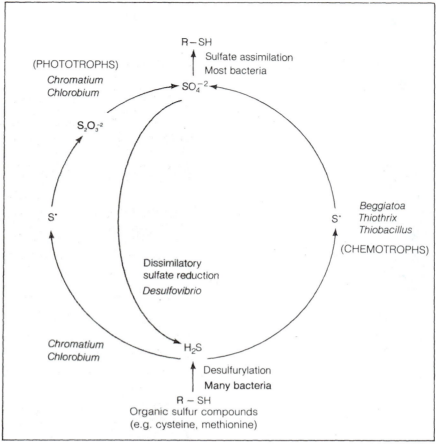

Figure 48-1. The Sulfur Cycle. The sulfur cycle is an important process that takes place in nature. It consists of a sequence of transformations of sulfur-containing compounds achieved by the metabolic activities of microorganisms.

3. Examine material from the column periodically, both macroscopically for the appearance of photosynthetic bacteria, and microscopically with hanging drops to observe the various organisms involved in the sulfur cycle.

4. The column may be used as a source of inoculum for the isolation of sulfur cycle organisms.

Questions

1. Why is calcium sulfate added to the column?

2. What is the function of the decomposing plant material in the column?

3. Why is the column kept in the dark for the first few days?

B. EXAMINATION OF PHOTOSYNTHETIC SULFUR BACTERIA

Observations on a Winogradsky column may reveal reddish-purple growth below the water-mud inter-face. You also may see a "bloom" of purple in a shallow pond. In both instances, what you are seeing is probably due to the growth of purple sulfur bacteria, microorganisms that grow in ponds that are rich in sulfur and in organic matter.

The organisms use H_2S as a source of reducing power to generate NADPH. This, together with ATP derived from radiant energy, can be used to fix CO_2, producing organic matter. Because photosynthesis as performed by these organisms is an anaerobic process, the bacteria must be at a level in the aquatic ecosystem where oxygen is limited. As this is at a level well below the surface of the water, light penetrating to that depth and not already absorbed by the surface algae is limited to wavelengths approaching the infrared (700 to 1100 nm). These are the wavelengths that are absorbed by the purple sulfur bacteriochlorophylls. These organisms are capable of storing sulfur granules as a reserve material for future oxidation and concomitant NADPH production. These sulfur granules are easily observed in wet mounts.

The purple sulfur bacteria belong to the family Chromatiaceae. The family, consisting of Gram-negative

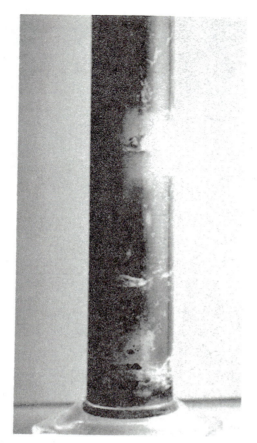

Figure 48-2. The Winogradsky Column

rods, cocci, and spiral-shaped bacteria, includes perhaps one of the largest bacterial species, *Chromatium okenii.*

Materials

Modified van Niel's medium in ground glass-stoppered bottles

Winogradsky column (or pond or river water)

Pasteur pipets with curved ends (the pipets may be flamed to obtain a curved tip)

Depression slides

Cover slips

Vaseline for hanging drops

Gram reagents

Light with 40-watt bulb

Procedure

First Period

1. With a curved Pasteur pipet, fish some of the reddish-purple growth from the Winogradsky column. If pond water is used, start with a 1 ml sample.

2. Inoculate the material into a bottle of van Niel's modified medium. Transfer 1 ml from this bottle to a second bottle. Mix well and transfer 1 ml from the second bottle to a third, and then from that to a fourth bottle. MAKE SURE ALL BOTTLES ARE COMPLETELY FULL OF MEDIUM. IF THEY ARE NOT, ADD SUFFICIENT MEDIA TO FILL.

3. Incubate on a countertop at room temperature near a 40-watt bulb.

Second Period

1. Periodically examine the bottles of medium for a period of 2 to 3 weeks. Should an obvious reddish cloudiness appear, perform hanging drops and Gram stains. Observe organisms in the hanging drops for motility, size, shape, and the presence of internal sulfur granules (which appear yellow).

2. Discard the bottles as indicated by your instructor.

Questions

1. Name the major groups of photosynthetic bacteria. Are they all oxygen-evolving organisms? Are they all involved in the sulfur cycle?

2. What purposes do the reduced sulfur serve for the photosynthetic sulfur bacteria?

3. What wavelengths of light do the anoxygenic photosynthetic bacteria absorb? What selective advantage does this provide?

C. ISOLATION OF *THIOBACILLUS*

The thiobacilli perform a series of steps in the sulfur cycle somewhat analagous to nitrification in the nitrogen cycle. Like the nitrifiers, the thiobacilli are chemoautotrophs (chemolithotrophs) and require an inorganic source of energy. The thiobacilli are capable, under aerobic conditions, of using reduced sulfides and elemental sulfur as energy sources, and liberating sulfates. Hence, more than one group of bacteria can oxidize reduced sulfides to sulfur and to sulfates; the sulfur purple bacteria under anaerobic conditions (exercise 48B) and the thiobacilli under aerobic conditions. The thiobacilli are found in nature in aerobic environments that contain sulfur or reduced sulfides (e.g., coal dust), and are capable of dropping the pH in some environments to less than 1 due to the production of excess sulfates.

Materials

Modified Starkey's medium, 4 ml/tube

Thiosulfate medium, pH 7 and 4.5, 25 ml/flask

Winogradsky column

Coal dust

Soil or pond water as a source of inoculum

Shaker

Procedure

First Period

1. Inoculate tubes containing modified Starkey's broth with 0.1 ml of material from the aerobic aquatic region of the Winogradsky column, a pinch of coal dust, a pinch of soil, or 0.1 ml pond water (or any desired combination of these).

2. Incubate 2 to 4 weeks at room temperature, observing the tubes during each laboratory period for color changes.

3. Inoculate the flasks of thiosulfate broth with either 1.0 ml from an aquatic source or 1.0 g of soil. If coal is available, a pinch of coal dust may be substituted. Inoculate replicate samples for both pH 7 and pH 4.5 media.

4. Incubate on a shaker at room temperature for 2 to 4 weeks, occasionally checking the pH to note changes.

Second Period

1. Look for growth in the Starkey's medium. This may take several forms, as given below:

a. Complete oxidation of sulfides to yield abundant acid (SO_4^{-2}) and possibly the accumulation of precipitated free sulfur. The reaction can be written as follows:

$$5Na_2S_2O_3 + H_2O + 4O_2 \rightarrow 5Na_2SO_4 + H_2SO_4 + S$$

b. Partial oxidation of the sulfides to neutral polythionates. This frees the Na ions to form NaOH and a strong alkaline reaction, which is noted by a blue color in the pH indicator. The reaction is as follows:

$$2Na_2S_2O_3 + H_2O + \frac{1}{2}O_2 \rightarrow Na_2S_4O_6 + 2NaOH$$

2. Prepare wet mounts and Gram stains using material from the tubes showing thiobacilli activity.

3. Check the pH of the thiosulfate media for changes. Also look for free elemental sulfur. Prepare Gram stains and wet mounts.

4. Discard all tubes and flasks as indicated by your instructor.

Questions

1. How might the thiobacilli contribute to acid mine waters?

2. Are the thiobacilli autotrophs or heterotrophs? Are they chemotrophs or phototrophs?

3. Do the thiobacilli store sulfur granules internally as a means of storing energy?

4. Why do the thiobacilli oxidize reduced sulfur compounds?

D. ISOLATION OF SULFATE REDUCERS

Sulfate reducers, such as the obligately anaerobic *Desulfovibrio* and *Desulfotomaculum*, are similar to nitrate reducers in that they respire anaerobically, using an external oxidized inorganic compound as a terminal electron acceptor. In this instance, however, the organisms use SO_4^{-2}, reducing it to H_2S.

The organisms are widely found in nature, in both fresh and marine waters and bottom ooze, in soils, and occasionally in foods such as fermenting olives. The liberated H_2S is odiferous and extremely toxic. In nature, when reduced iron is present, the liberated S atoms can combine with the metallic salts to produce ferrous sulfide (FeS), a black compound often responsible for the blackness of bottom ooze.

Materials

Desulfovibrio medium, screw-capped tubes

0.5% sterile ferrous ammonium sulfate

Winogradsky column

River or pond water or a soil suspension

9 ml dilution blanks

1 ml pipets

Procedure

First Period

1. Prepare dilutions of the mud from the Winogradsky column or the soil through 10^{-6}.

2. Melt tubes of *Desulfovibrio* medium (loosen the caps first) at 100°C to drive off the dissolved oxygen. Allow the tubes to cool to about 50°C.

3. While the medium is still liquid, transfer 1.0 ml aliquots from the appropriate dilutions to corresponding tubes of medium.

4. Pipet 1 ml of sterile ferrous ammonium sulfate to each of the tubes of medium. DO NOT SHAKE THE TUBES. Roll the tubes between the palms of your hands to mix the solutions.

5. Tighten the screw caps. Incubate the tubes at room temperature for 10 to 14 days, observing occasionally for distinct black colonies in the agar.

Second Period

1. Examine the tubes for black colonies embedded in the agar. They may be enumerated and the number present per ml or gram of the original sample may be calculated.

2. With an inoculating loop, remove a typical colony and perform a Gram stain on the organisms.

3. Discard all tubes as indicated by your instructor.

Questions

1. Why do the *Desulfovibrio* reduce sulfates?

2. What organisms comparable in energy metabolism are found involved in the nitrogen cycle?

3. In what area of the Winogradsky column might you expect to find *Desulfovibrio?*

4. How does sulfide liberation by this organism differ from sulfide liberation by other bacteria that metabolize the amino acid cysteine?

EXERCISE 49
LUMINESCENT BACTERIA

Some Gram-negative, polarly-flagellated marine bacteria, most notably species of the genera *Lucibacterium* (formerly *Beneckea*) and *Photobacterium,* emit light (bioluminescence) of a blue-green color. These organisms are widely distributed in marine waters and are often found associated with marine fish and cephalopods. Some marine animals even have special organs in which the **bioluminescent bacteria** grow.

Several compounds are required for bacterial bioluminescence, the enzyme luciferase, reduced flavin mononucleotide ($FMNH_2$), molecular oxygen, and a long chain saturated aldehyde. The bacteria will emit light as long as these requirements are present. The overall reaction may be written as follows:

$$FMNH_2 + O_2 + R\text{-}CHO$$
$$\downarrow \text{luciferase}$$
$$FMN + H_2O + R\text{-}COOH + hv,$$

where hv is the energy of the emitted photon.

An effective way to isolate the organisms is to place a dead marine fish in a pan with a small amount of salt water and incubate at 10° to 15°C for 2 to 3 days. The water should become faintly luminescent, and shortly thereafter, luminescent spots should appear on the fish. These may then be subcultured for isolation.

Materials

Dead marine fish (fresh or frozen)

Salt water

Incubator set at 10° to 15°C

Sterile 3% NaCl

Peptone-glycerol-phosphate-NaCl agar plates

Peptone-glycerol-phosphate-NaCl broth, 25ml/250ml flask

Test tubes

Inoculating loop

Procedure

First Period

1. Incubate a dead marine fish in a pan containing salt water at 10° to 15°C for 2 to 3 days. The fish should be lying in the water but not submerged, as the bioluminescent bacteria are aerobes.

Second Period

1. Examine the fish in a dark environment. Make sure your eyes become acclimatized to the dark. Look for the bioluminescence that should appear on the fish.

2. With a sterile inoculating loop, pick colonies and transfer them to tubes of 3% NaCl.

3. Streak from the 3% NaCl onto plates of peptone-glycerol-phosphate-salt agar.

4. Incubate the plates at 10° to 15°C for 2 to 4 days.

5. Inoculate a flask of broth from the 3% salt. Incubate as above.

Third Period

1. In the dark, examine the plates for bioluminescent colonies. There may be many colonies of competing nonbioluminescent bacteria on the plates, so a second transfer may be required.

2. Observe the flask for bioluminescence. Shake the flask. As the medium swirls, you can observe the light swirling also.

3. Discard all materials as indicated by your instructor.

Questions

1. Where in nature might you expect to find luminescent bacteria?

2. What other living organisms, often seen abundantly on warm summer nights, exhibit bioluminescence?

3. Is bioluminescence oxygen-dependent?

EXERCISE 50
PERIPHYTIC BACTERIA

It is often advantageous, especially in environments low in nutrients, for organisms to attach themselves to surfaces. The ability to attach is termed **periphytic;** slippery rocks in water are caused by periphytic organisms.

Nutrients tend to concentrate on surfaces and at interfaces. Consequently, periphytic microorganisms attaching to surfaces have a ready supply of nutrients in an environment which might otherwise not support their growth.

A variety of organisms are periphytic. These include the stalked bacterium *Caulobacter*, the sheathed bacterium *Sphaerotilus*, and the budding bacterium *Hyphomicrobium*, as well as encapsulated bacteria, diatoms, unicellular and filamentous algae, and protozoa. Representatives from all of these groups may be found on surfaces in aquatic habitats.

Materials

Small tank with pond or river water

Microscope slides

Test tube rack

Gentian violet, 1:50 in tap water

Procedure

First Period

1. Fill a small fish tank about ½ full with water from a pond or stream.

2. Arrange microscope slides in pairs, back to back,

in a test tube rack. Place the rack in the water so the slides are at least ⅔ submerged.

3. Keep the tank on a bench in the laboratory.

Second Period

1. After a five-day period, remove a pair of slides.

2. Let the slides air dry, then place in a beaker containing dilute gentian violet for about an hour.

3. Air dry and examine microscopically. Draw your observations. It is not necessary to identify the organisms.

Third Period

1. After another five-day period, remove a second pair of slides and observe as before. Note any changes in populations.

2. Discard all materials as indicated by your instructor.

Questions

1. Name three ways by which an organism might attach to a substrate or interface.

2. What advantages might periphytic microorganisms have in a nutritionally poor stream over their non-periphytic counterparts?

3. How might the ability to attach to a surface help strict aerobes growing in a test tube of broth? On wood shavings, in the production of vinegar?

4. Draw the life cycle of *Caulobacter*.

PART XII
WATER SANITARY ANALYSIS

A wide variety of microorganisms pathogenic to humans are transmitted by contaminated water. Usually these organisms are inhabitants of the intestinal tract and are introduced into **potable** (drinking) water by fecal contamination. These microorganisms, as well as others capable of causing skin, eye, ear, and throat infections, may also be transmitted by contaminated swimming or bathing waters.

Among the many microorganisms of fecal origin which cause water-borne disease outbreaks, *Salmonella typhi* (typhoid fever), *Shigella* spp. (shigellosis), and *Salmonella paratyphi* (salmonellosis) have been the most frequently encountered in the last three decades. *Vibrio cholerae* (cholera), *Campylobacter jejuni* (dysentery), and *Escherichia coli* (dysentery) are water-borne bacterial pathogens occasionally responsible for disease in the United States. In addition, viruses such as Hepatitis A virus (infectious hepatitis) and poliovirus (poliomyelitis), as well as protozoa such as *Entamoeba histolytica* (amoebic dysentary) and *Giardia lamblia* (giardiasis), have caused outbreaks due to consumption of waters contaminated with fecal pollutants.

Organisms that may cause disease to swimmers and bathers by entering the body through the mucous membranes or breaks in the skin include *Staphylococcus aureus, Pseudomonas aeruginosa,* and *Leptospira* spp., as well as the protozoan *Naeglaria fowleri* (primary amoebic meningoencephalitis) and the trematode *Schistosoma* spp. (swimmer's itch).

In order to ensure that waters are safe for consumption and recreational use, it is necessary to establish the absence of these organisms and related pathogens. Given the diversity of the microbial pathogens in question and the sporadic nature of their occurrence in aquatic environs, examination for their detection would be tedious, time-consuming, and costly. Hence, **indicator organisms,** organisms which indicate the possible presence of pathogens, are used to determine the sanitary quality of waters. For potable and recreational waters, these indicators are a group of bacteria known as **coliforms;** in recreational waters, *S. aureus, Streptococcus mitus-salivarius,* and/or *P. aeruginosa* may also be used—in this instance, to indicate agents of potential skin, eye, ear, or throat infections.

A complete analysis of methods and media required for the detection of these organisms may be found in *Standard Methods for the Examination of Water and Wastewater* (American Public Health Association, 15th ed.).

EXERCISE 51
MOST PROBABLE NUMBER METHOD
FOR COLIFORM ANALYSIS

In the **most probable number** (MPN) test for water quality, coliforms are defined as aerobic to facultatively anaerobic, Gram-negative, asporogenous rod-shaped bacteria that ferment lactose with the production of acid and gas within 48 hours at 35°C. These can be further separated into coliforms of fecal origin (fecal coliforms), which are capable of producing gas in E.C. broth within 24 hours at 44.5°C; and nonfecal coliforms (those originating from soils and vegetable matter).

The MPN test is a multiple tube test employing media designed to enrich and select for coliforms in order to establish an index of the number of coliform bacteria in the sample to be tested. It is not an actual enumeration. The test is initiated by inoculating 10 ml aliquots of the sample into double strength lauryl tryptose broth. This is the **presumptive test,** and any tube showing gas after the required incubation is presumed to contain coliforms. Positive tests are often referred to as "blows" due to the presence of gas in the Durham tubes.

Should any of the presumptive tubes "blow," they must be subjected to a **confirmed test** and, if desired, a **fecal coliform test.** These tests are performed to determine if the blow in the presumptive test was indeed due to coliforms (and not the synergistic action of several bacterial species) and to determine if the coliforms are of fecal origin. For the confirmed test, a tube of brilliant green bile lactose (BGBL) broth is inoculated with a loopful of culture medium from the positive presumptive tube and incubated at 35°C for up to 48 hours. This is performed for all positive presumptive tubes. If it is desirable to test also for fecal coliforms, tubes of E.C. broth are also inoculated, but are incubated at 44.5° ± 0.1°C for 24 hours. Gas production in any of these tubes constitutes a positive reaction, as these media and incubation conditions are highly selective for coliforms and fecal coliforms respectively. The appropriate MPN tables (table 51-1) are then consulted to determine the index of coliform density. The 1975 United States Environmental Protection Agency Drinking Water Standards have established an MPN of less than 2.2 coliforms/100 ml as an indication that the water sample in question meets safe drinking standards.

When necessary, a **completed test** may also be performed, using inocula from positive confirmed tubes.

The procedure for complete coliform MPN analysis is shown in figure 51-1.

Materials

Potable water sample containing approximately 10 coliforms/ml

Double strength lauryl tryptose broth (2×) with inverted Durham tubes

Brilliant green bile lactose broth with inverted Durham tubes

E.C. broth with inverted Durham tubes

Eosin methylene blue agar plates

Nutrient agar slants

Gram staining reagents

Incubator set at 35°C

Water bath set at 44.5°C

Slant cultures of *E. coli* as controls

Inoculating loop

Procedure

First Period

1. Pipet 10.0 ml aliquots of the potable water sample into each of 5 tubes of 2x lauryl tryptose broth.

2. Inoculate a separate tube of lauryl tryptose with the provided *E. coli* culture. This will be a positive control. Perform each succeeding test with this control.

3. Incubate the tubes at 35°C for 48 hours.

TABLE 51-1
MPN INDEX AND 95% CONFIDENCE LIMITS FOR VARIOUS COMBINATIONS OF POSITIVE AND NEGATIVE RESULTS WHEN FIVE 10 ml PORTIONS ARE USED.

NO. OF TUBES GIVING POSITIVE REACTION OUT OF 5 OF 10 ml EACH	MPN INDEX FOR 100 ml	95% CONFIDENCE LIMITS	
		LOWER	UPPER
0	< 2.2	0	6.0
1	2.2	0.1	12.6
2	5.1	0.5	19.2
3	9.2	1.6	29.4
4	16.0	3.3	52.9
5	>16.0	8.0	Infinite

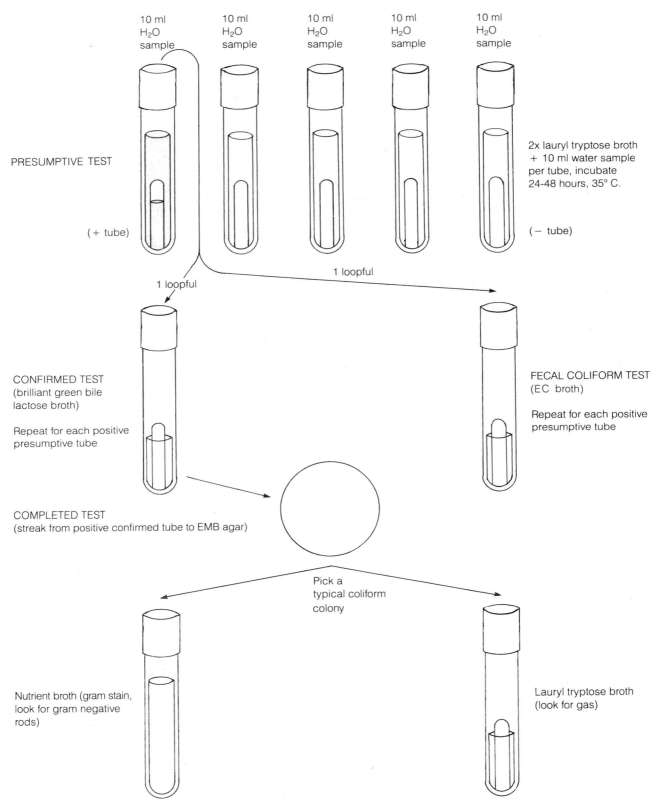

10 ml H₂O sample · 10 ml H₂O sample · 10 ml H₂O sample · 10 ml H₂O sample · 10 ml H₂O sample

PRESUMPTIVE TEST

2x lauryl tryptose broth + 10 ml water sample per tube, incubate 24-48 hours, 35° C.

(+ tube)

(− tube)

1 loopful

1 loopful

CONFIRMED TEST
(brilliant green bile lactose broth)

Repeat for each positive presumptive tube

FECAL COLIFORM TEST
(EC broth)

Repeat for each positive presumptive tube

COMPLETED TEST
(streak from positive confirmed tube to EMB agar)

Pick a typical coliform colony

Nutrient broth (gram stain, look for gram negative rods)

Lauryl tryptose broth (look for gas)

Figure 51-1. The Most Probable Number Method. A standard analysis of water consists of the presumptive test, the confirmed test, and the completed test. The most probable number (MPN) can be determined from the results of the confirmed test or fecal coliform test.

1. Observe the tubes of 2x lauryl tryptose. This medium enriches for coliforms. If any were present in each 10 ml sample added to the tubes, a positive reaction should have resulted. Discard all negative tubes.

2. Obtain one tube of brilliant green bile lactose broth and one tube of E.C. broth for each presumptive blow.

• These media and the incubation conditions are selective for coliforms and fecal coliforms respectively.

3. Using an inoculating loop, inoculate one set of each of these tubes with a loop of culture media from each respective presumptive blow. Remember to include the control.

4. Incubate the BGBL broth tubes at 35°C for 48 hours. Observe for gas.

5. Incubate the E.C. broth tubes in a 44.5 ± 0.1°C water bath for 24 hours. Observe for gas.

Third Period

1. From a positive BGBL tube, streak a plate of EMB agar. Incubate at 35°C for 24 hours.

2. Refer to the MPN table (table 51-1) for the fecal coliform and total coliform densities of the water sample tested, based on the number of positive BGBL and EC broth tubes respectively. Discard all tubes as indicated by your instructor.

Fourth Period

1. Observe the EMB agar plate for typical (nucleated, with or without metallic green sheen), atypical (opaque,

unnucleated, mucoid, pink), or negative (all others) colonies (see Appendix D).

2. Pick a typical colony (if none are present, use an atypical). Transfer to a nutrient agar slant and to a tube of lauryl tryptose broth. Incubate at 35°C for 24 hours.

3. Discard the plates

Fifth Period

1. Gram stain growth from the nutrient agar slant. Gram-negative rods should be observed since coliforms are Gram-negative.

2. Observe the lauryl tryptose tubes for gas. It should be observed.

3. Discard all tubes as indicated by your instructor.

Questions

1. What is the purpose of lauryl tryptose broth in the MPN tests?

2. What are coliforms?

3. Why, in the case of positive presumptive tests, might it be of interest to a water company to know if the coliforms are from fecal material instead of from soil?

4. How can you distinguish between fecal and nonfecal coliforms?

5. How does brilliant green bile lactose broth indicate coliform growth?

EXERCISE 52
MEMBRANE FILTER METHOD
FOR COLIFORM ANALYSIS

The **membrane filter** (MF) method provides a more rapid coliform analysis than the MPN method. However, this test should not be used to detect coliforms in waters high in turbidity or in noncoliform bacteria.

Media are available for the detection and enumeration of both total and fecal coliforms. The United States Environmental Protection Agency Drinking Water Standards for the MF method recommends, for potable waters, a quality limit of 1 coliform/100 ml and an action limit of 4 coliforms/100 ml. An **action limit**

means the water company must take immediate action to remedy the problems that are responsible for the presence of the coliforms.

Materials

Membrane filter apparatus (fig. 52-1)

Forceps

95% alcohol

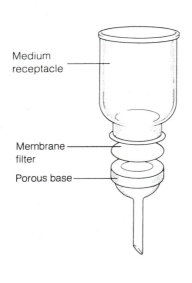

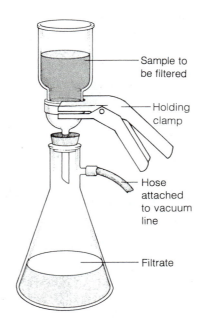

Figure 52-1. Filtration. Water samples are often checked for microorganisms by filtering. The waters are passed through a filter with a pore size of 0.47 μm that allows the passage of water but not of bacteria or larger organisms. The microorganisms stuck on the filters can be detected by placing the filter on a selective and differential agar medium. The resulting colonies can then be counted and identified.

Culture dishes, 60 × 15 mm or 50 × 12 mm

Membrane filters, 0.47 μm pore size

Absorbent pads, 48 mm diameter

Water bath, 44.5°C

Incubator, 35°C

m-Endo broth for total coliforms, 1.8 to 2.0 ml/tube

m-FC broth for fecal coliforms, 1.8 to 2.0 ml/tube

Potable water sample with approximately 10 to 50 coliforms/100 ml

Procedure

First Period

1. Using alcohol-sterilized forceps, place a sterile filter pad over the porous base of the filter apparatus. Clamp the medium receptacle in place.

2. Under partial vacuum, filter a 100 ml volume of water. Rinse the medium receptacle by filtering three 20 to 30 ml aliquots of sterile water to wash onto the filter any remaining organisms which may have become attached to the sides.

3. Pour a tube of m-Endo broth (containing 1.8 to 2.0 ml broth) into a 60 × 15 mm Petri dish. Repeat with m-FC broth into a fresh plate.

4. To each plate, aseptically add a sterile absorbent pad, using the alcohol-sterilized forceps.

5. Remove the clamp and the medium receptacle from the filter apparatus. Alcohol-sterilize the forceps and gently remove the filter pad. Place it on the absorbent pad in the Petri dish containing m-Endo broth. Avoid entrapment of air between the absorbent pad and the filter pad.

6. Repeat the process with m-FC broth and a second 100 ml aliquot of the water sample.

7. Incubate the m-Endo plates at 35°C for 24 hours.

8. Place the m-FC plates in a waterproof plastic bag, seal, and submerge in a 44.5°C water bath for 24 hours.

Second Period

1. Observe the m-Endo plates for typical colonies and enumerate.

• Typical coliform colonies will be pink to dark red with a metallic sheen (see Appendix D).

2. Count the number of blue colonies on the m-FC plates. These are typical fecal coliform colonies (see Appendix D).

3. Discard all plates as indicated by your instructor.

Questions

1. What advantages does the MF test have over the MPN? What disadvantages?

2. How is m-Endo broth effective in the isolation and detection of coliforms?

3. What is the function of the sterile absorbent pad in the Petri dish? What alternative to the absorbent pad might you use?

4. Why are the fecal plates with m-FC broth placed in waterproof bags in a water bath instead of in an air incubator?

EXERCISE 53
NONCOLIFORM INDICATORS OF RECREATIONAL WATER QUALITY

Swimmers and bathers may be subjected to a variety of potentially pathogenic microbes in addition to those of fecal origin. The absence of coliforms, either fecal or nonfecal, from swimming pools, spas, hot tubs, etc., does not preclude the possible presence of organisms capable of causing eye, ear, skin, throat, or other infections resulting from the passage of microorganisms through skin pores or cuts or through mucous membranes.

Unfortunately, no established, reliable indicator organisms are available to reflect possible danger to swimmers and bathers. Hence, routine analysis of recreational waters is usually restricted to coliform analysis and, on occasion, total plate counts.

The standard plate count is performed as an indicator of disinfection efficiency for recreational waters that are routinely disinfected. However, because a large percentage of swimming pool-associated illnesses are due to *Streptococcus mitis-salivarius* (found in saliva and sinus drainage), *Staphylococcus aureus* (a skin and upper respiratory tract pathogen), and *Pseudomonas aeruginosa* (the causative agent of many eye and ear infections), these organisms have been used as auxilliary indicators, supporting the coliform and standard plate counts. This exercise will examine the methods for the isolation of *S. aureus* and *P. aeruginosa*.

Materials

Membrane filter apparatus

Forceps

95% alcohol

Culture dishes, 60 × 15 mm or 50 × 12 mm

Membrane filters, 0.47 μm pore size

Absorbent pads, 48 mm diameter

m-*Staphylococcus* broth, 1.8 to 2.0 ml/tube

Double strength asparagine broth (2×)

Single strength asparagine broth (1×)

Black light (Woods light)

Acetamide broth

Water sample with 50 to 100 *P. aeruginosa* and 50 to 100 *S. aureus*/ml

Gram staining reagents

3% hydrogen peroxide

Slant culture of *P. aeruginosa* as a control

Incubator set at 35°C

Procedure

First Period

1. Pipet 10 ml aliquots of sample water into each of 5 tubes of 2× asparagine broth.

2. Pipet 1.0 ml aliquots into each of 5 tubes of 1× asparagine broth.

3. Pipet 0.1 ml aliquots into each of 5 tubes of 1× asparagine broth.

4. Incubate the 15 tubes at 35°C for 24 hours. This is the MPN test for *P. aeruginosa*.

5. Inoculate a tube of 1× asparagine broth with a loop of the control culture of *P. aeruginosa*. Use this control throughout the exercise.

6. Filter a 100 ml aliquot of sample water through a membrane filter. Rinse the medium receptacle several times with 25 to 30 ml sterile water to wash onto the filter any remaining organisms.

7. Aseptically place the filter into a 60 × 15 mm (or 50 × 12 mm) culture dish containing 1.8 to 2.0 ml m-staphylococci broth and an absorbent pad. Incubate at 35°C for 24 hours. This is an MF test for *S. aureus*.

Second Period

1. Examine the tubes of asparagine broth for presumptive positive tubes by exposing the tubes to the

long-wave ultraviolet light of a black light.

• A greenish fluorescence, due to metabolic byproducts of many pseudomonads, constitutes a positive test. **DO NOT LOOK DIRECTLY INTO THE LIGHT.**

2. Transfer a loop of culture medium from each positive presumptive tube into respective tubes of acetamide broth. Make sure the tubes are properly labeled to indicate the source of the inoculum (i.e., from the presumptive tubes initially inoculated with 10.0, 1.0, or 0.1 ml). Discard all presumptive tubes. Incubate 24 to 36 hours at 37°C. This will determine if the positive presumptive tube was due to *P. aeruginosa* or to other fluorescent pseudomonads.

3. Examine the m-*Staphylococcus* plates. Count all colonies except those that are extremely mucoid or rough (these are probably spore-formers).

• The high (7.5%) NaCl concentration of this medium makes it highly selective for *S. aureus* colonies, which are typically golden yellow in color (see Appendix D).

4. Perform a Gram stain and a catalase test on a staphylococcal colony. You should observe catalase-positive, Gram-positive cocci in clusters.

5. Determine the number of staphylococci/100 ml water sample.

6. Discard the plates as indicated by your instructor.

Third Period

1. Examine the tubes of acetamide broth.

• A positive reaction is the development of a purple red color (see Appendix D).

2. Determine the *P. aeruginosa* MPN/100 ml by consulting the appropriate MPN tables (table 53-1).

Questions

1. Why are coliforms alone not indicative of the sanitary quality of swimming pools or spas?

2. How can you detect the presence of *P. aeruginosa* in the presumptive media?

TABLE 53-1.
MPN INDEX AND 95% CONFIDENCE LIMITS FOR VARIOUS COMBINATIONS OF POSITIVE RESULTS WHEN FIVE TUBES ARE USED PER DILUTION (10 ml, 1.0 ml, 0.1 ml)

COMBINATION OF POSITIVES	MPN INDEX /100 ml	95% CONFIDENCE LIMITS LOWER	95% CONFIDENCE LIMITS UPPER	COMBINATION OF POSITIVES	MPN INDEX /100 ml	95% CONFIDENCE LIMITS LOWER	95% CONFIDENCE LIMITS UPPER
0-0-0	< 2	0	6	4-2-0	22	7	67
0-0-1	2	<0.5	7	4-2-1	26	9	78
0-1-0	2	<0.5	7	4-3-0	27	9	80
0-2-0	4	<0.5	11	4-3-1	33	11	93
				4-4-0	34	12	93
1-0-0	2	<0.5	7	5-0-0	23	7	70
1-0-1	4	<0.5	11	5-0-1	31	11	89
1-1-0	4	<0.5	11	5-0-2	43	15	110
1-1-1	6	<0.5	15	5-1-0	33	11	93
1-2-0	6	<0.5		5-1-1	46	16	120
				5-1-2	63	21	150
2-0-0	5	<0.5	13	5-2-0	49	17	130
2-0-1	7	1	17	5-2-1	70	23	170
2-1-0	7	1	17	5-2-2	94	28	220
2-1-1	9	2	21	5-3-0	79	25	190
2-2-0	9	2	21	5-3-1	110	31	250
2-3-0	12	3	28	5-3-2	140	37	340
3-0-0	8	1	19	5-3-3	180	44	500
3-0-1	11	2	25	5-4-0	130	35	300
3-1-0	11	2	25	5-4-1	170	43	490
3-1-1	14	4	34	5-4-2	220	57	700
3-2-0	14	4	34	5-4-3	280	90	850
3-2-1	17	5	46	5-4-4	350	120	1,000
4-0-0	13	3	31	5-5-0	240	68	750
4-0-1	17	5	46	5-5-1	350	120	1,000
4-1-0	17	5	46	5-5-2	540	180	1,400
4-1-1	21	7	63	5-5-3	920	300	3,200
4-1-2	26	9	78	5-5-4	1,600	640	5,800
				5-5-5	≥2,400	—	—

3. Would proper chlorination or iodination of swimming pools, spas, or hot tubs be effective against pathogens such as *P. aeruginosa* or *S. aureus*?

4. Name some other pathogens (or diseases they cause) that may be transmitted by swimming or bathing in contaminated waters.

EXERCISE 54
ISOLATION OF ACTINOMYCETES FROM WATER AND WASTEWATER TREATMENT PLANTS

Actinomycetes are a group of filamentous bacteria, distinguished from other bacteria by their branching filaments. Individual cells range from 0.5 to 2.0 μm in diameter. The cells may develop as nonseptated hyphae. The hyphae usually form cross walls and fragment into individual cells, with each cell then capable of forming branching filaments.

These prokaryotic organisms, which resemble fungi in morphology and spore formation, are of concern in water and wastewater treatment plants. In the former, where surface or well water is treated to yield potable water, the actinomycetes can cause taste and odor problems. In wastewater treatment plants, particularly those secondary plants known as activated sludge plants, the actinomycetes can cause a serious problem known as **bulking,** or poor settling of flocculent material in the treated water.

Among the various actinomycetes causing taste or odor problems (usually an earthy-musty odor), the genus most commonly encountered is *Streptomyces*, a common soil bacterium. In wastewater treatment plants where bulking is a problem, the genus *Nocardia* may be encountered.

The actinomycetes are a nutritionally diverse group of bacteria. No single medium will support the growth of all of these organisms. Most media used for the selection of these organisms are nutritionally poor, in order to inhibit the growth of other organisms.

Materials

Starch-casein agar, 15 ml/tube and 5 ml/tube

Sterile buffered dilution water, 99 ml/bottle and 9 ml/tube

Cyclohexamide stock solution (actidione)

Water sample from aeration basin of activated sludge plant, or garden soil

Incubator set at 28°C

Water bath set at 50°C

Procedure

First Period

1. If treated wastewater is used, prepare dilutions through 10^{-3} (see figure 54-1). If garden soil is used as a substitute, prepare dilutions through 10^{-6}.

2. Pour three plates of starch casein agar and allow them to solidify.

3. Melt three tubes of starch-casein agar (5 ml agar/tube). Allow the tubes to cool to approximately 46°C to 50°C.

4. Add to the melted and cooled tubes of agar 2 ml aliquots from the 10^{-1}, 10^{-2}, and 10^{-3} dilutions of wastewater, respectively (10^{-4}, 10^{-5} and 10^{-6} if garden soil is used). Then add 1.0 ml of cyclohexamide solution to each tube. Mix well.

5. Pour the contents of each tube onto labeled starch-casein agar plates from step 2. Swirl gently.

6. Invert and incubate the plates at 28°C for 7 days.

Second Period

1. Examine the plates for colonial appearance. Actinomycete colonies will be dull or chalky when covered with aerial hyphae. Colonies may adhere tightly to the agar and have a leatherlike texture. A magnification of $100\times$ will verify the filamentous growth.

- The cyclohexamide should inhibit fungal growth. Such growth, if it occurs, can be differentiated from actinomycete growth by the larger hyphal diameter of the fungi.

Questions

1. How are actinomycetes responsible for bulking in activated sludge sewage treatment plants?

2. Why is cyclohexamide added to the medium?

3. How can you differentiate between actinomycetes and filamentous fungi?

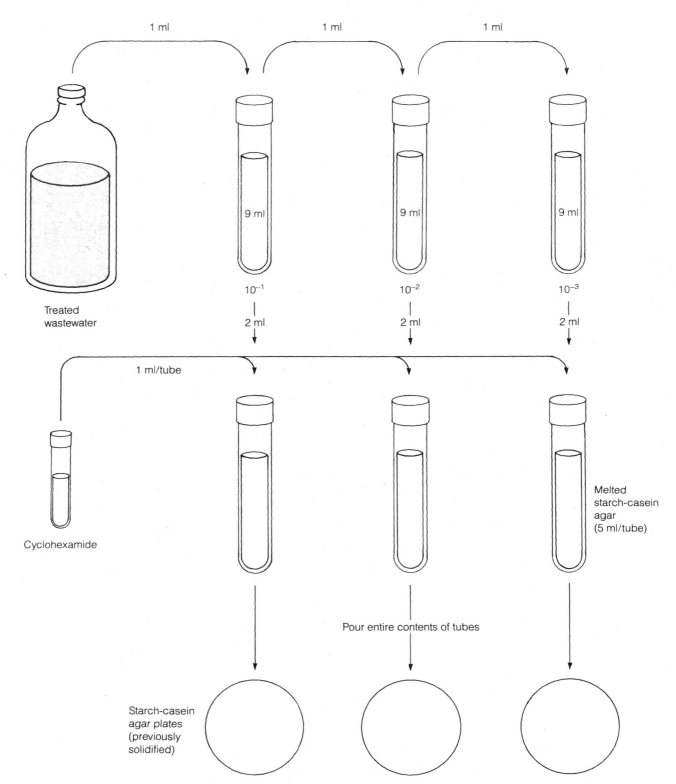

1 ml　　　　　　　1 ml　　　　　　　1 ml

9 ml　　　9 ml　　　9 ml

10^{-1}　　　10^{-2}　　　10^{-3}

2 ml　　　2 ml　　　2 ml

Treated
wastewater

1 ml/tube

Cyclohexamide

Melted
starch-casein
agar
(5 ml/tube)

Pour entire contents of tubes

Starch-casein
agar plates
(previously
solidified)

**Figure 54-1. Serial Dilutions and Plating Format for the Isolation
and Enumeration of Actinomycetes**

PART XIII
FOOD MICROBIOLOGY

The foods we consume are rarely sterile, unless they are cooked or treated in some manner to eliminate contaminating organisms. In many instances, these microorganisms have been introduced into the foods from the soil or from food handlers and equipment involved in food processing. In some cases, specific microorganisms are purposely added to a raw foodstuff as part of a process for the production of an edible food product.

Regardless of the source of the microorganisms or the reason for their being in the food, the microbes metabolize constituents of that food and excrete waste products. Some of the waste products may make food unpalatable, and consequently the food is considered to be spoiled. The microbes responsible for these changes are known as **food spoilage** organisms. Waste products from other contaminating organisms (or the organisms themselves) may not alter the taste or odor of the food in question, but may cause illness in the consumer. These organisms are responsible for **food poisonings** or **food intoxications.** In yet other instances, the waste products of the microorganisms may be desirable, and the organisms responsibile are considered to be involved in **food production.** Finally, there are organisms whose presence indicates the sanitary quality of the foods. These are the **indicator bacteria.**

In the following exercises, you will be introduced to some of the organisms involved in food spoilage, food poisonings and intoxications, food production, and sanitary analysis of foods. For a detailed review of all the available methods for food analysis, refer to *Compendium of Methods for the Microbiological Examination of Foods*, prepared by the American Public Health Association (Marvin L. Speck, ed.).

EXERCISE 55
FOOD SPOILAGE

The physical and chemical composition of foods differs greatly from one foodstuff to another. Imagine, for instance, the difference in texture and nutritional components of apple sauce as compared to hamburger. The organisms involved in food spoilage will tend to be those that are best able to use the molecules making up the foodstuff and that produce unpalatable waste products. Depending on the food in question, these organisms may be **lipolytic, proteolytic,** or **pectinolytic.** They may be, if salt or sugar is added as preservatives, **halophilic** or **saccharophilic,** respectively. If heat was used in prior treatment of the food, the surviving organisms may also be **thermophilic, thermoduric,** or **endosporogenous.**

Standard plate counts (SPC) are routinely performed in the food industry to determine the numbers of organisms in food. Obviously, the greater the number of spoilage organisms in the food, the more rapidly that food may deteriorate in quality. Standard plate counts may also be performed to enumerate specific spoilage organisms (e.g., proteolytics, flat-sour endospores, etc.) with the use of selective, differential, or selective-differential media.

A. ENUMERATION OF FOOD SPOILAGE ORGANISMS

The standard plate count is often referred to as the **aerobic plate count (APC).** This method is used to enumerate the number of bacteria in a variety of ecosystems. The method will be used here to count the number of bacteria per gram of hamburger. You should understand that only those microbes found in hamburger and which also are capable of growth on the medium and under the incubation conditions employed can be enumerated.

Materials

Hamburger sample, diluted 1:10

Plate count agar (PCA)

99 ml dilution blanks

9 ml dilution blanks

Sterile 1 ml pipets

Sterile Petri plates

Incubator set at 35°C

Procedure

First Period

1. Obtain a hamburger sample previously diluted 1:10 by blending 50 g hamburger with 450 ml sterile diluent.

2. Make dilutions through 10^{-5} as indicated in figure 55-1.

3. Label, in duplicate, sterile Petri plates 10^{-3} through 10^{-6}.

4. Aseptically pipet aliquots from the appropriate dilution blanks to the Petri plates as shown in figure 55-1.

5. Add melted and cooled PCA (also known as Standard Methods Agar) to the Petri plates. Swirl gently. Allow to solidify.

6. Incubate the plates at 35°C for 24 to 48 hours.

Second Period

1. Arrange the duplicate plates in order of lowest dilution to highest dilution.

2. Count the number of colonies on the duplicate plates that have between 30 and 300 colonies. Take the average.

3. Determine the APC/g hamburger by multiplying the average number determined in step 2 above by the inverse of the dilution factor of those plates.

4. Record your results and discard the plates as indicated by your instructor.

Questions

1. If foods such as hamburger are to be cooked, why are standards for total bacteria necessary?

2. Why is a 50 gram sample blended with 450 ml of diluent to yield a 1:10 dilution instead of the more traditional 1 gram to 9 ml diluent?

3. Does the number recorded by the standard plate count as you performed it accurately reflect the total bacterial count from that sample? Explain.

4. What advantages would the pour plate technique have over the spread plate technique? What disadvantages does the former method have?

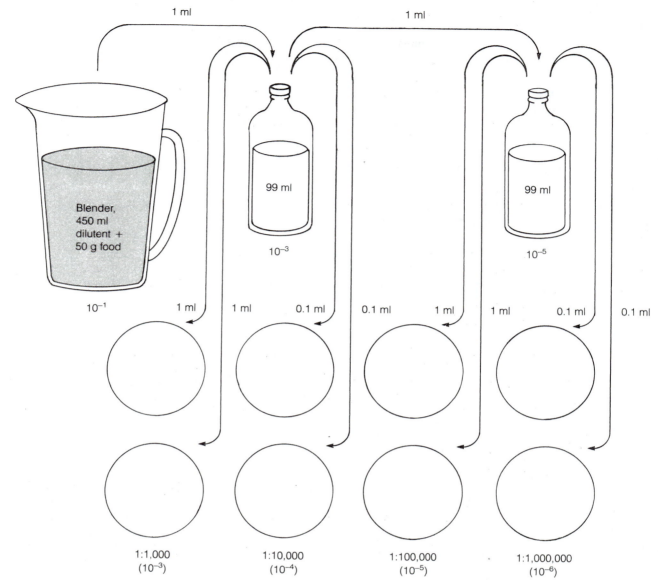

Figure 55-1. Dilution Series and Plating Format for Plate Counts of Hamburger Sample

B. ENUMERATION OF PROTEOLYTIC BACTERIA IN FOOD

Proteolytic bacteria can spoil food by breaking down the proteins and liberating unpalatable endproducts. These organisms can be enumerated by plating samples onto media containing sterile skim milk, which contains the protein casein. Organisms capable of using the casein are generally regarded as being proteolytic.

Materials

Plate count agar

Sterile skim milk. 2 ml/tube

Slants of *Bacillus subtilis* and *Escherichia coli*

Incubator set at 32°C

Procedure

First Period

1. Use the dilutions depicted in figure 55-1 to prepare samples of hamburger.

2. Label, in duplicate, sterile Petri plates 10^{-3} through 10^{-6}.

3. Aseptically pipet aliquots from the appropriate dilutions onto the plates as shown in figure 55-1.

4. Aseptically add 2 ml of sterile skim milk to each of the plates.

5. Add melted and cooled PCA to each of the plates. Swirl gently. Allow to solidify.

6. Make two additional plates of skim milk agar and allow them to solidify. On one plate, spot a culture of *E. coli*. On the other, spot a culture of *B. subtilis*. Incubate all the plates at 32°C for 24 to 48 hours.

Second Period

1. Arrange the plates in order of lowest to highest dilution.

2. Note which pair of duplicate plates contains a countable number (30 to 300) of colonies showing zones of clearing.

- Those colonies that show a zone of clearing around them usually consist of organisms that are proteolytic. When the casein (responsible for the whitish color of the skim milk agar) is hydrolyzed, a clear zone results. Use the *E. coli* and *B. subtilis* control plates to see positive and negative reactions.

3. Determine the number of proteolytic bacteria/g hamburger.

4. Record your results and discard the plates as indicated by your instructor.

Questions

1. Why are proteolytic organisms undesirable in foods such as hamburger?

2. Would saccharolytic organisms be of significant concern in hamburger?

3. Why are proteolytic organisms enumerated on skim milk (casein) agar, if there is no casein in hamburger?

4. How can you determine proteolysis on casein agar?

5. Colonies exhibiting clearing on skim milk agar usually indicate proteolysis. What is the exception?

C. ENUMERATION OF LIPOLYTIC BACTERIA

Lipolytic organisms are those capable of breaking down lipids or fats and liberating fatty acids. In foods such as butter, this accumulation of fatty acids contributes to the rancidity (and hence spoilage) of butter. In this exercise you will be looking for lipolytic organisms in butter.

Materials

Fat agar

Butter melted and kept at 45°C

9 ml dilution blanks, kept warm at 45°C

Sterile warm 1 ml pipets

Sterile Petri plates

Slant cultures of *Pseudomonas fluorescens* and *Escherichia coli*

Incubator set at 35°C

Procedure

First Period

1. Obtain a sample of warm, melted butter from the water bath.

2. Label, in duplicate, plates from 10^{-1} to 10^{-3}.

3. Pipet 1 ml of melted butter, using a warm pipet, into a 9 ml blank of warm diluent. Serially dilute in warm diluent and plate in duplicate as shown in figure 55-2.

4. Add melted and cooled fat agar to each of the plates. Swirl gently. Allow to solidify.

5. Pour two additional fat agar plates. Spot a culture of *P. fluorescens* on one plate and *E. coli* on the other.

6. Incubate the plates at 35°C for 24 to 48 hours.

Second Period

1. Arrange the duplicate sets of plate from lowest dilution to highest.

2. Determine the duplicate set of plates containing a countable number of reddish colonies with underlying pink-red zones of precipitation.

- The color is due to the liberation of fatty acids by the lipolytic bacteria. The resultant acidity causes the pH indicator to turn red. Check the plate with the control organisms for positive and negative reactions.

3. Determine the number of lipolytic bacteria/g butter.

4. Record your results and discard the plates as indicated by your instructor.

Questions

1. Is fat agar differential, selective, or both? Explain.

2. What microbial activities lead to rancidity in butter and similar foods?

3. How can you distinguish a fat hydrolyzing colony on a fat agar plate?

D. ENUMERATION OF MOLDS AND YEASTS FROM FRUIT

Fruits, both fresh and frozen, are typically very high in sugar and in acidity. Microbes responsible for spoilage of these fruits are usually those that can survive

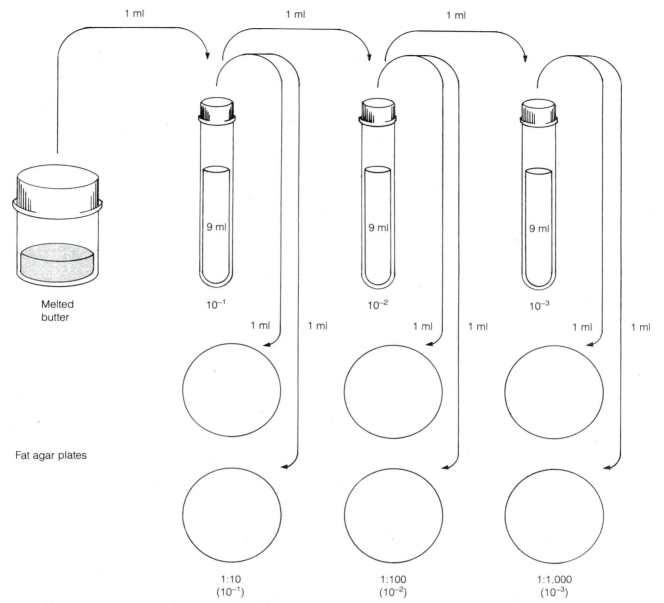

Figure 55-2. Flowchart for the Isolation and Enumeration of Li-polytic Bacteria from Butter

In the figure: "1 ml" (three times across top), "Melted butter", "9 ml" (three tubes), 10^{-1}, 10^{-2}, 10^{-3}, "1 ml" labels, "Fat agar plates", "1:10 (10^{-1})", "1:100 (10^{-2})", "1:1,000 (10^{-3})".

and grow under hypertonic and acidic conditions. Although some bacteria are capable of spoiling fruits, the organisms most often responsible are the molds and yeasts. In order to enumerate these organisms, media are used that are high in sugar content and/or low in pH. The media tend to inhibit bacterial growth and permit fungal growth.

Materials

Fruit, fresh or frozen, diluted 1:10 by blending 50 g fruit with 450 ml sterile diluent

Sabouraud agar

9 ml dilution blanks

Sterile pipets

Sterile Petri plates

Incubator set at 25°C

Procedure

First Period

1. Label sterile Petri plates 10^{-1} through 10^{-3} in duplicate.

2. Make dilutions of the fruit sample and plate as indicated in figure 55-3.

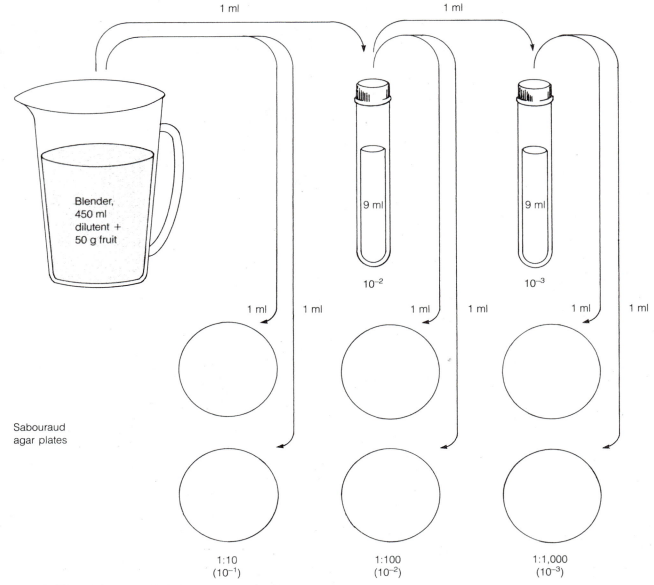

1 ml

1 ml

Blender,
450 ml
dilutent +
50 g fruit

9 ml

9 ml

10^{-2}

10^{-3}

1 ml

1 ml

1 ml

1 ml

1 ml

1 ml

Sabouraud
agar plates

1:10
(10^{-1})

1:100
(10^{-2})

1:1,000
(10^{-3})

Figure 55-3. Flowchart for the Isolation and Enumeration of Molds and Yeasts from Fruit

3. Aseptically add melted and cooled Sabouraud agar. Swirl gently. Allow to solidify.

4. Incubate the plates at 25°C for 7 days.

Second Period

1. Arrange the duplicate sets of plates from lowest to highest dilution.

2. Count the number of mold and yeast colonies.

3. Determine the number of molds and yeasts/g fruit. Discard the plates.

Questions

1. Why are molds and yeasts usually involved in the initial spoilage of fruits and vegetables?

2. How is Sabouraud agar a suitable medium for the growth of molds and yeasts?

3. How would a simple stain distinguish between a yeast colony and a bacterial colony that may grow on Sabouraud agar?

4. What foods do you commonly observe that are contaminated with molds?

EXERCISE 56
FOOD POISONING

A variety of microorganisms are capable of causing food-related diseases, due either to infections or to toxins they produce. Infectious and food poisoning organisms include bacteria (*Salmonella, Shigella, Staphylococcus aureus* and *Clostridium botulinum*), viruses (hepatitis A virus), and parasites *(Trichinella, Toxoplasma)*. These organisms may be introduced by food handlers, either directly or indirectly, or they may be associated with animal carcasses.

Foods are not routinely examined for the presence of microorganisms responsible for disease. Such examinations are usually limited to case studies performed on leftover foods following a food-related disease outbreak.

In this exercise you will be looking for organisms that cause food poisoning. It is important that you distinguish between organisms that cause food poisonings and those organisms that cause food-borne infections such as *Listeria monocytogenes, Salmonella typhi, Shigella* spp., hepatitis A, and *Trichinella*.

Materials

Food sample, blended and diluted to 1:10 and inoculated to yield 100 *S. aureus* per gram food

Phenol red mannitol salt agar

Sterile 9.0 ml dilution blanks

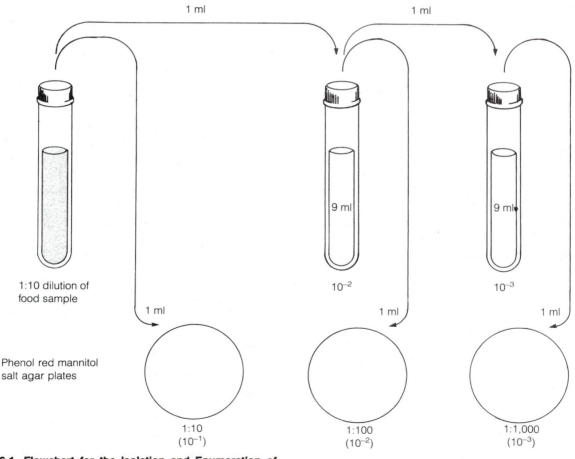

Figure 56-1. Flowchart for the Isolation and Enumeration of *S. aureus* from Food

Sterile Petri plates

Sterile 1.0 ml pipets and pipet bulbs

Gram stain reagents

Slant cultures of *S. aureus, S. epidermitis,* and *E. coli*

Incubator set at 37°C

Procedure

First Period

1. Label three sterile Petri plates 10^{-1} through 10^{-3} respectively.

2. Aseptically prepare dilutions as shown in figure 56-1.

3. Pipet aliquots from the appropriate dilution blanks to the plates as indicated.

4. Add melted and cooled phenol red mannitol salt agar to each of the plates. Swirl gently. Allow to solidify.

5. Divide a plate of phenol red mannitol salt agar into three sections. Inoculate each section with one of the control organisms.

6. Incubate the plates at 37°C for 24 to 48 hours.

Second Period

1. Examine the phenol red mannitol salt agar plates for typical colonies of *S. aureus.* (See Appendix D.)

2. Gram stain a typical colony of *S. aureus* and typical colonies that may have grown on any of the control plates.

3. Record your results and discard the plates as indicated by your instructor.

Questions

1. Is phenol red mannitol salt agar differential, selective, or both? Explain.

2. List those food poisonings that may be caused by food handlers who are carriers.

3. What food poisonings are most often due to improperly home-canned foods?

4. What food-borne infections are often due to pathogens introduced from a slaughtered animal?

5. How can differential and/or selective media be used in the rapid isolation and identification of bacterial food pathogens?

EXERCISE 57
EVALUATING THE
SANITARY QUALITY OF FOODS

It is impractical to analyze all foods as they are processed for the possible presence of pathogens. Historically, the sanitary quality of foods has been determined by looking for the presence of **indicator organisms**—bacteria which, if present in greater than allowable numbers, indicate the food is not of acceptable sanitary quality. The presence of these indicator organisms does not necessarily mean that food-borne pathogens are present in the food. Nor, unfortunately, does the absence of indicator bacteria indicate that the food is safe for consumption. However, the use of the indicator system, coupled with a standard plate count, does in most instances provide the consumer with foods of good sanitary quality.

In the food industry, two groups of indicator organisms are used. The first group is the **coliform** group, defined as Gram-negative, facultatively anaerobic asporogenous rods that ferment lactose with the production of acid and gas within 48 hours at 37°C. Some of the coliforms (e.g., *Enterobacter aerogenes*) are soil-borne and hence are found abundantly on vegatation. Other coliforms (primarily *E. coli*) are found in the intestinal tracts of humans and warm-blooded animals. The presence of these latter organisms (fecal coliforms) in foods poses a significant public health hazard, as pathogens of intestinal origin may also be present.

Obviously, it is imperative that indicator organisms survive storage of the food for an appreciable time (or at least longer than the survival period of pathogens whose presence the indicators suggest). In this respect, there is evidence that in frozen foods at least, another group of indicators function better than the coliforms. This group is called the **fecal streptococci,** or the **enterococci,** and are Gram-positive, catalase-negative cocci. They are primarily varieties of *Streptococcus faecalis* and *Streptococcus faecium* and are considered, in food microbiology, the most common of the group D streptococci encountered in the industry.

A. COLIFORM ANALYSIS OF FOODS

Materials

Violet red bile agar (VRBA)

Brilliant green bile lactose (BGBL) broth, with Durham tubes

Food sample (the instructor may spike the sample if desired)

Slant culture of *E. coli*

Blenders

Balances

9 ml dilution blanks

1 ml pipets

Colony counters

Incubator set at 35°C

Procedure

First Period

1. Weigh a 50 g sample of the food to be tested. Blend for 2 minutes with 450 ml sterile diluent in a clean blender. This is a 1:10 dilution.

2. Make dilutions as indicated in figure 57-1 through 10^{-3}.

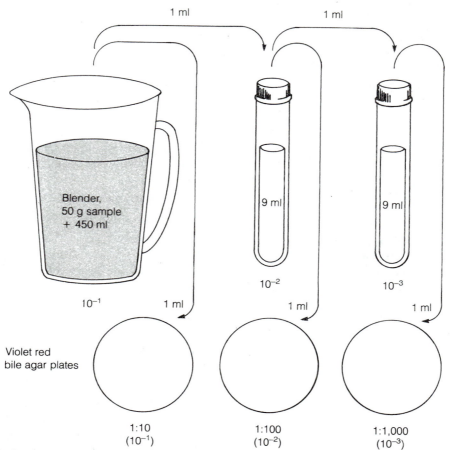

Blender, 50 g sample + 450 ml

1 ml   1 ml

9 ml   9 ml

10^{-2} 10^{-3}

10^{-1} 1 ml   1 ml   1 ml

Violet red bile agar plates

1:10   1:100   1:1,000
(10^{-1})   (10^{-2})   (10^{-3})

Figure 57-1. Dilution Series and Flowchart for Coliform Analysis of Foods

3. Label three sterile plates 10^{-1}, 10^{-2}, and 10^{-3} respectively.

4. Aseptically pipet appropriate aliquots from the dilutions to the plates as indicated in figure 57-1. Add melted and cooled violet red bile agar to each of the plates. Swirl gently. Allow to solidify.

5. After the agar has solidified, add approximately 5 ml of additional VRBA as an overlay. Again allow to solidify.

6. Streak an additional plate of VRBA with the control *E. coli* provided.

7. Incubate all plates at 35°C for 24 hours.

Second Period

1. Examine the test and control plates.
- Coliforms on this medium are purplish-red colonies surrounded by a reddish zone of precipitated bile, 0.5 mm in diameter or larger (see Appendix D).

2. Arrange the plates in order of lowest to highest dilution. Count the coliform colonies, using a colony counter, on the plate exhibiting between 30 and 150 colonies.

3. Determine the total coliforms/g food sample.

4. To confirm that these are indeed coliform colonies, fish out 5 to 10 typical colonies and transfer to separate tubes of brilliant green bile lactose broth. Incubate at 35°C for 24 to 48 hours. Include a colony from the control plate.

5. Record your results. Discard the plates as indicated by your instructor.

Third Period

1. Examine the tubes of BGBL for gas.
- This medium is highly selective for coliforms, and tubes showing gas are considered positive. If the tube also shows aerobic pellicle growth, a Gram stain should be performed since Gram-positive gas formers may be present.

2. If all the BGBL tubes were not positive, the number of coliforms/g food must be recalculated.
- Multiply the percentage of positive BGBL tubes (gas production, Gram-negative rods) by the original violet-red bile agar count, times the inverse of the dilution factor used on that plate.

3. Record your results. Discard all tubes as indicated by your instructor.

Questions

1. Define coliforms. How are they used in the analysis of the sanitary quality of foods?

2. Is brilliant green bile lactose selective? Why?

3. Is violet-red bile agar differential, selective, or both? Explain.

4. Why do you overlay VRBA?

B. FECAL STREPTOCOCCAL ANALYSIS OF FOODS

Materials

KF streptococcal agar

Sterile Petri plates

Hydrogen peroxide, 3%

Bile esculin agar slants

Brain heart infusion (BHI) + 6.5% NaCl

Colony counters

Food sample (the instructor may spike the sample if desired)

Slant culture of *S. faecalis* grown on BHI

BHI broth

Incubators set at 35°C and 45°C

Inoculation loop

9 ml dilution blanks

Procedure

First Period

1. Label three sterile Petri plates 10^{-1} through 10^{-3} respectively.

2. Weigh 50 g of food sample and blend with 450 ml of sterile diluent in a blender (the same sample used in exercise 57A may be used if both exercises are being performed).

3. Make a dilution series as indicated in figure 57-2.

4. Aseptically pipet the appropriate aliquots from the dilution blanks to the plates as indicated in figure 57-2.

5. Add melted and cooled KF agar to each of the plates. Pour an additional plate to streak the control *S. faecalis*.

6. Incubate all plates at 35°C for 48 hours.

Second Period

1. Arrange the KF plates in order from lowest to highest dilution. Select the plate which has between 15 and 150 typical colonies (see Appendix D).

2. Determine the number of fecal streptococci/g food.

3. Pick several typical colonies (including the control growth) and transfer to tubes of BHI broth. Incubate at 35°C for 18 to 24 hours.

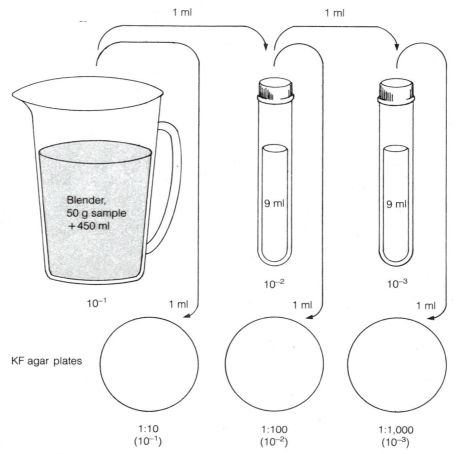

Figure 57-2. Dilution and Plating Scheme for the Isolation and Enumeration of Fecal Streptococci from Foods

Third Period

1. Observe the BHI broth for growth. Prepare Gram stains and observe for Gram-positive cocci, elongated, in pairs and short chains.

2. Pour about one half the contents from the tubes of BHI broth into clean test tubes. Test for catalase activity by adding 1 ml of 3% hydrogen peroxide to one set of culture tubes and look for the generation of oxygen bubbles. Streptococci are catalase-negative, so gas should not be observed. Discard these tubes.

3. Using the other set of tubes, inoculate a loop of culture into BHI broth + 6.5% NaCl, into bile esculin agar slants, and into BHI broth. Incubate the BHI + 6.5% NaCl at 35°C for 72 hours, the bile esculin at 35°C for 24 hours, and the BHI broth at 45°C for 24 hours.

Fourth Period

1. Examine the bile esculin agar (see Appendix D).

2. Examine the BHI broth incubated at 45°C for growth. Group D streptococci will grow at 45°C.

3. Record your results. Discard the bile esculin and BHI tubes as indicated by your instructor.

Fifth Period

Examine the BHI + NaCl for growth. The only enterococci that do not grow on this medium are *S. equinus* and *S. bovis*.

Questions

1. Define fecal streptococci. How are they used in the analysis of the sanitary quality of foods?

2. Is KF streptococcal agar differential, selective, or both? Explain.

3. In what types of foods might fecal streptococci be a more reliable indicator than coliforms?

4. Would the absence of these indicator organisms in foods demonstrate the absence of food-borne pathogens?

EXERCISE 58
EXAMINATION OF
EATING UTENSILS AND EQUIPMENT

In the home, in restaurants, and in processing plants, foods may become contaminated by disease causing organisms due to improperly cleaned utensils, drainboards, and equipment. Although a high bacterial count in these environments may in itself not represent an immediate health threat, it does indicate a general state of uncleanliness. Dried food particles, especially in hard-to-reach parts of food production equipment, serve as excellent nutrient sources for microbes.

A common cause of food poisoning and food-borne infections in the home is cross-contamination of foods. This can result from improper cleaning or disinfecting of food preparation areas. For example, raw poultry that is cut or stuffed on a cutting board may introduce potential pathogens such as *Salmonella* to the cutting board. While these bacteria that are indigenous to the poultry would be destroyed by proper cooking and hence pose no direct threat to the consumer, the organisms remaining on the cutting board may contaminate other foods that will not be cooked, such as a salad made with raw vegetables.

The United States Public Health Service requires, as a minimum standard, one of the following criteria in cleaning and dishwashing procedures: application of hot water (76.7°C for 30 seconds), hypochlorite at 50 ppm in water at least 23.8°C for 60 seconds, or an iodine sanitizer at 12 ppm iodine in water heated to at least 23.8°C with a pH less than 5 for 60 seconds.

On eating utensils, the maximum safe bacterial plate count recommended is 100 colonies/utensil, or 12.5 bacteria per square inch of significant surface. Significant surfaces constitute that portion of the utensil, dish, etc., which comes in contact with either the food or the mouth.

For equipment on which 5 areas of approximately 8 square inches each have been examined, residual bacterial counts not exceeding 500 (i.e., averaging 12.5/sq.in.) are satisfactory. When counts from a particular establishment are frequently or consistently in excess of the standards, a need for improvement in washing, sanitizing, handling, and/or storage is usually indicated.

Examination of eating utensils and equipment is usually conducted by swabbing significant surfaces with cotton swabs or by performing, where applicable, a **Rodac plate test.** In the swab method, the significant surface is wiped with a sterile swab. The swab is placed in sterile diluent to return it to the laboratory. If chemical sanitizers are used to wash the surfaces, neutralizing agents must be added to the diluent. A list of sanitizers and their neutralizing agents, is given in table 58-1.

Materials

Sterile swabs

9 ml of sterile diluent in screw-capped tubes

Plate count agar

1 ml pipets

Colony counters

Utensils, drainboards, etc. (the instructor may wish this exercise to be a take-home assignment)

Rodac plates

Incubator set at 37°C

Procedure

First Period (Swab Method)

1. Wet a sterile cotton swab in the tube of diluent. Squeeze out the excess fluid on the inside of the tube.

2. Swab the significant surfaces of the utensils, dishes, and equipment provided.

   a. Cups and glasses: upper ½ inch of the inner and outer rims

   b. Spoons and forks: entire inner and outer surface that enters the mouth

   c. Plates, bowls, saucers: 1 square inch of any part of the surface coming in contact with the food

TABLE 58-1.
SANITIZERS AND NEUTRALIZING AGENTS

SANITIZER	NEUTRALIZING AGENT
halogens	Na thiosulfate
quartenary ammonium	lecithin, stearate
phenolics	Tween 80, charcoal
hexachloraphene	Tween 80
formalin	ammonium
hydrogen peroxide	catalase
mercurials	Na thioglycollate
mercuric chloride	sodium sulfide

d. Food machinery, cutting boards, etc.: any food contact surfaces, approximately 8 square inch area

3. Place the swab back in the diluent. Squeeze out the excess fluid and repeat step 2 with another similar utensil or dish surface area. Repeat until 4 utensils or dish surface areas have been sampled. For food machinery, cutting boards, etc., repeat step 2 until 40 square inches have been sampled.

4. Break the swab *below* the finger contact area into the tube of diluent as shown in figure 58-1. If the sample tube must be returned to the lab, store it in a cool place until analysis can be performed.

5. Plate 1 ml of the sample in duplicate onto sterile Petri plates. Add melted and cooled PCA. Swirl gently. Allow to solidify.

6. Incubate the plates at 37°C for 48 hours.

First Period (Rodac Method)

1. Pour melted agar up to the surface of the Rodac plate, so that a slight convex surface ensues, as shown in figure 58-1. Allow the medium to solidify.

2. Press the agar on a surface to be sampled and gently rock the plate back and forth until the entire surface of the agar has come in contact with the equipment or cutting board.

3. Cover the plate with the lid. Place the plate in a plastic bag and incubate at 37°C for 48 hours.

Second Period

1. Count the colonies on the plates.
• For the swab method, report as the "average plate count/utensil or square inch of utensil." For the Rodac method, count the colonies. The standard is 50 col-

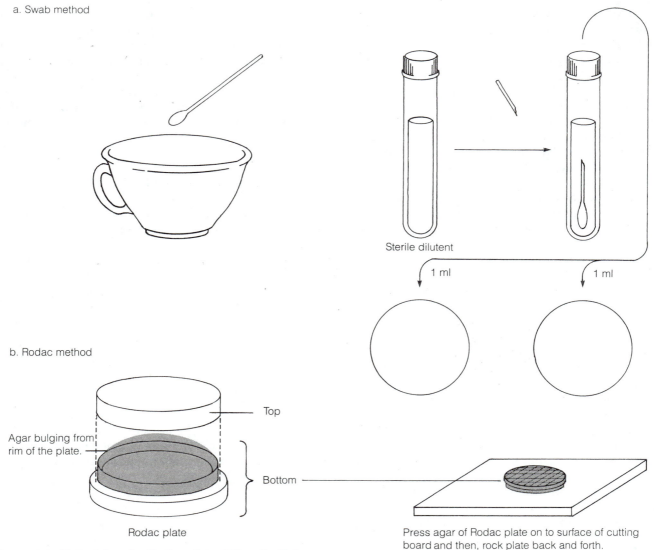

a. Swab method

b. Rodac method

Top

Agar bulging from rim of the plate.

Bottom

Rodac plate

Sterile dilutent

1 ml   1 ml

Press agar of Rodac plate on to surface of cutting board and then, rock plate back and forth.

Figure 58-1. Methodology for the Sampling of Utensils, Dishes, and Work Areas

onies per plate, as the plate covers a 4 square inch area.

2. Record your results. Discard the plates as indicated by your instructor.

Questions

1. Why, when sampling from dishes disinfected with halogens, should you add thiosulfate to the diluent to which the swab is added?

2. Why are cutting boards often responsible for cross contamination and subsequent food poisonings?

3. What are the criteria for dishwashing procedures in public establishments?

4. Give four sanitizing agents and their neutralizing agents.

EXERCISE 59
MICROBES AND FOOD PRODUCTION

It is difficult to determine the exact time at which humans were first aware of the presence and role of microorganisms in food production (or food spoilage for that matter). Certainly the time was well after van Leeuwenhoek's discovery of microorganisms. Pasteur recognized the role of microbes in the spoilage of wine, and Appert was the first to use heat to destroy organisms in the canning industry.

We do know that food processing has occurred for many thousands of years. For example, cereal cookery, brewing, and storage of foods dates back 6,000 to 8,000 years. It is likely, however, that foods produced by microbial activities were discovered by accident. The methods for the production of many foods have been refined through the centuries, resulting in today's modern food industry.

We know now that specific microorganisms can change the taste of raw materials and yield a desirable product. These same organisms also inhibit the growth of undesirable organisms, usually due to the byproducts formed as a result of metabolic actions on the raw materials. The most commonly produced byproducts are preservatives such as lactic and acetic acids. It is apparent then, that the activities of the food-producing microbes are in fact a means of preserving the microbial quality of the raw materials. This is achieved primarily by preventing the growth of undesirable organisms and their subsequent production of unpalatable or even toxic byproducts.

Most of the foods prepared using microorganisms are **fermented** foods. These include meat products such as cured hams and dried sausages; plant products such as kimchi, pickles, olives, and sauerkraut; beverages such as beer, ale, and wine; and breads such as sourdough and sour pumpernickel. A summary of dairy products can be found in Part XIV, Dairy Microbiology.

The predominant fermenting organisms in food production are the fermentative yeasts and the lactic acid bacteria, principally members of the genus *Saccharomyces*, and of the genera *Lactobacillus*, *Leuconostoc*, *Streptococcus*, and *Pediococcus*, respectively.

There are a few foods that are produced in part by fermentative organisms and in part by oxidative bacteria. Such foods include some sourdough breads and wine vinegar.

A. SAUERKRAUT PRODUCTION

Sauerkraut is typical of a fermented plant food product in which the predominant organisms responsible are the lactic acid bacteria. These organisms are widespread in nature and are also used extensively in the dairy industry (see Part XIV). Lactic acid bacteria are differentiated into two groups based on the end products of glucose metabolism. All members of the genera *Pediococcus* and *Streptococcus*, as well as some species of *Lactobacillus*, are homolactic; that is, they produce lactic acid as the sole fermentative product from glucose. The reaction is written as follows:

$$C_6H_{12}O_6 \rightarrow 2\ CH_3CHOHCOOH$$

The genus *Leuconostoc* and the remaining species of the lactobacilli are called heterolactic, as they yield equimolar amounts of lactic acid, ethanol, and carbon dioxide from glucose:

$$C_6H_{12}O_6 \rightarrow CH_3CHOHCOOH\ +\ CH_3CH_2OH\ +\ CO_2$$

The heterolactics are generally considered to be **flavor/aroma** bacteria, as they also produce small amounts of compounds that impart the flavor and aroma characteristic of some foods. These compounds are primarily **acetaldehyde** and **diacetyl.**

In preparing sauerkraut the outer leaves are removed from heads of cabbage. The heads are then rinsed to remove dirt and other extraneous material. They are then halved to remove the core, and shredded finely. Salt is added to give a concentration of 2.0 to 3.0% NaCl (this serves to inhibit some of the undesirable organisms and also to extract sugars from the cabbage). The salted and shredded cabbage is packed tightly into a suitable container and covered to prevent air from entering.

The resulting microbial activity is due to microorganisms that were initially present on the cabbage. As oxygen is limited, due to the manner in which the shredded cabbage is packed, the environment prevents oxidative yeasts and molds from growing, and favors fermentation instead. Undesirable fermentative organisms such as coliforms, which produce a variety of acids (some of which are unpalatable), are inhibited due to the salt concentration. Hence a favorable environment has been created for lactic fermentations.

Materials

Cabbage heads

Table salt

Sharp knives

Balance

Wide-mouth gallon jars

Tomato juice agar

9 ml dilution blanks

Sterile 1 ml pipets

Colony counters

Gram staining reagents

Hydrogen peroxide, 3%

Large polyethylene bags

Titration apparatus

0.1 N NaOH

phenolphthalein indicator

125 ml beakers

5 ml pipets

Incubator set at 30°C

Procedure

First Period

1. Remove and discard the outer leaves of the cabbage heads. Rinse the heads. Cut the heads in half and remove and discard the cores.

2. Weigh the cabbage, then shred finely. (Remember,

hemoglobin does not add to the flavor of the end product. If you want red sauerkraut, you should use red cabbage!)

3. Weigh enough salt to equal 2 to 3% of the weight of the cabbage. Sprinkle the salt over the shredded cabbage.

4. Pack the cabbage *tightly* into 1 gallon wide-mouthed jars so as to eliminate as many air spaces as possible. When the jar is roughly ¾ full, place a 11½ × 12½ inch × 1.01 mil food storage bag over the cabbage as shown in figure 59-1.

5. Fill the bag with water. This will act as a weight on the cabbage to maximize anaerobiosis. Incubate at room temperature for about 4½ weeks.

6. Pipet a 10 ml sample of cabbage juice into a 50 ml beaker.

7. Prepare dilutions through 10^{-6}, as shown in figure 59-2, using a 1 ml aliquot of the juice. Plate as indicated. Add melted and cooled tomato juice agar, swirl gently, and allow to solidify. Incubate at 30°C.

8. Pipet 5 ml of the remaining 9 ml of cabbage juice into a 125 ml Erlenmeyer flask and dilute with 5 ml distilled water. Boil gently for a few minutes to drive off carbon dioxide, cool rapidly, and add two drops of phenolphthalein indicator. Titrate with 0.1 N NaOH to a slightly pink end point. Calculate the percentage of acid, expressed as lactic acid, according to the formula

$$\% \text{ lactic acid} = \frac{\text{ml of alkali} \times \text{N of alkali} \times 9}{\text{ml sample}}$$

Figure 59-1. Sauerkraut Developing in a Jar

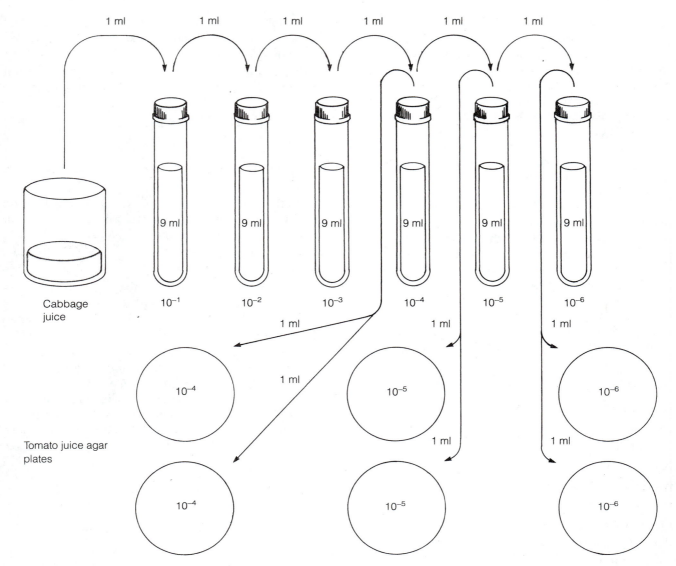

Figure 59-2. Dilution Series for Plate Count of Cabbage Juice

9. Using pH paper, record the pH of the remaining 4 ml of cabbage juice.

Second Period

1. Count the colonies from the paired series of plates exhibiting 30 to 300 colonies. Determine the numbers of colony-forming units/ml of sauerkraut juice at day 0.

2. Perform Gram stains and then catalase tests on the predominant colonies.

3. In each succeeding laboratory period, selected groups of students should repeat steps 6 through 9 performed in period one.

4. After a 4 to 5 week time period, the sauerkraut may be boiled and consumed. The final pH will be about 3 to 3.5.

5. Record the cumulative results from each sampling period for pH, titratable acidity, numbers of organisms; and the cellular morphology, Gram reactions, and catalase reactions of the predominant organisms enumerated each period. Discard all plates as indicated by your instructor.

Questions

1. What organisms are primarily responsible for sauerkraut production?

2. How are oxidative bacteria, molds, and yeasts eliminated or prevented from growing in the fermenting cabbage?

3. How are other fermenting organisms such as coli-forms inhibited?

4. What prevents endospores of *Clostridium botulinum* from germinating, growing, and producing toxins?

B. VINEGAR PRODUCTION

Vinegar is an excellent example of a microbially produced foodstuff that is the result of fermentation and subsequent oxidation. The first step in vinegar production is performed by a fermentative yeast such as *Saccharomyces cerevisiae*. The yeast ferments a sugar substrate (such as apple juice) and produces alcohol as follows:

$$C_6H_{12}O_6 \rightarrow 2\ CH_3CH_2OH\ +\ 2\ CO_2$$

Home production of vinegar uses the yeasts naturally occurring on the skin of grapes or apples. Commercially, *S. cerevisiae* is used as the starter culture. In either case, fermentation occurs.

Following fermentation, the ethanol produced by the yeast is metabolized aerobically by acetic acid bacteria such as *Acetobacter aceti* or other species. These oxidative organisms are capable of using the alcohol for carbon and energy, yielding acetic acid as the waste product, as follows:

$$CH_3CH_2OH\ +\ O_2 \rightarrow CH_3COOH\ +\ H_2O$$

Materials

Clean 250 ml Erlenmeyer flasks

Apple juice

Apple or raisins

Starter culture of *Saccharomyces cerevisiae* in apple juice

Gram staining reagents

Acetobacter culture

Inoculating loop

Aluminum foil

Procedure

First Period

1. Add approximately 50 ml of apple juice to each clean 250 ml flask.

2. Add a small piece of apple skin or 2 to 3 raisins to one of the flasks.

3. Add 1 ml of yeast starter culture to the other flask.

Cover both flasks with aluminum foil. Incubate at room temperature.

Second Period

1. In each succeeding lab period, examine the flasks for growth, odor, and gas. *Do not shake the flasks.* Reincubate the flasks.

2. When the odor of the ethanol becomes predominant in the flask to which the starter culture was added, prepare a Gram stain and observe for both yeasts and bacteria.

3. Add a loopful (or 1 ml) of *Acetobacter* culture to the flask (do not inoculate the flask which was initially spiked with the apple skin or raisins).

4. Incubate the flask at room temperature.

5. Allow the flask that contains the apple skin or raisins to progress naturally. The oxidizing organisms should be present in the flask, as they (as well as the yeasts) make up the normal flora of these fruits.

Third Period

1. Examine the flasks each lab period for the production of a pellicle, which would indicate the growth of the aerobic *Acetobacter*.

- DO NOT BREAK THE PELLICLE BY SHAKING THE FLASK.

2. When the pellicle is well-formed and the odor of vinegar is apparent, perform Gram stains. Compare the results to your observations of the Gram stains of the fermented substrate.

3. Record your results. Discard all flasks as indicated by your instructor.

Questions

1. What two major groups of organisms are responsible for the production of wine vinegar?

2. Which of these groups are fermentative? What is their fermentable substrate? What are their major waste products?

3. Which of the groups mentioned in question 1 are oxidative? What is their substrate? What is their major end product?

4. Why is wine vinegar more expensive, and tastier, than cider vinegar?

5. Why is vinegar pasteurized?

C. MICROORGANISMS IN SOURDOUGH PRODUCTION

The production of sourdough bread may involve some of the same processes and organisms noted in vinegar production, that is, the fermentative yeasts and oxidative acetic acid bacteria. However, some sourdough starters (such as those for the production of San Francisco sourdough) include the lactic acid bacterium *Lactobacillus sanfrancisco*. Although this organism is a lactic acid bacterium, it produces significant amounts of acetic acid in its fermentation of the sugar in the dough.

In those starters using yeasts and acetic acid bacteria, the yeasts produce CO_2 and alcohol (from the sugar present if the flour is unrefined or from added sugar if the flour used is refined). The gas (CO_2) causes the dough to rise and the alcohol provides a substrate for *Acetobacter*.

In starters employing yeasts and *Lactobacillus sanfrancisco*, the entire process is fermentative. Both organisms use the available sugars, again yielding primarily alcohol, acetic acid, and CO_2.

The acetic acid produced in either case contributes to the "sour" taste of sourdough breads.

Materials

Sourdough starter

Sabouraud agar

Ethanol-calcium carbonate agar

Flour, etc. (if the instructor wishes to make the bread)

Gram reagents

Inoculating loop

Incubator set at 30°C

Procedure

First Period

1. Prepare a Gram stain of the starter culture. Observe for the presence of both bacteria and yeast.

2. Streak a loop of starter culture onto both Sabouraud and ethanol-calcium carbonate agar. Incubate the plates at 30°C for 2 to 5 days.

3. Prepare sourdough bread if desired.

Second Period

1. Examine the Sabouraud agar for yeast colonies. Gram stain. Note the odor of the organisms. Discard the plates as indicated by your instructor.

2. Examine the ethanol-calcium carbonate plates for colonies exhibiting zones of clearing.

- Bacteria responsible for sourdough bread liberate acetic acid. The acid will in turn dissolve the calcium carbonate, leaving clear zones around those acid-producing colonies.

3. Gram stain and check the catalase reactions of a few of the organisms on the plate.

4. If sourdough bread was made, enjoy!

Questions

1. How are vinegar and sourdough production similar? Explain how they may differ.

2. Is ethanol-calcium carbonate agar differential, selective, or both? Explain.

3. Is the ethanol-calcium medium complex or chemically defined? Explain.

4. What types of bacteria did you find in the sourdough starter?

PART XIV
DAIRY MICROBIOLOGY

The dairy industry, with its manufacture of fermented milk products such as yogurt and buttermilk, and ripened cheeses such as Swiss and cheddar, is sufficiently different from other food processing industries (see Part XIII, Food Microbiology) that it can be treated as an entity in itself. It is similar to the other food industries in that it must also deal with **food spoilage organisms** and **food-borne pathogens.** Similarily, the dairy industry also uses **starter cultures** to produce the desired end products, and tests for the presence of **coliforms** to determine the sanitary quality of its products.

The factor that sets the dairy industry apart from other food industries is its use of a single substrate for all manufactured products: milk. In North America, the major source of milk is cows, although goat cheeses are reasonably popular. Humans also have used milk from reindeer, donkeys, zebras, and horses, both for direct consumption or as a substrate for natural fermentations.

The starter cultures used for dairy fermentations are lactic acid bacteria, both homolactic and heterolactic forms, discussed in Exercise 59. The starters are used for butter, cultured buttermilk, cottage cheese, cultured sour cream, yogurt, and other fermented dairy products. The starter cultures always include organisms that ferment the sugars to lactic acid (usually *Streptococcus lactis, S. cremoris,* or *S. diacetilactis).* If flavor and aroma compounds such as diacetyl are also desired in the end-product, **heterolactics** are included (such as *Leuconostoc citrovorum* or *Lactobacillus bulgaricus).*

In the production of cheeses, lactic acid fermentation of the milk sugar (primarily **lactose**) results in the curdling, or coagulating, of the milk protein (primarily **casein**). The **curd** is subsequently shrunk, pressed, and allowed to ripen in the presence of the desired microorganisms. In Swiss cheese production, the curd is inoculated with the bacterium *Propionibacterium;* in limburger cheese, it is the bacterium *Brevibacterium linens;* while in blue cheeses, it is the mold *Penicillium roquefortii.*

Although diseases are still transmitted by contaminated milk and milk products, the incidence of outbreaks has been drastically reduced since the introduction of **pasteurization.** Most outbreaks that occur today are due to the consumption of raw milk or products made from raw milk. Salmonellosis (*Salmonella*), brucellosis (*Brucella*), listeriosis (*Listeria*), tuberculosis (*Mycobacterium*), and viral diseases such as hepatitis have occurred due to consumption of unpasteurized dairy products. Pasteurization in milk is designed to expose the milk to a temperature high enough (and for long enough) to destroy all pathogens, but not to destroy the nutritional quality and taste of the milk.

The dairy industry uses coliform and total bacterial counts as indicators of the sanitary quality, nutritional quality, and shelf life of the milk or milk product. Table 62-1 lists some of the existing standards. Methods used in the analysis of milk and dairy products can be found in *Standard Methods for the Examination of Dairy Products*, published by the American Public Health Association (Elmer Marth, editor).

EXERCISE 60
ANALYSIS OF THE SANITARY QUALITY
OF MILK AND DAIRY PRODUCTS

A. COLIFORM ANALYSIS

In the dairy industry, as in other food industries, coliforms are the prime indicators of sanitary quality. In the dairy industry however, the temperature used for incubation in coliform analysis is 32°C, as opposed to the 35°C temperature required in food and water analysis.

Materials

Violet red bile agar (VRBA)

Raw milk (or milk spiked with approximately 100 coliforms/ml)

9 ml dilution blanks

Sterile Petri dishes

Sterile 1 ml pipets

Procedure

First Period

1. Make dilutions of the raw milk as indicated in fig. 60-1.

2. Label 3 sterile Petri plates as undiluted (10^0), 10^{-1} and 10^{-2}, respectively.

3. Aseptically pipet aliquots from the dilution blanks to the appropriate plates as indicated in fig. 60-1.

4. Add melted and cooled VRBA to the plates. Swirl gently. Allow to solidify.

5. After the medium has solidified, add another 5 ml to each plate. Swirl gently and allow it to solidify.

6. Incubate all plates at 32°C for 24 hours.

Second Period

1. Select the plate which has between 15 and 150 colonies that are subsurface, lenticular (lens-shaped), and deep red surrounded by a hazy pink halo (see Appendix D). Record as the coliform count/ml.

2. If confirmation is required, fish out several typical colonies and transfer them to separate tubes of brilliant green bile lactose broth. Incubate at 35°C for 24 to 48 hours.

3. Record your results. Discard the plates as indicated by your instructor.

Questions

1. Define coliforms.

2. How does this definition differ from that used in food or water analysis, with regard to incubation temperature?

3. Why is VRBA a useful medium for the detection of coliforms in the dairy industry?

4. Does the presence of coliforms in raw milk suggest fecal contaminants? Explain.

B. DYE REDUCTION TESTS

Dye reduction tests are not substitutes for coliform analyses, but can be used for a quick determination of the microbial load of the milk to be tested. The greater the microbial load, the more rapidly the milk will spoil. Should milk with high levels of microorganisms be used as a substrate for fermented products, the chances are also increased that these organisms may out-compete the desired starter cultures or otherwise have a deleterious effect on the desired production process.

There are two acceptable dye reduction tests, the **methylene blue** and **resazurin tests.** In these tests, standardized dye is added to the milk to be tested, which is held at 36°C. The microorganisms in the milk reduce the dye and so decolorize it. The reduction is due to their metabolic activities. The time required for decolorization of the dye-milk mixture by the microorganisms is an indication of the microbial load. Milk of poor quality (with a high microbial content) rapidly decolorizes the mixture, while high-quality milk requires several hours or more.

Standard dye solutions may be prepared by adding a methylene blue or resazurin tablet to 200 ml of boiling distilled water. The solution is stored in an ambered stoppered flask.

Materials

Standard resazurin solution

Raw milk (the instructor may use several samples that have been spiked with varying microbial loads)

36°C water bath

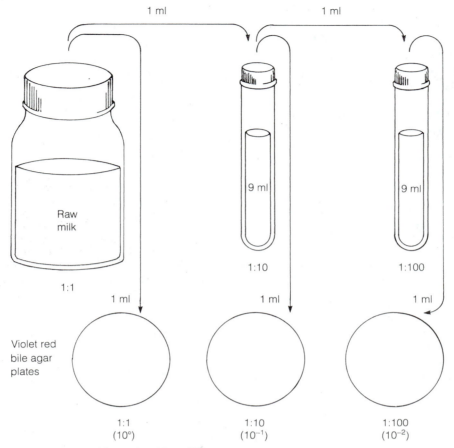

Figure 60-1. Dilution Series for Coliform Analysis of Raw Milk

Clean 16 × 125 mm screw-capped test tubes

Thermometer

Fluorescent light

1 ml pipets

Procedure

1. Pipet 1.0 ml standard resazurin dye into a clean screw-capped tube.

2. Add 10 ml of raw milk and tighten the screw cap. Invert the tube three times, but do not shake.

3. Place the tube in a 36°C water bath. Loosen the cap one turn. Include a second tube containing 10 ml raw milk and a thermometer. Start timing when this control tube reaches 36°C.

4. When the temperature has been reached (it should not take more than 10 minutes), tighten the screw cap on the sample tube. Invert three times.

5. Examine the milk-dye mixture 1 hour after the start time. Read color changes by both daylight and fluorescent light.

6. Compare the samples with the Munsell Color Standard 5P7/4 (colored insert). If reduction has occurred beyond this standard (the mixture is pink), the milk is of unsanitary quality.

$$\text{resazurin milk (purple)} \xrightarrow{[-O_2]} \text{resofurin milk (pink)} \xrightarrow{[-O_2]} \text{dihydroresofurin (colorless to white)}$$

Questions

1. Can dye reduction tests be used in place of coliform analysis in the examination of milk? Explain.

2. What is the value of a dye reduction test?

3. Describe how a dye reduction test works.

C. DIRECT MICROSCOPE COUNTS

The **direct microscopic count** (DMC) is an acceptable method for determining the **clump microscopic** count (CMC) of milk samples if the densities of microorganisms are relatively high. You can also use the procedure to detect abnormal milk products, to count leukocytes in the milk, and to identify specific morphological types of bacteria.

In order to carry out the DMC, you must first determine the area of the field of vision under oil immersion and calculate the **microscopic factor** (MF) of that microscope (see fig. 60-2).

Materials

Raw milk

Milk loops or a 0.01 ml metal syringe

40° to 45°C warming table

Levowitz-Weber stain

500 ml beakers for dip-rinsing, containing tap water at approximately 37°C

Procedure

1. Invert the milk sample gently 6 times.

2. With the milk loop or metal syringe, place 0.01 ml milk within the confines of a 1 sq. cm. area on a clean slide. Air dry.

3. Place the air-dried slide in Levowitz-Weber stain for 2 minutes. CAUTION: STAINING OF MILK SMEARS MUST BE DONE IN WELL-VENTILATED AREAS. Drain excess stain and air dry.

4. When the slide is dry, dip-rinse the slide in three successive changes of warm tap water. Air dry. DO NOT BLOT.

5. Observe the slide under oil immersion for the following:

   a) bacterial clumps (dark blue)

   b) leukocytes and somatic cells (light blue)

   c) milk protein (light blue background)

   d) fat globules (white or colorless areas)

   e) dirt, extraneous matter (brown-black)

6. Determine the number of clumps and somatic cells. A clump is one cell or a group of cells of the same type separated by a distance equal to or greater than twice the smallest diameter of the clumps or cells nearest each other (fig. 60-3).

7. Count the number of clumps and somatic cells in the recommended number of fields. Determine the average number per field. If the average number per field is 0.1 or less, count 30 to 40 fields; 0.1 to 1.0, count 20 fields; 1.0 to 10.0, count 10 fields.

8. Determine the CMC/ml as follows:

$$\text{average clump number/field} \times \text{MF} = \text{CMC/ml}$$

9. Classify the milk, using table 60-1.

TABLE 60-1.
CLASSES OF RAW MILK (DMC)

CLASSIFICATION	CLUMP COUNT RANGE	# FIELDS EXAMINED
1	under 30,000	30 to 40
2	30,000 to 300,000	20 to 30
3	300,000 to 600,000	20
4	600,000 and over	10

Stage micrometer

Stage micrometer scale magnified

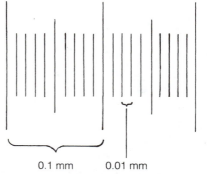

0.1 mm   0.01 mm

Figure 60-2. Calibration of the Microscopic Factor (MF). (*a*) Place a stage micrometer (0.01 mm scale) on the microscope. Locate under low power. Focus, then switch to oil immersion. (*b*) Count the number of 0.01 mm units you observe in one field. (*c*) Determine the diameter of the field, and from that, calculate the radius $\left(r = \dfrac{d}{2}\right)$. (*d*) Calculate the microscopic factor (MF) as follows: MF = $10,000/\pi r^2$, where π = 3.1416. The value 10,000 is used to determine the MF. This allows for corrections needed to determine the number of organisms per milliliter when only a 0.01 ml sample is used, and to adjust to the area of 1 square centimeter on the slide when the area of the field of vision is calculated in square millimeters.

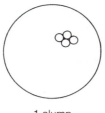
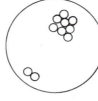
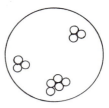

| 1 clump | 2 clumps | 3 clumps |

Figure 60-3. Determination of Clump Numbers

10. Note the morphological types predominant in the fields. Compare with the organisms in fig. 60-4.

Questions

1. What is the value of direct microscopic counts in the dairy industry?

2. What is a stage micrometer?

3. Why is it necessary to determine the microscopic factor?

4. Define a clump.

5. What, besides bacteria, can the direct microscopic count be used to enumerate?

Good quality milk. The vacuoles, white circular areas showing against a blue background, are caused by the removal of the fat droplets from the dried milk solids. Milk of good quality may contain an occasional leukocyte or bacterial cell but should never show large numbers of either.

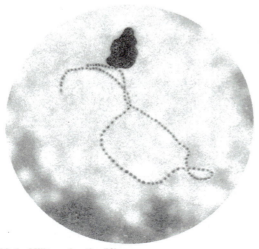

Mastitis milk from a cow shedding *Streptococcus agalactiae.* These long-chain streptococci and associated leukocytes are frequently seen in gargety milk. Use of a strip cup will frequently reveal flaky particles (garget) in such milk. The long chain of *S. agalactiae* plus large numbers of leukocytes are typical of mastitis. However, this is only one of the many organisms now found in milk from cows with a damaged udder condition.

Figure 60-4. Milk under the Microscope

Mastitis—a more serious case. Short-chain streptococci, as well as long-chain *Streptococcus agalactiae*, are shown with massive numbers of leukocytes. Here we see phagocytosis, the actual engulfing of bacterial chains as the leukocytes fight to destroy them. Frequently a cow thus infected will yield such milk in only one or two quarters while the remainder is normal. An entirely different organism, requiring a different drug for treatment, may also be found in the other quarters.

Dirty utensils—round bacterial cells. The irregular grouping of these cells, like a clump of grapes, is typical of the staphylococci. These organisms are associated with milk contaminated by contact with residues in open seams and crevices of utensils. These clumps are frequently derived from improperly cleaned milking machines or rubber parts. Leukocytes may or may not be present.

Dirty utensils—mixed cell growth. This is more typical of the flora associated with contaminated utensils. This milk contains clumps of staphylococci as well as diplococci, yeasts (the large cells), rods, and chains. This same picture is obtained when milk has been contaminated by dust.

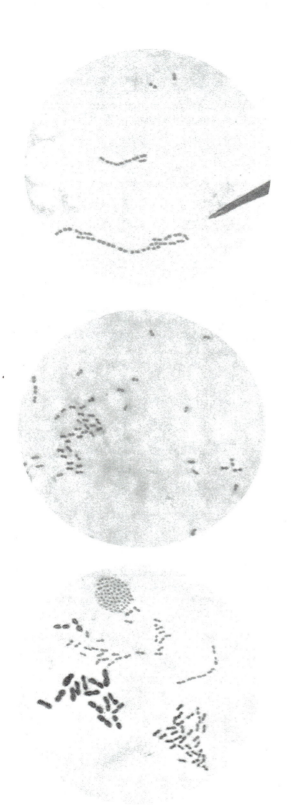

Poor cooling—paired round cells. The presence in the microscopic field of these paired cells (diplococci) or short chains of these round cells (streptococci) usually indicates poor cooling. This is the general appearance of *Streptococcus lactis*, the sour-milk organism. These organisms result from contamination.

Poor cooling—almost sour. When a sample of milk is on the verge of souring, the microscopic field shows many pairs of cells, double pairs, and short chains. The bacterial count is very high.

Poor production methods—mixed bacteria. When a field shows such a variety, it is a good bet that production methods are not what they should be. Here we see clumps of micrococci, some short chains, a group of paired bacilli (rods), as well as diplococci. This kind of contamination can easily ruin the quality of a large volume of milk. The lowered quality carries through to the pasteurized or manufactured product.

Poor production methods—mixed bacteria plus poor cooling. This field shows the same general mixture of bacterial types, as well as poor cooling. This is evidenced by the large number of paired cells or diplococci. Poor cooling, together with contamination, develops actively growing organisms and off-flavors. A culture of these organisms can quickly spoil a large supply of milk.

Dust contamination—hay and dirt. This group of heavy, paired, rod-shaped organisms is evidence of contamination by a culture of large bacilli, typical hay and soil organisms. These are spore-formers and quite heat resistant. Note that the pairs sometimes form chains.

Manure contamination—heavy sediment. This group of short rods probably came from manure. Filter pads remove visible particles of manure or manure-contaminated soil or bedding. But the organisms separate from the solid particles and go through the filter. They readily break down milk proteins to give off-flavors and sweet curdling.

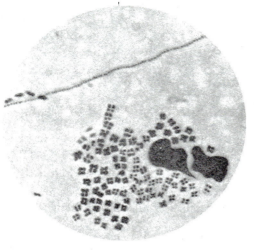

External contamination—tetrads. This typical arrangement of four micrococci is often associated with milkstone (accumulation of dried milk protein). Tetrads are heat loving. They sometimes come from the teat canal. They are relatively resistant to germicidal agents. Generally these organisms are considered "inert," that is, they cause little if any damage to the milk.

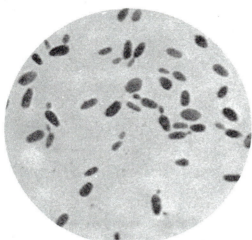

Yeast cells. These are much larger than bacteria. Normally spherical or slightly elongated, they frequently show small buds, their manner of reproduction. Yeasts normally are present in milk because of contamination by dirty rubber parts in milking machines. They may also be invaders in the case of mastitis, but this is not common. They produce a gassy, yeasty breakdown in cheese and, less frequently, in milk.

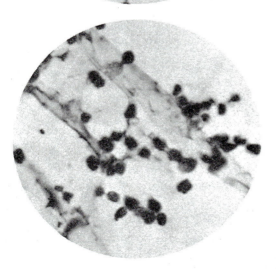

Molds—long, heavy threads or rods. Mold contamination is not commonly found in fresh milk unless it has come in contact with a moldy scum of milk solids. Occasionally, neglected rubber parts will cause this problem.

Prepared by H. V. Atherton and W. A. Dodge Animal and Dairy Science Department, University of Vermont.

EXERCISE 61
MASTITIS MILK

Pathogens of both humans and milking animals include *Mycobacterium* (tuberculosis), *Brucella* (brucellosis), *Coxiella* (Q-fever), and *Listeria* (listeriosis). The negative economic impact these organisms have as well as the illness and death they cause is well known. Many people, however, do not realize the importance of the organisms that cause mastitis.

Mastitis is an inflammation of the tissues of the udder. The predominant organisms causing bovine mastitis include species of the genera *Staphylococcus* and *Streptococcus*, although gram-negative rods are sometimes involved. Direct consumption of mastitis milk, or of dairy products made with such milk, poses a public health risk to the consumer. This is because the organisms involved are pathogens and they release disease causing products (enterotoxins) that are not destroyed by pasteurization.

Several methods are available for the detection of mastitis milk and are employed with relative success. These include rapid methods such as the **California** or **Wisconsin Mastitis Tests,** the **modified Whiteside test;** direct microscopic methods, including the **direct microscopic somatic cell count** (DMSCC); and the longer, more traditional **cultural methods** for the detection of specific organisms.

A. CALIFORNIA MASTITIS TEST

In the California Mastitis Test (CMT), a CMT reagent is added to the milk to be tested. The reagent contains a detergent and pH indicator (bromcresol purple). The detergent in the reagent combines with DNA in somatic cells in the milk, eliciting precipitate formation at a concentration of 150,000 to 200,000 somatic cells per ml. As the number of somatic cells increases, the precipitate becomes thicker, eventually forming a gel. Somatic cells are present in milk in high numbers due to infection. Thus, this test is used as a rapid screening for milk from diseased animals. The grading system (table 61-1) is based on the amount of gel formation and is basically subjective.

Materials

CMT reagent

Milk paddles

Mastitis milk

Procedure

1. Mix the milk sample to be tested by inverting the sample gently 6 times.

2. Place 2 ml of milk into a milk paddle.

3. Add 2 ml CMT reagent. Gently rotate for 10 seconds to allow mixing of the milk and reagent.

4. Grade the milk (see table 61-1).

5. Rinse the paddle and shake off excess moisture.

Questions

1. What is mastitis?

2. Name some organisms responsible for mastitis.

3. Explain how the California Mastitis Test works to determine milk quality.

4. Why should there be a large number of somatic cells in mastitic milk?

B. MODIFIED WHITESIDE TEST

The modified Whiteside test, like the CMT, is a subjective grading of a chemical reaction between somatic cells in the test milk sample and a reagent. In this test, the reagent is 4% NaOH. The reaction is one of increasing coagulation and decreasing opacity in milk as the somatic cell count increases.

Materials

Mastitis milk

4% NaOH

Test plates (slides over a black background with marked areas of 4 sq. cm.)

Eyedroppers

Applicator sticks

Procedure

1. Gently mix the milk by inverting the container 6 times. Place 5 drops of milk onto a slide held over a black background.

2. Add 2 drops of 4% NaOH.

3. Mix the milk and NaOH with an applicator stick over an area of 4 sq. cm. for 25 to 30 seconds.

4. Grade the milk (fig. 61-1).

5. Wash the slides.

TABLE 61-1.
GRADING AND INTERPRETING MILK SAMPLES (CMT)

SYMBOL	SUGGESTED MEANING	VISIBLE REACTION	INTERPRETATION (CELLS PER ML)
—	Negative	Mixture remains liquid with no evidence of precipitate formation	0 to 200,000
T	Trace	A slight precipitate forms, seen to best advantage by tipping paddle back and forth and observing mixture as it flows over bottom of cup. Trace reactions tend to disappear with continued movement of fluid.	150,000 to 500,000
1	Weak positive	A distinct precipitate forms, but with no tendency toward gel formation. In some milks, reaction is reversible with continued movement of the paddle and the precipitate may disappear.	400,000 to 1,500,000
2	Distinct positive	Mixture thickens immediately with some suggestion of gel formation. Upon swirling, mixture tends to move toward the center, leaving bottom of outer edge of cup exposed. When the motion is stopped, mixture levels out, once again covering bottom of cup.	800,000 to 5,000,000
3	Strong positive	A gel is formed which causes surface of mixture to become convex. Usually there is a central peak which remains projected above the main mass after motion of paddle has been stopped. Viscosity is greatly increased, so that the mass tends to adhere to bottom of cup.	Generally greater than 5,000,000
+	Alkaline milk	This symbol should be added to the CMT score whenever the reaction is distinctly alkaline, as indicated by a contrasting deeper purple color.	An alkaline reaction reflects depression of secretory activity, occurring as a result either of inflammation or of drying-off of the gland.
Y	Acid milk	Bromcresol purple is distinctly yellow at pH 5.2. This symbol should be added to the score when mixture is yellow.	Distinctly acid milk in the udder is rare. When encountered, it indicates fermentation of lactose by bacteria.

By permission. *Standard Methods for the Examination of Dairy Products*, 14th edition. American Public Health Association: Washington, D.C., 1978.

Questions

1. How are the modified Whiteside and the CMT tests similar?

2. What does the 4% NaOH do in the Whiteside test?

2. Why must you work in well ventilated areas with this stain?

3. Why is it important to observe more fields if the counts are low, yet observe less fields if the counts are high?

C. DIRECT MICROSCOPIC SOMATIC CELL COUNT

The direct microscopic somatic cell count (DMSCC) is an adaptation of the direct microscopic count. Levowitz-Weber stained milk smears are observed microscopically under oil immersion and leukocytes and somatic cells are counted. In this exercise, you should follow the procedure given previously in Exercise 60C.

Questions

1. How can you differentiate between somatic cells, bacteria, milk protein, and debris in milk when samples are stained with Lebowitz-Weber stain?

D. CULTURAL METHODS FOR MASTITIS MILK

Although tests such as the California mastitis test (CMT) and modified Whiteside test allow for rapid screening for mastitis milk, they do not identify the causative agent of the udder infection. Identification of these is usually performed by cultural methods using a variety of selective or differential media. Milk samples from suspected mastitis cases may be plated onto blood agar, eosin methylene blue agar, and phenol red mannitol salt agar. These media allow the growth or select for the most frequently encountered mastitis bacteria: hemolytic streptococci, coagulase-positive staphylococci, coliforms, and pseudomonads.

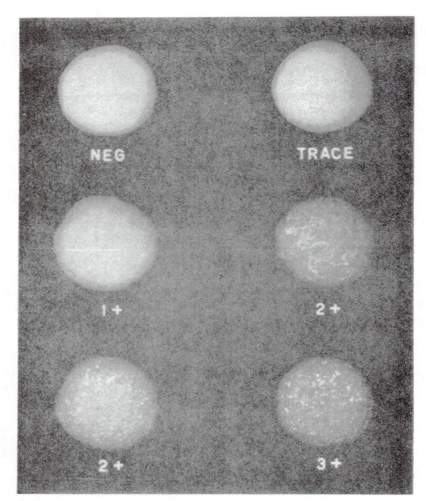

Figure 61-1. Appearance of Scoring Reactions in the Modified Whiteside Test under Magnification

Materials

Blood agar plates

Eosin methylene blue (EMB) agar plates

Phenol red mannitol salt (PRMS) agar plates

Stock cultures of *Escherichia coli*, a pseudomonad, *Staphylococcus aureus*, and a hemolytic streptococcus

Mastitis milk

Inoculating loop

Incubator set at 35°C

Procedure

First Period

1. Divide one plate each of blood, EMB, and PRMS agar into 5 sections each.

2. Into one section on each plate, streak the *E. coli*.

3. Streak the pseudomonad onto a second section of each plate.

4. Continue the process until all 4 control organisms have been streaked onto the three media.

5. Streak the mastitis milk sample onto the 5th section of each plate. Incubate all plates at 35°C for 24 to 48 hours.

Second Period

1. Examine each of the plates for growth and appearance of the various test organisms and for possible growth from the mastitis milk.

2. Record your results. Discard the plates as indicated by your instructor.

Questions

1. Explain why, of the organisms tested, only *S. aureus* grew on phenol red mannitol salt agar.

2. Explain why only the *E. coli* and the pseudomonad would grow on eosin methylene blue agar.

3. What are the advantages of cultural examination of suspected mastitis milk over the CMT or modified Whiteside tests? The disadvantages?

4. List the three types of hemolysis, the action of the organisms on red blood cells and the manifestations on the blood agar plate.

EXERCISE 62
PASTEURIZATION OF MILK

Pasteurization is a heating process that results in the reduction or elimination of spoilage organisms in foods such as beers, wines, and vinegar; or in the destruction of pathogens, in milk. In the dairy industry, the times and temperatures of heating are designed to eliminate all pathogens while not affecting the nutritional quality or taste of the milk. The treatments are sufficient to kill the most resistant of the asporogenous pathogens, *Mycobacterium tuberculosis* and *Coxiella burnetii*. Pasteurization also kills most of the nonpathogens in the milk. The surviving organisms are spore-formers and thermoduric/thermophilic organisms.

Three major pasteurization methods are used for milk: **low-temperature, long-time** (LTLT), or batch holding; **high-temperature, short-time** (HTST); and **ultra-high temperature** (UHT). These times and temperatures are, respectively, 63.5°C for 30 minutes, 72°C for 15 seconds, and 130°C or greater for 1 second or less.

The pasteurization of milk, in addition to removing pathogens, provides a product with an average refrigerated shelf life of 7 to 10 days. In addition, this pasteurized milk may be used to manufacture a variety of dairy products.

Materials

Raw milk in clean tubes, 10 ml/tube

1 ml pipets

9 ml dilution blanks

63.5°C water bath

Thermometer

Violet red bile agar (VRBA)

Plate count agar (PCA)

Colony counters

Incubator set at 32°C

Procedure

First Period

1. Remove a 1 ml aliquot of raw milk and prepare dilutions as shown in figure 62-1.

2. Pipet aliquots from the dilutions made in step 1 to the appropriately labeled sterile Petri plates as shown in figure 62-1.

3. Add melted and cooled VRBA and PCA to separate plates as directed. Allow to solidify.

4. When the VRBA plates have solidified, add an additional 5 ml melted and cooled VRBA as an overlay. Solidify.

5. Incubate the plates at 32°C for 24 hours.

6. Place the tube containing the raw milk in a water bath set at 63.5°C. Place a control tube with a thermometer in it beside the sample tube. When the control tube reaches 63.5°C, begin timing the ½ hour pasteurization time.

7. After the pasteurization time has elapsed, remove the sample milk from the water bath. Prepare dilutions as shown in figure 62-2 and plate VRBA and PCA as indicated. Remember to overlay the VRBA plates. Incubate at 32°C for 24 hours.

Second Period

1. Examine the PCA plates from the raw and pasteurized samples. Determine the number of colony-forming units per ml of raw and of pasteurized milk.

2. Examine the VRBA plate for typical coliforms, as described in Appendix D. Determine the coliform plate count per ml of both raw and pasteurized milk.

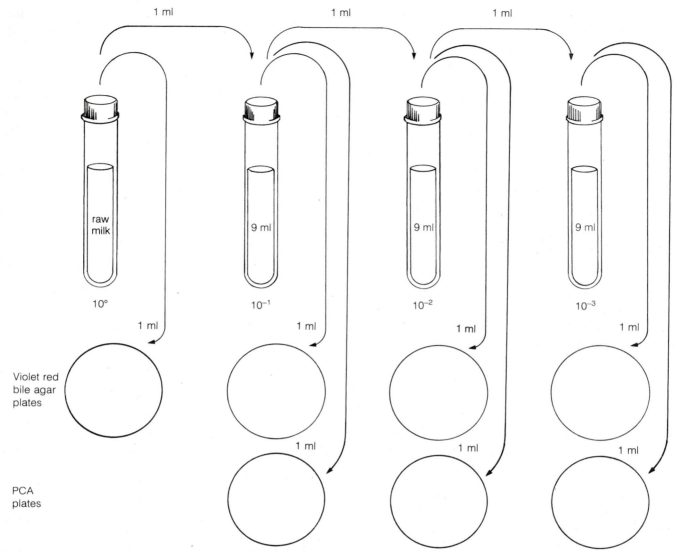

Figure 62-1. Dilution Series for Raw Milk Analysis

3. Compare your results to the standards given in table 62-1.

TABLE 62-1
BACTERIAL LIMITS FOR GRADE A RAW AND PASTEURIZED MILK

PRODUCT	LIMITS
Raw milk	200,000 to 400,000 CFU/ml 50 to 150 coliforms/ml
Grade A Pasteurized milk	less than 20,000 CFU/ml less than 5 coliforms/ml

Questions

1. Define pasteurization as it applies to the dairy industry.

2. What organisms tend to survive pasteurization? What organisms are killed?

3. Which microbes are the most heat-resistant of the asporogenous pathogens?

4. Differentiate between LTLT, HTST, and UHT as means of pasteurization.

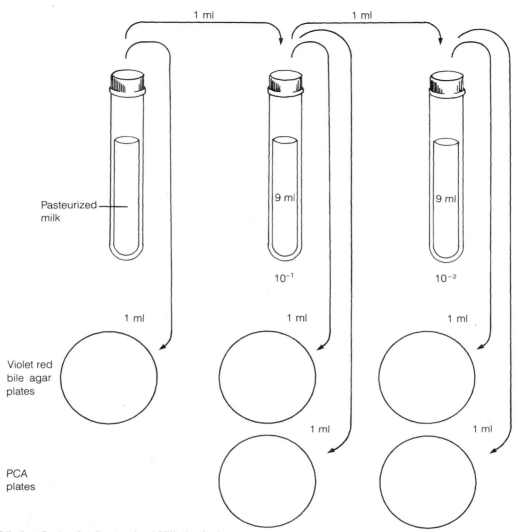

Figure 62-2. Dilution Series for Pasteurized Milk Analysis

EXERCISE 63
MICROBES AND THE
PRODUCTION OF DAIRY FOODS

Home and commercial production of dairy products is initiated by the appropriate lactic acid starter cultures. These organisms are necessary for the initial steps in cheese production as well as the production of strictly fermented products such as yogurt.

The lactic starters invariably include a strain that produces lactic acid (homolactics), such as *Strepto-*

coccus lactis, S. cremoris, or *S. diacetilactis.* If the desired end product requires flavor and aroma compounds such as diacetyl, the starter also includes heterolactics such as *Leuconostoc citrovorum* or *L. dextranicum.* Depending on the desired end product and incubation conditions employed in the manufacture of the product, the starter cultures may be single or mixed strains. In general, the streptococci make up around 90% of the mixed starter populations and convert most of the lactose to lactic acid. These organisms are capable of reducing the pH to 4.3 to 4.5 and produce a titratable acidity (calculated as lactic acid) of 0.8 to 1.0%.

A. YOGURT PRODUCTION

A simple and readily demonstrable example of dairy lactic fermentations is yogurt production. *Streptococcus thermophilus* and *Lactobacillus bulgaricus*, in a ratio of 1:1, are used as the starter cultures, although other lactic acid bacteria may be present and alter the flavor and aroma. During incubation, the acid-producing *Streptococcus* grows more rapidly than the flavor and aroma-producing *Lactobacillus*. A schematic outline of the end products is given as follows:

$$lactose \rightarrow glucose \rightarrow pyruvate \rightarrow \begin{cases} lactate\ (90\%) \\ acetate + CO2 \\ ethanol + CO2 \\ volatile\ fatty\ acids \\ flavor/aroma\ products \end{cases}$$

Materials

Milk (pasteurized)

Warm tap water

Instant powdered skim milk

Evaporated milk

Plain yogurt

46°C incubator

Magnetic stirrer and stirring bar

Plastic or styrofoam cups

Aluminum foil

Gram reagents

Balance

Procedure

First Period

1. Prepare a smear of commercial yogurt. Gram stain and observe for Gram-positive rods and cocci.

2. Prepare yogurt by one of the two methods given below.

Method A

   a. Add 125 ml warm tap water to a 1 liter flask.

   b. Add 135 g instant powdered milk and mix.

   c. Blend in 43 g yogurt.

   d. Add 275 ml warm tap water.

   e. Add 94 ml evaporated milk and stir.

   f. Decant into cups, cover with foil, and incubate at 46°C overnight. Refrigerate if not consumed immediately.

Method B

   a. Bring 1 liter of milk to a boil (a microwave oven is suitable).

   b. Add 5% by weight instant powdered milk and stir.

   c. When the milk cools to about 45°C, add 2% by volume yogurt. Blend.

   d. Decant and incubate as in step f above.

Second Period

1. Gram stain a smear of the prepared yogurt.

Questions

1. What group of bacteria is responsible for yogurt production? Which is the flavor and aroma culture? Which is responsible for the high acidity?

2. In making yogurt with pasteurized milk, why is the milk heated prior to adding the cultures?

3. Distinguish between homolactic and heterolactic bacteria in terms of their major end products from sugar fermentation.

B. MICROORGANISMS AND CHEESE PRODUCTION

Lactic fermentation of milk sugar is required for cheese production. The starter culture which is added differs depending upon the temperature at which the milk is to be incubated. The acid produced as well as added **rennin,** cause the curdling of the milk protein (curd formation). The **curd** is subsequently shrunk and pressed. It is then salted and allowed to ripen if a ripened cheese is desired.

In the ripening process, the cheese undergoes alterations in flavor, consistency, and texture. Under natural conditions, the ripening is due to the continued action of the rennin and of the lactic acid bacteria. The product is cheddar cheese.

Many cheeses, however, are ripened by the addition of specific organisms to the curd. In Swiss cheese, for example, the bacterium *Propionibacterium* is added. This organism converts the lactic acid to propionic acid, acetic acid, and CO_2. The propionic acid is primarily responsible for the taste and the carbon dioxide for the obvious holes in this cheese.

In producing soft and semisoft cheeses, ripening occurs because of bacterial activity on the surface of the cheese. The characteristic surface color and odor of Limburger cheese, for example, is due to the growth of *Brevibacterium*.

Not all cheeses are ripened with bacteria. Some, such as blue cheeses (e.g., roquefort and camembert), are produced by inoculating the curds with spores of the molds *Penicillium roquefortii* and *P. camembertii*, respectively. The characteristic blue veins appearing in these cheeses are due to the growth of fungal filaments.

Materials

Milk

Buttermilk

Cheesecloth

Samples of various cheeses such as blue, limburger, and Swiss

Gram staining reagents

Gelatin-peptone agar plates

Acidified potato dextrose agar plates

Inoculating loops

Incubators set at 30°C and 37°C

pH paper

Procedure

First Period

1. Prepare emulsions of the cheeses by grinding samples in sterile diluent.

2. Prepare smears of the emulsions and perform Gram stains.

3. Streak plates of gelatin-peptone agar with a loop of emulsion from the various cheeses. If a blue cheese is used, streak plates of acidified potato dextrose agar. Incubate the plates at 30°C for 2 to 7 days.

4. Inoculate 950 ml of milk in a beaker, heated to 37°C, with 30 ml buttermilk. Stir to obtain a good mixture.

5. Check the pH.

6. Cover the beaker with cheesecloth or a dishtowel. Incubate at 37°C for 48 hours.

Second Period

1. The unripened cheese in the beaker may be wrapped tightly in cheesecloth and allowed to ripen for a period of time as indicated by the instructor.

Questions

1. Why are most cheeses produced with pasteurized milk rather than raw milk?

2. What is the function of rennin? What is it and what is its source?

3. What end products are responsible for the flavor and appearance of Swiss cheese?

4. What is cottage cheese?

5. What would happen if you added vinegar to milk? Why?

PART XV
AGRICULTURAL MICROBIOLOGY

Microorganisms play diverse roles in agriculture, some of which are beneficial to humans and others detrimental. Beneficial activities include the cycling of nutrients such as nitrogen and sulfur (described previously in Part XI). Microorganisms improve soil fertility by forming compost and establish beneficial associations with plants (e.g., nitrogen fixation in legumes and alders) and animals (e.g., cellulose decomposition in rumens of ruminants).

In recent years some microorganisms have been used as substitutes for chemical insecticides in the continuing battle to control insect damage to valuable crops. Many destructive insects in agriculture, such as the cotton bollworm and cabbage looper, can now be controlled with microorganisms. These bioinsecticides thus present a viable alternative to chemical control of some insects.

A variety of microbial diseases affect farm animals. These include anthrax, fowl cholera, listeriosis, brucellosis, mastitis, foot and mouth disease, hog cholera and encephalitis. A variety of viruses, bacteria, fungi, and nematodes are also capable of causing diseases in plants. These plant pathogens are usually transmitted by water, wind, or insects. The symptoms usually include necrosis (death of all or part of the plant), wilt (loss of turgor), blight (death of foliage), galls (overgrowth or tumorlike warts), and chlorosis (lack of chlorophyll).

EXERCISE 64
PLANT PATHOGENS

Plant pathogens usually invade susceptible plants either by penetration through natural openings in the plant or through openings caused by mechanical abrasions. Bacterial pathogens such as *Pseudomonas*, *Agrobacterium*, *Xanthomonas*, and *Erwinia* cause a wide variety of diseases including galls, blights, wilts, and rots. Viruses typically cause mosaic diseases and leaf curls. The fungi, which are responsible for the majority of plant diseases, cause necrotic diseases such as root rot, soft rots, blights, and cankers; hypertrophies such as galls, warts, and leaf curls; as well as wilts, rusts, and mildew.

Plant diseases are varied and depend upon the activities of the invading organisms. Some plant pathogens inhibit or otherwise affect photosynthesis, others transform normal plant cells into tumor cells, still other pathogens inhibit the transfer of water through plants, and some break down the cementing material that holds plant cells together.

A. RUSTS AND SMUTS

Fungal diseases are often grouped as root rots, vascular wilts, rusts, smuts, or blights. The rusts and smuts are caused by obligate plant parasites and pathogens. The fungi causing smuts have a relatively simple life cycle, while those causing the rusts may have either a simple or complex life cycle.

Rust and smut diseases have been known to humans throughout history. Numerous Old Testament references are made to rusts destroying wheat. The diseases were called "rusts" because of the appearance of red-colored spores developing on the plants, giving them a rusty appearance.

A common rust disease is the wheat stem rust, caused by *Puccinia graminis*. This organism requires two different host species to complete its life cycle. Overwintering **teliospores** germinate in the spring, and the resultant wind-borne **basidiospores** infect only barberry leaves. After undergoing part of the life cycle in barberry, the resultant *Puccinia* **aeciospores** become wind-borne and are carried to wheat plants, where the disease can spread throughout an entire wheat-growing area (fig. 64-1).

Materials

Prepared slides of *P. graminis*
Barberry leaves infected with *P. graminis*

Plants bearing red rust (urediospores)
Prepared slides of *P. graminis* urediospores
Plants bearing teliospores (black rust stage)
Prepared slides of teliospores of *P. graminis*
Plants infected with smuts
Prepared slides of corn smut

Procedure

1. Examine the infected plants under a dissecting microscope. Note the colors.

2. Examine the provided prepared slides.

3. Observe the malformations on the smut infected corn.

4. Mount some smut material in detergent. Observe the smut spores.

5. Draw, using a simple diagram, the life cycle of *P. graminis*.

Questions

1. Distinguish between rots, wilts, rusts, smuts, cankers, and blights.

2. What kind of microorganisms cause rusts and smuts?

3. How might these organisms be transmitted in nature?

B. SOFT ROT

A soft rot, as its name implies, is a softening and discoloration of plant roots, stems, or storage organs primarily due to bacterial or fungal infection. In many instances, softening is caused by **pectinolytic enzymes** produced by these organisms. Plant pectin serves as the intracellular cementing material in plant tissues. A common organism responsible for some soft rots is the bacterium *Erwinia carotovora*.

Materials

Small, firm potatoes
Firm carrots
Broth cultures of *E. carotovora* and *E. atroseptica*
Sterile Petri dishes

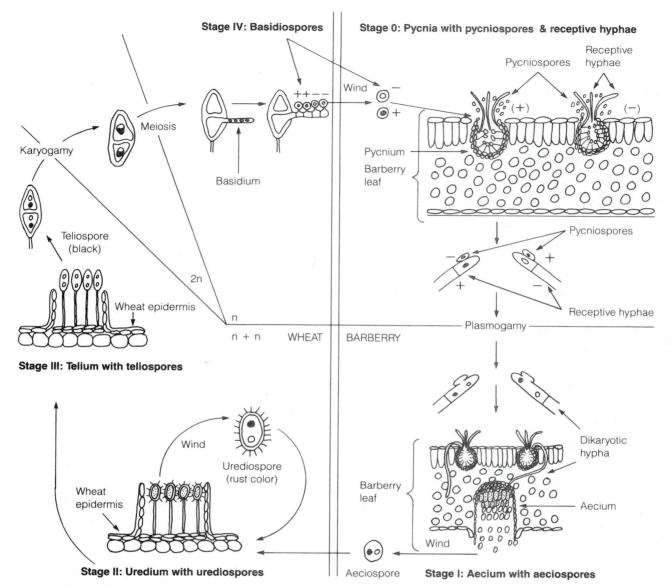

Figure 64-1. Life Cycle of *Puccinia graminis*

Labels in figure:

Stage IV: Basidiospores

Stage 0: Pycnia with pycniospores & receptive hyphae

Pycniospores

Receptive hyphae

Wind

Karyogamy

Meiosis

Basidium

Pycnium
Barberry leaf

(+)

(−)

Teliospore (black)

2n

Pycniospores

Receptive hyphae

Plasmogamy

n

Wheat epidermis

n + n   WHEAT   BARBERRY

Stage III: Telium with teliospores

Wind

Urediospore (rust color)

Dikaryotic hypha

Wheat epidermis

Barberry leaf

Aecium

Wind

Aeciospore

Stage II: Uredium with urediospores

Stage I: Aecium with aeciospores

Sterile distilled water

Alcohol

Scalpel, forceps

10 ml sterile pipets

Pasteur pipets

Sterile filter paper

Inoculating loop

Procedure

First Period

1. Wash the potatoes and carrots in tap water.

2. Dip the vegetables into 95% alcohol. Remove them and place in a sterile Petri dish to allow evaporation of the alcohol.

3. Alcohol-sterilize the forceps. Prepare slices of the potatoes and carrots (about 1 cm thick).

4. Place sterile filter paper into 6 sterile Petri plates. Moisten the filter paper with sterile distilled water. Place a potato slice into each of three dishes, and a carrot slice into each of the remaining three dishes.

5. Place a drop or two of *E. carotovora* onto one potato and one carrot slice, a drop or two of *E. atroseptica* onto a second slice of each, and sterile distilled water onto the third slice of each vegetable.

6. Place the lids on the Petri dishes. Incubate at room temperature for 10 to 14 days.

Second Period

1. Periodically examine the slices and look for discoloration. Remoisten the filter paper, if necessary, to prevent drying of the vegetable slices.

2. Sterilize an inoculating loop and touch the slices to determine firmness.

3. Note the odor of the vegetables.

4. Should soft rot symptoms appear, prepare a smear and perform a Gram stain.

5. Record your results. Discard the Petri plates and vegetables as indicated by your instructor.

Questions

1. Name four genera of bacterial plant pathogens.

2. Distinguish between galls, blights, wilts and rots.

3. What is the cellular morphology and Gram reaction of most bacterial plant pathogens?

4. Why were the carrot and potato slices dipped in alcohol?

C. CROWN GALL

Crown gall is caused by the soil bacterium *Agrobacterium tumefaciens*, a Gram-negative, motile organism. This bacterium carries with it a plasmid termed pTi that contains a sequence of genes responsible for crown gall. When roots or stems of susceptible plants (e.g., tomatoes) have been injured (by grafting, insects, or mechanical means), the pTi plasmid can penetrate the injured plant cells. The plasmid then induces a hormonal imbalance in the plant that ultimately leads to the proliferation of plant cells (**hyperplasia**) and the formation of tumors (**galls**). Figure 64-2 shows a diseased plant.

In this exercise, **Koch's postulates** will be introduced. To prove the etiological agent of a disease, these postulate must be fulfilled. One must first observe disease symptoms in an individual. The causative agent must then be isolated from the diseased individual and cultured in pure form in the laboratory. This pure culture, when inoculated into a healthy individual, must elicit the same symptoms as noted in the original diseased individual and must be reisolated from that second host.

Materials

Healthy and diseased (crown gall) tomato plants

Mortar and pestle

Sterile diluent

Trypticase soy agar plates

Gram stain reagents

Scalpel

Alcohol

Dissecting needles

Incubator set at 30°C

Procedure

First Period

1. Carefully remove some mature crown galls from a young tomato plant using an alcohol-sterilized scalpel. Sterilize the galls by dipping into alcohol. Allow the alcohol to evaporate.

2. Grind the galls using a mortar and pestle and sterile diluent.

3. Streak a loopful of the resulting pulp onto a plate of TSA.

4. Incubate at 30°C for 48 hours.

Second Period

1. Examine the TSA plates for *Agrobacterium*. Gram stain to look for Gram-negative rods. Keep the slides.

Figure 64-2. Genetic Engineering in Nature. A common soil bacterium, *Agrobacterium tumefaciens*, is able to infect certain plants like tomatoes, tobaccos, and begonias. During some of these infections, bacterial plasmids enter the plant cells and become part of the plant chromosomes. The bacterial plasmids disrupt the plant cells so that they reproduce and grow in an uncontrolled manner. Tumor-like growths on the stem are the result.

2. With a dissecting needle touch a colony and inoculate the bacterium along the stem of a healthy young tomato plant. Several inoculations may be necessary to initiate the infection.

3. Place the plants on a side bench and examine weekly for 2 to 3 weeks.

Third Period

1. When galls become evident, repeat the procedures as given in period 1, steps 1 and 2.

2. Perform a Gram stain from the pulp. Compare the morphology of the *Agrobacterium* cells with those of the cells stained from the TSA plates.

3. Record your results. Discard the plates as indicated by your instructor.

Questions

1. What is a plasmid? What role does the *Agrobacterium* plasmid play in crown gall?

2. How might you demonstrate that *Agrobacterium* causes crown gall?

3. How might crown gall be transmitted?

4. Is there any difference in the cellular morphology and size of *Agrobacterium* growing on TSA as compared to the pulp specimen? Explain.

D. TOBACCO MOSAIC VIRUS

Tobacco mosaic virus causes spotting or mottling of the leaves of a variety of susceptible plants including, of course, tobacco. The disease manifests itself as brown necrotic spots or white to yellow mottling on the leaves of infected plants (fig. 64-3).

Some of the mosaic viruses have been used in ornamental horticulture to produce variegations on leaves of plants, yielding plants with striking foliage and blossoms.

Materials

Tobacco plants with TMV

Healthy tobacco plants

Clean blenders

Sterile diluent

Funnel

Filter paper

Bacteriological filter apparatus with 0.45 um pore size

Carborundum

Cotton swabs

Figure 64-3. Viral Lesions. A leaf with light-colored zones that represent tobacco mosaic virus lesions. The leaf can be thought of as a lawn of plant cells.

Procedure

1. Remove several leaves from an infected tobacco plant. Mix them in a blender with about 100 ml sterile diluent.

2. Pour the resulting slurry through a funnel containing filter paper to remove particulate plant material.

3. Filter the resulting fluid through a bacteriological filter to remove all microorganisms except viruses, which will pass through the filter.

4. Rub the leaves of healthy tobacco plants with carborundum to cause abrasions.

5. Soak a cotton swab in the filtrate and rub onto the plants where the abrasions were made.

6. Put the plants on a side bench for 2 to 3 weeks, observing periodically for the appearance of mosaic disease.

Questions

1. Why is tobacco mosaic virus detrimental to some plant crops such as tobacco? What other plants are susceptible?

2. How might tobacco mosaic virus be used beneficially in the ornamental horticulture industry?

3. Would you expect that the virus could be isolated from commercial tobacco products? Explain.

EXERCISE 65
MYCORRHIZAE

Mycorrhiza means "fungal root," however, this term and the term **mycorrhizal** are used to describe a symbiotic relationship between plant roots and fungi. It has been suggested that most uncultivated plants and many crops are mycorrhizal. The fungi appear to aid almost all vascular plants by solubilizing inorganic nutrients, and by taking up water and organic materials from the rhizosphere. The more extensive the hyphal network, the better the fungus provides for the plant.

The influence of the mycorrhizal association can best be observed in poor soils, where nonmycorrhizal tree species grow poorly in comparison with mycorrhizal counterparts such as conifers, beeches, and oaks.

Fungi involved in these mycorrhizal relationships are widespread and belong to the imperfect fungi—ascomycetes, basidiomycetes, and zygomycetes. Most truffles and mushrooms found under specific tree species are mycorrhizal basidiomycetes.

Materials

Mycorrhizal demonstrations and prepared slides of mycorrhizal roots (e.g. orchid root)

Phloxine

Scalpels

Clean glass slides

Procedure

1. Examine the prepared slides of mycorrhizal roots.

2. Observe the many short lateral branched roots of mycorrhizal pines.

3. Make a thin cross-section of provided mycorrhizae and stain with phloxine. Examine microscopically.

Questions

1. Define mycorrhizae.

2. What are the benefits of mycorrhizal relationships to the fungus? To the plant?

3. What other symbiotic relationships are common between a microorganism and a plant?

EXERCISE 66
BIOINSECTICIDES

A variety of microorganisms are capable of causing disease or death in insects. These **insect pathogens** have been found, in some cases, to be effective in controlling insect infestation on plants, especially on edible crops. Perhaps the most commonly known **bioinsecticide** is *Bacillus thuringiensis*, which can be purchased commercially from local nurseries.

B. thuringiensis is an endospore-forming bacterium. During sporulation, the organism produces a crystalline protein, or **parasporal body,** that is liberated with the endospore. This parasporal body dissolves at an alkaline pH and can cause paralysis of the gut muscles of susceptible insects, specifically those with alkaline guts. This bioinsecticide has been successfully used against various insects such as cabbage worms, tent caterpillars, and gypsy moths.

Materials

Cabbage, brocolli or bean leaves

Trichoplusia ni (Hubner) first instar larvae

Glass vials

Dipel® (commercial preparation of *Bacillus thuringiensis*)

Procedure

First Period

1. Spray the leaves of the vegetable provided with a solution of Dipel as recommended by the manufacturer. Allow the leaves to dry. Keep a few unsprayed leaves as controls.

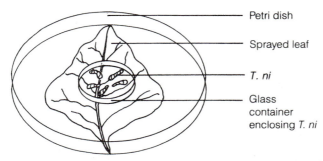

- Petri dish
- Sprayed leaf
- *T. ni*
- Glass container enclosing *T. ni*

Figure 66-1. *Trichoplusia ni* on vegetable leaves sprayed with bioinsecticide

2. Using clean scissors, cut leaves from both test and control plants (about 3 sq. cm. each). Place one leaf into each sterile Petri dish.

● You may wish to place a sterile filter paper on the bottom of the dish, then wet it with sterile water to maintain a moist environment.

3. Add 5 first instar larvae of *T. ni* onto each leaf, including the controls. Place an inverted glass vial over the leaves to prevent the larvae from escaping (fig. 66-1).

4. Place the plates on a side bench and observe daily for 5 days.

Second Period

1. Record the mortality rates on the control and test leaves.

2. Record your results. Discard the materials as indicated by your instructor.

Questions

1. What is a parasporal body?

2. Discuss the mode of action of the *B. thuringiensis* bioinsecticide.

3. What advantages might bioinsecticides have over chemical insecticides? What disadvantages?

EXERCISE 67
EXAMINATION OF RUMEN FLUID

Ruminants are herbivorous mammals in which the initial digestion of food takes place in a special organ called the **rumen.** Such animals include cows, sheep, goats, deer, elk, moose, and camels. These animals consume hay and grass as their principle food. Such plant material is rich in cellulose, which cannot be digested by most other higher animals.

Cellulose digestion by ruminants is due not to enzymatic action of the animal itself but to anaerobic, cellulolytic microbes present in the rumen. These organisms hydrolyze the cellulose to simple sugars that are then fermented, yielding organic acids. The acids pass through the rumen to the bloodstream and act as the main carbon and energy source for the animal.

The predominant cellulolytic organisms in the rumen are bacteria, primarily Gram-negative rods and vibrioid forms. Protozoa, especially ciliated forms, are also present in large numbers.

Materials

Rumen fluid

Depression slides

Cover slips and vaseline

Gram staining reagents

Minimal medium + 0.1% glucose + filter paper strips, 10 ml/tube

Anaerobe jars

Incubator set at 37°C

Procedure

First Period

1. Observe samples of the rumen fluid microscopically by preparing hanging drops and Gram stains.

2. Inoculate tubes of minimal media with 1.0 and 0.1 ml aliquots of rumen fluid.

3. Incubate the tubes in anaerobe jars at 37°C for 2 to 3 weeks. Observe routinely for cellulose digestion by shaking the anaerobe jar gently and noting the integrity of the filter paper strips.

Second Period

1. When disintegration of the paper occurs, remove the tubes from the jar.

2. Observe microscopically by preparing hanging drops and Gram stains.

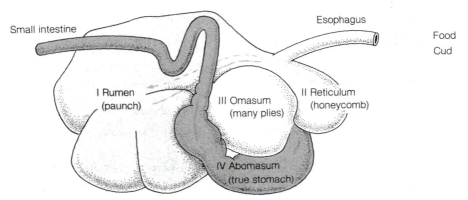

Figure 67-1. The Rumen. Food enters the rumen, where it is digested by various microorganisms. The partially digested food or cud is regurgitated, chewed, and swallowed again. The chewed cud enters the reticulum, and then passes through the omassum, abomassum, and finally the intestines. In the rumen, the cellulose is digested to fermentable sugars by the rumen microbiota. The sugars are utilized by the microbiota as carbon and energy sources. The ruminant obtains its nutrients as fermentation byproducts of microbial metabolism or as excess microbial mass.

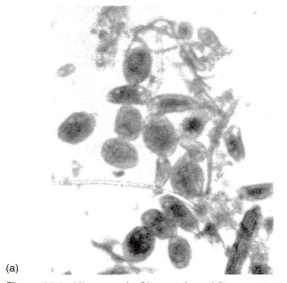

(a)

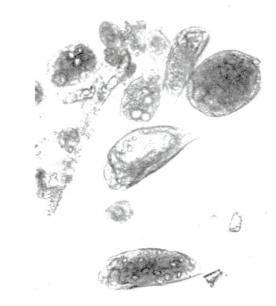

(b)

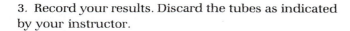

Figure 67-2. Microscopic Observation of Rumen Fluid

3. Record your results. Discard the tubes as indicated by your instructor.

Questions

1. What is a ruminant? What animals are included in this group?

2. What permits ruminants to digest cellulose?

3. What are the predominant cellulolytic organisms in the rumen?

4. What is a "fistulated" cow?

PART XVI
MEDICAL MICROBIOLOGY

Medical microbiology is a broad and rapidly changing field of applied microbiology. It deals with the cause, diagnosis, treatment, and prevention of infectious diseases. This area of microbiology, more than any other, has attracted considerable public attention because of the newsworthiness of epidemics or "new" diseases that occasionally appear in human populations. Much of the early work that took place during the Golden Age of Microbiology, pioneered by Louis Pasteur, Robert Koch, Paul Ehrlich, Elie Metchnikoff, and many others, dealt with various aspects of medical microbiology. During those early years, landmark discoveries were made in diagnostic microbiology, immunology, host-parasite interactions, and chemotherapy of infectious diseases. Many of these discoveries served as the bases for modern methods of treatment and control of infectious diseases.

In this section of the laboratory manual, you will perform some of the techniques that are in present use in the medical laboratory. You will first isolate various microorganisms that are part of your resident flora by using aseptic techniques and selective and differential media. In subsequent exercises, you will use various techniques to determine the susceptibility of pathogens to various antibiotics, determine your blood type, verify the identity of microorganisms, and finally, identify an unknown using a bacterial diagnostic kit commonly used in the clinical laboratory. As you perform all these exercises, keep in mind that similar (if not identical) procedures are routinely carried out in clinical laboratories all over the United States.

EXERCISE 68
NORMAL FLORA OF THE BODY

Microorganisms colonize all parts of the body that come into contact, either directly or indirectly, with the external environment. Microbes are found on the skin, in the oral cavity, the upper respiratory tract, the intestinal tract, and the urogenital tract.

The microbes that constitute the **normal body flora** are considered to be harmless under normal conditions, although some of the organisms are potential pathogens or opportunists. These latter organisms may cause disease if introduced to parts of the body that are generally free of microbial growth (e.g., the internal organs such as the bladder or kidneys) or to parts of the body where the normal flora that keeps invaders in check is altered.

The normal flora provides beneficial functions. Many of the organisms of the intestinal flora produce essential human growth factors (e.g., vitamin K) and aid in digestion. The normal flora also, by occupying space and using nutrients in the various habitats of the body, inhibit the invasion and growth of pathogens. Studies performed on germ-free animals also indicate that the normal flora enhances the functioning of the immune system.

The human fetus is normally free of microorganisms prior to passage down the birth canal. Colonization begins in the birth canal and continues after birth. Thus, by the time the infant is a few weeks old, it has a fairly well-developed flora.

A. THE ORAL CAVITY AND CARIOGENIC BACTERIA

The oral cavity (mouth) is a warm, moist environment that receives nutrients on a regular basis. It is, then, an ideal environment for microbial growth and does in fact have a diversified microflora. The organisms most often encountered are lactobacilli, streptococci, bacteroides, diphtheroids, filamentous forms and treponemes, as well as several species of protozoa, fungi, and viruses.

A variety of organisms constituting the indigenous microflora may cause diseases in the oral cavity. Such diseases include dental caries, cold sores (herpes), candidiasis (thrush), and Vincent's angina (trenchmouth). In most instances, the diseases of the oral cavity have their roots in poor dental hygiene.

Dental caries, or tooth decay, is a disease in which the hard, outer calcium layers of the teeth are broken down (decalcified) by acids produced primarily by cariogenic, plaque-forming bacteria. The causative agents appear to include *Streptococcus mutans* and *Streptococcus sanguis*. Although the susceptibility to dental caries is to some extent determined by an individual's genetic make-up, diet appears to play a significant role. The cariogenic streptococci, in the presence of sucrose, synthesize capsules that enable them and other of the oral microflora to attach to the teeth and to the gumlines. The streptococci are also fermentative, producing acids from the sugar. The acids are responsible for the decalcification of the hard outer surface of the teeth, thereby permitting bacterial attack on the pulp of the tooth which, if left untreated, can result in a variety of diseases.

Mitis-salivarius agar contains tellurite that selects for alpha and gamma hemolytic streptococci such as *S. mitis*, *S. salivarius* and *S. mutans*

Materials

Sterile toothpicks

Mitis-salivarius agar

Anaerobe jars and gas packs

Broth cultures of *Streptococcus mutans*, *Streptococcus mitis*, and *Streptococcus salivarius*

Gram stain reagents

Hydrogen peroxide

Incubator set at 35°C

Procedure

First Period

1. Using a sterile toothpick, scrape some dental plaque or tarter material from the buccal and lingual surfaces of your teeth.

2. Prepare a smear of the material. Gram stain and observe the morphological types.

3. Repeat step 1, but gently rub the toothpick onto a small section of a mitis-salivarius agar plate.

4. Streak for isolation on the mitis-salivarius plate.

5. Divide a second plate into three sections. Streak the control organisms, one to each section.

6. Incubate the plates anaerobically at 35°C for 24 hours, then remove from the jar until the next period.

Second Period

1. Compare the colonies of the control organisms with

the colonies that developed from the organisms in your mouth.

2. Perform Gram stains on the isolated colonies.

3. Perform catalase tests on the same colonies.

4. Record your results. Discard the plates as indicated by your instructor.

Questions

1. What are the predominant microbes in the oral cavity?

2. Why is good dental hygiene critical for disease prevention in the oral cavity?

3. Describe the process involved in the production of dental caries.

4. How might diet reduce the incidence of caries? Why?

B. SKIN FLORA

With the possible exception of those areas on the human skin that contain numerous hair follicles and sebaceous and sweat glands, the skin is a relatively dry ecosystem and is not a particularily favorable place for the growth of many microorganisms. Nutrients are limited, the concentration of salt may slow growth, and the acidic (pH 4 to 6) skin secretions are also somewhat inhibitory. The secretions of the sweat and sebaceous glands, however, promote the growth of the normal flora. The secretions provide water and liberate nutrients such as amino acids and lipids. Utilization of lipids by lipolytic flora results in the release of fatty acids that inhibit many organisms that might otherwise colonize the skin and subsequently cause disease.

The indigenous microflora of the human skin is primarily Gram-positive, and includes coccoid forms such as the micrococci and staphylococci, and diphtheroids such as the corynebacteria and propionibacteria, as well as some yeasts, such as *Pityrospum* sp., that require oily substances for growth.

Materials

Blood agar

Phenol red mannitol salt (PRMS) agar

Cotton swabs

Inoculating loop

Sterile diluent, 3 ml/screw cap tube

Gram staining reagents

Incubator set at 35°C

Procedure

First Period

1. Moisten a sterile swab in the diluent provided. Squeeze out the excess fluid by pressing the swab against the inner surface of the tube.

2. Rub the swab on a portion of skin (between the fingers; in the area between the nose and cheek).

3. Streak the material on the swab over a quadrant of a blood agar plate, carefully rolling the swab so that the entire cotton surface comes in contact with the medium. Discard the swab.

4. Using a sterilized inoculating loop, continue streaking for isolation.

5. Repeat the steps above with a fresh swab, substituting PRMS agar for the blood agar.

6. Incubate all plates at 35°C for 24 to 48 hours.

Second Period

1. Examine the blood agar plates. (Characterize the colonies and hemolytic activity (see Appendix D).

2. Examine the PRMS agar plates for staphylococci (see Appendix D).

3. Perform Gram stains on the predominant colonies.

4. Record your results. Discard the plates as indicated by your instructor.

Questions

1. Why is the skin a poor environment for the growth of most microbes?

2. What are the predominant bacterial genera found on the skin?

3. Where on the skin are these organisms most commonly found?

4. What are some skin diseases caused by the normal flora?

C. THROAT FLORA

The normal flora of the throat contains a variety of microorganisms. The most common of the indigenous populations include members of the staphylococci, streptococci, corynebacteria, spirochaetes, actinomycetes, and mycoplasmas, as well as yeasts and viruses.

Diagnosis of a common throat infection often may be achieved by taking a throat swab and plating the material on the swab onto appropriate media. Blood agar is one of the media routinely used, as it readily differentiates throat flora by hemolytic type.

Materials

Blood agar

Sterile swabs

Sterile tongue depressors

Inoculating loop

Hydrogen peroxide

Gram staining reagents

Incubator set at 35°C

Procedure

First Period

1. Perform a throat swab, using a sterile swab and, if needed to keep the tongue from interfering, a tongue depressor. When performing the throat swab, the area swabbed should be the tonsillar region, not the roof of the mouth.

2. Streak the material on the swab over one quadrant of a blood agar plate. Discard the swab.

3. Continue to streak for isolation, using a sterilized inoculating loop.

4. Incubate the plates at 35°C for 24 to 48 hours.

Second Period

1. Examine the plates. Characterize the colonies and hemolytic types.

2. Carry out Gram stains on the predominant organisms.

3. Perform a catalase test on the same colonies you Gram stained.

4. Record your results. Discard the plates as indicated by your instructor.

Questions

1. How might blood agar be useful in diagnosing the causative agent of a sore throat?

2. Is blood agar differential, selective, or both? Explain.

3. What is a "strep" throat? Are all sore throats "strep" throats? Explain.

4. Blood agar should be low in fermentable sugars. Why?

5. Blood agar should have 0.5% NaCl. Why?

EXERCISE 69
ANTIBIOTIC SENSITIVITY TESTING USING THE KIRBY-BAUER PROCEDURE

Antibiotics are substances produced by microorganisms (certain fungi and bacteria) that kill or inhibit the growth of other microorganisms. Antibiotics are used routinely in medicine to treat patients with infectious diseases. Modern medicine takes advantage of the many antibiotics that are commercially available to treat infectious diseases. Not all antibiotics, however, are effective against all infectious diseases. Before these drugs are administered to a patient, it must first be determined which one(s) to use. This is done by performing an **antibiotic sensitivity test** on the pathogen.

Antibiotic sensitivity assays involve testing selected antibiotics against a pathogen isolated from a diseased individual. The assay is carried out by inoculating a standard suspension of pathogen in broths containing various concentrations of an antibiotic. The lowest concentration that will inhibit the pathogen from re-producing in the broth can be measured by determining the turbidity of the broth after an incubation period. Inoculated broth tubes containing inhibitory concentrations of the antibiotic being tested will remain clear. Those tubes containing noninhibitory levels of antibiotics will become turbid, reflecting microbial growth. The efficiency of the antibiotic can then be determined by the lowest concentration of that antibiotic that will kill or inhibit the pathogen. This procedure for determining the effectiveness of an antibiotic, called the **minimum inhibitory concentration (MIC)** procedure, is very popular in clinical laboratories because it yields reliable results and is well-suited to automation. The dispensing of culture media, dilution of the various antibiotics, inoculation of microorganisms, and measuring of turbidity can all be preprogramed and computerized.

Another reliable procedure for determining the effectiveness of a drug, or the degree of sensitivity or resistance of a pathogen to various antibiotics, is the **Kirby-Bauer method.** The test, introduced by William Kirby and Alfred Bauer in 1966, consists of exposing a newly-seeded lawn of the bacterium to be tested, growing on a nutrient medium (Mueller-Hinton agar), to filter paper disks impregnated with various antibiotics. The culture is incubated for 16 to 18 hours and then examined for growth. If the organism is inhibited by one of the antibiotics, there will be a **zone of inhibition** around the disk, representing the area in which the microorganism was inhibited by that antibiotic.

The diameter of the zone of inhibition around an antibiotic disk is an indication of the sensitivity of the tested microorganism to that antibiotic. The diameter of the zone, however, is also related to the rate of diffusion of the antibiotic in the medium. This fact must be kept in mind when interpreting the zones of inhibition of various antibiotics (table 69-1). Figure 69-1 illustrates the basic procedure for the Kirby-Bauer test as well as some typical results.

The MIC and Kirby-Bauer tests are not restricted to the determination of antibiotic sensitivities of pathogenic organisms. These tests may be used to determine the sensitivity of any microorganism to a variety of antimicrobial agents, antibiotics, and synthetic chemotherapeutic drugs. This exercise, however, will be restricted to antibiotics.

You will determine the antibiotic sensitivity pattern for each of three different bacteria. At the end of the exercise, you will share your results with the rest of the class and reach conclusions as to the most effective way of treating infections caused by the three microorganisms.

Materials

Broth cultures of *Staphylococcus aureus, Escherichia coli,* and *Pseudomonas aeruginosa*

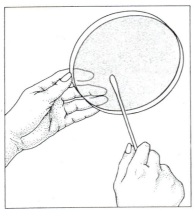

(a) Spread diluted culture over agar plate; allow to dry for a few minutes

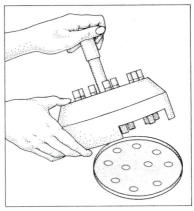

(b) Apply antibiotic impregnated disks to plate

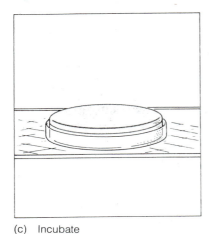

(c) Incubate

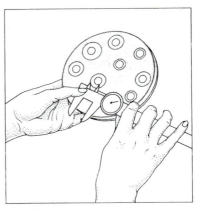

(d) Measure zones of inhibition

Figure 69-1. Basic Procedure for the Kirby-Bauer Test

TABLE 69-1.
INTERPRETATION OF ZONE OF INHIBITION

ANTIMICROBIAL AGENT	ABBREVIATION	CONCENTRATION	DIAMETER OF ZONE OF INHIBITION (mm)*		
			Resistant	Intermediate	Susceptible
Ampicillin	AM	10µg			
Gram negative			11	12–13	14
Staphylococci			20	21–28	29
Bacitracin	B	10 units	8	9–12	13
Carbenicillin	CB				
Escherichia coli		100 µg	17	18–22	23
Pseudomonas aeruginosa		50µg	12	13–14	15
Cephalothin	CR	30 µg	14		15
Chloramphenicol	C	30 µg	12	13–17	18
Clindamycin	CC	2 µg	14	15–16	17
Erythromycin	E	15 µg	13	14–17	18
Gentamycin	GM	10 µg	12	13–14	15
Kanamycin	K	30 µg	13	14–17	18
Methicillin	ME	5 µg	9	10–13	14
Neomycin	N	30 µg	12	13–16	17
Penicillin G	P	10 units			
Staphylococcus			20	21–28	29
Other organisms			11	12–21	22
Polymyxin B	PB	300 units	8	9–11	12
Streptomycin	S	10 µg	11	12–14	15
Tetracycline	T	30 µg	14	15–18	19
Vancomycin	VA	30 µg	9	10–11	12

*Zones of inhibition and their interpretation were obtained from: Bacto Sensitivity Discs for Use in the Antimicrobic Susceptibility Test. Technical information published by Difco Laboratories, Detroit, Michigan.

Plates of Mueller-Hinton agar

Antibiotic sensitivity disks and dispenser (BBL or Difco)

Forceps

Cotton- or dacron-tipped swabs

Alcohol jar

Calipers or ruler graduated in millimeters

Incubator set at 35°C

Procedure

First Period

1. Label one plate of Mueller-Hinton agar with your name, date, and organism assigned to you by your instructor.

2. Dip a cotton- or dacron-tipped swab into the broth culture of the bacterium and then roll the tip against the side of the broth tube to wring out excess culture from the tip.

3. Spread the bacterial culture over the entire surface of the Mueller-Hinton agar plate.

• To ensure an even distribution of the culture, swab the plate horizontally first, then vertically, and finally at a 45° angle.

4. Allow the plates to dry for about 5 minutes and then apply the antibiotic-impregnated disks to the surface of the agar.

• Record on your report form which antibiotics you used.

a. If you use an automatic dispenser, simply remove the cover from the dispenser, place over the uncovered Petri dish, and press firmly on the plunger to deliver the antibiotic-impregnated disks as illustrated in figure 69-1.

b. If individual cartridges are used, place the cartridge, antibiotic disk side down, directly over the portion of the agar plate you wish, and press on the lever to deliver the disk.

c. If loose disks are used, dip a pair of forceps in alcohol, flame to disinfect, and allow to cool. Pick a disk with the forceps and place it on the desired portion of the agar surface.

5. Use 6 to 8 different antibiotics to test the susceptibility pattern of your organism.

• Distribute the various disks with sterile forceps so that they are equally spaced from each other. This will minimize overlaps between the zones of inhibition of two different antibiotics.

6. With alcohol-flamed forceps, press gently on the disks so that there is good contact between the antibiotic disks and the agar surface.

• Do not press so hard that the disks become embedded in the agar.

7. Incubate the plate at 35°C overnight (or until the next laboratory period).

1. Measure the zones of inhibition for each of the antibiotics tested.

• Record the results obtained on the report form and interpret the significance of the results, using table 69-1 as an aid. For each antibiotic, determine whether your organism is susceptible or resistant.

2. Share your results with the rest of your class and record the class results on the report form.

• Interpret the class results in the same way as your own.

3. Discard all contaminated materials and put away supplies as indicated by your instructor.

Questions

1. Why is it necessary to test pathogens for sensitivity to a variety of antibiotics?

2. Why is it important that therapy not be initiated until a pathogen has been implicated?

3. Consider this statement: "*Bacillus anthracis* exhibits a zone of inhibition of 15 mm for both penicillin G and streptomycin." Does this mean that *B. anthracis* is equally sensitive to both antibiotics and either can be used to treat a disease caused by this pathogen? Explain your answer.

4. Of the antibiotics used to test the 3 different bacteria, which can be considered "broad-spectrum antibiotics" based on your results? Which are "narrow-spectrum antibiotics?"

5. Can the results obtained with the Kirby-Bauer test be used to determine the dose of antibiotic that should be given to the patient? Explain your answer.

EXERCISE 70
ABO BLOOD TYPING

Many surgical procedures routinely include blood transfusions to replace lost blood. This practice dates back to the fifteenth century, but is much more successful now. In past centuries more patients were lost than were saved by transfusions. Death was often due to clumping of the transfused blood in the patient's circulatory system. The reason for this, we now know, is that the recipient's antibodies in the serum reacts with incompatible donor's red blood cells and clumps them (agglutinates them). This reaction is now known as **hemagglutination** and serves as an indication of incompatibility between the recipient's and the donor's blood.

Before a patient receives blood from a donor, it is necessary to determine that the donor's blood is compatible with that of the recipient's. The compatibility is based on the presence of certain surface antigen groups on the plasma membrane of the red blood cell (erythrocyte). The principal antigen groups are the **ABO** and the **Rh** blood groups. In this exercise we will consider only the ABO blood groups.

The ABO blood groups were first discovered by

Karl Landsteiner in 1900 when he noted that human blood could be classified into four different groups based on the presence of agglutinating antigens (or agglutinogens on the blood cells). These antigens, now called A and B, are polysaccharides or glycolipids associated with the plasma membrane of red blood cells. The presence or absence of these antigens determines the blood group of the individual (table 70-1).

In this laboratory exercise we will carry out a typical blood typing procedure by the **slide agglutination test** (fig. 70-1), in which the patient's blood (in this case, yours) is mixed with antiserum directed against the A antigen and with antiserum directed against the B antigen. Each mixture is rocked back and forth for a minute or two and then examined for agglutination (clumping) of the red cells. The pattern of agglutination will determine to which blood type the patient belongs. Individuals with A type blood will have their blood agglutinated by anti-A antiserum but not by anti-B antiserum. Conversely, the blood of individuals with blood type B will be agglutinated by anti-B antiserum but not by anti-A antiserum. Blood

TABLE 70-1
MAJOR HUMAN BLOOD TYPES, ANTIGENS, AND ANTIBODIES

BLOOD TYPE	ANTIGENS ON ERYTHROCYTYES	ANTIBODIES IN SERUM
A	A	ANTI-B
B	B	ANTI-A
AB	A & B	NEITHER
O	NEITHER	BOTH ANTI-A & ANTI-B

type AB will be agglutinated by both antisera while blood type O will be agglutinated by neither anti-A nor anti-B antisera (fig. 70-1).

Materials

Anti-A and anti-B antisera

70% isopropyl alcohol

Disposable sterile lancets

Cotton balls and bandages

Clean glass slides

Grease pencil

Grease-cutting detergent

Toothpicks or applicators

Procedure

1. Prepare a slide for the agglutination test by drawing two separate circles on a clean glass slide, about 1.5 cm in diameter, with a grease pencil. Label one of the circles "A" and the other "B."

• To ensure good results, clean the slide (before drawing the circles) with a grease-cutting detergent to remove any grease.

2. Scrub your middle or index finger with a swab soaked in 70% isopropyl alcohol and then pierce the finger with a sterile lancet.

3. Allow one drop of blood to fall into each of the two circled areas. Wipe the blood off your finger with a clean cotton swab or gauze, and if you wish, put on an adhesive bandage.

4. Add a drop of anti-A antiserum to the drop of blood in area A and a drop of anti-B antiserum to the blood in area B. DO NOT CONTAMINATE THE DROPPERS WITH YOUR BLOOD.

5. Mix the antiserum with the blood using an applicator stick or a toothpick, using a DIFFERENT one for each antiserum.

6. Rock the slide back and forth for 2 to 3 minutes and then observe for the presence of clumps (agglutination). Use figure 70-1 as an aid.

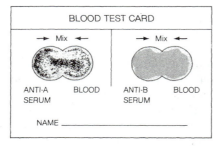

Group A

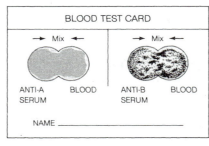

Group B

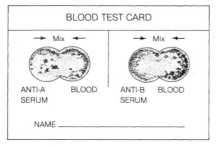

Group AB

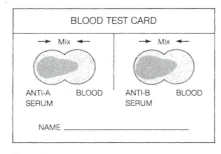

Group O

Figure 70-1. Agglutination Patterns of Human ABO Blood Groups. Blood typing is carried out by mixing known antisera against A and B antigens with blood samples. Blood that is agglutinated with anti-A antiserum only is A type blood. Blood agglutinated by anti-B antiserum only is type B blood. Blood agglutinated by both anti-A and anti-B antiserum is blood type AB. Blood that is not agglutinated by either antiserum is type O blood.

- View the slide in a well-illuminated area, preferably against a well-lit, white background.

7. Record your observations and your blood type on the report form provided.

8. Discard all contaminated materials and put away the supplies as indicated by your instructor.

Questions

1. Based on the results obtained with the ABO blood typing procedure, could you receive a blood transfusion from the student sitting at your left? Explain your answer briefly.

2. Explain why people with blood type O are considered universal donors. Are they also universal recipients? Explain your answer.

3. Could a type AB individual receive blood from a type B individual? How about from a type O individual? Explain your answer.

4. Is a type AB individual a universal donor? Explain your answer.

5. Could a person with type AB blood receive blood from a primate of the same ABO blood type? Explain your answer.

EXERCISE 71
SEROLOGICAL REACTIONS

Serology is the branch of immunology that studies *in vitro* antigen-antibody reactions. Serological procedures are based on the assumption that infections induce the host to produce specific antibodies and/or sensitized lymphocytes that react specifically with the infectious agent or its products. Serological procedures are designed to detect and/or quantify the immune response against such agents.

Serological reactions are used to diagnose infectious diseases because they are very specific and accurate. In addition, they can yield results in one day. This is in contrast to traditional methods of diagnosing infectious diseases that involve culturing and identifying the causative agent, which may take more than a week to yield results.

Typical serological tests involve mixing the patient's serum with known antigens. If the serum reacts with a particular antigen, the patient has been exposed to that antigen. Various techniques have been developed to detect antigen-antibody reactions, many of which have been used successfully in diagnosing infectious diseases.

In this exercise you will be introduced to some of the most common techniques presently in use in the clinical microbiology laboratory. These techniques are: the slide agglutination test, the tube precipitin test, the fluorescent antibody test, and the Rapid Plasma Reagin (RPR) test for the diagnosis of syphilis.

A. BACTERIAL AGGLUTINATION TEST

Agglutination reactions involve whole-cell antigens such as bacteria or red blood cells. These antigens, which contain many antigenic determinants on their surface, react with specific antibodies to form aggregates (or clumps) that are readily visible. The aggregates are particles joined to each other by antibody bridges (fig. 71–1).

In this exercise, you will perform a slide agglutination test, sometimes called a **Widal test,** to differentiate between *Escherichia coli* and *Salmonella typhimurium.* This procedure is routinely used in the laboratory to differentiate between closely related bacteria (e.g., those in the family Enterobacteriaceae) or between strains of the same species. The procedure involves mixing a suspension of an unknown bacterial isolate with a drop of specific antiserum. After 2 to 3 minutes, the slide is examined for evidence of clumping. Any clumping is considered to be an indication of agglutination.

Since you are dealing with an assay for an unknown bacterium, it is necessary that the reagents (the antiserum) and the procedure be properly tested. This is done with controls. A known positive control (*Salmonella*) and a known negative control (*Escherichia coli*) are tested along with the unknown. Results of the slide agglutination test are valid only if the positive

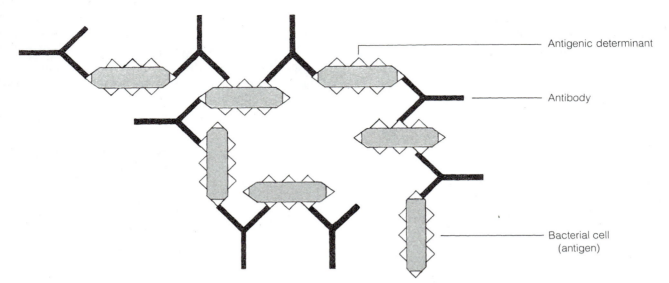

Figure 71-1. Agglutination Reaction

and negative controls give the anticipated results, indicating that the test system is working properly.

Materials

Suspension of *Escherichia coli* in phenol-normal saline (0.85% NaCl in water with 0.5% phenol)

Suspension of *Salmonella typhimurium* in phenol-normal saline

Salmonella O antiserum, polyvalent (Difco catalog #2264-47)

Spot plates

Phenol-normal saline solution

Pasteur pipets and bulbs

Beaker with disinfectant solution (for discarding pipets and used slides)

Procedure

1. Label 3 wells in a spot plate as follows: (a) saline control; (b) *E. coli*; and (c) *S. typhimurium* (fig. 71-2).

2. Using a different Pasteur pipet for each, add a drop of phenol-normal saline solution, a drop of *E. coli* suspension, and a drop of *S. typhimurium* suspension to the corresponding wells.
• Make sure that each drop is within the boundary of its well and that the drop does not fill any more than half the area of the well.

3. With a dropper or another Pasteur pipet, add a drop of *Salmonella* polyvalent antiserum to each of the three wells.

4. Rock the slide in an elliptical motion for approxi-

mately 2 minutes to mix the reagents, then examine each of the wells for clumping of cells.

5. Record your observations on the report form.

6. Discard all contaminated materials in a beaker with disinfectant or as indicated by your instructor.

Questions

1. Cite five examples in which serological reactions would be more useful than conventional methods of determining whether or not a person had been infected by a particular organism.

2. Describe briefly the principle involved in serological reactions.

3. How would you use the Widal test to demonstrate that a given isolate from a patient's blood is *Franciscella tularensis*?

4. Why is it necessary that a known positive and a known negative organism be tested concurrently with the unknown sample when using serological reactions?

B. THE PRECIPITIN TEST

Precipitin tests involve reactions between antigens and antibodies in which the antigens are soluble substances. When an antigen in solution is mixed with specific antiserum in the correct proportions, a precipitate will form because the bivalent antibodies will bind two or more molecules of antigen and cross-link them (fig. 71-3). Extensive cross-linking on soluble antigens and antibodies will form a lattice network that

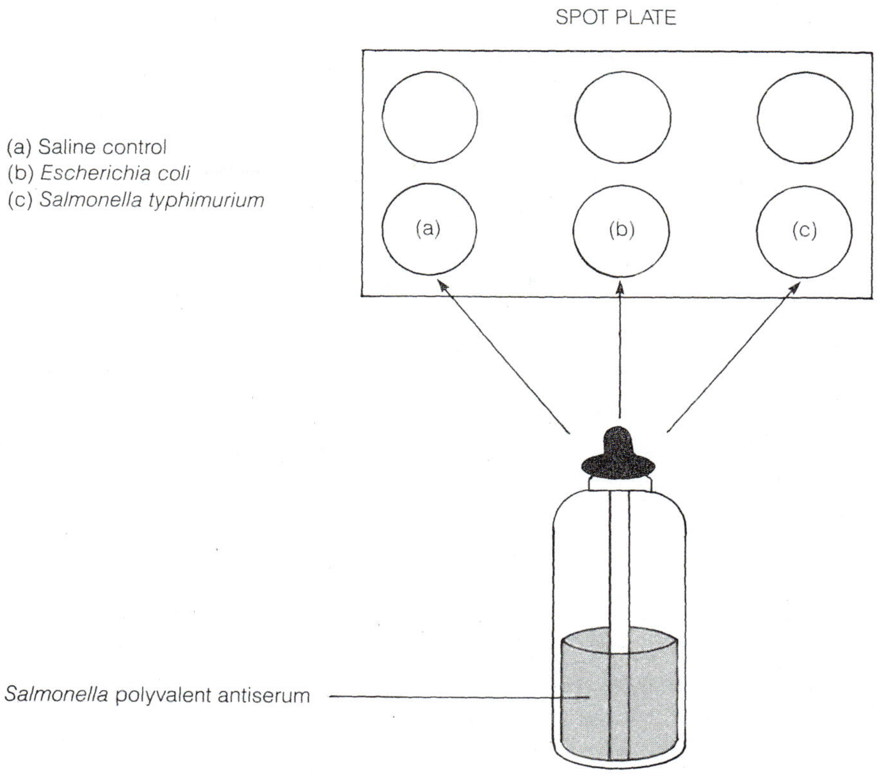

SPOT PLATE

(a) Saline control
(b) *Escherichia coli*
(c) *Salmonella typhimurium*

Salmonella polyvalent antiserum

Figure 71-2. Spot Plate System

will precipitate out of solution, indicating that an antigen-antibody reaction has taken place.

In the **classical precipitin reaction,** the same volume of antiserum is pipetted into several tubes and mixed with varying concentrations of an antigen. These tubes are then incubated for a prescribed amount of time, centrifuged at about 3,000 $\times$ G, and examined for the formation of a precipitate. Extensive precipitates will form in those tubes where antigen and antibodies are present in **optimal proportions.** In tubes where there is an excess of either antibody or antigen, little or no precipitate will form.

In this exercise you will perform a modification of the classical precipitin reaction called the **ring precipitation test.** In this test, the antiserum is pipetted carefully over the antigen in a tube so that the antiserum forms a layer on top of the antigen. The antiserum and the antigen should not mix (fig. 71-4b). During incubation, a cloudy ring of precipitation will form where the antigen and antibodies meet in optimal proportions. The antigen you will use is bovine serum and the antiserum consists of antibodies prepared by injecting goats (or any animal other than cows) with the serum from cows. Both antigen and antiserum are commercially available from a variety of different laboratories that specialize in preparing serological reagents.

Materials

Bovine serum albumin (1:500 dilution)

Goat antibovine serum albumin

Normal goat serum

Normal saline (0.85% NaCl in distilled water)

Durham tubes

Pasteur pipets and bulbs

1 ml pipets with bulbs

Test tube racks or molding clay in a wooden platform

Incubator set at 35°C

Procedure

1. Label 6 Durham tubes as follows (fig. 71-4):

   a. tube #1 as saline control

   b. tube #2 as normal serum control

   c. tubes #3 through 6 as 1:2, 1:4, 1:8, and 1:16

Place each of the tubes in a test tube rack or in molding clay so that they stand upright.

2. Add 0.4 ml of saline to all 6 tubes.

3. Add 0.4 ml of the bovine serum albumin solution to tubes 2 and 3. With the same pipet, mix the bovine

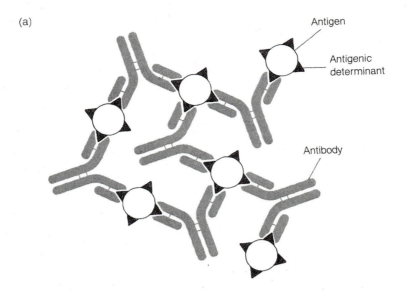

(a)

Antigen

Antigenic
determinant

Antibody

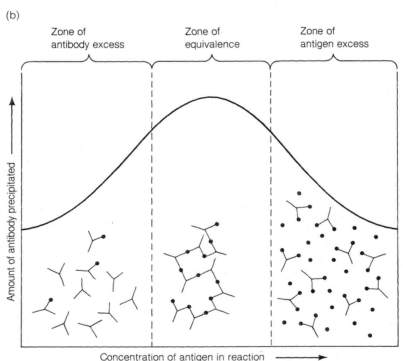

(b)

Zone of
antibody excess

Zone of
equivalence

Zone of
antigen excess

Amount of antibody precipitated

Concentration of antigen in reaction

Figure 71-3. The Classical Precipitin Reaction. *(a)* Cross-Linking in Antigen-Antibody Reactions. Antibodies are polyvalent molecules (i.e., they can bind to two or more antigen molecules). Antigen-antibody reactions can lead to the formation of lattice networks created by antibodies binding to antigens. The clumps can be so large that they settle out. Inside the host, these clumps can be phagocytosed by PMNs or macrophages. *(b)* The Classical Precipitin Reaction. The degree of cross-linking in antigen-antibody reactions is determined largely by the relative concentrations of antigen and antibody present. When increasing concentrations of antigen are re-acted with a standard concentration of antibody and the amount of precipitated antigen-antibody complexes measured, the results resemble those illustrated. At low antigen concentration, there is an excess of antibody, and very little cross-linking occurs. The precipitate (representing antigen-antibody complexes) is small and the supernatant fluid (containing unbound reagents) has a large concentration of antibody. At optimal concentrations of antigen, much cross-linking occurs, forming a large precipitate and little or no free antigen or antibody will be left in the supernatant fluid. At high concentrations of antigen, again little cross-linking takes place.

serum albumin well in the saline by taking up about 0.7 ml of the liquid into the pipet and then releasing it back into the tube. Repeat this procedure 5 to 7 times.

4. Discard 0.4 ml of the fluid in tube 2.

5. Using a different pipet, transfer 0.4 ml of the contents of tube 3 into tube 4. Using the same procedure

that you used in step 3 above, mix the albumin thoroughly in the saline. Repeat this procedure using a different pipet for each transfer to make the 1:8 and 1:16 dilutions. Discard 0.4 ml of the last dilution (so that all 6 tubes have the same amount of fluid in them.

6. Using a Pasteur pipet, carefully overlay 0.4 ml of normal goat serum over the bovine serum albumin in tube 2 (normal serum control).

• To avoid disruption of the upper layer, slant the tube so that it is almost horizontal (fig. 71-5) and let the serum run down the side of the tube until it layers over the albumin solution.

7. Using a different Pasteur pipet, overlay 0.4 ml of antibovine serum albumin over the saline in tube 1 and over the various dilutions of albumin in tubes 3 through 6.

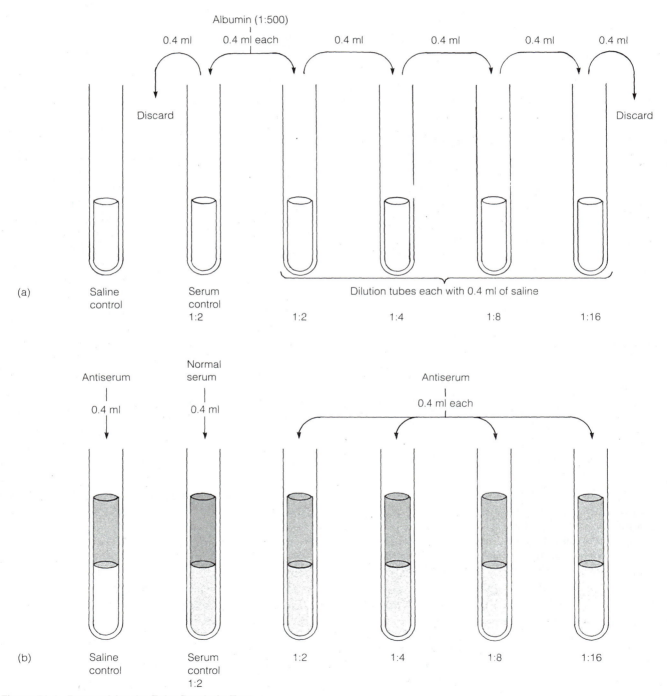

Figure 71-4. Protocol for the Tube Preciptin Test

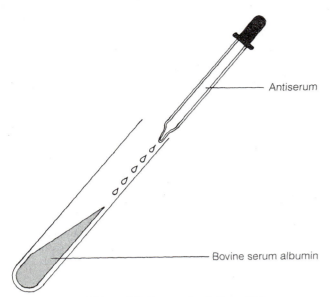

Antiserum

Bovine serum albumin

Figure 71-5. Addition of Antiserum to Antigen Tubes

8. Cover the tubes with aluminum foil or plastic wrap and place in an incubator set at 35°C for 2 to 4 hours.

9. Examine each of the tubes for the presence of a ring of precipitation at the interphase between the antiserum and albumin. Record the results and your observations on the report form.

10. Discard or put away all materials and supplies as indicated by your instructor.

Questions

1. Explain the following:
   a. why was the bovine serum albumin diluted?
   b. what is the function of the saline control?
   c. what is the purpose of the goat serum plus bovine serum albumin control?

2. Explain how you would perform an experiment using the precipitin test to determine whether a specific product was 100% beef or had pork filler added.

3. Suppose that you are a game warden on the trail of a poacher who has bagged a pronghorn antelope. Design an experiment that would allow you to determine that the meat in the suspected poacher's freezer is antelope meat, not deer meat as the poacher claims.

4. As a director in a clinical laboratory, you are asked to conduct a test to determine that a patient has a herpes infection, not a severe case of impetigo (a staphylococcal infection), as was originally suspected. The trouble is that the mere presence of antiherpes antibodies (or antistaphylococcal antibodies) is not an indication of active disease because the host is frequently exposed to these agents. How would you go

about determining the cause of the infection. HINT: It just so happens that there are two serum samples from the patient, one obtained at the time of admission to the hospital and the other a week after the patient was discharged from the hospital.

5. Explain why an antigen might have to be diluted before an antibody-antigen precipitate would be visible in a precipitin test.

C. IMMUNOFLUORESCENCE

Immunofluorescence techniques are antigen-antibody reactions in which antibodies are labeled by bonding a fluorescent dye to the Fc region of the antibody molecule. The resulting altered molecule, a "labelled antibody," is still able to bind to an antigen, but the reaction can be detected by examining the specimen with the aid of a fluorescence microscope (see exercise 1). One of the most common fluorescent dyes (or fluorochromes) is **fluorescein isothiocyanate (FITC).** This molecule absorbs ultraviolet wavelengths and emits a yellow-green light.

There are two common methods of using fluorescing antibodies in the clinical laboratory. These are the direct immunofluorescence method and the indirect immunofluorescence method (fig. 71-6). The **direct immunofluorescence method** involves the use of specific labeled antibodies to react with the antigen. For example, many laboratories employ the direct method to verify the identity of group A streptococci (e.g., *Streptococcus pyogenes*). The reagent, which consists of FITC-labeled antistreptococcal antibodies, may be purchased commercially from several laboratories specializing in serology. The procedure is carried out by preparing a smear of the suspected group A *Streptococcus* isolate, fixing it to a clean glass slide, and then flooding the fixed smear with FITC-labeled anti-Group A antiserum. The mixture is allowed to incubate in a moist chamber at 35°C for 30 minutes. The smear is then rinsed well with a saline solution to wash off unbound antibodies, covered with a drop of glycerol buffer and a coverslip, and viewed with the aid of a fluorescence microscope. During the incubation period, the FITC-labeled antibodies will react with Group A streptococci and coat their cell walls. When viewed with the fluorescence microscope, the cells will glow with a yellow-green fluorescence because the FITC-labeled molecules will absorb the ultraviolet light and emit a yellow-green light. If the isolate is not in Group A, the FITC-labeled antibodies will not bind to the cells and will be subsequently washing away during the saline rinse. These cells will not exhibit fluorescence.

The **indirect immunofluorescence method** is based on the same principles as the direct method, except that it adds an additional incubation step. Consider the **fluorescent Treponemal antibody (FTA) test** for syphilis as an example. The patient's serum (containing antitreponemal antibodies) is layered over a smear of *Treponema pallidum* (the agent that causes syphilis). This mixture is incubated in a moist chamber for 30 minutes at 35°C, after which any unbound antibody is removed by washing the smear several times with a saline solution. The rinsed smear is then flooded with FITC-labeled antihuman immunoglobulin antibodies (antibodies made in goats against human immunoglobulins), and incubated for an additional 30-minutes at 35°C. The labeled antibodies will react with the patient's antibodies attached to the treponemes. The smear is finally washed with saline, mounted in a glycerol solution, covered with a coverslip, and viewed with a fluorescence microscope. The treponemes will glow with a yellow-green fluorescence because they are coated with the patient's antibodies, which, in turn, are coated with FITC-labeled antihuman antibodies. If the patient has no antitreponemal antibodies in his or her blood, no fluorescence will be seen. Figure 71-6 illustrates the basic steps involved in the direct and indirect methods.

In this exercise you will use the direct fluorescence antibody test to differentiate a Group D *Streptococcus* (*Streptococcus faecalis*) from *Staphylococcus epidermidis*, to illustrate the usefulness of this technique in differentiating between two morphologically similar bacteria.

Materials

Broth cultures of *Streptococcus faecalis* and *Staphylococcus epidermidis* (alternatively, a mixture of both these cultures may be used)

FITC-labeled anti Group-D antibodies (Difco catalog #2321)

Gram staining reagents

Phosphate buffered saline (PBS), pH 7.2

50% glycerol in PBS

Glass slides and coverslips

Petri dish with a moistened filter paper (moist chamber)

China markers

Pasteur pipets and bulbs

Applicator sticks or toothpicks

Fluorescence microscope

Incubator set at 35°C

Procedure

1. Draw two circles with a China marking pencil on a clean, grease-free slide.
• Additionally, draw a double line between the two circles to prevent reagents on one side of the slide from spilling onto the other (see report form).

2. Prepare a smear of *Streptococcus faecalis* within one circle and *Staphylococcus epidermidis* within the other circle. Air dry and then heat-fix the smears by passing them 2 or 3 times over the flame on a Bunsen burner.

3. Using a Pasteur pipet, cover the smears with a few drops of the FITC-labeled anti-Group D antibodies, place the slide inside a moist chamber, and then incubate at 35°C for 30 minutes.

4. While the slides are incubating, prepare a Gram stain of each of the two broth cultures and examine

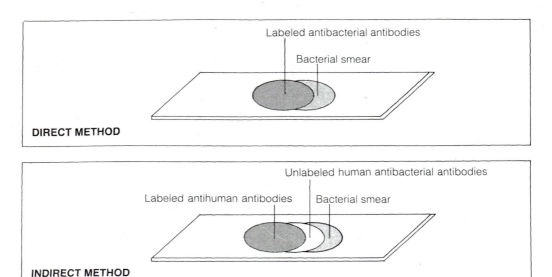

Figure 71-6. Fluorescence Antibody Methods

them at 1,000× magnification with the aid of a compound microscope. Record your observations on the report form.

5. After the incubation period, remove the slides from the incubator and rinse them several times with PBS.

• Do not squirt the PBS directly on the smear, but rather hold the slide at a slant and aim the stream of PBS at a point above the smear so that the fluid flows over the smear and rinses it gently. In order to achieve unequivocal results, the smear must be well rinsed so that all the unbound antibodies are washed off.

6. Add a drop of 50% buffered glycerol to each of the smears and cover with a clean coverslip.

7. Examine the smears with the aid of a fluorescence microscope.

• Turn to exercise 1 if you need additional information on the fluorescence microscope. Your instructor will show you the proper way to use the particular microscope model used in your laboratory.

8. Record your observations and conclusions on the report sheet.

9. Discard all contaminated materials and put away supplies as directed by your instructor.

Questions

1. Why is the procedure used in this exercise to identify Group D streptococci considered to be a direct method?

2. How would you modify this procedure to make it an indirect method?

3. Cite one advantage of the indirect over the direct method. HINT: many of the reagents used in direct methods are commercially purchased.

4. How would you use the immunofluorescence technique to detect a cancerous growth on a mouse spleen?

5. How would you use fluorescence methods to determine that a patient has tuberculosis? HINT: tuberculosis patients often have *Mycobacterium tuberculosis* in their sputum.

D. THE RAPID PLASMA REAGIN (RPR) TEST FOR SYPHILIS

Serological procedures for the diagnosis of syphilis are very popular because the etiological agent of the disease, *Treponema pallidum*, cannot be cultured in the laboratory and is difficult to see in clinical specimens unless the specimen is fresh and a dark field microscope is used (see exercise 1). Many states require that persons applying for marriage licences be tested for syphilis as a means of preventing the spread

of this sexually transmitted disease and of detecting new asymptomatic carriers. Also, family planning, premarital, prenatal, and venereal disease clinics, as a matter of routine procedure, test all participants for syphilis. This means that in many large communities, hundreds of specimens are tested for syphilis each week. The specimens are usually serum samples. The definitive serological test for syphilis is the FTA (fluorescent treponemal test), which was discussed briefly in exercise 71C. However, because this procedure is relatively expensive and takes time, it is not used as a screening test for syphilis. The **Rapid Plasma Reagin (RPR) test** was developed to screen many serum samples rapidly for the presence of antitreponemal antibodies.

The **RPR Card test** is a macroscopic testing procedure for the detection of syphilis, using unheated serum or plasma obtained from venous or finger puncture blood. This test makes use of a specially prepared, carbon-containing RPR antigen that is mixed with a specimen. The particle size of the carbon is such that the "nonreactive" (i.e., negative) specimens appear to have an even, light gray color. When a "reactive" (i.e., positive) specimen is encountered, flocculation occurs and there is a coagglutination of the carbon, which is readily visible without the aid of a microscope. The RPR Card test has a reactivity level similar to that of the VDRL (venereal diseases research laboratory) slide test for syphilis. Reactive sera are usually verified by the FTA test, as the RPR test alone is not suitable for an unequivocal diagnosis of syphilis. As with any serological procedure, the diagnosis of syphilis should not be made on a single reactive result without the support of a positive history or clinical evidence.

Materials

Reactive (4+) serum

Nonreactive (−) serum

RPR Card Test kit (available from Hynson, Wescott & Dunning, Baltimore, MD 21201, a division of Becton Dickinson and Company).

Rotating apparatus (optional)

Procedure

Each student will use only 2 circles on a card; one for the reactive serum and the other for the nonreactive serum. Since there are 10 circles on a card, 5 students may use a single card.

1. Label the card (fig. 71-7) with your name and type of serum used (+ or −).

• The cards are treated with a special coating and skin oils or perspiration within the circles will adversely affect the results.

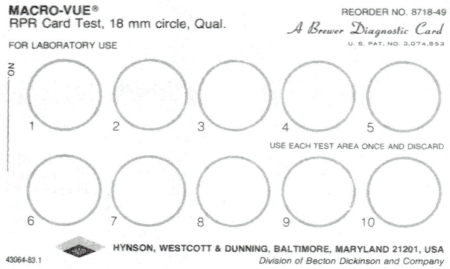

MACRO-VUE®
RPR Card Test, 18 mm circle, Qual.

A Brewer Diagnostic Card

REORDER NO. 8718-49

U. S. PAT. NO. 3,074,853

FOR LABORATORY USE

NO

1 2 3 4 5

USE EACH TEST AREA ONCE AND DISCARD

6 7 8 9 10

HYNSON, WESTCOTT & DUNNING, BALTIMORE, MARYLAND 21201, USA
Division of Becton Dickinson and Company

43064-83.1

Figure 71-7. The Rapid Plasma Reagin Card

2. Hold the Dispenstir between thumb and forefinger near the stirring or sealed end. Squeeze and do not release the pressure until the open end is below the surface of the specimen, then release the finger pressure to draw up the sample.

3. Hold the Dispenstir in a perpendicular position directly over the card test area to which the specimen is to be delivered (not touching the surface of the card). Squeeze the Dispenstir to allow one drop (approximately 0.05 ml) to fall within the circled area on the card.

4. Invert the Dispenstir, and with the sealed end, spread the specimen in the confines of the circle.

5. Shake the RPR antigen-dispensing bottle before use. Holding the dispensing bottle in a vertical position, dispense one or two drops in the upper corner of the card to make sure that the needle passage is clear and delivering 1/60 ml of antigen. Now place a "free-falling" drop on each of the two test areas. Do not mix the drop with the serum. This will be done by rotating the drops.

6. Rotate the card back and forth with your hands for 8 minutes.

• If a mechanical rotator is available, place the card on the rotator's platform under a humidifying lid and rotate at 100 RPMs for 8 minutes.

7. Following the 8 minutes of mixing, read the results macroscopically in the "wet" state with the aid of a high intensity lamp or bright daylight.

8. Record your results on the report form as follows:

Reactive: showing characteristic clumping ranging from slight but definite (minimally reactive) to marked and intense.

Nonreactive: showing no clumping or a slight roughness.

9. Put away all reagents and specimens as indicated by your instructor.

Questions

1. Give two reasons why the RPR card test is not a definitive test for the diagnosis of syphilis. Explain your answer. HINT: The RPR antigen is not derived from *Treponema pallidum* cells.

2. Why is it necessary that the proper volume of specimen and RPR antigen be used?

3. In view of the fact that the RPR antigen is a nontreponemal antigen, what must have been done in the laboratory that developed the procedure in order to ascertain that the card test is suitable for the diagnosis of syphilis?

4. What function do the carbon particles serve in the RPR test?

5. How would you incorporate this procedure in a comprehensive public health program in a family planning clinic?

EXERCISE 72
RAPID MULTITEST PROCEDURES FOR THE IDENTIFICATION OF ENTEROBACTERIACEAE

Rapid identification of disease-causing microorganisms is desirable so that proper early treatment of the patient can be initiated and suitable control measures can be implemented. In order to identify many pathogens, various tests have to be carried out. Many of these tests involve inoculating different types of selective or differential media with the isolated organism, incubating the media for 24–96 hours and evaluating the results after incubation. The culture media used must be prepared, stored, and checked for deterioration before cultivation so that the results obtained can be relied upon. All this requires time and can be very expensive.

It is often desirable, either for economy or expediency (or both), to use miniaturized, multitest systems or "rapid identification kits" to identify common pathogens. There are several of these kits on the market that are well suited for the identification of bacteria and yeasts. Most of these are made up of strips or tubes with compartments containing various types of culture media so that a large number of tests can be carried out. The results obtained using these miniaturized, multitest systems generally correlate well with those obtained using conventional methods (see Part VII). The kits can be stored for relatively long periods of time without deterioration, occupy little laboratory space, and the cost per isolate identified usually is much lower than the cost of conventional methods. Frequently, the results can be obtained in the same day or after overnight incubation. Many of the manufacturers of these kits provide computerized identification services at little or no extra cost.

In this exercise, you will identify an unknown bacterium in the family Enterobacteriaceae using the Enterotube II system and/or the API 20E strip system. Both of these systems are widely used in the clinical laboratory and have proven to be well suited for the identification of enteric bacteria.

A. THE ENTEROTUBE II SYSTEM

The Enterotube II is a multitest system developed and manufactured by Roche Diagnostics (Nutley, New Jersey). This system consists of a tube with 12 separate compartments (see figure 72-1 and the color insert section), each containing a different agar-solidified culture medium, for the identification of Gram negative, glucose fermenting, oxidase negative bacteria (Enterobacteriaceae). Using these 12 culture media, 15 different traits can be determined. The Enterotube II also contains an inoculating wire that extends through the middle of the tube (figure 72-2); it is used to pick a portion of a colony and to inoculate all 12 culture media in the compartments of the Enterotube II with the unknown organism. After inoculation, the Enterotube II is incubated for 24 hours before the results are read and interpreted. The final identification of the unknown can be made either by determining a five-digit **numerical code** (fig. 72-2) and consulting a Coding Manual, or by comparing the results obtained with those outlined in a differential chart.

Materials

Enterotube II systems

Broth culture of unknown (Enterobacteriaceae)

Plates of trypticase soy agar (TSA)

Gram-staining reagents

Kovac's reagent

α-naphthol (5% α-naphthol w/v in absolute ethanol)

Potassium hydroxide (20% aqueous solution)

Creatine solution (0.3% aqueous solution)

Oxidase reagent

Platinum inoculating loop

Filter paper squares (~1 cm²)

1 ml syringes with needles

Incubator set at 35°C

Procedure

First Period

1. Obtain an unknown culture from your instructor and streak a portion of it onto a plate of TSA so as to obtain well-isolated colonies.

2. Label the plate properly and incubate the culture, in an inverted position, at 35°C for 24–48 hours.

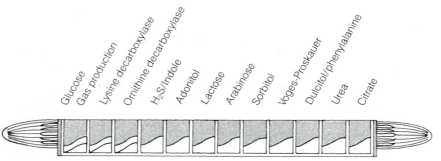

Figure 72-1. The Enterotube II

Second Period

1. Examine the TSA culture plate containing your unknown for evidence of well-isolated colonies. Select a well-isolated colony measuring 2–3 mm in diameter to carry out the identification procedure.

2. Perform a Gram stain on the unknown isolate using only a small portion of the colony.

3. Perform an oxidase test on the isolated colony.

• Use a platinum loop to pick up a small portion of the colony. Smear that portion of the colony over a

1. Remove caps at both ends and pick some organisms from a well-isolated colony measuring 2-3 mm in diameter.

2. Inoculate each compartment by drawing wire through all 12 compartments and then reinserting it through all the compartments again. Rotate the wire as it passes through the compartments.

3. Withdraw the wire a second time until the tip is in the H_2S compartment then break the wire off at the notch.

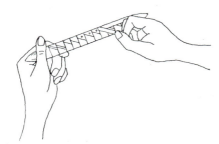

4. Replace caps at both ends and then strip off the plastic tape covering the holes on eight chambers. Incubate at 35°C for 24 hours.

Figure 72-2. How to Use the Enterotube II

5. After incubation, examine Enterotube II, interpret and record results with (+) or (−) on the report form, and perform indole and Voges-Proskauer tests.

6. Determine the ID code. Circle the numbers under the positive tests on the report form and add the circled numbers in each boxed group to determine the 5-digit code number.

small area of a filter paper square. Add the oxidase reagent to the smeared colony and note any color changes that take place.

4. Inoculate the Enterotube II* as recommended by the manufacturer (fig. 72-2) according to the following instructions:

   a. Remove the caps at both ends of the Enterotube II being careful not to contaminate the exposed tip of the inoculating wire. (DO NOT FLAME THE TIP OF THE WIRE SINCE IT IS ALREADY STERILE.)

   b. Pick up a small portion of the selected colony with the tip of the inoculating wire.

   c. Inoculate the Enterotube II with the culture by drawing the wire through the 12 compartments. DO NOT PULL THE WIRE OUT OF THE ENTEROTUBE II.

   • Rotate the inoculating wire as you withdraw it in order to ensure that all 12 compartments are sufficiently inoculated with organisms.

   d. Reinsert the wire through the Enterotube II so that the 12 chambers are reinoculated.

   e. Withdraw the inoculating wire once again until the tip of the wire is in the H₂S/indole compartment and then break the wire at the notch by bending it back and forth.

   • The portion of the wire remaining inside the Enterotube II helps maintain an anaerobic environment in those chambers where the wire remains in place.

   f. Replace the caps but not so tightly as to prevent some aeration.

   g. Remove the blue tape to expose the aeration vents (small holes) and slide the clear plastic band over the glucose compartment.

   • This is to avoid spilling wax in case there is extensive gas production by the bacteria.

   h. Incubate the Enterotube II at 35°C for 24 hours.

Third Period

1. Examine the Enterotube II, noting the changes that take place in each compartment. See table 72-1 as a guide to the interpretation of the results.

2. Perform the indole test by adding one or two drops of Kovac's reagent to the H₂S/indole compartment using a syringe with a needle. Read the results after 10 seconds.

• A positive indole test is indicated when the fluid in the chamber turns red.

3. Perform the Voges-Proskauer test by adding two drops of 20% KOH and three drops of α-naphthol

---

* The Enterotube II should be used only to identify Gram negative, oxidase negative bacteria (Enterobacteriaceae).

solution to the compartment labeled Voges-Proskauer. Use a new syringe for each solution. Read the results within 1 hour of adding the chemicals.

• A positive Voges-Proskauer test is indicated when the fluid in the chamber turns red.

4. Record the results on the report form provided.

5. Determine the five-digit identification number as follows:

   a. Consider only those tests that are positive. Add the numbers under the positive results within each test section.

   b. Enter the sum of the positive tests in the square labeled ID value. The five digits represent the identification number for your unknown.

6. Determine the identity of your unknown by comparing the five-digit identification number with the Enterotube II computer coding manual.

• Verify the identity of your unknown by comparing the results obtained with those outlined in table 72-2.

7. Enter the unknown number and the identity of the unknown on the report form provided.

8. Discard the Enterotube II and put away the syringes, needles, and reagents as indicated by your instructor.

Questions

1. What advantages does the Enterotube II have over the conventional method for the identification of unknown bacterial isolates?

2. Cite two shortcomings of the Enterotube II as a tool for the identification of unknown enteric bacteria.

3. Are the results obtained with the Enterotube II conclusive? Explain briefly your answer. HINT: Is an identification of *Salmonella* using the Enterotube II sufficient to establish the identity of the isolate?

4. Why are Gram negative, oxidase negative bacteria the only bacteria that are identifiable with the Enterotube II?

5. Design an Enterotube II-like system for the identification of Gram negative, oxidase positive, aerobic bacteria.

B. THE API 20E SYSTEM

The API 20E strip is a multitest system developed and manufactured by Analytab Products Inc. (Plainview, New York) that can be used to identify enteric and other Gram negative bacteria that are isolated from clinical specimens. The strip is comprised of 20 chambers, each consisting of a tube and a depression called a cupule (see figure 72-3 and the color insert section).

TABLE 72-1.
ENTEROTUBE II REACTIONS.

COMPART-MENT	MEDIUM/TEST	DESCRIPTION AND INTERPRETATION OF TESTS	RESULTS SUMMARY	
			Positive	Negative
1	Glucose (GLU)	Tests for glucose fermentation. A shift in pH is indicated by a change in color of medium from red to yellow, reflecting the production of acidic fermentation byproducts. A change in the color of the medium from red to yellow should be interpreted as a positive reaction. Orange should be interpreted as negative.	Yellow	Red/Orange
	Gas Production (GAS)	Gas from fermentation is indicated as a definite separation of the wax overlay from the surface of the culture medium. Bubbles in the culture medium should not be interpreted as evidence of gas production.	Separation of wax	No separation of wax
2	Lysine Decarboxylation (LYS)	Measures the ability of bacteria to decarboxylate lysine to produce the alkaline byproduct cadaverine. Any shift in the medium of the culture medium from yellow to purple should be interpreted as a positive reaction. The medium should remain yellow if decarboxylation does not take place.	Purple	Yellow
3	Ornithine Decarboxylation (ORN)	Measures the ability of bacteria to decarboxylate ornithine to produce the alkaline byproduct putresine. Any shift in the medium of the culture medium from yellow to purple should be interpreted as a positive reaction. The medium should remain yellow if decarboxylation does not take place.	Purple	Yellow
4	H_2S Production (H_2S)	H_2S is produced from the metabolism of sulfur-containing compounds (e.g., thiosulfate and amino acids) in the culture medium. Ferrous (Fe^{2+}) ions in the medium react with the H_2S to produce the black precipitate (FeS). Any blackening of the medium indicate that H_2S has been produced.	Black	No change
	Indole Formation (IND)	Indole is produced when tryptophan is degraded by the enzyme tryptophanase. After injection of Kovac's reagent into the medium (after 18–24 hours of incubation), any indole present will react with the reagent to produce a pink-red color.	Red	No change
5	Adonitol (ADON)	Tests for adonitol fermentation. A shift in pH is indicated by a change in color of medium from red to yellow, reflecting the production of acidic fermentation byproducts. A change in the color of the medium from red to yellow should be interpreted as a positive reaction. Orange should be interpreted as negative.	Yellow	Red/Orange
6	Lactose (LAC)	Tests for lactose fermentation. A shift in pH is indicated by a change in color of medium from red to yellow, reflecting the production of acidic fermentation byproducts. A change in the color of the medium from red to yellow should be interpreted as a positive reaction. Orange should be interpreted as negative.	Yellow	Red/Orange
7	Arabinose (ARAB)	Tests for arabinose fermentation. A shift in pH is indicated by a change in color of medium from red to yellow, reflecting the production of acidic fermentation byproducts. A change in the color of the medium from red to yellow should be interpreted as a positive reaction. Orange should be interpreted as negative.	Yellow	Red/Orange

TABLE 72-1.
CONTINUED

COMPART-MENT	MEDIUM/TEST	DESCRIPTION AND INTERPRETATION OF TESTS	RESULTS SUMMARY	
			Positive	Negative
8	Sorbitol (SORB)	Tests for sorbitol fermentation. A shift in pH is indicated by a change in color of medium from red to yellow, reflecting the production of acidic fermentation byproducts. A change in the color of the medium from red to yellow should be interpreted as a positive reaction. Orange should be interpreted as negative.	Yellow	Red/Orange
9	Voges-Proskauer (VP)	Tests for the production of acetoin, an intermediate in the 2,3 butanediol fermentation pathway. Acetoin is detected by the injection of 2 drops of solution containing 20% KOH and 0.3% creatine and 3 drops of alpha-napthol solution (5% wt/vol alpha-napthol in absolute ethanol). The development of a pink-red color 10 to 20 minutes after the addition of the alpha-naphthol solution indicates that acetoin was produced.	Pink	Colorless
10	Dulcitol (DUL)	Tests for dulcitol fermentation. A shift in pH is indicated by a change in color of medium from green to yellow, reflecting the production of acidic fermentation byproducts. A change in the color of the medium to yellow or pale yellow should be interpreted as a positive reaction.	Yellow	Green
	Phenylalanine Deaminase (PA)	Test for the formation of pyruvic acid from the deamination of phenylalanine. Pyruvic acid reacts with Fe^{3+} in the medium to cause a gray to black discoloration.	Black	Yellow
11	Urea (UREA)	Test for urease production. Hydrolysis of urea results in the production of ammonium, which makes the medium alkaline and causes a color change from yellow to red-purple. Light pink and other shades of red should be interpreted as positive.	Red-violet	Yellow
12	Citrate (CIT)	Test for the ability of certain bacteria to use citrate as the sole source of carbon. Utilization of citrate result in the production of alkaline metabolites, which turn the pH indicator in the culture medium from green to royal blue. Any intensity of blue should be interpreted as positive.	Blue	Green

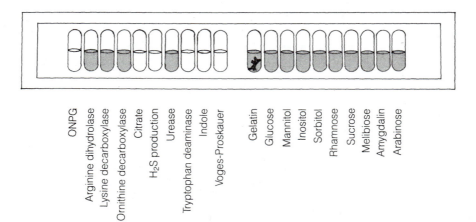

ONPG · Arginine dihydrolase · Lysine decarboxylase · Ornithine decarboxylase · Citrate · H₂S production · Urease · Tryptophan deaminase · Indole · Voges-Proskauer · Gelatin · Glucose · Mannitol · Inositol · Sorbitol · Rhamnose · Sucrose · Melibiose · Amygdalin · Arabinose

Figure 72-3. The API 20E Strip. The API 20E strip consists of 20 chambers, each consisting of a tube and a cupule. The tubes contain different types of dehydrated culture media. Twenty-six different traits can be determined using the strip.

TABLE 72-2. CHARACTERIZATION OF ENTEROBACTERIACEAE—THE ENTEROTUBE II SYSTEM

GROUPS			GLUCOSE	GAS PRODUCTION	LYSINE	ORNITHINE	H_2S	INDOLE	ADONITOL	LACTOSE	ARABINOSE	SORBITOL	VOGES-PROSKAUER	DULCITOL	PHENYLALANINE DEAMINASE	UREA	CITRATE
ESCHERICHIEAE		Escherichia	+ 100.0	+J 92.0	d 80.6	d 57.8	K 4.0	+ 96.3	5.2	+J 91.6	+ 91.3	± 80.3	0.0	d 49.3	0.0	0.1	0.2
		Shigella	+ 100.0	A 2.1	0.0	∓B 20.0	− 0.0	+ 37.8	0.0	−B 0.3	± 67.8	∓ 29.1	0.0	d 5.4	0.0	0.0	0.0
EDWARDSIELLEAE		Edwardsiella	+ 100.0	+ 99.4	+ 100.0	+ 99.0	+ 99.6	+ 99.0	− 0.0	− 0.0	∓ 10.7	0.2	− 0.0	− 0.0	− 0.0	− 0.0	− 0.0
SALMONELLEAE		Salmonella	+ 100.0	+C 91.9	+H 94.6	+I 92.7	+E 91.6	1.1	− 0.0	− 0.8	± 89.2	+ 94.1	− 0.0	dD 86.5	− 0.0	− 0.0	dF 80.1
		Arizona	+ 100.0	+ 99.7	+ 99.4	+ 100.0	+ 98.7	− 2.0	− 0.0	d 69.8	+ 99.1	+ 97.1	− 0.0	− 0.0	− 0.0	− 0.0	+ 96.8
	CITROBACTER	freundii	+ 100.0	+ 91.4	− 0.0	d 17.2	± 81.6	− 6.7	− 0.0	d 39.3	+ 100.0	+ 98.2	− 0.0	d 59.8	− 0.0	dw 89.4	+ 90.4
		amalonaticus	+ 100.0	+ 97.0	− 0.0	+ 97.0	− 0.0	+ 99.0	− 0.0	± 70.0	+ 99.0	+ 97.0	− 0.0	∓ 11.0	− 0.0	± 81.0	+ 94.0
		diversus	+ 100.0	+ 97.3	− 0.0	+ 99.8	− 0.0	+ 100.0	+ 100.0	d 40.3	+ 98.0	+ 98.2	− 0.0	± 52.2	− 0.0	dw 85.8	+ 99.7
PROTEEAE	PROTEUS	vulgaris	+ 100.0	±G 86.0	− 0.0	− 0.0	+ 95.0	+ 91.4	− 0.0	− 0.0	− 0.0	− 0.0	− 0.0	− 0.0	+ 100.0	+ 95.0	d 10.5
		mirabilis	+ 100.0	+G 96.0	− 0.0	+ 99.0	+ 94.5	− 3.2	− 0.0	− 2.0	− 0.0	− 0.0	∓ 16.0	− 0.0	+ 99.6	± 89.3	± 58.7
	MORGANELLA	morganii	+ 100.0	±G 86.0	− 0.0	+ 97.0	− 0.0	+ 99.5	− 0.0	− 0.0	− 0.0	− 0.0	− 0.0	− 0.0	+ 95.0	+ 97.1	−L 0.0
	PROVIDENCIA	alcalifaciens	+ 100.0	dG 85.2	− 0.0	− 1.2	− 0.0	+ 99.4	+ 94.3	− 0.3	− 0.7	− 0.6	− 0.0	− 0.0	+ 97.4	− 0.0	+ 97.9
		stuartii	+ 100.0	0.0	− 0.0	− 0.0	− 0.0	+ 98.6	∓ 12.4	3.6	4.0	3.4	− 0.0	− 0.0	+ 94.5	∓ 20.0	+ 93.7
		rettgeri	+ 100.0	+G 12.2	− 0.0	− 0.0	− 0.0	+ 95.9	+ 99.0	d 10.0	− 0.0	1.0	− 0.0	− 0.0	+ 98.0	+ 100.0	+ 96.0
KLEBSIELLEAE	ENTEROBACTER	cloacae	+ 100.0	+ 99.3	− 0.0	+ 93.7	− 0.0	− 0.0	∓ 28.0	± 94.0	+ 99.4	+ 100.0	+ 100.0	d 15.2	− 0.0	± 74.6	+ 98.9
		sakazakii	+ 100.0	+ 97.0	− 0.0	+ 97.0	− 0.0	∓ 16.0	− 0.0	+ 100.0	+ 100.0	− 0.0	+ 97.0	− 6.0	− 0.0	− 0.0	+ 94.0
		gergoviae	+ 100.0	+ 93.0	± 64.0	+ 100.0	− 0.0	− 0.0	− 0.0	∓ 42.0	+ 100.0	− 0.0	+ 100.0	− 0.0	− 0.0	+ 100.0	+ 96.0
		aerogenes	+ 100.0	+ 95.9	+ 97.5	+ 95.9	− 0.0	− 0.8	+ 97.5	+ 92.5	+ 100.0	+ 98.3	+ 100.0	− 4.1	− 0.0	− 0.0	+ 92.6
		agglomerans	+ 100.0	∓ 24.1	− 0.0		− 0.0	∓ 19.7	− 7.5	d 52.9	+ 97.5	d 26.3	± 64.8	d 12.9	∓ 27.6	d 34.1	d 84.2
	HAFNIA	alvei	+ 100.0	+ 98.9	+ 99.6	+ 98.6	− 0.0	− 0.0	− 0.0	d 2.8	+ 99.3	− 0.0	± 65.0	− 2.4	− 0.0	− 3.0	d 5.6
	SERRATIA	marcescens	+ 100.0	±G 52.6	+ 99.6	+ 99.6	− 0.0	−w 0.1	∓ 56.0	− 1.3	− 0.0	+ 99.1	+ 98.7	− 0.0	− 0.0	dw 39.7	+ 97.6
		liquefaciens	+ 100.0	d 72.5	± 64.2	+ 100.0	− 0.0	−w 1.8	d 8.3	d 15.6	+ 97.3	+ 97.3	∓ 49.5	− 0.0	d 0.9	dw 3.7	+ 93.6
		rubidaea	+ 100.0	dG 35.0	± 61.0	− 0.0	− 0.0	−w 2.0	± 88.0	+ 100.0	+ 100.0	− 8.0	+ 92.0	− 0.0	− 0.0	dw 4.0	± 88.0
	KLEBSIELLA	pneumoniae	+ 100.0	+ 96.0	+ 97.2	− 0.0	− 0.0	− 0.0	± 89.0	+ 98.7	+ 99.9	+ 99.4	+ 93.7	∓ 33.0	− 0.0	+ 95.4	+ 96.8
		oxytoca	+ 100.0	+ 96.0	+ 97.2	− 0.0	− 0.0	+ 100.0	± 89.0	∓ 98.7	+ 100.0	+ 98.0	+ 93.7	∓ 33.0	− 0.0	+ 95.4	∓ 96.8
		ozaenae	+ 100.0	d 55.0	∓ 35.8	− 1.0	− 0.0	− 0.0	+ 91.8	d 26.2	+ 100.0	± 78.0	− 0.0	− 0.0	− 0.0	d 14.8	d 28.1
		rhinoscleromatis	+ 100.0	− 0.0	− 0.0	− 0.0	− 0.0	− 0.0	+ 98.0	d 6.0	+ 100.0	+ 98.0	− 0.0	− 0.0	− 0.0	− 0.0	− 0.0
YERSINIAE	YERSINIA	enterocolitica	+ 100.0	− 0.0	− 0.0	+ 90.7	− 0.0	∓ 26.7	− 0.0	− 0.0	+ 98.7	+ 98.7	− 0.1	− 0.0	− 0.0	+ 90.7	− 0.0
		pseudotuberculosis	+ 100.0	− 0.0	− 0.0	− 0.0	− 0.0	− 0.0	− 0.0	− 0.0	± 55.0	− 0.0	− 0.0	− 0.0	− 0.0	+ 100.0	− 0.0

(continued on next page)

Courtesy of Roche Diagnostics, Nutley, NJ

E *S enteritidis* bioserotype Paratyphi A and some rare biotypes may be H₂S negative

F *S typhi*. *S enteritidis* bioserotype Paratyphi A and some rare biotypes are citrate-negative and *S cholerae-suis* is usually delayed positive.

G The amount of gas produced by *Serratia*, *Proteus* and *Providencia alcalifaciens* is slight, therefore, gas production may not be evident in the ENTER-OTUBE II.

H *S enteritidis* bioserotype Paratyphi A is negative for lysine decarboxylase.

I *S typhi* and *S gallinarum* are ornithine decarboxylase-negative.

J The Alkalescens-Dispar (A–D) group is included as a biotype of *E coli*. Members of the A–D group are generally anaerogenic, non-motile and do not ferment lactose.

K An occasional strain may produce hydrogen sulfide.

L An occasional strain may appear to utilize citrate.

The strip is used to determine 26 different traits of the unknown. Each tube contains a different type of dehydrated culture medium. The culture media are rehydrated with a saline (0.85% NaCl) suspension of the unknown bacterium when the tubes are filled. After all 20 chambers have been filled, the strip is incubated at 35°C. The results can be read after 5 hours of incubation, but more accurate results are obtained after 18–24 hours. The identification of the unknown is achieved by determining a seven-digit **profile index number** (see fig. 72-4) and consulting a Profile Recognition System (or the API Quick Index) or by using differential charts.

Materials

API 20E system strips

Broth culture of unknown (Enterobacteriaceae)

Plates of trypticase soy agar (TSA)

Screwcap tubes containing 5 ml of 0.85% (w/v) saline

Pasteur pipets

Sterile mineral oil

Gram-staining reagents

Kovac's reagent

Ferric chloride (10% aqueous solution)

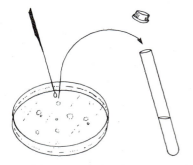

1. Pick colony and make a suspension in 5 ml of saline.

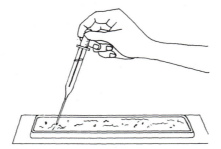

2. Dispense approximately 5 ml of water into the bottom of the API tray.

3. Place labeled API 20E strip into moistened tray.

4. Dispense saline suspension of organisms into the chambers of all twenty compartments.

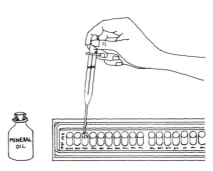

5. Fill cupules of ADH, LDC, ODC, H₂S, and URE chambers with sterile mineral oil using a sterile Pasteur pipette.

6. Circle the numbers over the positive results. Add the circled numbers in each boxed group to determine the 7-digit code.

Figure 72-4. Using the API 20E Strip

Sulfanilic acid (0.8% in 5N acetic acid)

N,N-dimethyl-α-naphthylamine (0.5% in 5N acetic acid)

α-naphthol (6% α-naphthol w/v in absolute ethanol)

Potassium hydroxide (40% aqueous solution)

Oxidase reagent

Platinum inoculating loop

Filter paper squares (~1 cm²)

Incubator set at 35°C

Procedure

First Period

1. Obtain an unknown culture from your instructor and streak a portion of it onto a plate of TSA to obtain well-isolated colonies.

2. Label the plate properly and incubate the culture, in an inverted position, at 35°C for 24–48 hours.

Second Period

1. Examine the TSA plate containing your unknown for evidence of colonial growth. Select a well-isolated colony measuring 2–3 mm in diameter to carry out the identification procedure.

2. Perform a Gram stain on the unknown isolate using only a small portion of the colony.

3. Perform an oxidase test on the isolated colony.

- Use a platinum loop to pick up a small portion of the colony (aseptically) and smear a portion of the colony over a small area of a filter paper square. Add the oxidase reagent to the smeared colony and note any color changes that take place.

4. Inoculate the API 20E strip as indicated in numbers 5–7. Use figure 72-4 as a guide. Record the results on the report form.

5. Prepare a bacterial suspension as follows:

   a. Flame a bacteriological loop, allow it to cool for a few seconds, and then touch (gently) the central portion of the selected colony.

   b. Mix the portion of the colony thoroughly with 5 ml of sterile saline in a screwcap tube to give a light milky appearance to the fluid.

6. Prepare the API 20E system as follows:

   a. Place an incubation tray and its lid on a disinfected bench top.

   b. Add 5 ml of water to the incubation tray to provide a humidified incubation environment for the API 20E strip.

   c. Remove an API 20E strip from the pouch, write your name and unknown number on it, and place the strip inside the incubation tray.

7. Inoculate the API 20E strip (fig. 72-4) as follows:

   a. Using a Pasteur pipet, fill the tube and cupule sections of the citrate (CIT), Voges-Proskauer (VP), and gelatin (GEL) chambers with the saline suspension of the unknown bacterium.

   - Filling the tubes is done best by slightly tilting the incubation tray with the strip and placing the tip of the Pasteur pipet against the side of the cupule.

   b. Fill ONLY the tube portion of the remaining chambers with the saline suspension.

   - Slightly underfill the ADH, LDC, ODC, H₂S, and URE tubes. This will facilitate reading the results.

   c. Fill the cupule of the ADH, LDC, ODC, H₂S, and URE chambers with sterile mineral oil.

8. Place the lid over the incubation chamber containing the API 20E strip.

   a. Streak a plate of TSA with a portion of the remaining unknown suspension to ascertain the purity of the suspension.

   b. Incubate the API 20E strip (in its incubation chamber) and the TSA plate at 35°C for 24 hours.

   - If you cannot read the results at 24 hours, incubate the strip until you can read it.

Third Period

1. Read and record in the report form the results of all of the tests, except those that require the addition of reagents (TDA, IND, VP, nitrate reduction, and catalase. Use tables 72-3 and 72-4 to interpret the results.

- Glucose should be positive (yellow with or without gas bubbles) because all Enterobacteriaceae are able to ferment glucose. If glucose is negative, consult figure 72-5 as a guide for the suggested protocol for proceeding with the identification of the unknown. If you have any questions regarding which procedure to follow, consult with your instructor.

2. Perform the tests in the sequence listed below and record the results obtained on the report form (use table 72-3 as a guide to the interpretation of the results):

   a. Add one drop of 10% FeCl₃ to the chamber labeled TDA.

   - A brown-red color indicates a positive reaction while a yellow color denotes a negative reaction.

   b. Perform the Voges-Proskauer test by adding one drop of 40% KOH and one drop of 6% α-naphthol to the chamber labeled VP.

   - A color change indicates a positive reaction while no color change denotes a negative reaction.

   c. Test for indole production by adding one drop of Kovac's reagent to the chamber labeled IND.

- A red ring indicates a positive reaction while a yellow ring denotes a negative reaction.

d. Test for nitrate reduction by adding two drops of sulfanilic acid (0.8%) and two drops of 0.5% N,N-dimethyl-α-naphthylamine to the chamber labeled GLU.

- A red color or gas bubbles indicate a positive reaction while yellow denotes a negative reaction.

e. Perform the catalase test by adding one drop of 1.5% H_2O_2 to the chamber labeled MAN.

- Bubbles indicate a positive catalase test while no bubbles indicate a negative test.

3. Identify your unknown using the Profile Recognition System provided by your instructor. Determine the seven-digit profile number as follows:

a. Add the numbers within each test section. Count only those tests that are positive.

TABLE 72-3.
SUMMARY OF RESULTS (18–24 HOUR PROCEDURE)

TUBE		POSITIVE	NEGATIVE	COMMENTS
				INTERPRETATION OF REACTIONS
ONPG		Yellow	Colorless	(1) Any shade of yellow is a positive reaction (2) VP tube, before the addition of reagents, can be used as a negative control.
ADH	Incubation 18–24 h	Red or Orange	Yellow	Orange reactions occurring at 36–48 hours should be interpreted as negative.
	36–48 h	Red	Yellow or Orange	
LDC	18–24 h	Red or Orange	Yellow	Any shade of orange within 18–24 hours is a positive reaction. At 36–48 hours, orange decarboxylase reactions should be interpreted as negative.
	36–48 h	Red	Yellow or Orange	
ODC	18–24 h	Red or Orange	Yellow	Orange reactions occurring at 36–48 hours should be interpreted as negative.
	36–48 h	Red	Yellow or Orange	
CIT		Turquoise or Dark Blue	Light Green or Yellow	(1) Both the tube and cupule should be filled. (2) Reaction is read in the aerobic (cupule) area.
H₂S		Black Deposit	No Black Deposit	(1) H₂S production may range from a heavy black deposit to a very thin black line around the tube bottom. Carefully examine the bottom of the tube before considering the reaction negative. (2) A "browning" of the medium is a negative reaction unless a black deposit is present. "Browning" occurs with TDA positive organisms.
URE	18–24 h	Red or Orange	Yellow	A method of lower sensitivity has been chosen. *Klebsiella*, *Proteus* and *Yersinia*, routinely give positive reactions.
	36–48 h	Red	Yellow or Orange	
TDA		Add 1 drop 10% Ferric chloride		(1) Immediate reaction. (2) Indole positive organisms may produce a golden orange color due to indole production. This is a negative reaction.
		Brown-Red	Yellow	
IND		Add 1 drop Kovacs' Reagent		(1) The reaction should be read within 2 minutes after the addition of the Kovacs' reagent and the results recorded. (2) After several minutes, the HCl present in Kovacs' reagent may react with the plastic of the cupule resulting in a change from a negative (yellow) color to a brownish-red. This is a negative reaction.
		Red Ring	Yellow	
VP		Add 1 drop of 40% Potassium hydroxide, then 1 drop of 6% alpha-naphthol		(1) Wait 10 minutes before considering the reaction negative. (2) A pale pink color (after 10 min.) should be interpreted as negative. A pale pink color which appears immediately after the addition of reagents but which turns dark pink or red after 10 min. should be interpreted as positive. Motility maybe observed by hanging drop or wet mount preparation.
		Red	Colorless	
GEL		Diffusion of the pigment	No Diffusion	(1) The solid gelatin particles may spread throughout the tube after inoculation. Unless diffusion occurs, the reaction is negative. (2) Any degree of diffusion is a positive reaction.

TABLE 72-3.
CONTINUED

TUBE	POSITIVE	INTERPRETATION OF REACTIONS NEGATIVE	COMMENTS
GLU	Yellow or Gray	Blue or Blue-Green	**Fermentation** *(Enterobacteriaceae, Aeromonas, Vibrio)* (1) Fermentation of the carbohydrates begins in the most anaerobic portion (bottom) of the tube. Therefore, these reactions should be read from the bottom of the tube to the top. (2) A yellow color at the bottom of the tube only indicates a weak or delayed positive reaction.
MAN INO SOR RHA SAC MEL AMY ARA	Yellow	Blue or Blue-Green	**Oxidation** (Other Gram-negatives) (1) Oxidative utilization of the carbohydrates begins in the most aerobic portion (top) of the tube. Therefore, these reactions should be read from the top to the bottom of the tube. (2) A yellow color in the upper portion of the tube and a blue in the bottom of the tube indicates oxidative utilization of the sugar. This reaction should be considered positive **only** for non-*Enterobacteriaceae* gram-negative rods. This is a negative reaction for fermentative organisms such as *Enterobacteriaceae*.

GLU	After reading GLU reaction, add 2 drops 0.8% sulfanilic acid and 2 drops 0.5% N, N dimethyl-alpha-naphthylamine		(1) Before addition of reagents, observe GLU tube (positive or negative) for bubbles. Bubbles are indicative of reduction of nitrate to the nitrogenous (N_2) state. (2) A positive reaction may take 2–3 minutes for the red color to appear. (3) Confirm a negative test by adding zinc dust or 20 mesh granular zinc. A pink-orange color after 10 minutes confirms a negative reaction. A yellow color indicates reduction of nitrates to nitrogenous (N_2) state.
Nitrate Reduction	NO_2 N_2 gas	Red Bubbles. Yellow after reagents and zinc	Yellow Orange after reagents and zinc	

MAN INO SOR Catalase	After reading carbohydrate reaction, add 1 drop 1.5% H_2O_2		(1) Bubbles may take 1–2 minutes to appear. (2) Best results will be obtained if the test is run in tubes which have no gas from fermentation.
	Bubbles	No bubbles	

Analytab Products, Division of Sherwood Medical, October 1985. API 20E® System. Package Product Insert. Plainview, NY. Used with permission.

b. Enter the sum of the positive tests in the square labeled profile number. The seven digits represent the profile number for your unknown.

4. Verify your identification with the aid of the differential charts provided in tables 72-5 and 72-6.

5. Turn in your unknown report form to your instructor with the genus and species of your unknown.

6. Discard the inoculated API 20E strips and put away all the reagents and materials used as indicated by your instructor.

Questions

1. The API 20E strip is used routinely to identify pathogens in the clinical laboratory. The results are very reliable and correspond well with those obtained using conventional identification methods. What must have been done during the design and testing of the API 20E strip to make it such a reliable rapid identification procedure?

2. Why was mineral oil dispensed on the LDC and ODC cupules? In your answer, include the necessity of using sterile mineral oil.

3. Could Gram negative, oxidase positive bacteria be identified using the API 20E strip?

4. Why isn't it possible to identify such bacteria using the Enterotube II?

5. Why are the results obtained with the API 20E strip after 5 hours of incubation not as reliable as those obtained after 24 hours?

TABLE 72-4.
SUMMARY OF CHEMICAL AND PHYSICAL PRINCIPLES OF THE TESTS ON THE API 20E

| TUBE | CHEMICAL/PHYSICAL PRINCIPLES | COMPONENTS | |
		REACTIVE INGREDIENTS	QUANTITY
ONPG	Hydrolysis of ONPG by beta-galactosidase releases yellow orthonitrophenol from the colorless ONPG; ITPG (isopropylthio-galactopyranoside) is used as inducer.	ONPG ITPG	0.2 mg 8.0 µg
ADH	Arginine dihydrolase transforms arginine into ornithine, ammonia and carbon dioxide. This causes a pH rise in the acid-buffered system and a change in the indicator from yellow to red.	Arginine	2.0 mg
LDC	Lysine decarboxylase transforms lysine into a basic primary amine, cadaverine. This amine causes a pH rise in the acid-buffered system and a change in the indicator from yellow to red.	Lysine	2.0 mg
ODC	Ornithine decarboxylase transforms ornithine into a basic primary amine, putrescine. This amine causes a pH rise in the acid-buffered system and a change in the indicator from yellow to red.	Ornithine	2.0 mg
CIT	Citrate is the sole carbon source. Citrate utilization results in a pH rise and a change in the indicator from green to blue.	Sodium Citrate	0.8 mg
H_2S	Hydrogen sulfide is produced from thiosulfate. The hydrogen sulfide reacts with iron salts to produce a black precipitate.	Sodium Thiosulfate	80.0 µm
URE	Urease releases ammonia from urea; ammonia causes the pH to rise and changes the indicator from yellow to red.	Urea	0.8 mg
TDA	Tryptophane deaminase forms indolepyruvic acid from tryptophane. Indolepyruvic acid produces a brownish-red color in the presence of ferric chloride.	Tryptophane	0.4 mg
IND	Metabolism of tryptophane results in the formation of indole. Kovacs' reagent forms a colored complex (pink to red) with indole.	Tryptophane	0.2 mg
VP	Acetoin, an intermediary glucose metabolite, is produced from sodium pyruvate and indicated by the formation of a colored complex. Conventional VP tests may take up to 4 days, but by using sodium pyruvate, API has shortened the required test time. Creatin intensifies the color when tests are positive.	Sodium Pyruvate Creatine	2.0 mg 0.9 mg
GEL	Liquefaction of gelatin by proteolytic enzymes releases a black pigment which diffuses throughout the tube.	Kohn Charcoal Gelatin	0.6 mg
GLU MAN INO SOR RHA SAC MEL AMY ARA	Utilization of the carbohydrate results in acid formation and a consequent pH drop. The indicator changes from blue to yellow.	Glucose Mannitol Inositol Sorbitol Rhamnose Sucrose Melibiose Amygdalin (I +) Arabinose	2.0 mg 2.0 mg 2.0 mg 2.0 mg 2.0 mg 2.0 mg 2.0 mg 2.0 mg 2.0 mg
GLU Nitrate Reduction	Nitrites form a red complex with sulfanilic acid and N, N-dimethyl-alpha-naphthylamine. In case of negative reaction, addition of zinc confirms the presence of unreduced nitrates by reducing them to nitrites (pink-orange color). If there is no color change after the addition of zinc, this is indicative of the complete reduction of nitrates through nitrites to nitrogen gas or to an anaerogenic amine.	Potassium Nitrate	80.0 µg
MAN INO SOR Catalase	Catalase releases oxygen gas from hydrogen peroxide.		

Analytab Products, Division of Sherwood Medical, October 1985. API 20E® System. Package Product Insert. Plainview, NY. Used with permission.

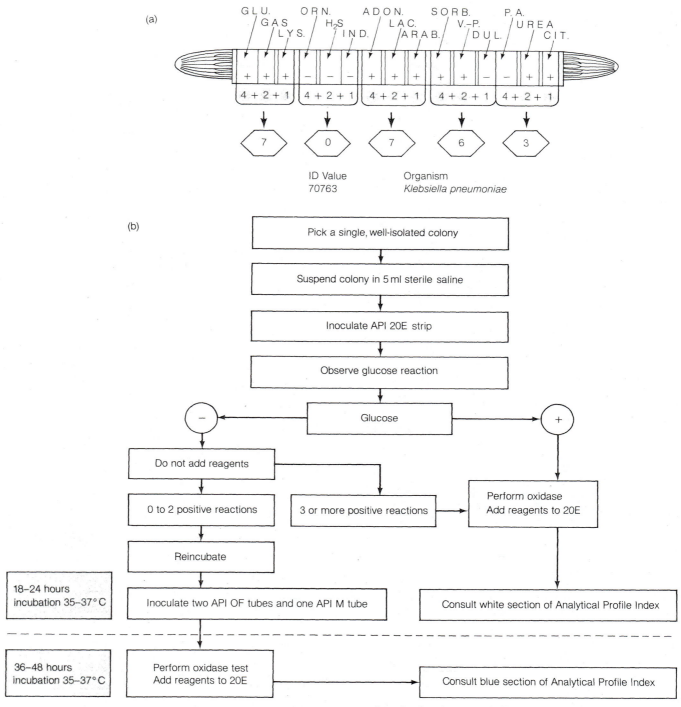

Figure 72-5. Outline of Procedure for Use and Interpretation of the API 20E Strip. Identification of bacteria using the API 20E strip. The strip consists of 20 wells each containing a different type of medium. A colony is suspended in 5 ml of saline and used to inoculate the 20 wells. The strip is then incubated for 18–24 hours. The results are assessed. See color insert of the API 20E strip.

TABLE 72-5

DIFFERENTIATION OF *ENTEROBACTERIACEAE* BY BIOCHEMICAL TESTS PROVIDED ON API 20E®
(FIGURES INDICATE THE PERCENTAGE OF POSITIVE REACTIONS AFTER 18–24 HOURS OF INCUBATION AT 35–37°C.)

		API 20E									
		ONPG	ADH	LDC	ODC	CIT	H₂S	URE	TDA	IND	VP
Escherichieae	*Escherichia coli*	98.7	3.0	82.8	75.7	0.0	1.0	0.0	0.0	92.6	0.0
	E. fergusonii	93.3	0.0	100.0	100.0	0.0	0.0	0.0	0.0	100.0	0.0
	E. hermannii	99.9	0.0	0.1	99.9	3.2	0.0	0.0	0.0	99.9	0.0
	E. vulneris	100.0	22.2	45.0	0.0	0.0	0.0	0.0	0.0	0.0	0.0
	Shigella dysenteriae	33.3	0.0	0.0	0.0	0.0	0.0	0.0	0.0	33.0	0.0
	Sh. flexneri	0.9	0.0	0.0	0.9	0.0	0.0	0.0	0.0	36.2	0.0
	Sh. boydii	10.4	0.0	0.0	9.1	0.0	0.0	0.0	0.0	50.0	0.0
	Sh. sonnei	87.6	0.0	0.0	92.9	0.0	0.0	0.0	0.0	0.0	0.0
	Edwardsiella tarda	0.0	0.0	99.9	99.0	0.0	94.4	0.0	0.0	99.0	0.0
Proteeae	*Citrobacter freundii*	92.3	21.3	0.0	52.4	65.1	60.7	0.0	0.0	10.0	0.0
	C. diversus	97.5	63.6	0.0	98.8	91.0	0.0	0.0	0.0	98.8	0.0
	C. amalonaticus	95.0	40.0	0.0	98.0	77.0	0.0	0.0	0.0	100.0	0.0
	Salmonella enteritidis	1.9	69.2	96.2	95.7	75.4	85.7	0.0	0.0	3.0	0.0
	Sal. typhi	0.0	5.8	99.0	0.0	0.0	8.3	0.0	0.0	0.0	0.0
	Sal. cholerae suis	0.0	18.4	98.0	98.0	4.1	65.3	0.0	0.0	0.0	0.0
	Sal. paratyphi A	0.0	0.0	0.0	100.0	0.0	0.6	0.0	0.0	0.0	0.0
	Sal. subgroup 3	98.7	48.1	96.1	97.4	50.7	98.1	0.0	0.0	0.0	0.0
Klebsielleae	*Klebsiella pneumoniae*	99.0	0.2	74.8	0.9	74.9	0.0	58.6	0.0	0.0	87.4
	Kl. oxytoca	100.0	0.0	86.7	0.0	60.0	0.0	60.0	0.0	100.0	92.4
	Kl. ozaenae	90.0	23.3	32.2	1.1	40.0	0.0	6.7	0.0	0.0	0.0
	Kl. rhinoscleromatis	0.0	0.0	0.0	0.0	0.0	0.0	0.0	0.0	0.0	0.0
	Enterobacter aerogenes	99.5	0.0	98.8	99.0	83.7	0.0	0.3	0.0	0.0	93.6
	Ent. cloacae	99.0	93.5	0.1	97.3	85.5	0.0	0.4	0.0	0.0	96.6
	Ent. agglomerans	98.3	1.0	0.0	0.0	55.2	0.0	3.4	3.7	50.0	33.2
	Ent. gergoviae	94.0	0.0	28.8	99.9	82.0	0.0	100.0	0.0	0.0	95.2
	Ent. sakazakii	100.0	99.0	0.0	85.7	85.7	0.0	0.0	0.0	5.0	73.7
	Ent. taylorae	98.1	86.8	0.0	92.5	88.7	0.0	0.0	0.0	0.0	92.1
	Ent. amnigenus 1	78.6	0.0	0.0	78.6	5.0	0.0	0.0	0.0	0.0	28.6
	Ent. amnigenus 2	100.0	95.0	0.0	100.0	33.3	0.0	0.0	0.0	0.0	10.0
	Ent. intermedium	100.0	0.0	10.0	90.0	0.0	0.0	0.0	0.0	0.0	10.0
	Serratia liquefaciens	98.1	0.6	82.4	99.0	88.7	0.0	5.0	0.0	0.0	57.8
	Ser. marcescens	94.2	0.0	98.5	95.7	74.9	0.0	29.0	0.0	0.0	65.9
	Ser. rubidaea	100.0	0.0	68.5	0.0	81.5	0.0	3.3	0.0	0.0	63.5
	Ser. odorifera 1	95.0	0.0	99.0	100.0	95.0	0.0	0.0	0.0	99.9	50.0
	Ser. odorifera 2	95.0	0.0	99.0	0.0	90.9	0.0	0.0	0.0	99.9	54.5
	Ser. fonticola	99.0	0.0	66.7	97.6	21.4	0.0	0.0	0.0	0.0	0.0
	Ser. plymuthica	99.0	0.0	0.0	0.0	16.7	0.0	0.0	2.0	0.0	75.0
	Hafnia alvei	71.2	1.8	100.0	99.0	10.6	0.0	1.0	0.0	0.0	15.4
Salmonelleae	*Proteus penneri*	0.5	0.0	0.0	0.0	41.2	73.2	98.9	99.6	0.0	0.0
	Proteus vulgaris	0.5	0.0	0.0	0.0	41.2	83.2	98.9	99.6	100.0	0.0
	Prot. mirabilis	0.2	0.6	2.0	98.4	57.8	83.4	99.0	98.7	1.9	2.4
	Providencia alcalifaciens	0.0	0.0	0.0	0.0	97.5	0.0	0.0	100.0	99.0	0.0
	Prov. stuartii Ure −	0.0	0.0	0.8	0.0	85.1	0.0	0.0	91.7	97.1	0.0
	Prov. stuartii Ure +	1.0	0.0	0.0	0.0	68.6	0.0	100.0	75.7	88.6	0.0
	Prov. rettgeri	1.0	0.0	0.0	0.0	70.7	0.0	99.0	99.0	97.4	0.0
	Morganella morganii	0.5	0.0	0.5	99.0	2.2	0.0	99.0	91.8	97.2	0.0
Yersiniae	*Yersinia enterocolitica*	81.1	0.0	0.0	85.1	0.0	0.0	93.7	0.0	69.4	0.0
	Y. intermedia	95.1	0.0	0.0	100.0	0.0	0.0	97.6	0.0	97.6	2.4
	Y. fredericksenii	95.6	0.0	0.0	100.0	0.0	0.0	100.0	0.0	99.0	0.0
	Y. kristensenii	87.5	0.0	0.0	87.5	0.0	0.0	100.0	0.0	99.0	0.0
	Y. pseudotuberculosis	77.1	0.0	0.0	0.0	13.3	0.0	96.2	0.0	0.0	0.0
	Presumptive Y. pestis	68.6	0.0	0.0	0.0	0.0	0.0	0.0	0.0	0.0	8.6
	Y. ruckeri AN	73.5	0.0	93.9	95.9	0.0	0.0	0.0	0.0	0.0	0.0
	Cedecea davisae	50.0	86.4	0.0	95.5	36.4	0.0	0.0	0.0	0.0	36.4
	Cedecea lepagei	83.3	66.7	0.0	0.0	66.7	0.0	0.0	0.0	0.0	25.0
	Cedecae sp. 3	90.0	90.0	0.0	0.0	75.0	0.0	0.0	0.0	0.0	0.0
	Cedecea neteri	99.0	99.0	0.0	0.0	80.0	0.0	0.0	0.0	0.0	66.7
	Cedecea sp. 5	90.0	90.0	0.0	90.0	25.0	0.0	0.0	0.0	0.0	75.0
	Tatumella ptyseos	0.0	0.0	0.1	0.0	1.0	0.0	0.0	1.0	0.0	50.0
	Kluyvera sp.	91.7	0.0	56.7	90.0	91.7	0.0	0.0	0.0	99.0	0.0
	Moellerella wisconsensis	75.0	0.0	0.0	0.0	37.5	0.0	0.0	0.0	0.0	12.5
	Ewingella americana	90.9	0.0	0.0	0.0	27.7	0.0	0.0	0.0	0.0	95.5
	CDC Enteric Group 17	99.9	5.3	0.0	99.9	26.1	0.0	0.0	0.0	0.1	21.1
	CDC Enteric Group 41	100.0	0.0	0.0	0.0	0.0	0.0	0.0	0.0	88.9	0.0

TABLE 72-5
CONTINUED

					API 20E									SUPPLEMENTARY TESTS**			
GEL	GLU	MAN	INO	SOR	RHA	SAC	MEL	AMY	ARA	OXI	NO$_2$	GAS	MOT	MAC	OF-O	OF-F
0.0	99.9	99.0	0.5	89.2	87.9	41.8	67.1	9.2	90.8	0.0	99.7	0.0	62.1	100.0	100.0	100.0
0.0	100.0	100.0	0.0	0.0	80.0	0.0	0.0	100.0	100.0	0.0	100.0	0.0	93.3	100.0	100.0	100.0
0.0	99.9	99.9	0.0	0.0	96.8	58.1	0.0	96.8	99.9	0.0	99.9	0.0	99.0	99.0	99.0	99.0
0.0	100.0	100.0	0.0	0.0	83.3	5.6	94.4	94.4	100.0	0.0	100.0	0.0	100.0	100.0	100.0	100.0
0.0	95.9	2.5	0.0	18.6	22.2	0.0	0.0	0.0	16.7	0.0	99.7	0.0	0.0	100.0	99.9	99.0
0.0	99.0	91.8	0.0	24.0	2.6	0.0	24.9	0.0	61.6	0.0	99.8	0.0	0.0	100.0	100.0	100.0
0.0	99.4	94.9	0.0	60.0	0.6	1.3	21.0	0.0	70.8	0.0	99.0	0.0	0.0	100.0	100.0	100.0
0.0	99.9	99.0	0.0	2.2	80.0	0.0	0.0	0.0	93.8	0.0	100.0	0.0	0.0	100.0	100.0	100.0
0.0	99.9	0.0	0.0	0.0	0.0	0.0	0.0	0.0	1.1	0.0	100.0	0.0	98.2	100.0	100.0	100.0
0.0	99.9	99.9	10.5	96.0	91.3	53.5	65.7	59.1	96.5	0.0	99.0	0.0	95.7	100.0	100.0	100.0
0.0	99.0	99.0	4.5	92.6	95.1	24.7	1.2	96.3	95.1	0.0	100.0	0.0	92.9	100.0	100.0	100.0
0.0	99.0	99.0	0.0	99.0	99.0	24.4	28.0	95.0	97.0	0.0	99.0	0.0	99.0	100.0	100.0	100.0
0.0	99.9	98.7	33.6	93.2	93.0	2.3	78.2	0.0	94.6	0.0	100.0	0.0	94.6	100.0	100.0	100.0
0.0	99.9	99.0	0.0	100.0	0.0	0.0	94.4	0.0	0.0	0.0	100.0	0.0	100.0	100.0	100.0	100.0
0.0	99.9	99.0	0.0	89.8	95.9	0.0	20.4	0.0	0.0	0.0	100.0	0.0	100.0	100.0	100.0	100.0
0.0	99.9	99.0	0.0	99.0	99.0	0.0	97.0	0.0	100.0	0.0	100.0	0.0	94.6	100.0	100.0	100.0
0.0	100.0	99.0	0.0	99.0	96.1	0.0	64.9	0.0	99.0	0.0	100.0	0.0	100.0	100.0	100.0	100.0
0.2	99.0	99.0	96.5	99.0	97.9	99.0	99.0	99.0	99.0	0.0	99.9	0.0	0.0	100.0	100.0	100.0
1.0	99.0	99.0	96.7	99.0	96.7	99.0	96.7	99.0	96.7	0.0	99.9	0.0	0.0	100.0	100.0	100.0
0.0	97.8	92.2	61.1	44.4	66.7	21.1	83.3	90.0	67.8	0.0	92.0	0.0	0.0	100.0	100.0	100.0
0.0	96.2	99.0	83.0	64.2	39.6	39.6	18.9	92.5	1.9	0.0	100.0	0.0	0.0	100.0	100.0	100.0
0.4	99.0	99.0	92.8	97.0	99.0	98.1	99.0	99.0	99.0	0.0	100.0	0.0	97.3	100.0	100.0	100.0
0.6	99.0	99.0	13.4	92.6	80.3	99.0	91.3	99.0	99.0	0.0	100.0	0.0	94.5	100.0	100.0	100.0
2.1	98.7	98.4	16.4	41.5	82.3	73.1	68.1	85.7	96.2	0.0	85.9	0.0	89.4	100.0	100.0	100.0
0.0	100.0	98.2	19.3	0.0	100.0	98.2	100.0	100.0	100.0	0.0	100.0	0.0	100.0	100.0	100.0	100.0
0.0	100.0	100.0	72.0	0.0	99.0	99.0	99.0	99.0	99.0	0.0	100.0	0.0	94.0	100.0	100.0	100.0
0.0	99.9	99.9	0.0	0.0	99.9	1.9	0.0	99.9	99.9	0.0	99.0	0.0	83.0	99.0	99.0	99.0
0.0	100.0	100.0	0.0	28.6	92.9	78.6	71.4	85.7	85.7	0.0	100.0	0.0	100.0	100.0	100.0	100.0
0.0	100.0	100.0	0.0	100.0	100.0	0.0	100.0	100.0	100.0	0.0	100.0	0.0	100.0	100.0	100.0	100.0
0.0	100.0	90.0	0.0	80.0	90.0	40.0	100.0	100.0	100.0	0.0	100.0	0.0	90.0	100.0	100.0	100.0
60.4	99.0	99.0	71.7	98.7	22.4	99.0	77.5	99.0	93.8	0.0	100.0	0.0	93.3	100.0	100.0	100.0
85.5	99.0	99.0	75.0	91.3	0.0	98.5	63.1	97.1	15.9*	0.0	95.8	0.0	98.6	100.0	100.0	100.0
62.9	96.7	98.9	42.4	4.3	2.3	84.8	82.6	94.6	85.8	0.0	100.0	0.0	88.0	100.0	100.0	100.0
99.0	99.0	99.0	99.0	99.0	99.0	100.0	99.0	99.0	99.0	0.0	99.0	0.0	99.0	100.0	100.0	100.0
99.0	99.0	99.0	99.0	99.0	99.0	0.0	99.0	99.0	99.0	0.0	99.0	0.0	87.5	100.0	100.0	100.0
0.0	99.0	99.0	88.1	97.6	95.2	0.0	99.0	97.6	92.9	0.0	99.0	0.0	99.0	100.0	100.0	100.0
66.7	99.0	99.0	50.0	95.0	0.5	100.0	66.7	99.0	99.0	0.0	99.0	0.0	95.0	99.0	100.0	100.0
0.0	99.0	95.2	0.0	1.0	71.5	0.0	2.9	11.5	90.2	0.0	100.0	0.0	93.0	100.0	100.0	100.0
52.8	97.3	0.4	1.3	0.0	2.7	89.6	0.0	65.2	0.5	0.0	100.0	0.0	94.7	100.0	100.0	100.0
52.8	97.3	0.4	1.3	0.0	2.7	89.6	0.0	65.2	0.4	0.0	100.0	0.0	94.7	100.0	100.0	100.0
76.6	96.3	0.4	0.0	0.4	0.0	0.7	0.1	1.0	0.2	0.0	93.8	0.0	95.9	100.0	100.0	100.0
0.0	99.0	2.5	2.5	0.0	0.0	2.5	0.0	0.0	2.9	0.0	100.0	0.0	96.5	100.0	100.0	100.0
0.4	99.0	0.8	99.0	0.0	0.0	3.7	0.0	0.8	3.3	0.0	100.0	0.0	87.0	100.0	100.0	100.0
0.0	99.0	14.0	85.7	0.0	0.0	62.9	0.0	0.0	0.0	0.0	100.0	0.0	87.0	100.0	100.0	100.0
0.4	99.0	84.1	78.8	0.0	41.2	34.4	0.0	33.4	1.5	0.0	98.8	0.0	94.4	100.0	100.0	100.0
0.0	97.0	0.2	0.0	0.0	0.0	0.3	0.0	0.0	1.2	0.0	88.5	0.0	87.7	100.0	100.0	100.0
0.0	99.0	99.1	25.2	98.2	5.4	100.0	0.9	87.8	76.6	0.0	98.7	0.0	0.0	100.0	100.0	100.0
0.0	99.0	99.0	63.4	95.1	100.0	100.0	100.0	99.0	46.3	0.0	98.7	0.0	0.0	100.0	100.0	100.0
0.0	99.0	99.0	11.1	95.6	100.0	100.0	0.0	97.8	57.8	0.0	98.7	0.0	0.0	100.0	100.0	100.0
0.0	99.0	99.0	62.5	99.0	0.0	0.0	0.0	99.0	87.5	0.0	98.7	0.0	0.0	100.0	100.0	100.0
0.0	98.1	97.1	0.0	0.0	77.1	0.0	9.5	0.0	29.5	0.0	95.0	0.0	0.0	100.0	100.0	100.0
0.0	99.0	97.1	0.0	71.4	0.0	0.0	0.0	11.4	0.0	0.0	47.9	0.0	0.0	100.0	100.0	100.0
0.0	83.7	95.9	0.0	0.0	0.0	0.0	0.0	0.0	0.0	0.0	50.0	0.0	0.0	100.0	100.0	100.0
0.0	99.0	99.0	27.3	0.0	0.0	100.0	0.0	99.0	1.0	0.0	99.0	0.0	99.0	99.0	99.0	99.0
0.0	99.0	99.0	0.0	0.0	0.0	0.0	0.0	99.0	0.0	0.0	99.0	0.0	80.0	99.0	99.0	99.0
0.0	99.0	99.0	0.1	0.1	0.1	99.0	90.0	99.0	0.0	0.0	90.0	0.0	90.0	90.0	99.0	99.0
0.0	99.0	99.0	0.0	99.0	0.0	99.0	0.0	99.0	0.0	0.0	99.0	0.0	90.0	90.0	99.0	99.0
0.0	100.0	99.9	0.0	95.0	0.0	99.0	50.0	99.0	0.0	0.0	90.0	0.0	90.0	90.0	99.0	99.0
0.0	100.0	0.0	0.0	0.0	0.0	80.0	90.0	80.0	0.0	0.0	98.0	0.0	0.1	99.0	99.0	99.0
0.0	99.0	91.7	0.0	5.0	85.0	66.7	87.0	99.0	90.0	0.0	95.0	0.0	97.0	100.0	100.0	100.0
0.0	100.0	5.0	0.0	0.0	0.0	100.0	100.0	0.0	0.0	0.0	100.0	0.0	0.0	100.0	100.0	100.0
31.8	100.0	81.8	0.0	0.0	0.0	0.0	0.0	13.6	4.5	0.0	90.9	0.0	40.9	100.0	100.0	100.0
0.0	99.0	99.9	21.1	84.2	0.1	99.9	0.0	99.9	99.9	0.0	99.0	0.0	5.0	99.0	99.0	99.0
0.0	100.0	100.0	0.0	11.1	100.0	66.7	100.0	100.0	100.0	0.0	100.0	0.0	100.0	100.0	100.0	100.0

**These tests may be required for differentiation of *Enterobacteriaceae* from other Gram-negative bacteria.
*Positive by oxidative metabolism.

Analytab Products, Division of Sherwood Medical, October 1985. API 20E® System. Package Product Insert. Plainview, NY. Used with permission.

TABLE 72-6. DIFFERENTIATION OF SOME NON-*ENTEROBACTERIACEAE* GRAM-NEGATIVE BACTERIA BY BIOCHEMICAL REACTIONS PROVIDED ON API 20E® (FIGURES INDICATE THE PERCENTAGE OF POSITIVE REACTIONS AFTER INCUBATION AT 35–37°C.)

ORGANISM		ONPG	ADH	LDC	ODC	CIT	H₂S	URE	TDA	IND	VP
Pseudomonas aeruginosa	24h	0.0	75.8	0.0	0.0	78.8	0.0	24.0	0.0	0.0	2.0
	48h	0.0	99.2	0.0	0.0	98.8	0.0	49.3	0.0	0.0	6.4
Ps. fluorescens	24h	0.0	51.0	0.0	0.0	48.0	0.0	1.9	0.0	0.0	9.4
	48h	0.0	91.3	0.0	0.0	93.5	0.0	2.2	0.0	0.0	26.1
Ps. putida	24h	0.0	59.0	0.0	0.0	63.6	0.0	2.9	0.0	0.0	20.6
	48h	0.0	88.0	0.0	0.0	92.8	0.0	6.3	0.0	0.0	28.1
Ps. cepacia	24h	61.1	0.0	16.3	8.1	75.0	0.0	0.0	0.0	0.0	4.6
	48h	76.2	0.0	63.1	30.4	96.4	0.0	1.2	0.0	0.0	16.0
Ps. maltophilia	24h	73.8	0.0	76.3	0.0	76.2	0.0	0.0	0.0	0.0	0.0
	48h	71.9	0.0	73.0	0.0	93.3	0.0	1.1	0.0	0.0	2.3
Ps. putrefaciens	24h	0.0	0.0	0.0	87.5	90.6	90.6	0.0	0.0	0.0	6.3
	48h	0.0	0.0	0.0	90.6	96.9	93.8	0.0	0.0	0.0	6.3
Ps. stutzeri	24h	0.0	0.0	0.0	0.0	18.0	0.0	0.0	0.0	0.0	9.4
	48h	0.0	1.9	0.0	0.0	59.3	0.0	0.0	0.0	0.0	11.1
Ps. pseudomallei	24h	0.0	0.0	0.0	0.0	0.0	0.0	0.0	0.0	0.0	0.0
	48h	0.0	70.8	0.0	0.0	29.2	0.0	0.0	0.0	0.0	4.2
Ps. paucimobilis	24h	80.0	0.0	0.0	0.0	10.0	0.0	5.0	0.0	0.0	15.0
	48h	86.4	0.0	0.0	0.0	54.6	0.0	5.0	0.0	0.0	31.8
Other *Pseudomonas* spp.	24h	0.4	0.7	0.0	0.0	18.0	0.0	0.7	0.0	0.0	9.7
	48h	1.9	1.9	0.0	0.0	60.9	0.0	1.9	0.0	0.0	15.2
Acinetobacter calcoaceticus var. anitratus	24h	0.0	0.0	0.0	0.0	28.2	0.0	0.0	0.0	0.0	14.9
	48h	0.0	0.0	0.0	0.0	54.0	0.0	0.0	0.0	0.0	22.0
Acinetobacter calcoaceticus var. lwoffi	24h	0.0	0.0	0.0	0.0	7.8	0.0	2.2	0.0	0.0	11.1
	48h	0.0	0.0	0.0	0.0	19.1	0.0	2.4	0.0	0.0	11.9
Flavobacterium spp. (11B)	24h	25.0	0.0	0.0	0.0	10.0	0.0	78.3	0.0	77.5	0.0
	48h	33.9	0.0	0.0	0.0	85.4	0.0	92.7	0.0	81.9	0.0
Flav. meningosepticum	24h	64.0	0.0	0.0	0.0	20.9	0.0	0.0	0.0	81.1	0.0
	48h	86.0	0.0	0.0	0.0	84.9	0.0	2.7	0.0	84.9	0.0
Flav. odoratum	24h	0.0	0.0	0.0	0.0	23.6	0.0	45.6	0.0	0.0	0.0
	48h	0.0	0.0	0.0	0.0	82.0	0.0	53.5	0.0	0.0	0.0
Flav. breve	24h	0.0	0.0	0.0	0.0	0.0	0.0	0.0	0.0	100.0	0.0
	48h	0.0	0.0	0.0	0.0	90.0	0.0	0.0	0.0	100.0	0.0
Flav. multivorum	24h	87.0	0.0	0.0	0.0	0.0	0.0	79.2	0.0	0.0	50.0
	48h	96.2	0.0	0.0	0.0	65.4	0.0	96.2	0.0	0.0	75.0
Flav. spiritivorum	24h	75.0	0.0	0.0	0.0	0.0	0.0	0.0	0.0	0.0	0.0
	48h	100.0	0.0	0.0	0.0	0.0	0.0	0.0	0.0	0.0	6.2
Bordetella bronchiseptica	24h	0.0	0.0	0.0	0.0	10.6	0.0	63.0	0.0	0.0	17.0
	48h	0.0	0.0	0.0	0.0	69.4	0.0	86.3	0.0	0.0	36.1
Alcaligenes spp.	24h	0.0	0.0	0.0	0.0	37.0	0.0	3.0	0.0	0.0	8.9
	48h	0.0	0.0	0.0	0.0	74.8	0.0	5.8	10.0	0.0	40.3
Moraxella spp.	24h	0.0	0.0	0.0	0.0	5.2	0.0	7.0	0.0	0.0	2.9
	48h	0.0	0.0	0.0	0.0	11.5	0.0	11.5	0.0	0.0	2.6
Pasteurella multocida	24h	3.7	0.0	0.0	13.0	0.0	0.0	0.0	0.0	88.9	0.0
	48h	3.9	0.0	0.0	13.0	0.0	0.0	0.0	0.0	96.2	0.0
Past. aerogenes	24h	70.0	0.0	0.0	95.0	0.0	0.0	95.0	0.0	0.0	0.0
	48h	70.0	0.0	0.0	95.0	0.0	0.0	95.0	0.0	0.0	0.0
Pasteurella-Actinobacillus spp.	24h	39.3	0.0	0.0	0.0	0.0	0.0	19.7	0.0	6.5	6.6
	48h	44.4	0.0	0.0	0.0	0.0	0.0	36.1	0.0	5.6	6.9
Chromobacterium	24h	0.0	96.2	0.0	0.0	57.7	0.0	0.0	0.0	19.2	0.0
	48h	0.0	96.2	0.0	0.0	95.8	0.0	0.0	0.0	19.2	0.0
Achromobacter xylosoxidans	24h	0.0	0.0	0.0	0.0	32.7	0.0	0.0	0.0	0.0	0.0
	48h	0.0	0.0	0.0	0.0	89.1	0.0	0.0	0.0	0.0	9.1
Achromobacter spp. (Vd)	24h	0.0	0.0	0.0	0.0	24.3	0.0	62.2	0.0	0.0	0.0
	48h	0.0	0.0	0.0	0.0	75.9	0.0	91.3	4.4	0.0	3.1
Agrobacterium radiobacter	24h	69.7	0.0	0.0	0.0	18.6	0.0	11.6	0.1	0.0	0.0
	48h	95.4	0.0	0.0	0.0	69.8	0.0	46.5	2.3	0.0	0.0

TABLE 72-6
CONTINUED

| | API 20E | | | | | | | | | | | | | SUPPLEMENTARY TESTS | | | |
GEL	GLU	MAN	INO	SOR	RHA	SAC	MEL	AMY	ARA	OXI	NO₂	N₂ GAS	MOT	MAC	OF-O	OF-F
75.5	57.8	0.0	0.0	0.0	0.0	0.0	2.7	0.0	10.9	98.6	12.8	56.8	86.4	99.2	98.4	0.0
87.2	61.6	0.0	0.0	0.0	0.0	0.8	12.0	0.0	29.6	99.2	12.8	87.2	86.4	99.2	98.4	0.0
37.7	28.3	0.0	0.0	0.0	0.0	0.0	7.6	0.0	13.2	100.0	50.0	0.0	78.3	99.0	97.8	0.0
60.9	43.5	0.0	0.0	0.0	2.2	0.0	19.6	0.0	26.1	100.0	50.0	0.0	78.3	99.0	97.8	0.0
0.0	41.2	0.0	0.0	0.0	2.9	0.0	11.8	0.0	8.8	97.9	9.4	0.0	93.8	96.9	93.8	0.0
0.0	46.9	0.0	0.0	0.0	3.1	0.0	31.3	0.0	18.8	100.0	9.4	0.0	93.8	96.9	93.8	0.0
46.5	70.5	2.3	0.0	1.2	0.0	13.2	0.0	24.4	13.8	90.7	40.5	0.0	67.9	88.1	97.6	0.0
64.3	97.6	3.6	1.2	1.2	0.0	26.2	2.4	36.9	39.2	90.7	40.5	0.0	67.9	88.1	97.6	0.0
79.0	1.9	0.0	0.0	0.9	0.0	0.0	0.0	0.9	0.0	4.8	27.0	0.0	88.8	91.0	49.4	0.0
92.1	1.9	0.0	0.0	1.1	0.0	0.0	0.0	1.1	0.0	4.8	27.0	0.0	88.8	91.0	49.4	0.0
93.8	3.1	0.0	0.0	0.0	0.0	3.1	0.0	0.0	0.0	100.0	96.9	0.0	93.8	96.9	9.4	0.0
96.9	3.1	0.0	0.0	0.0	0.0	9.4	0.0	0.0	0.0	100.0	96.9	0.0	93.8	96.9	9.4	0.0
22.6	7.6	0.0	0.0	0.0	0.0	0.0	3.8	0.0	1.9	98.1	44.4	51.9	68.5	98.1	81.5	0.0
25.9	13.0	0.0	0.0	0.0	0.0	0.0	5.6	0.0	1.9	100.0	8.2	91.9	68.5	98.1	81.5	0.0
91.7	37.5	79.2	75.0	79.2	0.0	54.2	0.0	12.5	0.0	100.0	0.0	100.0	95.8	100.0	100.0	0.0
100.0	100.0	100.0	91.7	100.0	0.0	100.0	0.0	79.2	41.7	100.0	0.0	100.0	95.8	100.0	100.0	0.0
0.0	5.0	0.0	0.0	0.0	0.0	5.0	0.0	0.0	5.0	50.0	0.0	0.0	40.9	0.0	95.5	0.0
9.1	9.1	0.0	0.0	0.0	4.6	18.2	0.0	4.6	5.0	59.1	0.0	0.0	40.9	0.0	95.5	0.0
9.0	0.4	0.0	0.0	0.0	0.4	0.0	0.7	0.4	0.7	99.3	68.4	9.4	62.1	85.9	25.1	0.0
28.9	1.6	0.0	0.0	0.0	0.4	0.0	0.7	0.8	0.7	99.6	68.4	9.4	62.1	85.9	25.1	0.0
12.6	87.4	0.0	0.0	0.0	0.0	1.2	88.5	0.0	79.3	0.0	3.2	0.0	0.0	90.5	98.4	0.0
12.6	87.4	0.0	0.0	0.0	0.0	1.2	89.1	0.0	79.7	0.0	3.2	0.0	0.0	90.5	98.4	0.0
3.3	0.0	0.0	0.0	0.0	0.0	0.0	0.0	0.0	0.0	0.0	7.1	0.0	0.0	70.2	0.0	0.0
11.9	0.0	0.0	0.0	0.0	0.0	0.0	0.0	0.0	0.0	0.0	7.1	0.0	0.0	70.2	0.0	0.0
85.0	0.0	0.0	0.0	0.0	0.0	0.0	0.0	0.0	0.0	100.0	0.0	0.0	0.0	57.5	90.0	10.0
100.0	0.0	0.0	0.0	0.0	0.0	0.0	0.0	0.0	0.0	100.0	4.0	22.0	0.0	57.5	90.0	10.0
87.9	0.0	0.0	0.0	0.0	0.0	0.0	0.0	0.0	0.0	100.0	6.1	0.0	0.0	48.5	93.9	6.1
98.8	0.0	0.0	0.0	0.0	0.0	0.0	0.0	0.0	0.0	100.0	6.1	0.0	0.0	48.5	93.9	6.1
58.6	0.0	0.0	0.0	0.0	0.0	0.0	0.0	0.0	0.0	100.0	0.0	0.0	0.0	84.4	0.0	0.0
96.0	0.0	0.0	0.0	0.0	0.0	0.0	0.0	0.0	0.0	100.0	0.0	0.0	0.0	84.4	0.0	0.0
20.0	0.0	0.0	0.0	0.0	0.0	0.0	0.0	0.0	0.0	100.0	0.0	0.0	0.0	99.0	20.0	0.0
99.0	0.0	0.0	0.0	0.0	0.0	0.0	0.0	0.0	0.0	100.0	0.0	0.0	0.0	100.0	20.0	0.0
10.7	46.4	0.0	0.0	0.0	0.0	25.0	0.0	7.1	17.9	96.4	0.0	0.0	15.4	84.6	96.2	0.0
10.7	61.5	0.0	0.0	0.0	15.4	84.6	0.0	38.5	23.1	96.4	0.0	0.0	15.4	84.6	96.2	0.0
0.0	0.0	0.0	0.0	0.0	0.0	0.0	0.0	0.0	0.0	100.0	0.0	0.0	0.0	0.0	99.9	31.2
0.0	0.0	0.0	0.0	0.0	0.0	0.0	0.0	0.0	0.0	100.0	0.0	0.0	0.0	0.0	99.9	31.2
0.0	0.0	0.0	0.0	0.0	0.0	0.0	0.0	0.0	0.0	95.7	77.8	0.0	88.9	97.2	0.0	0.0
2.8	0.0	0.0	0.0	0.0	0.0	0.0	0.0	0.0	0.0	100.0	77.8	0.0	88.9	97.2	0.0	0.0
1.5	0.0	0.0	0.0	0.0	0.0	0.0	0.0	0.0	0.0	99.3	43.9	8.6	84.9	91.4	0.0	0.0
10.1	0.0	0.0	0.0	0.0	0.0	0.0	0.0	0.0	0.0	100.0	43.9	8.6	84.9	91.4	0.0	0.0
2.2	0.0	0.0	0.0	0.0	0.0	0.0	0.0	0.0	0.0	100.0	9.7	0.0	0.0	23.9	0.0	0.0
11.5	0.0	0.0	0.0	0.0	0.0	0.0	0.0	0.0	0.0	100.0	9.7	0.0	0.0	23.9	0.0	0.0
0.0	29.6	74.1	0.0	68.5	0.0	77.8	0.0	0.0	0.0	81.5	52.4	0.0	0.0	2.0	19.6	19.6
0.0	30.8	74.1	0.0	65.4	0.0	78.9	0.0	0.0	0.0	82.7	52.4	0.0	0.0	2.0	19.6	19.6
0.0	99.0	0.0	90.0	0.0	10.0	99.0	0.0	0.0	80.0	85.0	100.0	0.0	0.0	100.0	100.0	100.0
0.0	100.0	0.0	90.0	0.0	10.0	100.0	0.0	0.0	80.0	85.0	100.0	0.0	0.0	100.0	100.0	100.0
1.6	13.1	0.0	0.0	0.0	0.0	3.3	0.0	0.0	1.6	91.8	59.7	0.0	0.0	9.7	25.0	25.0
2.8	13.1	0.0	0.0	0.0	0.0	3.3	0.0	0.0	1.6	97.2	59.7	0.0	0.0	9.7	25.0	25.0
92.3	96.2	0.0	0.0	0.0	0.0	11.5	0.0	0.0	0.0	95.8	75.0	0.0	95.8	99.0	99.0	99.0
100.0	96.2	0.0	0.0	0.0	0.0	16.7	0.0	0.0	0.0	95.8	75.0	0.0	95.8	99.0	99.0	99.0
0.0	0.0	0.0	0.0	0.0	0.0	0.0	0.0	0.0	0.0	100.0	65.5	43.6	65.5	99.0	47.3	0.0
1.8	1.8	0.0	0.0	0.0	0.0	0.0	0.0	0.0	0.0	100.0	65.5	43.6	65.5	99.0	47.3	0.0
0.0	0.0	2.7	2.7	2.7	2.7	5.4	0.0	5.4	0.0	100.0	15.6	71.9	99.0	99.0	53.1	0.0
0.0	0.0	12.5	12.5	12.5	12.5	18.8	0.0	5.4	15.6	100.0	28.1	91.3	99.0	99.0	53.1	0.0
0.0	0.0	0.0	0.0	0.0	0.0	0.0	0.0	2.3	4.6	100.0	18.2	27.2	86.0	100.0	51.0	0.0
0.0	0.0	0.0	0.0	0.0	7.0	9.3	2.3	14.0	20.9	100.0	23.3	37.2	86.0	100.0	62.8	0.0

TABLE 72-6
CONTINUED

ORGANISM		API 20E									
		ONPG	ADH	LDC	ODC	CIT	H$_2$S	URE	TDA	IND	VP
Brucella spp.	24h	0.0	0.0	0.0	0.0	0.0	0.0	88.9	0.0	0.0	0.0
	48h	0.0	0.0	0.0	0.0	0.0	0.0	94.4	0.0	0.0	0.0
Eikenella corrodens	24h	0.0	0.0	33.3	76.3	0.0	0.0	0.0	0.0	0.0	0.0
	48h	0.0	0.0	37.0	94.0	0.0	0.0	0.0	0.0	0.0	0.0
CDC Group II F	24h	0.0	0.0	0.0	0.0	8.6	0.0	0.0	0.0	71.4	0.0
	48h	0.0	0.0	0.0	0.0	71.4	0.0	0.0	0.0	71.4	0.0
CDC Group II J	24h	0.0	0.0	0.0	0.0	0.0	0.0	86.0	0.0	50.0	0.0
	48h	0.0	0.0	0.0	0.0	4.4	0.0	98.0	0.0	50.0	0.0
CDC Group IV C-2	24h	0.0	0.0	0.0	0.0	32.4	0.0	17.7	0.0	0.0	2.9
	48h	0.0	0.0	0.0	0.0	84.9	0.0	84.9	0.0	0.0	9.1
CDC Group IV E	24h	0.0	0.0	0.0	0.0	10.0	0.0	75.0	0.0	0.0	0.0
	48h	0.0	0.0	0.0	0.0	38.0	0.0	87.0	0.0	0.0	0.0
CDC Group V E-1	24h	86.5	59.5	0.0	0.0	77.0	0.0	0.0	0.0	0.0	43.2
	48h	86.5	80.6	0.0	0.0	98.0	0.0	2.8	0.0	0.0	58.3
CDC Group V E-2	24h	0.0	0.0	0.0	0.0	79.0	0.0	0.0	0.0	0.0	60.6
	48h	0.0	0.0	0.0	0.0	96.0	0.0	0.0	0.0	0.0	78.6
Aeromonas hydrophila	24h	97.8	90.6	50.0	0.0	45.0	0.0	0.0	0.0	85.6	61.7
Aeromonas salmonicida (25c)	24h	11.1	88.9	100.0	0.0	0.0	0.0	0.0	0.0	0.0	0.0
Plesiomonas shigelloides	24h	95.5	95.5	100.0	100.0	0.0	0.0	0.0	0.0	100.0	0.0
Vibrio cholerae	24h	94.4	1.9	94.4	96.3	63.0	0.0	0.0	0.0	100.0	40.7
V. alginolyticus	24h	0.0	0.0	97.5	62.5	41.0	0.0	2.5	0.0	100.0	15.0
V. parahemolyticus	24h	0.0	0.0	100.0	89.7	58.8	0.0	8.6	0.0	100.0	5.2
V. vulnificus	24h	100.0	0.0	75.0	90.0	71.4	0.0	0.0	0.0	100.0	18.2
V. fluvialis	24h	99.0	92.8	0.0	0.0	21.4	0.0	0.1	0.0	78.4	0.0
V. mimicus	24h	99.0	0.2	90.4	90.3	63.0	0.0	0.0	0.0	90.0	0.0
V. damsela	24h	7.1	100.0	71.4	0.0	0.0	0.0	85.7	0.0	0.0	0.0
V. hollisae	24h	0.0	0.0	0.0	0.0	0.0	0.0	0.0	0.0	85.7	0.0

TABLE 72-6
CONTINUED

				API 20E										SUPPLEMENTARY TESTS		
GEL	GLU	MAN	INO	SOR	RHA	SAC	MEL	AMY	ARA	OXI	NO$_2$	N$_2$ GAS	MOT	MAC	OF-O	OF-F
0.0	0.0	0.0	0.0	0.0	0.0	0.0	0.0	0.0	0.0	100.0	85.0	0.0	0.0	0.0	0.0	0.0
0.0	0.0	0.0	0.0	0.0	0.0	0.0	0.0	0.0	0.0	100.0	85.0	0.0	0.0	0.0	0.0	0.0
0.0	0.0	0.0	0.0	0.0	0.0	0.0	0.0	0.0	0.0	100.0	3.7	0.0	0.0	0.0	0.0	0.0
0.0	0.0	0.0	0.0	0.0	0.0	0.0	0.0	0.0	0.0	100.0	3.7	0.0	0.0	0.0	0.0	0.0
88.6	0.0	0.0	0.0	0.0	0.0	0.0	0.0	0.0	0.0	99.0	0.0	0.0	0.0	2.0	0.0	0.0
89.8	0.0	0.0	0.0	0.0	0.0	0.0	0.0	0.0	0.0	99.0	0.0	0.0	0.0	2.0	0.0	0.0
4.1	0.0	0.0	0.0	0.0	0.0	0.0	0.0	0.0	0.0	100.0	0.0	0.0	0.0	2.4	0.0	0.0
78.3	0.0	0.0	0.0	0.0	0.0	0.0	0.0	0.0	0.0	100.0	0.0	0.0	0.0	2.4	0.0	0.0
0.0	0.0	0.0	0.0	0.0	0.0	0.0	0.0	0.0	0.0	100.0	0.0	0.0	72.7	90.9	0.0	0.0
0.0	0.0	0.0	0.0	0.0	0.0	0.0	0.0	0.0	0.0	100.0	0.0	0.0	72.7	90.9	0.0	0.0
0.0	0.0	0.0	0.0	0.0	0.0	0.0	0.0	0.0	0.0	100.0	0.0	66.0	29.0	73.1	0.0	0.0
0.0	0.0	0.0	0.0	0.0	0.0	0.0	0.0	0.0	0.0	100.0	8.0	66.0	29.0	73.1	0.0	0.0
13.5	86.5	0.0	13.5	0.0	16.2	2.7	13.5	2.7	78.4	0.0	30.6	0.0	63.9	91.7	94.4	0.0
77.8	86.5	0.0	13.9	0.0	30.6	8.3	16.1	5.6	91.7	2.7	63.0	0.0	63.9	91.7	94.4	0.0
6.1	45.5	0.0	15.2	0.0	0.0	0.0	3.0	0.0	81.8	0.0	7.1	0.0	96.4	99.0	99.0	0.0
64.3	50.0	0.0	14.3	3.6	0.0	0.0	3.0	0.0	92.9	0.0	7.1	0.0	96.4	99.0	99.0	0.0
94.4	98.9	96.7	0.0	12.7	10.0	82.2	5.6	61.1	61.1	99.0	98.7	0.0	96.0	99.0	99.0	99.0
100.0	88.9	100.0	0.0	11.1	0.0	0.0	0.0	0.0	0.0	100.0	100.0	0.0	0.0	99.0	100.0	100.0
0.0	100.0	0.0	100.0	0.0	0.0	0.0	0.0	0.0	0.0	100.0	99.0	0.0	95.5	99.0	99.0	99.0
92.6	98.2	98.2	0.0	0.0	0.0	100.0	0.0	5.6	0.0	100.0	96.2	0.0	99.0	96.2	99.0	99.0
55.0	100.0	100.0	0.0	0.0	0.0	100.0	0.0	5.0	7.5	100.0	47.4	0.0	97.4	99.0	94.7	94.7
68.8	100.0	96.6	0.0	0.0	3.5	1.7	0.0	12.1	41.2	100.0	63.8	0.0	98.3	98.3	99.0	99.0
100.0	100.0	36.4	0.0	0.0	0.0	0.0	0.0	93.0	0.0	100.0	54.6	0.0	99.0	99.0	99.0	99.0
21.4	100.0	100.0	0.0	7.1	0.0	99.0	0.0	50.0	99.9	100.0	100.0	0.0	50.0	100.0	100.0	100.0
92.6	98.1	95.1	0.0	0.0	0.0	0.0	0.0	0.6	0.0	100.0	92.0	0.0	95.0	96.2	100.0	100.0
0.0	100.0	0.0	0.0	0.0	0.0	7.1	0.0	0.0	0.0	100.0	92.9	0.0	100.0	0.0	100.0	100.0
0.0	0.0	0.0	0.0	0.0	0.0	0.0	0.0	0.0	0.0	100.0	14.3	0.0	100.0	0.0	7.1	7.1

Analytab Products, Division of Sherwood Medical, October 1985. API 20E® System. Package Product Insert. Plainview, NY. Used with permission.

Other *Pseudomonas spp.*
Ps. vesicularis
Ps. pickettii
Ps. alcaligenes
Ps. acidovorans
Ps. testosteroni
Ps. diminuta
Ps. pseudoalcaligenes
CDC Group V a-1

Moraxella spp.
M. osloensis
M. phenylpyruvica
M. non-liquefaciens
M. lacunata
M. atlantae
CDC Group M5

Brucella spp.
B. abortus
B. suis
B. melitensis
B. canis

Flavobacterium spp.
CDC II B

Pasteurella - Actinobacillus spp.
Past. haemolytica
Past. urea
Past. pneumotropica
A. suis
A. equuli
A. lignieresi

Chromobacterium
C. violaceum

fluorescent Pseudomonas group
Ps. aeruginosa
Ps. fluorescens
Ps. putida

Ps. stutzeri
CDC Group Vb - 1, 3
Ps. mendocina

Alcaligenes spp.
Al. faecalis
Al. odorans
Al. denitrificans

Achromobacter spp.
CDC Vd - 1, 2

APPENDIX A
HOW TO CALCULATE DILUTIONS

Standard plate counts of microbial populations are carried out frequently in microbiology laboratories. Samples such as water, milk, shellfish, meats, and soils are analyzed to determine the numbers of microorganisms present. Often, the microbial populations in these samples are so large that the samples have to be diluted before plating in order to obtain plates that will yield between 30 and 300 colonies. More than 300 colonies on a plate of agar medium may be difficult to enumerate due to crowding; less than 30 colonies may yield statistical errors. Dilutions are carried out by mixing a known amount of sample (either in ml or g) with a known volume of sterile buffered water or saline. This is called the **diluent.** The figure below illustrates a series of dilutions of a sample.

To calculate the dilution value of a sample you use the following formula:

$$\text{Dilution} = \frac{\text{Volume of sample}}{\text{Volume of sample} + \text{Volume of diluent}}$$

In the dilution below, 1 ml of sample was mixed with 9 ml of diluent so that the total dilution is $1/1 + 9 = 1/10$ or 10^{-1}. To calculate the **total dilution** of a series of dilutions, determine the dilution value for each step and then multiply them together. In the example below, each dilution was a 1/10 dilution. The total dilution value is: $1/10 \times 1/10 \times 1/10 = 1/1,000$ or 10^{-3}. Practice calculating dilutions by solving the problems given below.

To determine the number of **colony-forming units/ml** in the sample, multiply the number of CFU/ml on the plate by the reciprocal of the total dilution. For example, assume 0.1 ml of the 10^{-3} dilution was plated onto TSA, and after an incubation period of 24 hours, 45 colonies developed. Since 0.1 ml was plated, there are 450 colonies/ml in a 10^{-3} dilution of the sample. In order to obtain the number of CFU/ml in the undiluted sample, multiply the 450 colonies by 1000 to obtain the number of CFU/ml of sample. This should yield 450,000 CFU/ml of sample, or 4.5×10^5 CFU/ml.

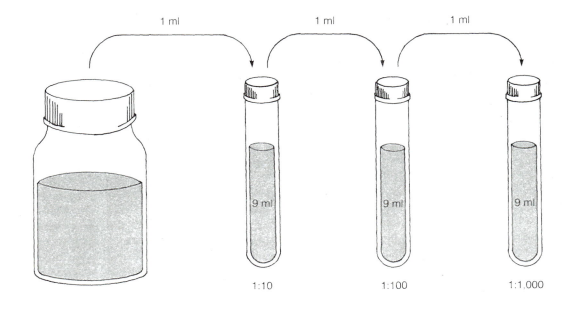

Suppose that instead of plating 0.1 ml of the 10^{-3} dilution, 0.5 ml is plated and 50 colonies appeared on the plate. To convert the number of colonies on the plate to colonies/ml, divide the number of colonies by the volume plated and solve. In the example: CFU/ml = 50/0.5 = 100 CFU/ml. Now multiply the number of CFU/ml by the reciprocal of the dilution value to obtain the number of CFU/ml of sample. In the example: 100 × 1000 = 100,000 or 1×10^5 CFU/ml. Practice doing this with the problems given below.

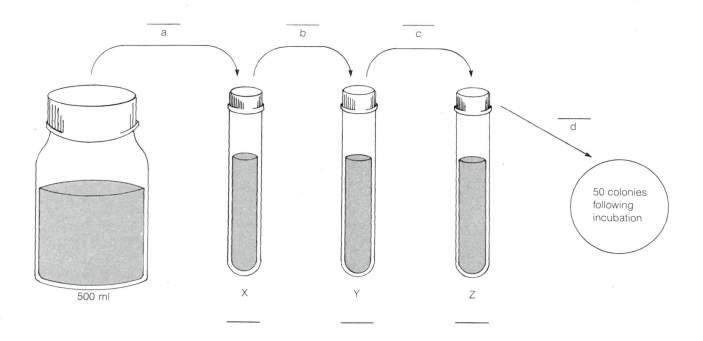

USING THE DIAGRAM ABOVE AND THE VALUES IN THE CHART BELOW, DETERMINE THE NUMBER OF CFU/ML OF SAMPLE FOR EACH PROBLEM.

PROBLEM NUMBER	VOLUME DIS-PENSED (ml) INTO TUBES			VOLUME OF DILUENT IN TUBES			DILUTION FACTOR OF TUBES			d VOLUME PLATED (ML) FROM LAST DILUTION TUBE	CFU/ML
	a	b	c	X	Y	Z	X	Y	Z		
1	1	1	1	9	9	9				1	
2	1	1	0.1	9	9	9.9				1	
3	0.1	1	0.1	9.9	9	9.9				1	
4	3	5	1	7	5	4				0.5	
5	0.4	0.6	0.8	3.6	5.4	7.2				0.5	
6	0.1	11	3	9.9	99	17				0.1	

NOTES AND CALCULATIONS

APPENDIX B
RANGES AND COLOR CHANGES
OF COMMON pH INDICATORS

		COLOR OF SOLUTION	
pH INDICATOR	pH RANGE	ACIDIC	BASIC
Brilliant green	0.0-2.6	Yellow	Green
Bromcresol green	3.8-5.4	Yellow	Blue-green
Bromcresol purple	5.2-6.8	Yellow	Purple
Bromthymol blue	6.0-7.6	Yellow	Blue
Congo red	3.0-5.0	Blue	Red
Cresol red	2.3-8.8	Orange	Red
2,4-dinitrophenol	2.8-4.0	Colorless	Yellow
Ethyl violet	0.0-2.4	Yellow	Blue
Litmus	4.5-8.3	Red	Blue
Malachite green	0.2-1.8	Yellow	Blue-green
Methyl green	0.2-1.8	Yellow	Blue
Methyl Red	4.8-6.0	Red	Yellow
Neutral Red	6.8-8.0	Red	Amber
Phenol Red	6.8-8.4	Yellow	Red
Phenolphthalein	8.2-10.0	Colorless	Pink
Resazurin	3.8-6.4	Orange	Violet

APPENDIX C
ORGANISMS USED
IN THE LABORATORY MANUAL

Algae

Organism: *Anabaena* species
Strain Number: 15-1225 Carolina Biological
Used in exercises 31A, 31B.

Organism: *Ceratium* species
Strain Number: 15-3245 Carolina Biological
Used in exercises 31A, 31B.

Organism: *Chlamydomonas* species
Strain Number: 15-1245 Carolina Biological
Used in exercises 31A, 31B.

Organism: *Chlorella* species
Strain Number: 15-1250 Carolina Biological
Used in exercises 31A, 31B.

Organism: *Cyclotella* species
Strain Number: 15-3020 Carolina Biological
Used in exercises 31A, 31B.

Organism: *Nostoc* species
Strain Number: 15-1230 Carolina Biological
Used in exercises 31A, 31B.

Organism: *Oscillatoria* species
Strain Number: 15-1235 Carolina Biological
Used in exercises 31A, 31B.

Organism: *Spirogyra* species
Strain Number: 15-1320 Carolina Biological
Used in exercises 31A, 31B.

Bacteria

Organism: *Acetobacter aceti*
Strain Number: ATCC 15973
Special Characteristics: 26°C recommended growth temperature.
Used in exercise 59B.

Organism: *Acinetobacter calcoaceticus*
Strain Number: ATCC 33306, 33307, or 33308
Special Characteristics: Auxotrophic strain (trp⁻), 37°C recommended growth temperature.
Used in exercise 39.

Organism: *Acinetobacter calcoaceticus*
Strain Number: ATCC 33305
Special Characteristics: Prototrophic strain (trp⁺), 37°C recommended growth temperature.
Used in exercise 39.

Organism: *Bacillus cereus*
Strain Number: ATCC 21768, 14893
Special Characteristics: 30°C recommended growth temperature.
Used in exercises 11B, 44B.

Organism: *Bacillus megaterium*
Strain Number: ATCC 12872
Special Characteristics: 30°C recommended growth temperature.
Used in exercises 11A, 11B.

Organism: *Bacillus stearothermophilus*
Strain Number: ATCC 7953
Special Characteristics: 55°C recommended growth temperature.
Used in exercises 16, 21B, 21D, 22A.

Organism: *Bacillus subtilis*
Strain Number: ATCC 605/a
Special Characteristics: 30°C recommended growth temperature.
Used in exercises 7A, 7B, 7C, 8, 9A, 9B, 10B, 11C, 23A, 23B, 27B, 55B.

Organism: *Brevibacterium linens*
Strain Number: ATCC 8377
Special Characteristics: 30°C recommended growth temperature.
Used in exercise 12.

Organism: *Citrobacter freundii*
Strain Number: ATCC 6750, 8090
Special Characteristics: Produce H_2S under certain conditions. 37°C recommended growth temperature.
Used in exercise 26C.

Organism: *Clostridium sporogenes*
Strain Number: ATCC 3584
Special Characteristics: 37°C recommended growth temperature. Anaerobic.
Used in exercise 10B.

Organism: *Clostridium butyricum*
Strain Number: ATCC 19398
Special Characteristics: 37°C recommended growth temperature. Anaerobic.
Used in exercise 10B.

Organism: *Clostridium perfringens (C. welchii)*
Strain Number: ATCC (Order a nonpathogenic strain).
Special Characteristics: Many strains produce lethal toxins. 37°C recommended growth temperature. Anaerobic.
Used in exercises 19A, 19B, 19C, 19D.

Organism: *Corynebacterium xerosis*
Strain Number: ATCC (Order a nonpathogenic strain).
Special Characteristics: 37°C recommended growth temperature.
Used in exercise 11A.

Organism: *Enterobacter aerogenes*
Strain Number: ATCC e13048
Special Characteristics: 30°C recommended growth temperature.
Used in exercises 24B, 24C, 26JA, 26C, 27A.

Organism: *Erwinia atroseptica*
Strain Number: ATCC 4446
Special Characteristics: USDA permit required for distribution. 26°C recommended growth temperature.
Used in exercise 64B.

Organism: *Erwinia carotovora*
Strain Number: ATCC 25270
Special Characteristics: USDA permit required for distribution. 26°C recommended growth temperature.
Used in exercise 64B.

Organism: *Escherichia coli*
Strain Number: CGSC strain 5581 (Dr. Barbara Bachman)
Special Characteristics: F⁻ auxotrophic strain (*gal⁻, lac⁻, leu⁻, strʳ*).
Used in exercise 41B.

Organism: *Escherichia coli*
Strain Number: CGSC strain 5461 (Dr. Barbara Bachman)
Special Characteristics: Hfr prototrophic strain (*strˢ*).
Used in exercise 41B.

Organism: *Escherichia coli* K12.
Strain Number: ATCC e25254 (*lac⁻*)
Special Characteristics: F⁻ lac deletion mutant.
Used in exercise 41A.

Organism: *Escherichia coli* K12.
Strain Number: ATCC e23724 (*lac y⁻, thr⁻ leu⁻*), ATCC e23722 (*lac z⁻*), ATCC e25253 (*lac z⁻*)
Special Characteristics: F⁻ lac point mutants. Require vitamin B1 (thiamine).
Used in exercise 41A.

Organism: *Escherichia coli* K12.
Strain Numbers: e2521 (*lac y⁻*), ATCC e25255 (*lac z⁻*)
Special Characteristics: F⁻ lac point mutants. Require vitamin B1 (thiamine).
Used in exercise 41A.

Organism: *Escherichia coli* B.
Strain Number: ATCC 23226
Special Characteristics: B strain, host to T-even bacteriophage.
Used in exercises 32, 33, 35.

Organism: *Escherichia coli* K12.
Strain Number: ATCC e23725
Special Characteristics: K12 strain (lac⁺).
Used in exercise 38.

Organism: *Escherichia coli*
Strain Number: ATCC 25922
Special Characteristics: Reference strain for monitoring the accuracy and precision of antimicrobial disk diffusion tests.
Used in exercise 69.

Organism: *Escherichia coli*
Strain Number: ATCC 11229
Special Characteristics: 37°C recommended growth temperature.
Used in exercises 7A, 7B, 7C, 7E, 8, 9A, 9B, 15, 16, 17, 18A, 18b, 19A, 19B, 19C, 19D, 21C, 21D, 22C, 23A, 23B, 23C, 24A, 24B, 24C, 25A, 25C, 26A, 26B, 26D, 27A, 27B, 27C, 43, 44A, 44B, 45, 46B, 46C, 47A, 47C, 51, 55B, 55C, 56, 57A, 61C, 61D, 71A.

Organism: *Streptococcus agalactiae*
Strain Number: ATCC e13813 or 27956
Special Characteristics: 37°C recommended growth temperature.
Used in exercise 61C.

Organism: *Klebsiella pneumoniae*
Strain Number: ATCC e13883
Special Characteristics: 37°C recommended growth temperature.
Used in exercise 10A.

Organism: *Lactobacillus bulgaricus*
Strain Number: ATCC 11842
Special Characteristics: 45°C recommended growth temperature.
Used in exercise 17.

Organism: *Micrococcus luteus*
Strain Number: ATCC 9341
Special Characteristics: 30°C recommended growth temperature.
Used in exercises 8, 10A, 19A, 19B, 19C, 19D, 24A.

Organism: *Mycobacterium smegmatis*
Strain Number: ATCC 19420
Special Characteristics: 37°C recommended growth temperature.
Used in exercise 9B.

Organism: *Mycobacterium phlei*
Strain Number: ATCC 354
Special Characteristics: 37°C recommended growth temperature.
Used in exercises 9B, 16.

Organism: *Neisseria sicca*
Strain Number: ATCC 29256
Special Characteristics: 37°C recommended growth temperature.
Used in exercises 19A, 19B, 19C, 19D.

Organism: *Neisseria subflava*
Strain Number: ATCC 14221
Special Characteristics: 37°C recommended growth temperature.
Used in exercise 9A.

Organism: *Proteus mirabilis*
Strain Number: ATCC 14273
Special Characteristics: 37°C recommended growth temperature.
Used in exercise 23D.

Organism: *Proteus vulgaris*
Strain Number: ATCC e13315
Special Characteristics: 37°C recommended growth temperature.
Used in exercises 7E, 10C, 26B, 26D, 27C, 45.

Organism: *Pseudomonas aeruginosa*
Strain Number: ATCC 10145
Special Characteristics: Neotype strain. 37°C recommended growth temperature.
Used in exercises 25A, 53.

Organism: *Pseudomonas aeruginosa*
Strain Number: ATCC 27853
Special Characteristics: Reference for monitoring the accuracy and precision of disk diffusion tests. 37°C recommended growth temperature.
Used in exercise 69.

Organism: *Pseudomonas denitrificans*
Strain Number: ATCC 13867
Special Characteristics: 30°C recommended growth temperature.
Used in exercises 47C, 53.

Organism: *Pseudomonas fluorescens*
Strain Number: ATCC e13525
Special Characteristics: 26°C recommended growth temperature.
Used in exercises 10C, 16, 17, 25C, 44B, 46B, 47A, 55C, 61C.

Organism: *Salmonella typhimurium*
Strain Number: ATCC e23595 (*trpD⁻ cysB⁻*)
Special Characteristics: Auxotrophic strain. 37°C recommended growth temperature.
Used in exercises 40, 71A.

Organism: *Salmonella typhimurium*
Strain Number: ATCC e2356
Special Characteristics: Prototrophic strain. 37°C recommended growth temperature.
Used in exercise 40.

Organism: *Salmonella typhimurium*
Strain Number: Ames' strain TA-1538
Special Characteristics: Auxotrophic strain (his⁻). 37°C recommended growth temperature.
Used in exercise 42.

Organism: *Serratia marcescens*
Strain Number: ATCC e13880
Special Characteristics: Strain should actively produce the pigment prodigiosin (red) at 25°C but not at 37°C.
Used in exercises 21D, 23C, 37A, 37B, 37C, 37D.

Organism: *Staphylococcus aureus*
Strain Number: ATCC 27697, 27693, 27691
Special Characteristics: Bacteriophage host.
Used in exercise 36.

Organism: *Staphylococcus aureus*
Strain Number: ATCC 25923
Special Characteristics: Reference strain for monitoring the accuracy and precision of antimicrobial disk diffusion tests. 37°C recommended growth temperature.
Used in exercise 69.

Organism: *Staphylococcus aureus*
Strain Number: ATCC 25923
Special Characteristics: 37°C recommended growth temperature.
Used in exercises 7E, 22A, 22B, 22D, 24A, 44A, 45, 46B, 46C, 53, 56, 61C, 61D.

Organism: *Staphylococcus epidermidis*
Strain Number: ATCC 14990
Special Characteristics: 37°C recommended growth temperature.
Used in exercises 7A, 7B, 7C, 7E, 9A, 9B, 18A, 18B, 22A, 23D, 25B, 25C, 44B, 56, 71C.

Organism: *Streptococcus faecalis*
Strain Number: ATCC e19433
Special Characteristics: 37°C recommended growth temperature.
Used in exercises 7E, 25B, 57B, 71C.

Organism: *Streptococcus lactis*
Strain Number: ATCC 19435
Special Characteristics: 37°C recommended growth temperature.
Used in exercises 7E, 19A, 19B, 19C, 19D.

Organism: *Streptococcus mitis*
Strain Number: ATCC 9895

Special Characteristics: 37°C recommended growth temperature.
Used in exercises 7E, 68A.

Organism: *Streptococcus mutans*
Strain Number: ATCC 25175
Special Characteristics: 37°C recommended growth temperature.
Used in exercise 68A.

Organism: *Streptococcus salivarius*
Strain Number: ATCC 25975
Special Characteristics: 37°C recommended growth temperature.
Used in exercise 68A.

Organism: *Streptomyces sp.*
Special Characteristics: 25°C recommended growth temperature.
Used in exercises 46B, 46C.

Organism: *Vibrio fischeri*
Strain Number: ATCC 15381 (Morita's *Vibrio marinus*, strain MP-1)
Special Characteristics: True marine psychrophile, isolated by Morita and co-workers. Maximum growth temperature, 20°C; optimum growth at 15°C; will grow at −5.5°C; sensitive to temperatures above 20°C. Grows on Marine Agar/Broth 2216.
Used in exercise 16.

Fungi

Organism: *Alternaria* sp.
Used in exercises 29A, 29B.

Organism: *Aspergillus* sp.
Used in exercises 29A, 29B.

Organism: *Geotrichum* sp.
Used in exercises 29A, 29B.

Organism: *Penicillium* sp.
Used in exercises 17, 18A, 18B, 29A, 29B, 44B.

Organism: *Rhizopus stolonifer (nigricans)*
Strain Numbers: ATCC 12938 (plus strain) and 12939 (minus strain).
Special Characteristics: 24°C recommended growth temperature.
Used in exercises 17, 18A, 18B, 29A, 29B, 29C.

Organism: *Saccharomyces cerevisiae*
Used in exercises 4, 8, 11C, 13, 17, 18A, 18B, 22D, 24A, 29A, 29B, 59B.

Insects

Organism: *Trichoplusia ni* (Hubner)
Special Characteristics: First instar larvae of the cabbage looper, for demonstration of effects of a bioinsecticide (Dipel - active ingredient: *Bacillus* thuringiensis).
Used in exercise 66.

Protozoa

Organism: *Amoeba* sp.
Used in exercises 30A, 30C.

Organism: *Chilomonas*
Strain number: L8 Carolina Biological
Used in exercise 30A.

Organism: *Didinium*
Strain number: L13 Carolina Biological
Used in exercise 30A.

Organism: *Euglena* sp.
Used in exercises 30A, 30C.

Organism: *Paramecium* sp.
Used in exercises 30A, 30C.

Organism: *Vorticella*
Strain number: L6 Carolina Biological
Used in exercise 30A.

Viruses

Organism: Bacteriophage P22 (host: *S. typhimurium*)
Strain Number: ATCC 19585-B1
Special Characteristics: Temperate transducing phage for certain strains of *S. typhimurium*.
Used in exercise 40.

Organism: Bacteriophage 81 (host: *S. aureus*)
Strain Number: ATCC e27701-B1
Special Characteristics: Phage virulent for certain strains of *S. aureus*; used in phage typing of isolates of staphylococci.
Used in exercise 36.

Organism: Bacteriophage type 71 (host: *S. aureus*)
Strain Number: ATCC e 27697-B1
Special Characteristics: Phage virulent for certain strains of *S. aureus*; used in phage typing of isolates of staphylococci.
Used in exercise 36.

Organism: Bacteriophage type 52A (host: *S. aureus*)
Strain Number: ATCC e27693-B1
Special Characteristics: Phage virulent for certain strains of *S. aureus*; used in phage typing of isolates of staphylococci.
Used in exercise 36.

Organism: Bacteriophage type 47 (host: *S. aureus*)
Special Characteristics: Phage virulent for certain strains of *S. aureus*; used in phage typing of staphylococcal isolates.
Used in exercise 36.

Organism: Bacteriophage T4 (host: *E. coli* strain B)
Strain Number: ATCC e11303-B4
Special Characteristics: Phage virulent for *E. coli*.
Used in exercises 32, 35.

Organism: Bacteriophage T2 (host: *E. coli* strain B)
Strain Number: ATCC e11303-B2
Special Characteristics: Phage virulent for *E. coli*
Used in exercises 32, 35.

Prepared Slides

Specimen: *Balantidium* sp.
Use: Provides a good quality prepared slide of a parasitic protozoan (ciliate) for microscopic study.
Used in exercise 30A.

Specimen: *Borrelia recurrentis*
Use: Provides a good quality, prepared specimen of a spirochete for microscopic study.
Used in exercise 4A.

Specimen: Corn smut
Use: Provides a good quality prepared slide from plant material affected by the agent of corn smut for microscopic study.
Used in exercise 64A.

Specimen: Endotrophic mycorrhizae
Use: Provides a good quality prepared slide for microscopic study of symbiotic mycorrhizal fungus-plant relationships.
Used in exercise 65.

Specimen: *Entamoeba* sp.
Use: Provides a good quality prepared slide of a parasitic protozoan (amoeba) for microscopic study.
Used in exercise 30A.

Specimen: *Giardia lamblia*
Use: Provides a good quality prepared slide of a parasitic protozoan (flagellate) for microscopic study.
Used in exercise 30A.

Specimen: Known bacterium stained with a fluorescent dye or a fluorescent antibody.
Special characteristics: Several different organisms are available in this type of preparation.
Use: Provides a good quality prepared slide for use in study of fluorescent microscopy and immunofluorescent technique.
Used in exercises 4C, 71C.

Specimen: *Morchella* sp.
Use: Provides a good quality prepared slide of a basidiomycetous fungus.
Used in exercises 29A, 29B.

Specimen: Orchid roots
Use: Provides a good quality prepared slide for microscopic study of symbiotic mycorrhizal fungus-plant relationships.
Used in exercise 65.

Specimen: *Plasmodium* sp.
Use: Provides a good quality prepared slide of a parasitic protozoan (sporozoan) for microscopic study.
Used in exercise 30A.

Specimen: *Puccinia graminis*
Special characteristics: Should exhibit teliospores.
Use: Provides a good quality prepared slide of the basidiomycetous fungal agent of wheat stem rust for microscopic study.
Used in exercise 64A.

Specimen: *Puccinia graminis*
Special characteristics: Should exhibit uredospores.
Use: Provides a good quality prepared slide of the basidiomycetous fungal agent of wheat stem rust for microscopic study.
Used in exercise 64A.

Specimen: *Puccinia graminis*
Use: Provides a good quality prepared slide of the basidiomycetous agent of wheat stem rust for microscopic study.
Used in exercises 29A, 29B, 64A.

Specimen: Representative microorganisms (bacteria, algae, protozoa, fungi) and microscopic animals (nematodes, water fleas)
Use: Introduces students to microbial world and provide good quality prepared specimens for development of skill in use of light microscope and microscopic measurements.
Used in exercises 1A, 1B, 2.

Specimen: *Trypanosoma* sp.
Use: Provides a good quality prepared slide of a parasitic protozoan (flagellate) for microscopic study.
Used in exercise 30A.

Sources of Cultures

American Type Culture Collection
12301 Parklawn Drive
Rockville, MD 20852

Dr. Bruce Ames,
Department of Biochemistry,

University of Calif., Berkeley, CA 94720
(Source of *S. typhimurium* TA1538 *his⁻*)

Dr. Barbara Bachman
Curator *E. coli* Genetic Stock,
Yale University Center, School of Medicine
333 Cedar St., P.O. Box 3333,
Department of Human Genetics,
New Haven, CT 06510
(Source of *E. coli* CGSC 5461 Hfr str^s and CGSC 5581 F⁻ *gal⁻ lac⁻ leu⁻ str^r*)

Collection Nationale d'Cultures de Microorganismes
Institut Pasteur
28 Rue du Docteur Roux
75724 Paris Cedex 15, France

Culture Centre of Algae and Protozoa
36 Storey's Way
Cambridge CB3 0DT
UK

Dico Laboratories
P.O. Box 1058A
Detroit, MI 48201

Midwest Culture Collection
1924 North 17th Street
Terre Haute, Indiana 47804

National Collection of Dairy Organisms
National Institute for Research in Dairy
Shinfield, Reading RG2 9AT
UK

National Collection of Industrial Bacteria
Torry Research Station
135 Abbey Road
Aberdeen AB9 8DG, UK

National Collection of Type Cultures
Central Public Health Laboratory
Colindale Avenue
London NW9 5HT, UK

Presque Isle Cultures
P.O. Box 8191
Presque Isle, PA 16505

University Micro Reference Laboratory
7885 Jackson Road
Ann Arbor, MI 48103

APPENDIX D
CULTURE MEDIA

Name of Medium: Acetamide Broth.

Ingredients:

Acetamide	10.0	g
Sodium chloride	5.0	g
Dipotassium phosphate	1.39	g
Monopotassium phosphate	0.73	g
Phenol red	0.012	g
Distilled water	1000.0	ml

Autoclave to sterilize.

Final pH: 6.9–7.2.

Type of Medium: Chemically defined (synthetic), differential and selective enrichment broth.

Use of Medium: Used for confirmation of an identification of *Pseudomonas aeruginosa* in the MPN test. *P. aeruginosa* is the only fluorescent pseudomonad capable of growth on acetamide as its sole carbon and nitrogen source at temperatures above 37°C. Deamination of the acetamide results in the accumulation of ammonium hydroxide ions in the medium, with the concurrent change of color of the pH indicator. Incubation at 37°C will inhibit *P. fluorescens* growth.

Medium used in exercise 53.

Name of Medium: Ammonium Broth.

Ingredients:

Ammonium sulfate	2.0	g
Magnesium sulfate · 7H$_2$O	0.5	g
Ferrous sulfate · 7H$_2$O	0.03	g
Sodium chloride	0.3	g
Magnesium carbonate	10.0	g
Dipotassium phosphate	1.0	g
Tap water	1000.0	ml

Final pH: 7.3.

Type of Medium: Chemically defined (synthetic), selective enrichment broth.

Use of Medium: For the enrichment and isolation of nitrifying chemolithotrophic bacteria. This medium lacks an organic carbon and energy source. The sole energy source is ammonia nitrogen, with atmospheric carbon dioxide and the carbonate ions in the broth providing the only carbon sources.

Medium used in exercise 47B.

Name of Medium: Asparagine Broth.

Ingredients:

DL-asparagine	3.0	g
Dipotassium phosphate	1.0	g
Magnesium sulfate · 7H$_2$O	0.5	g
Distilled water	1000.0	ml

Autoclave to sterilize.

Final pH: 6.9–7.2.

Type of Medium: Chemically defined (synthetic), differential and selective enrichment broth.

Use of Medium: Selective enrichment for the detection and isolation of fluorescent pseudomonads. Pseudomonads are capable of using asparagine as their sole carbon and nitrogen source. Fluorescent pseudomonads can be distinguished by the presence of a bluish-green fluorescent pigment in the medium which can be detected under long wave ultraviolet light with a Wood's lamp.

Medium used in exercise 53.

Name of Medium: *Azotobacter* Nitrogen-Free Broth.

Ingredients:

Bacto mannitol	15.0	g
Dipotassium phosphate	0.5	g
Magnesium sulfate · 7H$_2$O	0.2	g
Calcium sulfate	0.1	g
Sodium chloride	0.2	g
Calcium carbonate	5.0	g
Tap water	1000.0	ml

Final pH: 8.3.

Type of Medium: Chemically defined (synthetic), selective enrichment broth.

Use of Medium: For the enrichment and isolation of *Azotobacter* sp. (aerobic, nonsymbiotic nitrogen fixing bacteria). This medium contains no available fixed nitrogen and organisms capable of growth must be able to use atmospheric nitrogen.

Medium used in exercise 47D.

Name of Medium: Bile Esculin Azide Agar.

Ingredients:

Bacto beef extract	5.0	g
Bacto proteose peptone #3	3.0	g
Bacto oxgall	10.0	g
Bacto esculin	1.0	g
Ferric ammonium citrate	0.5	g
Bacto tryptone	17.0	g
Sodium chloride	5.0	g
Sodium azide	0.15	g
Bacto agar	15.0	g
Distilled water	1000.0	ml

Autoclave to sterilize. If plates are desired, cool to 50 to 55C and pour plates. It tubes are desired, allow medium in tubes to solidify in slanted position. Medium if reheated will darken due to esculin breakdown.

Final pH: 7.1 at 25°C.

Type of Medium: Complex, differential and selective agar medium.

Use of Medium: Group D streptococci tolerate the sodium azide and oxgall (bile salts) and hydrolyze esculin, which produces a dark brown colored product around their colonies. Gram negative bacteria are inhibited by the azide and Gram positive bacteria other than the fecal and other Group D streptococci are inhibited by the bile salts.

Medium used in exercise 57B.

Name of Medium: Blood Agar.

Ingredients:

Sterile, molten blood agar base medium*, cooled to 50–55°C	1 liter	
Sterile, defibrinated sheep blood**, added aseptically	5%	(v/v)

Immediately mix blood and basal medium and dispense in Petri plates and/or tubes as needed.

*Any of several complex infusion agar media may be suitable for use as a blood agar base, including heart infusion agar, Columbia blood agar base, neopeptone infusion agar, tryptic soy agar, etc. All such basal media should be nutritionally rich, have a final pH of 7.2–7.4, and lack any added fermentable carbohydrate (e.g., dextrose).

**Sheep blood is commonly used in the United States, but not worldwide. The source species of the blood can affect patterns of hemolysis of some bacteria.

Final pH: 7.2 to 7.4 at 25°C.

Type of Medium: Complex, enriched and differential.

Use of Medium: For the cultivation of nutritionally fastidious microorganisms. Microorganisms can be differentiated as to their "hemolytic" type on this medium, based on observed changes in the blood agar surrounding each organism's colonies. Blood agar is normally an opaque rich red color. Beta-hemolytic organisms will produce a zone of clearing around their colonies, due to an actual lysis of red cell membranes and cell destruction by bacterial exoenzymes known as hemolysins. Alpha-hemolytic organisms will produce a zone of green, olive, or brown discoloration around their colonies, due to oxidative effects of peroxide wastes on heme (pigmented portion of hemoglobin). Gamma-hemolytic organisms will exhibit no detectable change in the blood agar around their colonies.

Medium used in exercises 7E, 61D, 68B.

Name of Medium: Brain Heart Infusion (BHI) Broth.

Ingredients:

Infusion from calf brains	200.0	g
Infusion from beef heart	250.0	g
Proteose peptone, Difco	10.0	g
Bacto dextrose	2.0	g
Sodium chloride	5.0	g
Disodium phosphate	2.5	g
Distilled water	1000.0	ml

Autoclave to sterilize.

Final pH: 7.4 at 25°C.

Type of Medium: Complex infusion agar.

Use of Medium: Medium is highly nutritive and suitable for the culture of many fastidious heterotrophic microorganisms.

Medium used in exercise 57B.

Name of Medium: Brilliant Green Bile Lactose (BGBL) Broth.

Ingredients:

Bacto peptone	10.0	g
Bacto lactose	10.0	g
Bacto oxgall	20.0	g
Brilliant green	0.0133	g
Distilled water	1000.0	ml

Dispense in tubes containing inverted Durham tubes as gas traps. Autoclave to sterilize. Allow autoclave to cool to 75°C before opening the chamber to avoid entrapment of air bubbles in the Durham tubes.

Final pH: 7.2 at 25°C.

Type of Medium: Complex, selective broth.

Use of Medium: For confirmed and completed tests for coliform analysis in foods, dairy products, and waters. Inoculate medium and incubate at 35–37°C for 48 hours. Brilliant green is inhibitory to Gram positive bacteria and bile (oxgall) is inhibitory to non-coliform Gram negative bacteria. Lactose fermentation by the organisms growing in this medium, as indicated by gas formation, indicates coliform bacilli are present.

Medium used in exercises 51, 57A.

Name of Medium: Bromcresol Purple Fermentation Broth.

Ingredients:

Bacto casitone	10.0	g
Bromcresol purple (0.2%*)	5.0	ml
Distilled water	1000.0	ml

*Bromcresol purple (0.2%) stock solution should be made up separately and filter sterilized.

Adjust pH to 7.1–7.2, dispense in 8 ml quantities in 16 × 125 mm tubes with inverted Durham tube inserts (as gas traps), and autoclave the complete basal medium to sterilize.

Filter sterilized, concentrated (5 to 10%, w/v) aqueous stock solutions of carbohydrates are aseptically added to the previously sterilized basal medium to achieve a final concentration of 0.5 to 1.0%. Glucose and mannitol may be added to the medium before sterilization.

Final pH: 7.1–7.2 at 25°C.

Type of Medium: Complex, differential broth medium.

Use of Medium: Used to determine the abilities of nonfastidious heterotrophic bacteria (e.g., enterics, staphylococci, micrococci, etc.) to ferment various carbohydrates. A richer basal medium (such as cystine trypticase soy agar) may be necessary to support the growth of more nutritionally fastidious bacteria (e.g., most streptococci, neisseriae, and corynebacteria, species) during fermentation tests. Inoculate the broth, incubate and observe daily for up to 7 days for signs of fermentation. A positive reaction is indicated by the appearance of a definite yellow color throughout the medium (acidification), with or without the development of a definite gas bubble in the inverted Durham tube (production of gaseous wastes). A negative reaction is indicated by the persistence of the original color of the medium (lavender) or development of a deeper purple color (alkalinization of the medium as a result of utilization of the peptone in the medium as an energy source).

Medium used in exercise 24A.

Name of Medium: *Citrate Agar (Simmons).*

Ingredients:

Magnesium sulfate	0.2	g
Ammonium dihydrogen phosphate	1.0	g
Dipotassium phosphate	1.0	g
Sodium citrate	2.0	g
Sodium chloride	5.0	g
Bacto agar	15.0	g
Bacto brom thymol blue	0.08	g
Distilled water	1000.0	ml

Tube in 8 to 10 ml amounts and autoclave to sterilize. Allow tubes of medium to cool in a slanted position.

Final pH: 6.8 at 25°C.

Type of Medium: Chemically defined (synthetic), selective and differential, agar slant medium.

Use of Medium: Used to determine the ability of microorganisms to grow with citrate salts as the sole organic source of carbon. Cultures are incubated and examined daily for up to 4 days. A positive reaction is indicated by a color change in the medium from the original deep green to deep blue (alkalinization of the medium). Growth on the slant may or may not be obvious. A yellow color change at the top of the slant is an artifact which sometimes results from dessication of the medium (if prepared medium has been stored too long before use or culture has been incubated under conditions of inadequate humidity).

Medium used in exercise 27.

Name of Medium: *Corn Meal Agar*.

Ingredients:

Corn meal, infusion from	50.0	g
Bacto agar	15.0	g
Distilled water	1000.0	ml

Mix ingredients with water and heat to boiling to dissolve. Autoclave to sterilize, cool to 50 to 55°C and dispense in plates or tubes (for slants) as desired.

*1%(v/v) Tween 80 may be added before sterilization to promote chlamydospore formation by *Candida albicans*. More luxuriant growth of some fungi is obtained if 0.2% glucose (Bacto dextrose) is added before sterilization.

Final pH: 6.0 at 25°C.

Type of Medium: Complex, differential agar medium.

Use of Medium: For the cultivation of fungi and for the production of chlamydospores by *Candida albicans*. Cultures are inoculated and incubated at 23 to 25°C for up to 4 days.

Medium used in exercise 29B.

Name of Medium: Decarboxylase Broth (Moeller).

Ingredients:
Basal medium:

Bacto peptone	5.0	g
Bacto beef extract	5.0	g
Bacto dextrose	0.5	g
Bacto brom cresol purple	0.01	g
Cresol red	0.005	g
Pyridoxal	0.005	g
Distilled water	1000.0	ml

To make the various amino acid-specific media, add one of the following amino acids, dispense to tubes in 3 to 4 ml amounts and autoclave at 121°C for 10 min:

L-lysine.dihydrochloride, or
L-arginine.monohydrochloride, or
L-ornithine.dihydrochloride .... 10.0 g/l*

*If using DL-amino acids, add 20.0 g/liter.

Final pH: 6.0 at 25°C (ornithine medium may need to have pH readjusted).

Type of Medium: Complex, differential broth medium.

Use of Medium: Used to detect presence of specific amino acid decarboxylase activities in microorganisms. The inoculated cultures should be overlayed with a 4 to 5 mm layer of sterile mineral oil or paraffin before incubation. A tube of the basal medium must always be inoculated to serve as a control in the media containing 1% of a specific L-amino acid. Cultures are incubated and examined daily for up to 4 days. A positive decarboxylase reaction is indicated by the development of a purple (alkaline pH) color in the amino acid-containing medium, while the decarboxylase base culture remains yellow to yellow-green. If the basal medium culture turns purple, it may be inferred that some of the amino acids present in small amounts in the peptone or beef extract of the basal medium are being decarboxylated. In this situation, it is not possible to conclude that the amino acid added (at 1% of the L-form) to the basal medium is specifically being decarboxylated.

Medium used in exercise 26C.

Name of Medium: DNase (Deoxyribonuclease) Agar.

Ingredients:

Bacto tryptose	20.0	g
Sodium chloride	5.0	g
Deoxyribonucleic acid	2.0	g
Agar	15.0	g
Distilled water	1000.0	ml

Autoclave to sterilize. Cool to 50–55°C, aseptically dispense to sterile Petri dishes, and allow to solidify.

Final pH: 7.3 at 25°C.

Type of Medium: Complex, differential, agar plating medium.

Use of Medium: Used for the detection of deoxyribonuclease activity by bacterial colonies. Spot inoculate the medium (if desired, several cultures may be tested on one plate) and incubate the culture for 18 to 24 hours. To test for DNase activity, flood the plate with 0.1 N hydrochloric acid and observe for a zone of clearing around the culture(s) indicating hydrolysis of the DNA. The acid will precipitate intact highly polymerized DNA in the medium around the culture(s) if hydrolysis has not occurred.

An alternative approach to this test uses DNase Test Agar with Methyl green (0.05 g per liter of medium). Methyl green combines only with highly polymerized DNA at a pH of 7.5. Therefore, a positive DNase reaction in this medium is indicated by a clearing of the methyl green color around the culture. This approach has the advantage of not requiring the addition of hydrochloric acid, thereby permitting direct subculture of colonies for further tests.

Medium used in exercise 23C.

Name of Medium: *Desulfovibrio* Medium.

Ingredients:

Yeast extract	5.0	g
Ammonium phosphate, dibasic	0.3	g
Sodium lactate	3.0	g
Sodium sulfite	2.0	g
Dipotassium phosphate	1.7	g
Sodium thioglycollate	0.5	g
Bacto agar	3.0	g
River/pond water	1000.0	ml

Tube in 8 ml volumes in 10 ml screw cap tubes. Autoclave. Before using, melt medium, cool and add 1 ml of 0.5% sterile ferrous ammonium sulfate to each tube.

Final pH: Do not adjust the pH.

Type of Medium: Complex, differential enrichment agar medium.

Use of Medium: For the enrichment and isolation of anaerobic dissimilatory sulfide producers (e.g., *Desulfovibrio*). *Desulfovibrio* is able to use lactate as a carbon and energy source and will respire anaerobically, using oxidized sulfur compounds as the terminal electron acceptors. The medium is kept in a largely reduced form by thioglycollate. Sterile ferrous ammonium sulfate may be added after inoculation of the medium. With this addition, a black zone will appear around colonies of *Desulfovibrio*, due to the formation of ferrous sulfide, following production of sulfide ions from this organism's energy metabolism.

Medium used in exercise 48D.

Name of Medium: Eosin Methylene Blue (EMB) Agar.

Ingredients:

Bacto peptone	10.0	g
Bacto lactose	5.0	g
Bacto sucrose	5.0	g
Dipotassium phosphate	2.0	g
Bacto agar	13.5	g
Bacto eosin Y	0.4	g
Bacto methylene blue	0.065	g
Distilled water	1000.0	ml

Autoclave to sterilize.

Final pH: 7.2 at 25°C.

Type of Medium: Complex, selective and differential plating medium.

Use of Medium: For the detection and isolation of lactose- or sucrose-fermenting, Gram negative, enteric bacilli. Colonies which ferment either disaccharide take up various combinations of the two

dyes and are colored either dark purple with a greenish metallic surface sheen, or have dark purple centers with transparent peripheries, or are pink with no sheen. Nonfermenter colonies are colorless.

Medium used in exercises 7E, 44A, 51, 61C.

Name of Medium: EC (*Escherichia coli*) Broth.

Ingredients:

Bacto tryptose	20.0	g
Bacto lactose	5.0	g
Bacto bile salts #3	1.5	g
Dipotassium phosphate	4.0	g
Monopotassium phosphate	1.5	g
Sodium chloride	5.0	g
Distilled water	1000.0	ml

Dispense medium to tubes containing inverted Durham tubes as gas traps. Autoclave to sterilize. Allow the autoclave to cool to 75°C before opening chamber to avoid entrapment of air bubbles in the Durham tubes.

Final pH: 6.9 at 25°C.

Type of Medium: Complex, selective enrichment broth.

Use of Medium: This medium selects for coliforms of fecal origin, when cultures are incubated at 44.5°C. The bile salts inhibit fecal streptococci and endospore formers from growth. Lactose fermentation with the production of gas is indicative of the presence of coliforms when incubated at 37°C and of fecal coliforms at 44.5°C.

Medium used in exercise 51.

Name of Medium: Ethanol-Calcium Carbonate (EtOH-CaCO$_3$) Agar.

Ingredients:

Ethanol	20.0	ml
Yeast extract	5.0	g
Calcium carbonate (powder)	5.0	g
Agar	15.0	g
Tap water	1000.0	ml

Sterilize all but the ethanol in an autoclave. Add the alcohol just before pouring the plates.

Final pH: Do not adjust the pH.

Type of Medium: Complex, differential agar medium.

Use of Medium: Isolation of *Acetobacter*. *Acetobacter* oxidizes ethanol to acetic acid, which dissolves the CaCO$_3$ in the medium, creating a clear zone around *Acetobacter* colonies.

Medium used in exercise 59C.

Name of Medium: Fat Agar.

Ingredients:

Trypticase soy agar	40.0	g
Cottonseed oil	50.0	g
Tween 80	1.0	ml
Neutral red	0.1	g
Distilled water	1000.0	ml

Dispense to tubes and autoclave to sterilize.

Final pH: 7.2.

Type of Medium: Complex, differential agar.

Use of Medium: For detection of lipolytic bacteria. On this medium, lipolytic bacteria liberate free fatty acids from hydrolysis of the cottonseed oil; the fatty acids cause precipitation of the neutral red to yield a red area below (and sometimes surrounding and including) the lipolytic colonies.

Medium used in exercise 55C.

Name of Medium: Gelatin Peptone Agar*.

Ingredients:

Trypticase	17.0	g
Phytone	3.0	g
Sodium chloride	5.0	g
Dipotassium phosphate	2.5	g
Glucose	2.5	g
Gelatin	30.0	g
Agar	15.0	g
Distilled water	1000.0	ml

Adjust pH to 7.3 and autoclave to sterilize. Cool to 50 to 55°C, dispense to sterile disposable** Petri dishes, and allow to solidify.

*The above medium, without gelatin, is available commercially as trypticase soy agar.

**Since mercury ions are highly microbistatic and adsorb strongly to glass, avoid reuse of glassware contaminated by this reagent for microbiological cultures.

Final pH: 7.3 at 25°C.

Type of Medium: Complex, differential plating medium.

Use of Medium: Used for the detection of gelatin hydrolysis by bacterial colonies on an agar-containing gelatin medium. Spot inoculate the medium, incubate the culture for 48 hours and test for gelatin hydrolysis by flooding the culture with 1.0 ml of acidified mercuric chloride *(reagent is highly poisonous; never mouth pipet)*. A positive reaction is indicated by the appearance of a zone of clearing around the growth. If gelatin hydrolysis has not occurred, the mercury ions form a white precipitate with the gelatin throughout the medium.

Medium used in exercise 63B.

Name of Medium: Glucose Minimal Salts Broth.

Ingredients:

Glucose	0.2	g
Dipotassium phosphate	0.1	g
Ammonium chloride	0.05	g
Magnesium sulfate · 7H$_2$O	0.02	g
Ferric chloride · 6H$_2$O	0.0005	g
Distilled water	100.0	ml

Dispense medium to 10 ml amounts in tubes and autoclave to sterilize.

Final pH: 7 at 25°C.

Type of Medium: Chemically defined, minimal medium for some heterotrophs.

Use of Medium: Used to test ability of heterotrophs to grow in a minimal medium with glucose as the sole available organic carbon and energy source.

Medium used in exercise 5.

Name of Medium: Heart Infusion (HI) Agar.

Ingredients:

Infusion from beef heart	500.0	g
Bacto tryptose	10.0	g
Sodium chloride	5.0	g
Bacto agar	15.0	g
Distilled water	1000.0	ml

Autoclave to sterilize.

Final pH: 7.4 at 25°C.

Type of Medium: Complex infusion agar.

Use of Medium: For the cultivation of nutritionally fastidious microorganisms. Suitable as basal medium for preparation of blood agar (see Blood Agar).

Medium used (as blood agar base) in exercises 7E, 61, 68B.

Name of Medium: KF *Streptococcus* Agar*.

Ingredients:

Proteose peptone #3, Difco	10.0	g
Bacto yeast extract	10.0	g
Sodium chloride	5.0	g
Sodium glycerophosphate	10.0	g
Maltose	20.0	g
Lactose	1.0	g
Sodium azide	0.4	g
Bacto bromcresol purple	0.015	g
Bacto agar	20.0	g
Distilled water	1000.0	ml

Heat to boiling to dissolve ingredients. Continue to heat for an additional 5 minutes. Do not autoclave. Cool to 50 to 55°C, add the following solution to complete the medium, and pour plates (or use for making pour plates):

Bacto TTC Solution 1% (triphenyl-tetrazolium chloride 1%)	10.0	ml

Final pH: 7.2 at 25°C.

Type of Medium: Complex, differential and selective agar medium.

Use of Medium: For the detection and enumeration of fecal streptococci. Incubate cultures for 48 hours at 34–36°C. With the aid of a dissecting microscope (15×), count all colonies showing a red or pink center as streptococci. Sodium azide is the selective agent in the medium. Fecal streptococci tolerate the azide and can be differentiated from most other bacteria on the basis of their acid production from maltose and lactose. The acid reduces triphenyltetrazolium to an acid azo dye which causes fecal streptococcus colonies to turn deep red.

Medium used in exercise 57B.

Name of Medium: Lauryl Tryptose Broth*.

Ingredients:

Bacto Tryptose	20.0	g
Bacto Lactose	5.0	g
Dipotassium phosphate	2.75	g
Monopotassium phosphate	2.75	g
Sodium chloride	5.0	g
Sodium lauryl sulfate	0.1	g
Distilled water	1000.0	ml

*Double strength broth, required if 10 ml samples of water are inoculated, requires double concentration of lactose, to ensure the final concentration is adequate.

Dispense to tubes containing inverted Durham tubes as gas traps. Autoclave to sterilize. Allow autoclave to cool to 75°C before opening the chamber to avoid entrapment of air bubbles in the Durham tubes.

Final pH: 6.8 at 25°C.

Type of Medium: Complex, enrichment broth.

Use of Medium: For the detection of coliform bacteria in water, food and dairy products. Sodium lauryl sulfate selects for coliform bacteria, which ferment the lactose to produce acidic and gaseous wastes. The latter wastes are trapped in the inverted Durham tubes.

Medium used in exercise 51.

Name of Medium: m-Endo Broth.

Ingredients:

Bacto yeast extract	1.5	g
Bacto casitone	5.0	g
Bacto thiopeptone	5.0	g
Bacto tryptose	10.0	g
Bacto lactose	12.5	g
Sodium desoxycholate	0.1	g
Dipotassium phosphate	4.375	g
Monopotassium phosphate	1.375	g
Sodium chloride	5.0	g
Sodium lauryl sulfate	0.05	g
Sodium sulfite	2.1	g
Bacto basic fuchsin	1.05	g
Distilled water	1000.0	ml

Dissolve ingredients in distilled water containing 20 ml of ethanol by boiling briefly. Do not autoclave. Cool, dispense to sterile pads or into tubes, 1.8–2.0 ml/tube, and use the same day.

Final pH: 7.2 at 25°C.

Type of Medium: Complex, differential and selective broth.

Use of Medium: For coliform analysis of waters by membrane filter method. Inoculated membrane filters are placed on sterile absorbent pads soaked with freshly prepared medium. Cultures are incubated at 35°C for 24 hours. Gram positive microorganisms are inhibited by the desoxycholate. Coliform (lactose fermenters) colonies appear pink, possibly with a metallic sheen (due to the effects of the acidic wastes from fermentation on the sodium sulfite-basic fuchsin indicator system). Non-lactose fermenting enteric colonies are colorless to slightly pink.

Medium used in exercise 52.

Name of Medium: m-FC (Fecal Coliforms) Broth.

Ingredients:

Bacto tryptose	10.0	g
Proteose peptone #3, Difco	5.0	g
Bacto yeast extract	3.0	g
Sodium chloride	5.0	g
Bacto lactose	12.5	g
Bacto bile salts #3	1.5	g
Aniline blue	0.1	g
Distilled water	1000.0	ml

Suspend ingredients in distilled water. Add 10 ml of 1% Bacto Rosolic Acid in 0.2 N NaOH solution and heat to boiling for 1 minute. Do not autoclave. Cool to room temperature and add 2.0 ml broth to sterile tubes or to each sterile absorbent pad (in Petri dishes) on which the inoculated membrane filter(s) are to be placed.

Final pH: 7.4 at 25°C.

Type of Medium: Complex, differential and selective broth.

Use of Medium: For the detection, enumeration, and isolation of fecal coliforms at 44.5°C by the membrane filter method. The inoculated membrane filters are placed on the sterile pads that have been soaked with 2.0 ml of this broth. The Petri dishes containing the pads are covered tightly and submerged in a 44.5°C water bath for 24 hours, before examination for (and enumeration of) coliform colonies. The bile salts and dye are inhibitory to non-enteric forms. Fecal coliforms will grow at 44.5°C and produce blue colonies, due to absorption of the aniline blue from the medium. All other microorganisms capable of growth produce grey colonies. Fecal streptococci will be inhibited.

Medium used in exercise 52.

Name of Medium: m-*Staphylococcus* Broth.

Ingredients:

Bacto tryptone	10.0	g
Bacto yeast extract	2.5	g
Bacto lactose	2.0	g
Bacto mannitol	10.0	g
Dipotassium phosphate	5.0	g
Sodium chloride	75.0	g
Distilled water	1000.0	ml

Autoclave to sterilize. Use 1.8 to 2.0 ml of this medium to saturate the sterile paper pads on which the inoculated membrane is placed or dispense in 2 ml aliquots into sterile tubes.

Final pH: 7.0 at 25°C.

Type of Medium: Complex, differential and selective broth.

Use of Medium: For the detection and isolation of pathogenic staphylococci by the membrane filter method. This medium is highly inhibitory for many bacteria, due to the salt concentration. Golden yellow colonies detected on this medium are considered to be *S. aureus.*

Medium used in exercise 53.

Name of Medium: Mannitol Salt Agar.

Ingredients:

Proteose peptone #3, Difco	10.0	g
Bacto Beef extract	10.0	g
D-Mannitol	10.0	g
Sodium chloride	75.0	g
Bacto Agar	15.0	g
Phenol red	0.025	g
Distilled water	1000.0	ml

Autoclave to sterilize.

Final pH: 7.4 at 25°C.

Type of Medium: Complex, differential and selective.

Use of Medium: For the detection and isolation of staphylococci, in particular, *Staphylococcus aureus.* This medium inhibits many organisms, due to its high (7.5%) NaCl content; it tends to select for micrococci and staphylococci. Mannitol acidifiers (e.g., *S. aureus*) have yellow zones around their colonies, due to acid production and subsequent change in indicator color.

Medium used in exercises 7E, 44A, 56, 61D, 68B.

Name of Medium: Marine Agar.

Ingredients:

Bacto peptone	5.0	g
Bacto yeast extract	1.0	g
Ferric citrate	0.1	g
Sodium chloride	19.45	g
Magnesium chloride	8.8	g
Sodium sulfate	3.24	g
Calcium chloride	1.8	g
Potassium chloride	0.55	g
Sodium bicarbonate	0.16	g
Potassium bromide	0.08	g
Strontium chloride	0.034	g
Boric acid	0.022	g
Sodium silicate	0.004	g
Sodium fluoride	0.0024	g
Ammonium nitrate	0.0016	g
Disodium phosphate	0.008	g
Bacto agar	15.0	g

Final pH: 7.6 +/− 0.2 at 25C.

Type of Medium: Complex agar medium (available from Difco Labs, Detroit, Mi., as Bacto Marine Agar 2216).

Use of Medium: This medium can be used for the isolation, enumeration, and maintenance of heterotrophic marine bacteria. When used for the cultivation of psychrophilic marine bacteria (eg., *Vibrio fischeri* (V. *marinus*), strain MP-1), it should be prepared as plates, slants, or deeps in advance, cooled below 20C, and inoculated. Pour plates or shake cultures of most such psychrophiles are not feasible, since these organisms are usually very sensitive to environmental temperatures above 20C. V. marinus MP-1 cultures have been reported to die when exposed to 42C for as little as 1 min.!

Medium used in exercise 16.

Name of Medium: Marine Broth

Ingredients:

Bacto peptone	5.0	g
Bacto yeast extract	1.0	g
Ferric citrate	0.1	g
Sodium chloride	19.45	g
Magnesium chloride dried	5.9	g
Sodium sulfate	3.24	g
Calcium chloride	1.8	g
Potassium chloride	0.55	g
Sodium bicarbonate	0.16	g
Potassium bromide	0.08	g
Strontium chloride	0.034	g
Boric acid	0.022	g
Sodium silicate	0.004	g
Sodium fluoride	0.0024	g
Ammonium nitrate	0.0016	g
Disodium phosphate	0.008	g

Final pH: 7.6 +/− 0.2 at 25C.

Type of Medium: Complex broth medium (available from Difco Labs, Detroit, Mi., as Bacto Marine Broth 2216).

Use of Medium: This medium can be used for the culture of heterotrophic marine bacteria. When used for the cultivation of marine psychrophiles like V. *fischeri* (V. *marinus*) strain MP-1, it should be prepared in advance, cooled below 20C, and inoculated. Marine psychrophiles tend to be quite sensitive to temperatures above 20C.

Medium used in exercise 16.

Name of Medium: Methyl Red-Voges Proskauer (MRVP) Broth.

Ingredients:

Buffered peptone	7.0	g
Glucose	5.0	g
Dipotassium phosphate	5.0	g
Distilled water	1000.0	ml

Dispense medium in 8–10 ml amounts to tubes and autoclave to sterilize.

Final pH: 6.9 at 25°C.

Type of Medium: Complex, differential broth medium.

Use of Medium: Used for the determination of the Methyl Red (MR) and Voges-Proskauer (VP) reactions of heterotrophic fermentative bacteria. Incubate the culture in this medium for 48 hours, then divide the growth medium evenly into two clean test tubes.

Perform the MR test. It indicates the approximate final pH of the culture. Add 5 drops of MR reagent to one of the tubes of medium. A positive MR reaction is indicated by a distinct pink to red color and shows the presence of sufficient acid wastes to lower the pH to 5.0 or less. A negative MR reaction is indicated by a yellow color.

To perform the VP test, add 0.5 ml of 40% KOH-0.3% creatine to the other tube of medium, and mix vigorously. Then add 0.5 ml of alpha-naphthol and agitate vigorously for at least 30 seconds. A positive VP reaction is indicated by the development of a pink to red color within 30 min.

Medium used in exercise 24B.

Name of Medium: Minimal 0.4% Galactose Agar.

Ingredients: Solutions A, B, and D are prepared and sterilized as for Minimal 0.4% Glucose Media, unsupplemented. Solution C is prepared as follows and sterilized by autoclaving.

Solution C:

Galactose	20.0	g
Distilled water	100.0	ml

Solution E is prepared as follows and sterilized by filtration:

Solution E:

DL-leucine	100.0	mg
Streptomycin	10.0	mg
Distilled water	100.0	ml

To prepare complete medium, aseptically mix solutions as follows:

Solution A	400.0	ml
Solution B	180.0	ml
Solution C	20.0	ml
Solution D	300.0	ml
Solution E	100.0	ml

Final pH: 7.0 at 25°C.

Type of Medium: Chemically defined medium.

Use of Medium: Used in genetic studies of certain auxotrophic mutants.

Medium used in exercise 41B.

Name of Medium: Minimal 0.4% Glucose Media, unsupplemented.

Ingredients: See recipes for Minimal 0.4% Glucose Medium, solutions A, B, C, and D. To prepare complete media, aseptically combine the sterile solutions as follows:

Solution A	400.0	ml
Solution B	180.0	ml
Solution C	20.0	ml*

For Minimal 0.4% Glucose Agar, unsupplemented, add:

Solution D	300.0	ml
Sterile distilled water	100.0	ml

For Minimal 0.4% Glucose Broth, unsupplemented, add:

Sterile distilled water	400.0	ml

The supplemented Minimal Broth (and Agar) Media used in this manual employ the above Solutions A, B, C (usually), and D (for agar media) in these same proportions; the various supplements are then added aseptically as solutions E, etc.

*For exercise 41B, the Minimal Broth (without carbon or energy source) should be prepared with 20.0 ml of sterile distilled water substituted for solution C.

Final pH: 7.0 at 25°C.

Type of Medium: Minimal, chemically defined broth (agar) media.

Use of Medium: Minimal glucose media are used in studies of physiologic requirements and genetic characteristics of heterotrophic bacteria. These defined media are very commonly modified by substitution of alternate carbon and energy sources (for glucose) or supplementation with specific growth factors, inhibitors, or other compound to determine these characteristics.

Medium used in exercises 40, 42 (see also 38, 39, 40, and 41 for examples of use of modified minimal media).

Name of Medium: Minimal 0.4% Glucose Media, with supplements*.

Ingredients: Solutions A, B, C, and D are prepared and sterilized as for Minimal 0.4% Glucose Media, unsupplemented. The composition of solution E depends on its intended use; the recipes for solution E for several exercises follow:

*For exercise 40 (Minimal 0.4% Glucose Agar, with 0.1% Tryptophan), Solution E is prepared as follows and sterilized by autoclaving:

Solution E (Ex. 40):

Tryptophan	1.0	g
Distilled water	100.0	ml

*For exercise 41A (Minimal 0.4% Glucose Broth, with 0.1 μg/ml vitamin B1 (thiamine), 0.1 mg/ml DL-leucine, and 0.1 mg/ml DL-threonine), Solution E is prepared as follows and sterilized by filtration:

Solution E (Ex. 41A):

Thiamine (vitamin B1) hydrochloride, (1 mg/ml aq. sln.)	0.1	ml
DL-leucine	100.0	mg
DL-threonine	100.0	mg
Distilled water	100.0	ml

For exercise 41B (Minimal 0.4% Glucose Agar, with 0.01 mg/ml streptomycin), Solution E is prepared as follows and sterilized by autoclaving:

Solution E (Ex. 41B):

Streptomycin	10.0	g
Distilled water	100.0	ml

To prepare complete medium, aseptically mix solutions as follows:

Solution A	400.0	ml
Solution B	180.0	ml
Solution C	20.0	ml

Solution D	300.0	ml
Solution E	100.0	ml*

Final pH: 7.0 at 25°C.

Type of Medium: Chemically defined media (except for agar).

Use of Medium: Used in genetic studies of various bacterial auxotrophic mutants.

Medium used in exercises 40, 41A, 41B.

Name of Medium: Minimal 0.4% Glucose Medium (Solution A*).

Ingredients:

Monopotassium phosphate	5.0	g
Dipotassium phosphate	14.0	g
Ammonium chloride	4.0	g
Distilled water	400.0	ml

Autoclave.

*When cooled to room temperature, solution A is mixed aseptically with solutions B, C (usually), and (for minimal agar media) D, to make complete minimal broth (agar) media (e.g., see Minimal 0.4% Glucose Broth (Agar). Various supplemented Minimal media will also require further aseptic additions.

Final pH: 7.0 at 25°C.

Type of Medium: Solution A contributes a source of phosphorus and nitrogen, as well as the buffer, to minimal media.

Use of Medium: Solution A must be mixed after sterilization, in appropriate amounts, with other solutions (B, C, [usually], D [for agar], etc.) in order to produce various types of minimal media for use in physiologic and/or genetic studies of heterotrophic microorganisms.

Medium used in exercises 38, 39, 40, 41, 42.

Name of Medium: Minimal 0.4% Glucose Medium (Solution B*).

Ingredients:

Ferrous sulfate · 7H₂O	0.010	g
Magnesium sulfate · 7H₂O	0.100	g
Distilled water	180.0	ml

Autoclave. *When cooled to room temperature, Solution B is mixed aseptically with solutions A, C (usually), and (for minimal agar media) D, to make complete minimal broth (agar) media (e.g., see Minimal 0.4% Glucose Broth (Agar)). Various supplemented Minimal media will also require further aseptic additions.

Final pH: 7.0 at 25°C.

Type of Medium: Solution B contributes a source of ferrous and magnesium ions to minimal media.

Use of Medium: Solution B must be mixed after sterilization, in appropriate amounts, with other solutions (A, C, D [for agars], etc.) in order to produce various types of minimal media for use in physiologic and/or genetic studies of heterotrophic microorganisms.

Medium used in exercises 38, 39, 40, 41, 42.

Name of Medium: Minimal 0.4% Glucose Medium (Solution C*).

Ingredients:

Glucose	20.0	g
Distilled water	100.0	ml

Autoclave. *When cooled to room temperature, Solution C is mixed aseptically with solutions A, B and (for minimal agar media) D, to make complete minimal glucose broth (or agar) media (e.g., see Minimal 0.4% Glucose Broth/Agar). Various supplemented Minimal media will also require further aseptic additions.

Final pH: 7.0 at 25°C.

Type of Medium: Solution C contributes glucose as the carbon and energy source to minimal media for heterotrophs.

Use of Medium: Solution C (or an alternative carbon and energy source) must be mixed after sterilization, in appropriate amounts, with other solutions (A, B, D [for agar], etc.) in order to produce various types of minimal media for use in physiologic and/or genetic studies of heterotrophic microorganisms.

Medium used in exercises 40, 41, 42 (see also 38, 39, and 41 for examples of minimal media with substitutions for glucose).

Name of Medium: Minimal 0.4% Glucose Medium (Solution D*).

Ingredients:

Agar-agar	15.0	g
Distilled water	300.0	ml

Autoclave. *When cooled to room temperature, solution is mixed aseptically with solutions A, B, and C (usually) to make complete minimal agar media (e.g., Minimal 0.4% Glucose Agar). Various supplemented Minimal media will require further aseptic additions.

Final pH: 7.0 at 25°C.

Type of Medium: Solidification agent for minimal media.

Use of Medium: Solution D must be mixed after sterilization, in appropriate amounts, with other solutions (A, B, and C [usually], etc.) in order to produce various types of minimal agar media for use in physiologic and/or genetic studies of heterotrophic microorganisms.

Medium used in exercises 38, 39, 40, 41, 42.

Name of Medium: Minimal 0.4% Glucose Broth (with 0.05% Yeast Extract).

Ingredients: Solutions A and B are prepared as for Minimal 0.4% Glucose Media, unsupplemented. Solution C is prepared as follows:

Solution C:

Glycerol	20.0	g
Distilled water	100.0	ml

Solution D is prepared as follows:

Solution D:

Bacto yeast extract	0.5	g
Distilled water	400.0	ml

Solutions A, B, C, and D are sterilized separately by autoclaving.

To prepare complete medium, aseptically mix solutions as follows:

Solution A	400.0	ml
Solution B	180.0	ml
Solution C	20.0	ml
Solution D	400.0	ml

Final pH: 7.0 at 25°C.

Type of Medium: Chemically defined, minimal broth medium (except for undefined yeast extract supplement).

Use of Medium: This medium is used in genetic studies with certain bacterial auxotrophic mutants.

Medium used in exercise 38.

Name of Medium: Minimal 0.4% Lactose Agar, with supplements*.

Ingredients: Solutions A, B, and D are prepared and sterilized as for Minimal 0.4% Glucose Media, unsupplemented. Solution C is prepared as follows and sterilized by autoclaving:

Solution C:

Lactose	20.0	g
Distilled water	100.0	ml

The composition of solution E depends on the supplemented medium's particular use; the recipes for solutions E for 2 different exercises follow:

*For exercise 41A (Minimal 0.4% Lactose Agar, with 0.1 μg/ml vitamin B1 (thiamine), 0.1 mg/ml DL-leucine, and 0.1 mg/ml DL-threonine), Solution E is prepared, sterilized and used exactly as described for Minimal 0.4% Glucose Broth, with these same supplements.

*For exercise 41B (Minimal 0.4% Lactose Agar, with 0.1 mg/ml DL-leucine and 0.01 mg/ml streptomycin), Solution E is prepared, sterilized and used exactly as described for Minimal 0.4% Galactose Agar, with these same supplements.

To prepare complete medium, aseptically mix solutions as follows:

Solution A	400.0	ml
Solution B	180.0	ml
Solution C	20.0	ml
Solution D	300.0	ml
Solution E	100.0	ml*

Final pH: 7.0 at 25°C.

Type of Medium: Chemically defined media (except for agar).

Use of Medium: Used in genetic studies of various bacterial auxotrophic mutants.

Medium used in exercises 41A, 41B.

Name of Medium: Minimal 0.2% Sodium Acetate Agar.

Ingredients: Solutions A, B, and D are prepared and sterilized as for Minimal 0.4% Glucose Media, unsupplemented. Solution C is prepared as follows and sterilized by autoclaving:

Solution C:

Sodium acetate	2.0	g
Distilled Water	120.0	ml

To prepare complete medium, mix solutions as follows:

Solution A	400.0	ml
Solution B	180.0	ml
Solution C	120.0	ml
Solution D	300.0	ml

Final pH: 7.0 at 25°C.

Type of Medium: Chemically defined minimal agar medium.

Use of Medium: Used in genetic studies with certain bacterial auxotrophic mutants.

Medium used in exercise 39.

Name of Medium: Minimal Medium + Glucose + Cellulose.

Ingredients:

Ammonium chloride	20.0	g
Ammonium nitrate	4.0	g
Sodium sulfate, anhydrous	8.0	g
Dipotassium phosphate	12.0	g
Monopotassium phosphate	4.0	g

Magnesium sulfate · 7H$_2$O	0.4	g
Glucose	0.1	g
Filter paper strip		1/tube
Tap water	1000.0	ml

Final pH: Do not adjust pH.

Type of Medium: Selective enrichment.

Use of Medium: For the selective enrichment and isolation of cellulolytic microorganisms. Cellulose is the sole organic carbon and energy source, except for a minimal amount of glucose. Cellulolytic organisms will attach (usually) to the filter paper strips and gradually cause the paper to disintegrate.

Medium used in exercise 67.

Name of Medium: Mitis-Salivarius Agar.

Ingredients:

Bacto tryptose	10.0	g
Proteose peptone #3, Difco	5.0	g
Proteose peptone, Difco	5.0	g
Bacto dextrose	1.0	g
Saccharose, Difco	50.0	g
Dipotassium phosphate	4.0	g
Trypan blue	0.075	g
Bacto crystal violet	0.0008	g
Bacto agar	15.0	g
Distilled water	1000.0	ml

After sterilization in autoclave, cool to 50–55°C, and add 1.0 ml/liter of Bacto Chapman tellurite solution. Mix well and immediately pour plates. Do not heat medium after addition of tellurite.

Final pH: 7.0 at 25°C.

Type of Medium: Complex, differential and selective.

Use of Medium: For isolation and differentiation of nonhemolytic streptococci, especially from saliva, fecal, body cavity exudate and similar specimens. *Streptococcus mitis* produces small blue colonies, *S. salivarius* produces blue, smooth or rough "gumdrop" colonies, 1 to 5 mm in diameter, enterococci form dark blue to black shiny raised colonies, 1 to 2 mm in diameter. Most bacteria and molds are inhibited by the tellurite.

Medium used in exercise 68A.

Name of Medium: Modified Starkey's Medium.

Ingredients:

Sodium thiosulfate · 5H$_2$O	10.0	g
Dipotassium phosphate	2.0	g
Monopotassium phosphate	2.0	g
Calcium chloride	0.1	g
Magnesium sulfate · 7H$_2$O	0.1	g
Ammonium sulfate	0.1	g
Ferric chloride · 6H$_2$O	0.02	g
Manganese sulfate	0.02	g
Bromthymol blue	0.01	g
Pond/river water	1000.0	ml

Final pH: 6.6.

Type of Medium: Selective enrichment broth.

Use of Medium: For the selective enrichment and isolation of the thiobacilli (chemolithotrophs). This medium contains no organic

sources of carbon or energy. The thiosulfate ions provide a suitable energy source for the thiobacilli.

Medium used in exercise 48C.

Name of Medium: Modified van Niel's Medium.

Ingredients:

Ammonium chloride	1.0	g
Dipotassium phosphate	1.0	g
Magnesium chloride · 7H$_2$O	1.0	g
Pond/river water	1000.0	ml

Autoclave, then add:

Sodium bicarbonate 5% solution (filter sterilized)	20.0	ml
Sodium sulfide · 9H$_2$O 10% solution (autoclaved)	10.0	ml

Final pH: 8–8.5 (adjust with sterile phosphoric acid).

Type of Medium: Inorganic selective enrichment broth.

Use of Medium: For the selective enrichment of photosynthetic purple sulfur bacteria (photolithotrophs). Sulfide ions are available as an electron donor and bicarbonate ions provide a source of inorganic carbon.

Medium used in exercise 48B.

Name of Medium: Nitrate Broth.

Ingredients:

Bacto beef extract	3.0	g
Bacto peptone	5.0	g
Potassium nitrate	1.0	g
Distilled water	1000.0	ml

Insert an inverted Durham tube in each tube of medium before sterilization, if detection of denitrification by signs of gas production is intended. Autoclave to sterilize.

Final pH: 7.0 ± 0.2, at 25°C.

Type of Medium: Complex, differential, infusion broth medium.

Use of Medium: To detect signs that a specimen (e.g., soil) or a culture of a heterotrophic bacterium is able to reduce nitrate ions to nitrite or to nitrogen gas. Inoculate broth and incubate culture for (at least) 1 week. Test for signs of reduction in the following order: first look for appearance of an obvious gas bubble in Durham tube as a sign of reduction of nitrate ions to nitrogen or other insoluble gaseous products. If no obvious bubble appears after incubation, test for reduction of nitrate to nitrite ions: either add 5 drops of sulfamic acid to culture and look for evolution of bubbles within 5 minutes or add 12 to 15 drops each of alpha-naphthylamine (weak carcinogen) and sulfanilic acid, mix and look for development of a pink to red color. If no bubbles have appeared after incubation or after adding sulfamic acid, confirm that nitrate ions are still present by the addition of several grains of powdered zinc and the appearance of bubbles.

Medium used in exercise 47C.

Name of Medium: Nitrite Broth.

Ingredients:

Sodium nitrite	1.0	g
Magnesium sulfate · 7H$_2$O	0.5	g
Ferrous sulfate · 7H$_2$O	0.03	g
Sodium chloride	0.3	g

Sodium carbonate .... 1.0 g
Dipotassium phosphate .... 1.0 g
Tap water .... 1000.0 ml

Final pH: 7.3.

Type of Medium: Inorganic, selective enrichment broth.

Use of Medium: For the enrichment and isolation of nitrifying chemolithotrophic bacteria. This medium lacks an organic carbon and energy source. The sole energy source is nitrite, with atmospheric carbon dioxide and the carbonate ions in the broth serving as the only carbon sources.

Medium used in exercise 47B.

Name of Medium: Nutrient Agar.

Ingredients:
Beef extract .... 3.0 g
Tryptone .... 5.0 g
Agar.. 15.0 g
Distilled water .... 1000.0 ml

Adjust pH to 7.0, dispense to tubes and autoclave to sterilize.

Final pH: 7.0 at 25°C.

Type of Medium: Complex infusion agar.

Use of Medium: General purpose infusion agar medium, supports the growth of nonfastidious heterotrophic microorganisms.

Medium used in exercises 6, 7A, 7C, 44B, 51.

Name of Medium: Nutrient Broth.

Ingredients:
Beef extract .... 3.0 g
Tryptone .... 5.0 g
Distilled water .... 1000.0 ml

Adjust pH to 7.0 and autoclave to sterilize.

Final pH: 7.0 at 25°C.

Type of Medium: Complex infusion broth medium.

Use of Medium: General purpose infusion broth medium, supports the growth of nonfastidious heterotrophic microorganisms.

Medium used in exercises 7B, 7D.

Name of Medium: Nutrient Broth with 0.1% Glucose.

Ingredients:
Bacto Beef extract .... 3.0 g
Bacto Peptone................................ 5.0 g
Glucose .... 1.0 g
Distilled water .... 1000.0 ml

Autoclave to sterilize.

Final pH: 6.8 at 25°C.

Type of Medium: Complex infusion broth.

Use of Medium: General purpose medium for cultivation of nonfastidious heterotrophic microorganisms.

Medium used in exercises 43, 44A.

Name of Medium: Nutrient Gelatin.

Ingredients:
Bacto beef extract .... 3.0 g
Bacto peptone .... 5.0 g
Bacto gelatin .... 120.0 g
Distilled water .... 1000.0 ml

Dispense to tubes in 8 to 10 ml amounts and autoclave to sterilize. Allow tubes to cool in an upright position.

Final pH: 6.8 at 25°C.

Type of Medium: Complex, differential semisolid medium.

Use of Medium: Used to determine the ability of microorganisms to hydrolyze gelatin (proteolytic activity). Medium is stab-inoculated and culture incubated and examined for evidence of hydrolysis (liquefaction of part or all of the medium) for up to 6 weeks. If the culture must be incubated at temperatures above 20°C, the tube must be cooled to 20°C or below before observation for liquefaction (unhydrolyzed gelatin itself liquefies, reversibly, as the environmental temperature rises above 20°C).

Medium used in exercise 27.

Name of Medium: Peptone (4%, w/v) Broth.

Ingredients:
Bacto peptone .... 40.0 g
Distilled water .... 1000.0 ml

Autoclave to sterilize.

Final pH: Approximately 7.0.

Type of Medium: Complex broth.

Use of Medium: For the detection of ammonifying activity in soil microorganisms.

Medium used in exercise 47A.

Name of Medium: Peptone Glycerol Phosphate Sodium Chloride Medium.

Ingredients:
Peptone .... 10.0 g
Dipotassium phosphate .... 2.0 g
Sodium chloride.............................. 30.0 g
Glycerol .... 5.0 ml
Tap water .... 1000.0 ml
Agar.. 17.5 g

Final pH: Adjust to pH 7.0 before autoclaving.

Type of Medium: Complex.

Use of Medium: Enrichment medium for luminescent bacteria.

Medium used in exercise 49.

Name of Medium: Phage Top Agar.

Ingredients:
Tryptone .... 8.0 g
Sodium chloride............................. 5.0 g
Agar.. 7.5 g
Distilled water .... 1000.0 ml

Dispense to screw cap tubes in desired quantities (3–5 ml) and autoclave to sterilize.

Final pH: 7.0 at 25°C.

Type of Medium: Complex soft agar medium.

Use of Medium: Used to deposit mixtures of bacteriophage and host bacteria on the surface of prepoured agar plates.

Medium used in exercises 32, 33, 35, 40, 42.

Name of Medium: Phenylalanine Agar.

Ingredients:

Bacto yeast extract	3.0	g
Dipotassium phosphate	1.0	g
Sodium chloride	5.0	g
DL-phenylalanine	2.0	g
Bacto agar	12.0	g
Distilled water	1000.0	ml

Heat ingredients in boiling water to dissolve completely. Dispense in tubes and autoclave to sterilize. Cool medium in slanted position.

Final pH: 7.3 at 25°C.

Type of Medium: Complex differential agar slant.

Use of Medium: Used to detect phenylalanine deaminase activity in heterotrophic microorganisms. Inoculate the entire slant with a pure culture and incubate at 35 to 37°C for 18 to 24 hours. Test for activity by adding 3 to 5 drops each of an 8 to 12% ferric chloride solution and a 0.1 N HCl solution to the culture. Rotate the tubes to wet and loosen the growth. A positive test is indicated by the appearance of a green color. *Proteus* and *Providencia* species will give a positive reaction within 1 to 5 minutes after addition of the reagents. Other members of the Enterobacteriaceae are ordinarily negative for this activity.

Medium used in exercise 26D.

Name of Medium: Plate Count Agar*.

Ingredients:

Tryptone	5.0	g
Yeast extract	2.5	g
Dextrose (glucose)	1.0	g
Agar	15.0	g
Distilled water	1000.0	ml

Autoclave to sterilize.

*This medium is also known as Standard Methods Agar or as Tryptone Glucose Yeast Extract Agar.

Final pH: 7.0 at 25°C.

Type of Medium: Complex infusion agar.

Use of Medium: For the enumeration of bacteria from water, food, and dairy products.

Medium used in exercises 14A, 14B, 15, 43, 46, 46A, 46B, 55A, 55B, 58, 62.

Name of Medium: Potassium Chloride (KCl) Broth.

Ingredients:

Potassium chloride	5.0	g
Tryptone	40.0	g
Calcium chloride, 1.0M*	0.5	ml
Distilled water	1000.0	ml

Autoclave to sterilize.

*CaCl₂, 1.0M, is prepared separately, as follows:

Calcium chloride, anhydrous	11.1	g
Distilled water	100.0	ml

Final pH: 7.2.

Type of Medium: Complex peptone broth.

Use of Medium: This broth is used to dilute the bacteriophage lysate in the transduction experiment.

Medium used in exercise 40.

Name of Medium: Potato Dextrose Agar (Acidified).

Ingredients:

Infusion from potatoes	200.0	g
Bacto dextrose	20.0	g
Bacto agar	15.0	g
Distilled water	1000.0	ml

Autoclave to sterilize.

Final pH: 5.6 at 25°C (or 3.5, using sterile 10% tartaric acid, added after autoclaving).

Type of Medium: Complex, selective infusion agar.

Use of Medium: For the enumeration, isolation, and growth of fungi from foods, including dairy products. The infusion from the potatoes provides suitable nutrients for fungi, while the lowered pH is inhibitory for most bacteria.

Medium used in exercise 63B.

Name of Medium: Sabouraud Dextrose (Dextrox) Agar.

Ingredients:

Neopeptone, Difco	10.0	g
Bacto dextrose	40.0	g
Bacto agar	15.0	g
Distilled water	1000.0	ml

Final pH: 5.6 at 25°C.

Type of Medium: Complex, selective.

Use of Medium: This medium is selective for molds, yeasts and other aciduric microorganisms due to the lowered pH of the medium and (to a lesser extent) the elevated sugar content.

Medium used in exercises 46, 55D, 59C.

Name of Medium: Sabouraud Dextrose Broth.

Ingredients:

Neopeptone, Difco	10.0	g
Glucose	20.0	g
Distilled water	1000.0	ml

Autoclave to sterilize.

Final pH: 5.6 at 25°C.

Type of Medium: Complex, selective infusion broth.

Use of Medium: Used for the cultivation of molds and yeasts. Acid pH is inhibitory to most heterotrophic bacteria.

Medium used in exercise 4A.

Name of Medium: Salt (6.5% NaCl) Broth.

Ingredients: Same as for brain heart infusion broth*, but additional sodium chloride should be added to increase the final concentration to 6.5%:

Brain heart infusion broth . . . . as described
Sodium chloride . . . . 60.0 g

Dispense medium to tubes and autoclave to sterilize.

*Nutrient broth may be used as the basal medium in place of brain heart infusion broth, but some nutritionally fastidious bacteria may find this formulation inadequate.

Final pH: 7.0 at 25°C.

Type of Medium: Complex, selective infusion broth.

Use of Medium: To determine the ability of a microorganism to tolerate and grow in an elevated concentration of NaCl. Inoculate medium lightly (because heavy inocula may produce apparent signs of growth immediately after inoculation), incubate and examine culture daily for up to 7 days. A positive reaction is indicated by appearance of obvious signs of growth (turbidity, sediment, a pellicle, or some combination of these).

Medium used in exercise 57B.

Name of Medium: Skim Milk Agar.

Ingredients: Prepare, dispense in tubes and sterilize an infusion agar medium like trypticase soy agar or plate count agar. Cool to 50–55°C and aseptically add sterile skim milk in the following proportion:

Infusion agar base (sterile) . . . . 20.0 ml
Sterile skim milk . . . . 3.0 ml

Thoroughly mix the molten medium and immediately pour into petri plates. Alternatively, the milk and the infusion agar base may be added directly to the plates and mixed therein.

Final pH: Do not pH.

Type of Medium: Complex, differential agar plating medium.

Use of Medium: Used for the detection of casein hydrolysis (proteolytic activity) by microbial colonies. Spot inoculate the medium, incubate and examine daily for signs of hydrolysis. A positive reaction is indicated by the appearance of a clear zone surrounding the growth in the originally cloudy white medium. In milk at a neutral pH, unhydrolyzed casein is a large molecular weight protein in a stable colloidal state; when casein is digested, the ensuing mixture of amino acids and peptides go into true solution and the milk in the medium clears.

Medium used in exercise 23B, 55B.

Name of Medium: Spirit Blue Agar.

Ingredients:
Bacto tryptone . . . . 10.0 g
Bacto yeast extract . . . . 5.0 g
Bacto agar . . . . 20.0 g
Spirit blue . . . . 0.15 g

Sterilize by autoclaving and cool to 50–55°C. To complete the medium, add (slowly, with agitation) either of the following, dispense to petri plates and allow to solidify:

Bacto lipase reagent . . . . 30.0 ml

Or: Prepare a cottonseed/olive oil emulsion (25%, v/v), in 400 ml of warm distilled water containing 1.0 ml of Tween 80. Add 30.0 ml of this to Spirit Blue agar base.

Final pH: 6.8 at 25°C.

Type of Medium: Complex, differential, plating medium.

Use of Medium: Used for the detection and enumeration of lipolytic microorganisms. Spot-inoculate or streak the medium (may also be inoculated as a pour plate), incubate for up to 72 hours, and observe culture for the appearance of a halo surrounding the growth, which is indicative of lipolysis. A negative reaction is indicated by the persistence of a pale lavender color and an opalescence throughout the medium.

Medium used in exercise 23D.

Name of Medium: Sporulating Agar.

Ingredients:
Bacto beef extract . . . . 1.5 g
Bacto yeast extract . . . . 3.0 g
Bacto casitone . . . . 4.0 g
Bacto peptone . . . . 6.0 g
Bacto dextrose . . . . 1.0 g
Bacto agar . . . . 15.0 g
Distilled water . . . . 1000.0 ml

Autoclave to sterilize.

Final pH: 6.5 at 25°C.

Type of Medium: Complex.

Use of Medium: For the promotion of endospore formation by *Bacillus subtilis* (for use in preparation of endospore suspensions).

Medium used in exercise 46B.

Name of Medium: Starch Agar.

Ingredients:
Bacto beef extract . . . . 3.0 g
Soluble starch, Difco . . . . 10.0 g
Bacto agar . . . . 12.0 g
Distilled water . . . . 1000.0 ml

Autoclave to sterilize, cool to 50–55°C, dispense in sterile Petri dishes and allow to solidify.

Final pH: 7.5 at 25°C.

Type of Medium: Complex, differential infusion agar medium.

Use of Medium: Used as a plating medium for the detection of starch hydrolysis by bacterial colonies. Spot inoculate the medium and incubate the culture for 48 hours. To test for starch hydrolysis, flood the culture with Gram's iodine. A positive reaction is indicated by the appearance of a colorless zone (the iodine solution itself is a pale yellow color) around the growth. A negative reaction is indicated by the development of a blue or purple zone around the culture. Partially hydrolyzed starch will produce a brownish complex with iodine (tri-iodide ions).

Medium used in exercise 23A.

Name of Medium: Starch-Casein Agar.

Ingredients:
Soluble starch . . . . 10.0 g
Vitamin free casein . . . . 0.3 g

Sodium nitrate	2.0	g
Sodium chloride	2.0	g
Dipotassium phosphate	2.0	g
Magnesium sulfate · 7H$_2$O	0.05	g
Calcium carbonate	0.02	g
Ferrous sulfate · 7H$_2$O	0.01	g
Bacto agar	15.0	g
Tap water	1000.0	ml

Final pH: 7.0 at 25°C.

Type of Medium: Complex.

Use of Medium: With cycloheximide added as a fungal inhibitor, this medium is suitable for the isolation of filamentous bacteria.

Medium used in exercises 46A, 54.

Name of Medium: Sulfide-Indole-Motility (SIM) Agar.

Ingredients:

Bacto peptone	30.0	g
Bacto beef extract	3.0	g
Peptonized iron, Difco	0.2	g
Sodium thiosulfate	0.025	g
Bacto agar	3.0	g
Distilled water	1000.0	ml

Dissolve and dispense medium in 10 ml amounts to 16 × 125 mm tubes. Autoclave to sterilize and allow medium to solidify in a vertical position to produce deeps.

Final pH: 7.3 at 25°C.

Type of Medium: Complex, differential, agar tubed medium.

Use of Medium: Used for the detection of hydrogen sulfide production, indole formation, and motility by heterotrophic bacteria. Medium is carefully stab-inoculated and incubated for 18 to 24 hours. If free sulfide ions are produced during incubation, they react with the iron to produce a black precipitate of iron sulfide in the previously clear, amber-colored medium. Motility can be detected as a diffuse cloud of turbidity (growth) away from the line of inoculation. To test for indole production, add an overlay of 1 ml of Kovac's reagent to the culture. Indole's presence is indicated by the development of a pink to deep red color in the top (Kovac's) layer. In the absence of indole, the top layer remains colorless. Always include an uninoculated tube of the medium, as a visual negative control, when these tests are performed.

Medium used in exercises 26A, 26B.

Name of Medium: Thiosulfate Medium.

Ingredients:

Sodium thiosulfate · 5H$_2$O	10.0	g
Ammonium chloride	1.0	g
Maganesium chloride	0.5	g
Dipotassium phosphate	0.6	g
Monopotassium phosphate	0.4	g
Trace salts solution*	10.0	ml
Tap water	1000.0	ml

*The trace salts solution (described below) is prepared in 0.1 N HCl and is added to the above medium before autoclaving.

Final pH: 7 or 4.5.

Type of Medium: Selective, inorganic, chemically defined enrichment broth.

Use of Medium: For the selective enrichment and isolation of the chemolithotrophs of the genus *Thiobacillus*. The medium contains no organic carbon or energy sources. The thiosulfate ions in the medium provide a suitable energy source for the thiobacilli.

Medium used in exercise 48C.

Name of Medium: Thiosulfate Medium—Trace Salts Solution.

Ingredients:

Ferrous sulfate · 7H$_2$O	0.300	g
Manganese chloride · 4H$_2$O	0.180	g
Cobalt nitrate · 6H$_2$O	0.130	g
Zinc sulfate · 7H$_2$O	0.040	g
Molybdic acid	0.020	g
Cupric sulfate · 5H$_2$O	0.001	g
Calcium chloride	1.0	g
Hydrochloric acid (0.1 N)	1000.0	ml

Final pH: Do not adjust pH.

Type of Medium: Concentrated source of trace metal ions.

Use of Medium: Added as a supplement to thiosulfate basal medium, before autoclaving. Ten ml of this supplement is sufficient to supply one liter of medium with adequate amounts of essential trace metal ions.

Medium used in exercise 48C.

Name of Medium: Tomato Juice Agar.

Ingredients:

Tomato juice (400 ml)	20.0	g
Bacto peptone	10.0	g
Bacto peptonized milk	10.0	g
Bacto agar	11.0	g
Distilled water	1000.0	ml

Autoclave to sterilize.

Final pH: 6.1 at 25°C.

Type of Medium: Complex, selective enrichment for aciduric heterotrophs.

Use of Medium: For the enrichment and enumeration of lactobacilli. Lactobacilli grow poorly on commonly used complex organic culture media, but many are enhanced by the tomato juice and will produce large colonies on this medium. The acidic pH tends to inhibit many other bacteria that are encountered in saliva.

Medium used in exercise 59A.

Name of Medium: Top Agar (see Phage Top Agar).

Name of Medium: Trypticase Soy Agar.

Ingredients:

Trypticase	17.0	g
Phytone	3.0	g
Sodium chloride	5.0	g
Dipotassium phosphate	2.5	g
Glucose	2.5	g
Agar	15.0	g
Distilled water	1000.0	ml

Adjust pH to 7.3, dispense to tubes, and autoclave to sterilize.

Final pH: 7.3 at 25°C.

Type of Medium: Complex infusion agar medium.

Use of Medium: General purpose infusion agar medium which supports the growth of a wide variety of heterotrophic microorganisms.

Medium used in exercises 7A, 7C.

Name of Medium: Trypticase Soy Broth.

Ingredients:

Trypticase	17.0	g
Phytone	3.0	g
Sodium chloride	5.0	g
Dipotassium phosphate	2.5	g
Glucose	2.5	g
Distilled water	1000.0	ml

Adjust pH to 7.3 and autoclave to sterilize.

Final pH: 7.3 at 25°C.

Type of Medium: Complex infusion broth.

Use of Medium: General purpose infusion broth medium which supports the growth of a wide variety of heterotrophic microorganisms.

Medium used in exercise 7B.

Name of Medium: Trypticase Soy Broth with 50 Units/ml Mycostatin.

Ingredients:

Trypticase	17.0	g
Phytone	3.0	g
Sodium chloride	5.0	g
Dipotassium phosphate	2.5	g
Glucose	2.5	g
Distilled water	800.0	ml

Adjust pH to 7.3 and autoclave to sterilize. Prepare a mycostatin solution separately as follows:

Mycostatin	50,000	Units
Distilled Water	200.0	ml

The mycostatin solution is filter sterilized and then aseptically added to the cooled, sterile trypticase solution to form the complete medium.

Final pH: 7.3 at 25°C.

Type of Medium: Complex selective broth medium.

Use of Medium: Used in studies of the effects of antimicrobial agents on microorganisms.

Medium used in exercises 22C, 22D.

Name of Medium: Trypticase Soy Broth with 200 Units/ml Penicillin.

Ingredients:

Trypticase	17.0	g
Phytone	3.0	g
Sodium chloride	5.0	g
Dipotassium phosphate	2.5	g
Glucose	2.5	g
Distilled water	800.0	ml

Adjust pH to 7.3 and autoclave to sterilize. Prepare a penicillin solution separately as follows:

Penicillin	200,000	Units
Distilled water	200.0	ml

The penicillin solution is filter sterilized and then aseptically added to the cooled, sterile trypticase solution to form the complete medium.

Final pH: 7.3 at 25°C.

Type of Medium: Complex selective broth medium.

Use of Medium: Used to determine the sensitivity of microorganisms to penicillin.

Medium used in exercises 22C, 22D.

Name of Medium: Trypticase Soy Broth with 400 micrograms/ml Streptomycin.

Ingredients:

Trypticase	17.0	g
Phytone	3.0	g
Sodium chloride	5.0	g
Dipotassium phosphate	2.5	g
Glucose	2.5	g
Distilled water	800.0	ml

Adjust pH to 7.3 and autoclave to sterilize. Prepare a streptomycin solution separately as follows:

Streptomycin	400.0	mg
Distilled water	200.0	ml

The streptomycin solution is filter sterilized and then aseptically added to the cooled, sterile trypticase solution to form the complete medium.

Final pH: 7.3 at 25°C.

Type of Medium: Complex selective broth medium.

Use of Medium: Used in studies of the effects of antimicrobial agents on microorganisms.

Medium used in exercises 22C, 22D.

Name of Medium: Trypticase Soy Broth with 0.08M Sulfanilamide

Ingredients:

Trypticase	17.0	g
Phytone	3.0	g
Sodium chloride	5.0	g
Dipotassium phosphate	2.5	g
Glucose	2.5	g
Distilled water	800.0	ml

Adjust pH to 7.3 and autoclave to sterilize. Prepare a sulfanilamide solution separately as follows:

Sulfanilamide	13.8	g
Distilled water	200.0	ml

The sulfanilamide solution is filter sterilized and then aseptically added to the cooled, sterile trypticase solution to form the complete medium. 1.38 g/100 ml = 0.08 M.

Final pH: 7.3 at 25°C.

Type of Medium: Complex selective broth medium.

Use of Medium: Used to determine the sensitivity of microorganisms to sulfanilamide.

Medium used in exercises 22C, 22D.

Name of Medium: Tryptose Phosphate Agar*.

Ingredients:

Bacto tryptose	20.0	g
Bacto dextrose	2.0	g
Sodium chloride	5.0	g
Disodium phosphate	2.5	g
Agar	15.0	g
Distilled water	1000.0	ml

Autoclave to sterilize.

*Tryptose phosphate medium is commercially available as a broth.

Final pH: 7.3 at 25°C.

Type of Medium: Complex infusion agar.

Use of Medium: Used for cultivation of nutritionally fastidious microorganisms.

Medium used in exercise 11A.

Name of Medium: Urea Agar (Christensen)*.

Ingredients:

Bacto peptone	1.0	g
Bacto dextrose	1.0	g
Sodium chloride	5.0	g
Potassium phosphate, monobasic	2.0	g
Urea, Difco	20.0	g
Bacto phenol red	0.012	g
Distilled water	100.0	ml

Filter sterilize the above Urea Agar Base. Do not boil or autoclave. Also prepare the following agar solution which is autoclaved to sterilize:

Bacto agar	15.0	g
Distilled water	900.0	ml

Cool the sterile agar solution to 50–55°C and aseptically add the filter sterilized Urea Agar Base. Mix well and dispense to tubes. Slant tubes to produce solid media with a butt about 2 cm deep and a slant about 3 cm long.

Final pH: 6.8 at 25°C.

Type of Medium: Complex, differential agar slant.

Use of Medium: Weakly buffered medium used to rapidly detect urease (urea hydrolysing) activity in bacteria. Medium is inoculated on the entire slant surface from isolated colonies and incubated at 35°C. Strongly urease positive organisms such as *Proteus* species will usually cause the entire medium to alkalinize and shift in color from pale yellow or colorless to pink within about 4 to 6 hours. Negative cultures should continue to be incubated and examined daily for 4 to 6 days for any sign of weak or delayed urease activity.

Medium used in exercise 27C.

Name of Medium: Violet Red Bile Agar.

Ingredients:

Bacto yeast extract	3.0	g
Bacto peptone	7.0	g
Bacto bile salts #3	1.5	g
Bacto lactose	10.0	g
Sodium chloride	5.0	g
Agar	15.0	g
Neutral red	0.03	g
Crystal violet	0.002	g
Distilled water	1000.0	ml

Heat to boiling for no more than 2 minutes to dissolve completely. Do not autoclave. Cool to 45 to 50°C and pour into sterile plates.

Final pH: 7.4 at 25°C.

Type of Medium: Complex, differential, and selective agar.

Use of Medium: For the isolation and enumeration of coliforms from food and dairy products. Use pour plate technique with freshly prepared (not autoclaved) medium. After inoculated medium has solidified, add 4 ml of medium to form an overlay. Incubate at 35°C for 18 to 24 hours. Bile salts and crystal violet inhibit the growth of Gram positive bacteria. Coliforms form purple-red or pink submerged colonies, 1–2 mm in diameter, usually surrounded by a reddish zone of precipitated bile. The color of the coliform colonies and the precipitated bile are due to the action on the neutral red of acids from lactose fermentation. Gram negative non-lactose fermenting rods which grow on this medium form colorless colonies.

Medium used in exercises 57A, 60, 62.

APPENDIX E
STAINS, INDICATORS, AND REAGENTS

Stains and Indicators

Stain: Carbolfuchsin (Ziehl-Neelsen)

Composition:

Basic fuchsin	0.3	g
Ethyl alcohol, 95%	10.0	ml
Phenol, heat-melted crystals	5.0	ml
Distilled water	95.0	ml

Dissolve the basic fuchsin in the alcohol. Heat phenol crystals to about 45°C to melt them, then manually transfer 5.0 ml of molten phenol crystals to the water and mix to dissolve. Mix the basic fuchsin solution with the phenol solution and allow the mixture to stand for several days. Filter the reagent before use.

Special Instructions: Handle phenol crystals and concentrated phenol solutions with caution. See comments under phenol, 5.0%, w/v, in the Reagents section of this appendix.

Use: This solution may be used alone as a simple, positive stain, or as the primary stain in the "hot" acid-fast stain procedure (Ziehl-Neelsen, 1882–1883).

Used in exercises 8A, 9B, 12.

Indicator: Bromcresol purple, 0.04%, w/v

Composition:

Bromcresol purple	0.1	g
NaOH, 0.1 N	1.85	ml
Ethyl alcohol, 50%, q.s. to	250.0	ml

Grind bromcresol purple in 0.1 N NaOH, then add 50% ethanol to 250 ml. Generally, a pH indicator is added to fermentation media, before sterilization, to achieve a final concentration of 18 to 20 mg per liter of medium.

Use: This pH indicator is added to fermentation broth media to allow visual detection of fermentation activity by a color shift in the indicator due to acid waste production. Its pK value is lower than that of phenol red; more acid must be produced by a culture to cause a visible color shift (purple at pH 6.8, yellow at pH 5.2) than is necessary with phenol red. It is present in Moeller's decarboxylase medium, where a shift from yellow-green to purple indicates decarboxylase activity.

Used in exercises 24A, 26C.

Indicator: Bromthymol blue, 0.04%, w/v

Composition:

Bromthymol blue	0.4	g
Ethyl alcohol, 95%	500.0	ml
Distilled water	500.0	ml

Dissolve the bromthymol blue in ethyl alcohol and then dilute the solution with water.

Use: This pH indicator is used in Simmons' citrate agar medium to allow visual detection of microbial growth on citrate as its sole organic carbon and energy source. A positive finding is indicated by the development of a blue color on the originally green slant, due to alkalinization of the medium above a pH of 7.5. This indicator is yellow below a pH of 6.0.

Used in exercise 27A.

Stain: Congo red

Composition:

Congo red	0.5	g
Distilled water	100.0	ml

Special Instructions: If one omits the counterstain step with the modified Maneval's stain, the procedure using Congo red is capable of producing a good negative stain, which may be of real value in evaluating the actual shape and size of some bacterial cells.

Use: This stain is used as the initial reagent in the modified Maneval's method for the detection of bacterial capsules. Congo red won't bind to the cells' structures or capsule (if present), but dries on the glass slide to form a colored background which outlines the physical border of the cells (with or without capsules) in the mixture. When internal negatively-charged structures of the cell are stained by the modified Maneval's stain, capsules will be revealed in unstained relief.

Used in exercises 10A, 12.

Stain: Crystal violet (modified Hucker)

Composition:

Solution A:

Crystal violet (certified)	2.0	g
Ethyl alcohol, 95%	20.0	ml

Solution B:

Ammonium oxalate	0.8	g
Distilled water	80.0	ml

Combine solutions A and B. Store at room temperature for 24 hours before use, then filter through paper into stain bottle.

Use: This solution may be used alone as a simple, positive stain or as the primary reagent in the gram stain.

Used in exercises 8A, 9A, 12, 47D, 51, 53, 56, 59A, 59B, 59C, 63A, 63B, 64C, 67, 68A, 68B, 71C, 72A, 72B.

Stain: Eosin Y

Composition:

Eosin Y	1.0	g
Distilled water	100.0	ml

Use: This pink stain may be used as the counterstain in the stain procedure (Gohar, 1944) for the detection of metachromatic granules (volutin).

Used in exercises 11A, 12.

Stain: Gentian violet

Composition: 1:50, in tap water

Use: This reagent is used to lightly stain slides for the detection of periphytic bacteria.

Used in exercise 50.

Reagent/Stain: Gram's iodine

Composition:

Iodine	1.0	g
Potassium iodide	2.0	g
Distilled water	300.0	ml

Grind iodine and potassium iodide in mortar. Add water in small amounts with further grinding until dissolved. Add remainder of the water, mix, and store in amber glass bottle.

Special Instructions: Elemental iodine is intensely irritating to eyes, skin and mucous membranes, and can be poisonous if ingested in large quantities. Prepare this reagent in a well-ventilated chemical hood, wear gloves and goggles.

Use: This solution is the second reagent (mordant) in the gram stain. It forms an insoluble chemical complex with the crystal violet which is more firmly bound to cellular structures. It may also be similarly used in the stain procedure to detect metachromatic granules (volutin). In the stain procedure to detect starch and/or glycogen inclusions, this solution directly forms colored complexes with starch, glycogen, or dextrans. It is also used to test for signs of starch hydrolysis.

Used in exercises 9A, 11A, 11C, 12, 23, 47D, 51, 53, 56, 59A, 59B, 59C, 63A, 63B, 64C, 67, 68A, 68B, 71C, 72A, 72B.

Stain: Gray's flagella stain

Composition:

Solution A (mordant):

Potassium alum (aluminum potassium sulfate, KAl (SO$_4$)$_2$ · 12H$_2$O), saturated solution*	5.0	ml
Mercuric chloride, saturated solution**	2.0	ml
Tannic acid, 20%, w/v, in distilled water	2.0	ml

*At 20C, 11.4 g of potassium alum dissolves in 100.0 ml of distilled water.

**At 20C, 6.9 g of mercuric chloride dissolves in 100.0 ml of distilled water.

Solution B (saturated solution of the dye):
Basic fuchsin, certified by the Biological

Commission, for flagella stains	6.0	g
Ethyl alcohol, absolute	100.0	ml

Staining solution (prepare and filter each day):

Solution A (mordant)	9.0	ml
Solution B (saturated basic fuchsin)	0.4	ml

Use: This complex solution is used to stain and demonstrate bacterial flagella by the method of Gray (1926).

Used in exercises 10C, 12.

Reagent/Stain: India ink

Composition: A high quality India ink, in which the carbon particles are small (ca. 0.02 μm to 0.1 μm) and not aggregated.

Special Instructions: Not all commercially available India ink preparations are suitable for use in this procedure. Pelikan Drawing Ink, 17 Black, Gunther Wagner, Germany is recommended. Higgins waterproof India ink is acceptable, if the coarse particles are first removed by low speed centrifugation.

Use: This suspension of carbon particles may be used alone, in place of an acidic stain (such as nigrosin, or congo red), in a simple, negative (relief) stain procedure to reveal the shape and size of microbial cells and in a wet capsule stain procedure (Duguid, 1951).

Used in exercises 8B, 12.

Stain: Lactophenol cotton blue

Composition:

"Cotton blue"*	0.05	g
Phenol, crystals	20.0	g
Lactic acid (specif. grav. 1.21)	20.0	g
Glycerol	40.0	g
Distilled water	20.0	ml

*Cotton blue is available as China Blue or Poirrier's Blue.

Dissolve the cotton blue in the distilled water. Then individually dissolve the other ingredients in order.

Special Instructions: Avoid direct contact of skin with concentrated phenol crystals or solutions, which may cause severe, prolonged chemical burns.

Use: This staining solution is used for the microscopic study of stained wet mounts of fungi. Cotton blue will stain chitin which is commonly present in the walls of fungi.

Used in exercises 29A, 29B, 29C.

Stain: Levowitz-Weber stain

Composition:

Methylene blue	0.6	g
95% ethyl alcohol	52.0	ml

Add the alcohol to the dye slowly, with stirring, until the dye is completely dissolved. Continue stirring and add:

Tetrachloroethane	44.0	ml

Place in a closed 100 ml container and let stand 12 to 24 hr in a refrigerator (4 to 5C). Warm to room temperature and carefully add:

Glacial acetic acid	4.0	ml

Filter stain through Whatman no. 2 filter paper or its equivalent. Store in a tightly closed container in a cool, dark place, but do not refrigerate.

Special Instructions: This reagent should be prepared in a chemical hood. Wear nonabsorbent gloves. Avoid inhalation of fumes from the ingredients or the final stain.

Use: This stain is used in the direct microscopic count and evaluation of raw milk. The reagent fixes the raw milk specimen, and stains the microorganisms to permit their microscopic detection and enumeration.

Used in exercises 60C, 61C.

Stain: Loeffler's alkaline methylene blue

Composition:

Solution A:

Methylene blue chloride	0.3	g
Ethyl alcohol, 95%	30.0	ml

Solution B:

Potassium hydroxide	0.01	g
Distilled water	100.0	ml

Mix solutions A and B.

Special Instructions: Exercise caution when handling potassium hydroxide pellets. These pellets are very hygroscopic, so weigh them quickly.

Use: This solution may be used alone as a simple, positive stain, as the counterstain in the acid fast stain procedure (Ziehl-Neelsen, 1882–1883), or it may be used in the stain procedure for detection of metachromatic granules (volutin, Gohar, 1944).

Used in exercises 8A, 9B, 11A, 12.

Reagent/Stain: Lugol's iodine

Composition:

Potassium iodide	10.0	g
Iodine	5.0	g
Distilled water	100.0	ml

Dissolve the potassium iodide in the distilled water. Next, slowly add the iodine crystals and shake the mixture frequently until the iodine dissolves. Filter the solution and transfer to a tightly stoppered amber glass bottle for storage.

Special Instructions: Elemental iodine is intensely irritating to eyes, skin and mucous membranes, and can be poisonous if ingested in large quantities. Prepare reagent in a well ventilated chemical hood, wear gloves and goggles.

Use: Lugol's iodine (strong iodine solution) may be used as an alternative to Gram's iodine (weak iodine solution) in the procedure to detect starch and/or glycogen inclusions. The iodine solution directly forms distinctly colored complexes with starch, glycogen, and dextrans.

Used in exercises 11C, 12.

Stain: Malachite green

Composition:

Malachite green	5.0	g
Distilled water	100.0	ml

Use: This solution of a basic dye is used as the primary stain in the procedure of Schaeffer and Fulton (1933) to detect bacterial endospores.

Used in exercises 10B, 12, 46B.

Stain: Modified Maneval's stain

Composition:

Fuchsin*	0.05	g
Ferric chloride	3.0	g
Acetic acid, glacial	5.0	ml
Phenol, liquified (89–90%)	3.9	ml
Distilled water	95.0	ml

*Fuchsin is included in place of fast green FCG or light green.

Special Instructions: Prepare stain in a well ventilated chemical hood. Avoid direct contact with or ingestion of concentrated phenol or glacial acetic acid. Also, avoid inhalation of vapors from glacial acetic acid.

Use: This staining solution is used as the modified counterstain in Maneval's procedure to detect bacterial capsules. It stains the negatively charged internal structures of bacteria red but will ordinarily not bind to (or stain) the typically uncharged capsule layer present on the cellular surfaces of certain bacteria.

Used in exercises 10A, 12.

Stain: Methyl green

Composition:

Methyl green, certified Biological Stain	0.5	g
Distilled water	100.0	ml

Dissolve the dye in the water. Extract the solution with equal volumes of chloroform until the chloroform is colorless (6 to 8 times). The extracted aqueous solution may be added to DNase test agar before it is autoclaved or it may be filter sterilized and aseptically added to the sterile molten agar at 50 to 55C.

Special Instructions: Methyl green dye is available as a certified Biological Stain, C.I. no. 43590, from Fisher Scientific Co., Pittsburgh, Pa., catalog no. M295. Perform the chloroform extraction in a well ventilated chemical hood. Avoid direct contact, ingestion or inhalation of chloroform. Do not dispose of waste chloroform in sink drain as it is denser than water. See notes under "Chloroform" in this appendix. Store the reagent at 4C after chloroform extraction.

Use: Methyl green reagent may be added to DNase test agar medium as an indicator to aid in the detection of DNA hydrolysis following incubation. Dehydrated DNase test agar media with and without added methyl green are available commercially from Difco Laboratories, Detroit, Mi.

Used in exercise 23C.

Indicator: Methyl red

Composition:

Methyl red	0.1	g
Ethyl alcohol, 95%	300.0	ml

Dissolve dye in the alcohol and add distilled water to make 500 ml.

Use: This pH indicator is added to grown cultures in MRVP broth (glucose-peptone broth) to allow visual estimation of the final pH (MR test). If the pH of the culture has dropped to ca. 4.4 or below, the added indicator will show a definite pink to red color; this is a positive MR test finding. If the pH is between ca. 4.4 and 6.4, an orange color will develop, which is interpreted as a "doubtful" MR test. At a pH above 6.4 a yellow color occurs, which is a negative MR test finding.

Used in exercise 24B.

Stain: Nigrosin

Composition:

Nigrosin	10.0	g
Distilled water	100.0	ml

Mix the nigrosin with the water and heat in a boiling water bath for 30 min. Add 0.5 ml formalin as a preservative and filter twice through double thickness filter paper.

Special Instructions: Be sure to manually pipet formalin using disposable pipets. Avoid ingestion, inhalation of (or direct contact with) formalin.

Use: This solution may be used alone as a simple, negative stain.

Used in exercises 8B, 12.

Indicator: Phenol red, 0.02%, w/v

Composition:

Phenol red	0.1	g
NaOH, 0.1 N	2.82	ml
Ethyl alcohol, 50%, q.s. to	250.0	ml

Grind phenol red in 0.1 N NaOH, then add 50% ethyl alcohol to 250 ml. Generally, a pH indicator is added to fermentation media, before sterilization, to achieve a final concentration of 18 to 20 mg per liter of medium.

Use: Phenol red is added to carbohydrate broths to allow visual detection of fermentation activity in a culture as a color shift (orange at pH 7.4, to yellow at 6.8) in the indicator signifying acid waste production. Phenol red is a more sensitive indicator of an acid shift than bromcresol purple. Phenol red performs a similar function in mannitol-salts agar plates. It is also used in Christensen's urea agar slants where a shift from pale yellow (pH 6.8) to pink (pH 8.4) indicates urease activity.

Used in exercises 24A, 27B.

Stain: Phloxine

Composition:

Phloxine B	6.5	g
Distilled water	100.0	ml

Use: This stain is used to aid in the microscopic detection and study of mycorrhizae.

Used in exercise 65.

Indicator: Resazurin solution

Composition:

Resazurin tablet, certified by the Biological Stain Commission, dye content approximately 11 mg		1 tablet
Distilled water, autoclaved or boiled	200.0	ml

Autoclave or boil distilled water (final volume, 200 ml ± 2 ml) in a dark, stoppered glass container. While still hot, add 1 resazurin tablet, which should completely dissolve before the solution cools. Keep in a cool dark place. Solution must be made fresh each week.

Use: This reagent is used in the dye reduction test to determine the sanitary quality of raw milk.

Used in exercise 60B.

Stain: Safranin

Composition:

Safranin 0 (certified)	2.5	g
Ethyl alcohol, 95%	100.0	ml

Working solution:

Stock solution	10.0	ml
Distilled water	90.0	ml

Use: This solution may be used alone as a simple, positive stain or as the counterstain in the gram stain procedure, the endospore stain procedure, and the stain procedure to detect poly-beta-hydroxybutyric acid (PHB) inclusions.

Used in exercises 8A, 9A, 10B, 11B, 12, 46B, 47D, 51, 53, 56, 59A, 59B, 59C, 63A, 63B, 64C, 67, 68A, 68B, 71C, 72A, 72B.

Stain: Sudan black B

Composition:

Sudan black B	0.3	g
Ethyl alcohol, 70%, v/v	100.0	ml

Shake the mixture thoroughly periodically during the day of preparation. Allow the mixture to stand overnight before use.

Use: This solution of a neutral (uncharged, lipid-soluble) dye is used as the primary stain in the procedure (Burdon, 1946) to detect poly-beta-hydroxybutyric acid (PHB) inclusions.

Used in exercises 11B, 12.

Stain: Toluidine blue

Composition:

Toluidine blue 0	1.0	g
Distilled water	100.0	ml

Use: This solution may be used, in place of Loeffler's alkaline methylene blue, as the primary stain in the procedure for the detection of metachromatic granules (volutin).

Used in exercises 11A, 12.

Reagents

Reagent: Acid alcohol

Composition:

Hydrochloric acid, concentrated, reagent grade	3.0	ml
Ethyl alcohol, 95%	97.9	ml

Special Instructions: Prepare reagent in a well ventilated chemical hood. Avoid direct contact, inhalation, or ingestion of concentrated hydrochloric acid or of this reagent.

Use: This solution is used as the decolorizing agent (or "differentiator") in the "hot" acid-fast stain procedure (Ziehl-Neelsen, 1882–1883).

Used in exercises 9B, 12.

Reagent: Alpha-naphthol, 5%, w/v

Composition:

Alpha-naphthol*	5.0	g
Ethyl alcohol, absolute	100.0	ml

Store reagent in a tightly stoppered, amber glass bottle in the refrigerator.

In exercise 72B, development of the VP reaction in the API 20E system, specifies the use of a 6% (w/v) solution of alpha-naphthol in absolute ethyl alcohol.

Use: This organic solution is used as a reagent in the VP test to detect production of the fermentation waste product acetylmethylcarbinol (acetoin) by organisms growing in MRVP broth (glucosepeptone broth). Under alkaline conditions and in the presence of a small amount of creatine, acetoin is oxidized to diacetyl, which interacts with alpha-naphthol and creatine, to produce a pink to red product, usually within 30 minutes.

Used in exercises 24C, 72A, 72B.

Reagent: API 20E Strip

Composition: Commercial multitest system for the identification of members of the family Enterobacteriaceae.

Special Instructions: This system is available commercially from Analytab Products Inc., Plainview, New York. Handle aseptically; they are sterilized commercially. Store at 4°C. Observe specified expiration date.

Use: This multitest system is used for the identification of members of the family Enterobacteriaceae and certain other Gram-negative bacteria.

Used in exercise 72B.

Reagent: Bleach, 5.25%

Composition:
Active ingredient:
Sodium hypochlorite 5.25%
Inert ingredients 94.75%

Special Instructions: This solution is available commercially under a variety of trade names (Clorox, Purex, etc).

Use: This solution is used in the study of the factors which make an antimicrobial agent effective.

Used in exercise 22A.

Reagent: Blood Typing Kit, Carolina Biological

Composition: Kit contains anti-A serum (hemagglutinating antibodies specific for Group A antigenic determinants on human erythrocytes) and anti-B serum (hemagglutinating antibodies specific for Group B antigenic determinants on human erythrocytes).

Special Instructions: Commercially available as cat. no. 70-4057. Store at 4°C. Observe specified expiration date.

Use: Blood typing kit reagents used in the test to identify the ABO blood group of students.

Used in exercise 70.

Reagent: Bovine serum albumin, 1:500

Composition:
Borate buffer:
Boric acid . . . .   6.184   g
Borax (sodium tetraborate) . . . .   9.536   g
NaCl . . . .   4.384   g
Distilled water . . . .   1000.0   ml

Dissolve ingredients in about 800 ml of distilled water with shaking; add distilled water to 1000.0 ml.

Then add and dissolve, without shaking contents:
Bovine serum albumin . . . .   2.0   g
Borate buffer, pH 8.5 (prepared above) . . . .   1000.0   ml

Use: This preparation is used as the soluble antigen in the exercise on the ring precipitin test.

Used in exercise 71B.

Reagent: Calcium sulfate

Composition: Calcium sulfate, anhydrous reagent; approximately 50 g needed for a Winogradsky column in a 250 ml graduate cylinder.

Use: This chemical in its pure, dry form is added to Winogradsky columns to contribute a source of oxidized sulfur (sulfate ions) for microbes capable of performing anaerobic respiration in the depth of the column with sulfate ions as their terminal electron acceptor.

Used in exercise 48A.

Reagent: California Mastitis Test Reagent

Composition: Commercial product ready to use . . . .1 pt liquid

Special Instructions: Available from Dairy Research Products, Inc., Spencerville, Indiana 46788.

Use: The CMT reagent is added to milk samples to be screened for evidence of mastitis. The amount of gel formation correlates with the number of somatic cells.

Used in exercise 61A.

Reagent: Chloroform

Composition: Chloroform (also known as trichloromethane) is a dense, volatile, colorless liquid with a characteristic odor. Since chemically pure chloroform is light sensitive, reagent grade chloroform usually contains 0.75% ethanol as a stabilizer. Store in a cool, dark area.

Special Instructions: Use carefully only in a well-ventilated chemical hood, avoid inhalation, ingestion, or skin contact, and handle with nonporous gloves and manual pipetting or transfer devices; chloroform is listed as a carcinogen by the EPA and has demonstrable serious toxic side effects if inhaled or ingested in large doses! Since chloroform is denser than water, it should not be disposed of down sink drains, but should be collected in a special waste container.

Use: Chloroform is used in the exercise on bacterial transduction to enhance lysis of infected bacterial cells by bacteriophage P22. It may also be used in the extraction of methyl green indicator reagent before this indicator is added to DNase test agar medium. Chloroform is sometimes added in small amounts to nonsterile reagents susceptible to contamination (e.g., nigrosin) as a preservative.

Used in exercises 8B, 23C, 40.

Reagent: cAMP (Cyclic adenosine monophosphate)

Composition: 10^{-1} M cAMP is made by mixing 0.329g cAMP with 10 ml distilled H_2O.

Special Instructions: The solution does not have to be sterilized since it is added late in the experiment. It may be filter sterilized but should not be autoclaved.

Use: This solution is used in the exercise on the activation of the lactose operon (an inducible enzyme system).

Used in exercise 38A.

Reagent: Cycloheximide stock solution

Composition: Commercial preparation known as *Actidione.*

Special Instructions: Poisonous! Avoid physical contamination with this substance. Observe all precautions indicated by the commercial producer. Available from The Upjohn Co., Kalamazoo, Mi.

Use: This reagent is aseptically added to starch-casein agar to prepare a selective plating medium for the isolation of the actinomycetes from soil. This medium should inhibit growth of fungi.

Used in exercises 46A, 54.

Reagent: N,N-Dimethyl-1-naphthylamine

Composition:

N,N-Dimethyl-1-naphthylamine	6.0	ml
Acetic acid, 5 N	1000.0	ml

Prepare 5 N acetic acid by mixing:

Acetic acid, glacial	300.0	ml
Distilled water	750.0	ml

Use 1000.0 ml of the last solution dissolve the N,N-Dimethyl-1-naphthylamine.

Special Instructions: This reagent has not been listed as a carcinogen by the Occupational Safety and Health Administration, Department of Labor, as has alpha-naphthylamine. However, in view of these two compounds' structural similarity, users should observe safety precautions (i.e., avoid mouth pipetting, aerosol production, and contact with the skin).

Use: This reagent is a substitute for alpha-naphthylamine (a listed carcinogen) in the test to detect the formation of nitrite ions by microbial reduction of nitrate ions. If nitrate broth/agar cultures contain nitrite ions, a pink to red color reaction will develop within a few minutes after addition of suitable amounts of sulfanilic acid and N,N-Dimethyl-1-naphthylamine. The same reagents may also be used to determine whether or not a culture has utilized nitrite ions in a nitrite agar/broth.

Used in exercises 25C, 72B.

Reagent: Diphenylamine

Composition:

Diphenylamine	0.5	g
Sulfuric acid, concentrated reagent	100.0	ml
Distilled water	20.0	ml

Carefully and slowly add the concentrated sulfuric acid to the distilled water, with stirring. Then slowly add the diphenylamine with stirring to dissolve.

Special Instructions: Prepare reagent only in a well ventilated chemical hood. Wear protective clothing, gloves and goggles. Avoid inhalation or other physical contact. May be irritating to mucous membranes. Remember to add concentrated acid slowly to water with stirring, not water to acid! Store reagent in a cool dark container.

Use: This reagent may be used to test ammonium broth cultures for the production of nitrate ions, if the test for nitrite ion formation is negative. When one drop of diphenylamine, one drop of concentrated sulfuric acid and one drop of culture are mixed in a spot plate, the appearance of a blue-black color is indicative of the presence of nitrate or nitrite ions in the medium.

Used in exercise 47B.

Reagent: Distilled water, sterile

Composition:

Distilled water	***.*	ml

Place in appropriate vessel, in measured volumes as needed (common volumes, when used as a diluent, would be 9.0 ml, 9.9 ml, 99.0 ml, etc.). Add an appropriate closure (cotton plugs, screw caps, Morton caps, etc.) and autoclave to sterilize. If vessels containing large volumes are to be sterilized, the time of sterilization will need to be set beyond 15 min, to ensure that the center of the liquid (like the periphery) has attained a sterilizing temperature.

Use: Commonly used in known volumes as a diluent for certain types of cultures/specimens before some type of quantitative enumeration is to be performed. Also used in the preliminary dilution of antimicrobial agents for the determination of the phenol coefficient.

Used in exercises 22B, 39, 64B.

Reagent: Enterotube II

Composition: Commercial multitest system for the identification of members of the family Enterobacteriaceae.

Special Instructions: This system is commercially available from Roche Diagnostics, Nutley, New Jersey. Handle aseptically; they arrived presterilized. Store at 4°C. Observe specified expiration date.

Use: This multitest kit is used for the indentification of Gram-negative, oxidase-negative, glucose-fermentative bacteria (members of the Enterobacteriaceae).

Used in exercise 72A.

Reagent: Ethyl alcohol, 95%

Composition:

Ethyl alcohol, 95%	reagent grade

Use: This solution is used as the decolorizing agent (or "differentiator") in the gram stain procedure.

Used in exercises 9A, 12, 47D, 51, 52, 53, 56, 59A, 59B, 59C, 63A, 63B, 64B, 64C, 67, 68A, 68B, 69, 71C, 72A, 72B.

Reagent/Stain: Ferric chloride, 10%, w/v

Composition:

Ferric chloride (FeCl₃ · 6H₂O)	10.0	g
Distilled water	100.0	ml

Use: This solution is used in the test for deamination of the amino acid phenylalanine by microorganisms with phenylalanine deaminase activity. The product of such activity, phenylpyruvic acid, will react with this reagent to form a green-colored complex.

Used in exercises 26D, 72B.

Reagent: Ferrous ammonium sulfate, 0.5%, w/v

Composition:

Ferrous ammonium sulfate	5.0	g
Distilled water	1000.0	ml

Dispense solution to screw capped tubes or bottles as desired and autoclave to sterilize.

Use: This sterile solution is added aseptically to *Desulfovibrio* medium before use to provide a reduced source of iron to react with any free sulfide ions released by metabolism and yield a visible black precipitate.

Used in exercise 48D.

Reagent: FITC-labeled anti-*Streptococcus* group D antibody

Composition: Antibody preparation, purified from an antiserum produced in an animal species against *Streptococcus* group D cells; after purification, the antibodies were chemically conjugated (linked) to the fluorochrome (fluorescent dye) FITC (fluorescein isothiocyanate).

Special Instructions: Commercially available from Difco Labs. as cat. no. 2321. Handle aseptically; sterilized by commercial vendor. Store at 4°C. Observe any specified expiration date.

Use: This antibody preparation is used in the exercise on direct immunofluorescence. The antibodies are chemically "conjugated" (bound) to a fluorescent dye (FITC = fluorescein isothiocyanate). The antibodies will specifically attach the fluorescent dye to any cells on a slide smear which possess the streptococcal group D antigen. When the slide is examined with suitable illumination under a fluorescence microscope, cells that have bound the antibody will fluoresce blue-green.

Used in exercise 71C.

Reagent: Glucose, 10%, w/v, sterile

Composition:

Glucose	100.0	g
Distilled water	1000.0	ml

Filter to sterilize.

Use: This sterile solution is used in the exercise on the activation of the lactose operon (an inducible enzyme system).

Used in exercise 38A.

Reagent: Glycerol in PBS (1:1, v/v)

Composition:

Glycerol, reagent grade	500.0	ml

Add to:

PBS (see phosphate buffered saline)	500.0	ml

Use: This solution is used as the mounting medium for smears in the exercise on the direct immunofluorescence technique.

Used in exercise 71C.

Reagent: Hydrochloric acid (HCl), 1.0N

Composition:

Hydrochloric acid, concentrated, reagent grade (ca. 38% HCl, w/v)	83.0	ml
Distilled water, make up to	1000.0	ml

Slowly and carefully add the concentrated acid to about 400 ml of water with continuous stirring and cooling. When the solution is cool, transfer it to a 1 liter volumetric flask and add sufficient distilled water to the mark.

Special Instructions: Concentrated hydrochloric acid solutions are volatile and can be very poisonous if mishandled! Prepare the solution in a well ventilated chemical hood. Avoid direct contact between the concentrated acid or this solution and skin or mucous membranes. Avoid ingestion—never mouth pipette. Remember, always add concentrated acids to water, not water to acids! Store the reagent in a cool area, away from the containers of strong mineral bases.

Use: Acid solution is used to adjust the pH of sterile trypticase soy broth (TSB) to various levels to test the effect of environmental pH on microbial growth.

Used in exercise 17.

Reagent: Hydrogen peroxide, 3%, w/v

Composition:

Hydrogen peroxide	3.0	%
Distilled water*	97.0	%

*Some commercially available versions of this reagent may have small amounts of certain organic compounds (e.g., acetanilide) added as stabilizers.

Special Instructions: Avoid ingestion, contamination of mucous membranes, or prolonged contact with skin. This preparation will deteriorate with exposure to heat or bright light, as well as with age. Store the solution in a cool, dark place and observe any expiration dates stamped on the bottle. This solution is readily available commercially from a wide variety of sources.

Use: This solution is used as the substrate in tests of cultures for catalase activity and in the procedure to test the factors that make an antimicrobial agent effective.

Used in exercises 22A, 25B, 53, 57B, 59A, 68A.

Reagent: Isopropyl alcohol, 70%, v/v

Composition:

Isopropyl alcohol	700.0	ml
Distilled water	300.0	ml

Special Instructions: Poisonous if ingested.

Use: This solution is used in the study of the factors that make an antimicrobial agent effective. It is also used as an antiseptic in the exercise on blood grouping.

Used in exercises 22A, 70.

Reagent: Kovac's reagent

Composition:

Para-dimethylaminobenzaldehyde	50.0	g
Amyl (or butyl) alcohol	750.0	ml
Hydrochloric acid, concentrated	250.0	ml

Dissolve the aldehyde in the alcohol. Then add the acid to the alcoholic solution.

Special Instructions: Prepare reagent in a well ventilated chemical hood. Avoid direct contact with solution or its ingredients.

Use: This reagent is used in sulfide-indole-motility (SIM) agar deeps for evidence of indole formation from the amino acid tryptophan (tryptophanase activity). When added as an overlay, the development of a pink to red color in the top layer is indicative of the presence of indole as a waste product. This test may be performed on any culture that has been grown in a rich animal peptone-containing medium (e.g., tryptone broth).

Used in exercises 26A, 72A, 72B.

Reagent: Lactose, 10%, w/v, sterile

Composition:

Lactose	100.0	g
Distilled water	1000.0	ml

Filter to sterilize.

Use: This sterile solution is used in the exercise on the activation of the lactose operon (an inducible enzyme system).

Used in exercise 38A.

Reagent: Lysis mixture, sterile

Composition:

NaCl	0.1	M
Sodium citrate	0.015	M
Sodium dodecylsulfate	0.05	%
Distilled water		

Filter to sterilize.

Use: This sterile lysis mixture is used in the exercise on bacterial transformation.

Used in exercise 39.

Reagent: Lysol® deodorizing cleaner (Available in stores)

Composition:

Active ingredients:

Alkyl (50% C14, 40% C12, 10% C16) dimethyl benzyl ammonium chloride	2.7	%
Ethyl alcohol	0.34	%
Tetrasodium ethylenediamine tetraacetate	0.13	%
Inert ingredients	96.83	%

Use: This solution is used in the study of the factors that make an antimicrobial agent effective.

Used in exercise 22A.

Reagent: Lysol® disinfectant (Available in stores)

Composition:

Active ingredients:

Soap	16.5	%
Ortho-phenylphenol	2.8	%
Ortho-benzyl-para-cholophenol	2.7	%
Ethyl alcohol	1.8	%
Xylenols	1.5	%
Isopropyl alcohol	0.9	%
Tetrasodium ethylenediamine tetraacetate	0.7	%
Inert ingredients	73.1	%

Special Instructions: Handle with care. Avoid ingestion or prolonged contact with this disinfectant.

Use: This solution is used in the study of the factors that make an antimicrobial agent effective and as the "unknown" antimicrobial agent in the determination of the phenol coefficient.

Used in exercises 22A, 22B.

Reagent: Mineral oil, sterile

Composition:

Mineral oil	chemically pure

Autoclave to sterilize.

Use: This reagent is used aseptically as an overlay to exclude air (oxygen) from certain types of cultures (such as Moeller's amino acid decarboxylase media) during incubation.

Used in exercises 26C, 72B.

Reagent: Nessler's reagent

Composition:

Solution A:

Potassium iodide	35.0	g
Mercuric iodide	50.0	g
Distilled water	200.0	ml

Prepare this solution in a 500.0 ml volumetric flask.

Solution B:

Sodium hydroxide	50.0	g
Distilled water	250.0	ml

Dissolve and cool; add solution B to solution A with continuous shaking. Add distilled water to bring the volume to 500.0 ml.

Allow fine reddish-brown precipitate to settle for one week. Decant the clear supernatant fluid and store the reagent in dark, well stoppered bottles.

Special Instructions: Prepare in a well ventilated chemical hood. Avoid contamination with components as well as final reagent. Wear gloves, protective clothing and goggles.

Use: This reagent is used for the detection of free ammonium ions in peptone broth cultures as a sign of ammonification. When several drops of Nessler's reagent are mixed with several drops of culture in a spot plate, the development of a distinct deeper yellow or orange color, than the color that develops when a sterile control is tested, indicates that the organism is showing ammonifying activity. This reagent is also used to test for residual ammonium ions the test for nitrification of ammonium broth.

Used in exercises 47A, 47B, 47C.

Reagent: ONPG (ortho-nitrophenyl-beta-D-galactoside), 1%, w/v

Composition:

ONPG	10.0	g
Distilled water	1000.0	ml

Filter to sterilize.

Use: This sterile, colorless solution is used in the exercise on the activation of the lactose operon (an inducible enzyme system).

Used in exercise 38A.

Reagent: Oxidase reagent

Composition:

Tetramethyl-para-phenylenediamine dihydrochloride (or dimethyl-para-phenylenediamine hydrochloride)	0.1	g
Distilled water	10.0	ml

Solution is unstable and chemical is spontaneously oxidized by exposure to air (oxygen). Avoid bubbling the solution and store in the refrigerator.

Special Instructions: The tetramethyl reagent is more sensitive, less toxic, and less stable than the dimethyl reagent. The reagent should be prepared fresh or stored in the refrigerator for no longer than 1 week. The dried chemical in pure form should be crystalline white and exhibit little or no signs of discoloration.

Use: This reagent forms the substrate in the test of cultures for the respiratory enzyme cytochrome oxidase. If a significant amount of this enzyme activity is present, it will oxidize the colorless substrate in oxidase reagent (dimethyl- or tetramethyl-para-phenylenediamine) to form a purple colored product, usually within 1 minute or less (Kovacs method, using a paper strip moistened with the reagent and rubbing the cells in with a platinum loop, is read at 10 seconds).

Used in exercises 25A, 72A, 72B.

Reagent: Phenol, 5.0%, w/v

Composition:

Phenol crystals, reagent grade*	5.0	g
Distilled water	100.0	ml

*Liquified phenol (89–90%, w/v) may be used in place of phenol crystals to achieve the same final concentration.

Special Instructions: Use nonporous gloves when handling phenol crystals or concentrated solutions of phenol! In concentrated form, this chemical is capable of causing a severe chemical burn if it is allowed to penetrate human tissues. If skin contact is made with the chemical it is helpful to immediately rinse the affected area with 95% ethanol, followed immediately by rinsing with copious amounts of water.

Use: This reagent is used in the experimental determination of the phenol coefficient for other antimicrobial agents.

Used in exercise 22B.

Reagent: Phenol-normal saline solution

Composition:

NaCl	8.5	g
Phenol, crystalline*	5.0	g
Distilled water	1000.0	ml

*Alternatively, one may add liquified phenol (ca. 85–90% phenol, w/v, in water) to achieve a final phenol concentration of 0.5%. In either case, handle concentrated phenol preparations with nonabsorbent gloves because of their ability to cause serious burns of human tissue.

Special Instructions: Handle concentrated phenol preparations with gloves.

Use: Phenol-normal saline solution is used for the preparation of killed suspensions of bacteria to be used as antigens in slide agglutination tests. Killed suspensions of a wide variety of Gram-negative bacterial species and serotypes that are suitable for slide agglutination tests are readily available from several commercial sources.

Used in exercise 71A.

Reagent: Phosphate buffered saline (PBS), pH 7.2

Composition:
Solution A:

NaCl	8.5	g
$NaH_2PO_4 \cdot H_2O$	1.38	g
Distilled water	1000.0	ml

Solution B:

NaCl	8.5	g
Na_2HPO_4	1.42	g
Distilled water	1000.0	ml

To prepare PBS, mix solutions in following proportions:

Solution A	280.0	ml
Solution B	720.0	ml

Check pH and readjust if necessary. PBS may be autoclaved to sterilize.

Use: This solution is used as a rinse and as part of the mounting medium in the exercise on the direct immunofluorescence technique.

Used in exercise 71C.

Reagent: Potassium carbonate, 2%, w/v, sterile

Composition:

Potassium carbonate	20.0	g
Distilled water	1000.0	ml

Filter to sterilize.

Use: This sterile solution is used in the exercise on the activation of the lactose operon (an inducible enzyme system).

Used in exercise 38A.

Reagent: Potassium hydroxide (40%, w/v)-creatine (0.3%, w/v)*

Composition:

KOH	40.0	g
Creatine	0.3	g
Distilled water	100.0	ml

Dissolve the potassium hydroxide in the distilled water carefully, with stirring and cooling. Then dissolve the creatine.

*In exercise 72A, development of the VP reaction in the Enterotube II system specifies the use of a 20% (w/v) potassium hydroxide-0.3% (w/v) creatine reagent.

Special Instructions: Avoid direct contact with potassium hydroxide. See additional comments under "Sodium hydroxide, 1.0 N" in this section.

Use: This reagent is used in the Voges-Proskauer (VP) test to detect the fermentation waste product acetylmethylcarbinol (acetoin), to alkalinize the final MRVP broth (glucose-peptone broth) culture. Under alkaline conditions acetoin is readily oxidized to diacetyl, which is detected by the development of a pink to red colored complex. This reaction develops more rapidly if creatine is added to the KOH and a second reagent, alpha-naphthol is used.

Used in exercises 24C, 72A, 72B.

Reagent: Pyrogallic acid crystals

Composition: Also known as pyrogallol. Reagent grade chemical in crystalline form. White, odorous crystals darken to a gray upon exposure to air and light. Keep in a well closed dark container.

Special Instructions: Poisonous! Use with caution only in a well ventilated, chemical hood and handle with nonabsorbent gloves. Avoid ingestion, inhalation or skin contact.

Use: Used with 4% NaOH solution to generate a low oxygen space above an inoculated agar deep to promote growth of anaerobes.

Used in exercise 19D.

Reagent: Rapid Plasma Reagin (RPR) Card Test Kit

Composition: Commercially available test kit for screening human sera for reaginic antibodies as a presumptive finding indicative of syphilis.

Special Instructions: This kit is commercially available from Hynson, Wescott, and Dunning, Baltimore, Md, 21201, a Division of Becton, Dickinson and Co. Store at 4°C. Observe expiration date on kit.

Use: This kit is used in the exercise on the detection of reaginic antibodies by the screening procedure known as the RPR Card Test.

Used in exercise 71D.

Reagent: Saline (0.9% NaCl, w/v)

Composition:

NaCl	9.0	g
Distilled water	1000.0	ml

Dispense in known volumes to tubes/flasks, add closures and autoclave to sterilize.

Use: Saline (0.9% NaCl solution) is used in known volumes as the diluent in procedures for the enumeration of microorganisms by direct microcounts (DMC), by standard plate counts (SPC—also known as the "viable count"), and by optical density (OD, also known as "turbidimetric") measurements. It is also used in small volumes in procedures where moistening of a sterile item or a specimen with a nontoxic, aqueous, sterile fluid is necessary.

Used in exercises 13, 14A, 14B, 15A, 15B, 21A, 32, 33, 34, 35, 43, 44, 46A, 46B, 55A, 55B, 55C, 55D, 56, 57A, 57B, 58, 59A, 60A, 62, 64C, 64D, 68B, 71B, 72B.

Reagent: Saline (3% NaCl, w/v)

Composition:

NaCl	30.0	g
Distilled water	1000.0	ml

Dispense in small quantities (1.0 to 2.0 ml) to screw capped tubes and autoclave to sterilize.

Use: This reagent is used in the experiment to isolate luminescent bacteria from marine specimens.

Used in exercise 49.

Reagent: Serum, anti-*Salmonella* O, polyvalent

Composition: This diagnostic reagent consists of a pool of antibodies specific for *Salmonella* O antigenic determinants typical of the commonly encountered serogroups, including groups A, B, C, D, and E. Antibody populations specific for other *Salmonella* antigens of diagnostic or epidemiologic significance are also present; the specificities present depend upon the commercial source from which this reagent is obtained. Consult the circular which usually accompanies this product.

Special Instructions: This antiserum is commercially available from Difco Labs as cat. no. 2264-47. Suitable substitute antiserum preparations are available from a variety of other commercial sources. Store at 4°C. Observe specified expiration date.

Use: This antiserum is used in the exercise to illustrate the slide agglutination test.

Used in exercise 71A.

Reagent: Serum, goat, anti-bovine serum albumin

Composition: Goat serum, demonstrated to contain specific antibodies for bovine serum albumin, detectable by the precipitin test.

Special Instructions: Commercially available from a variety of sources that specialize in the preparation of serological reagents.

Handle aseptically, the serum has been presterilized by the commercial vendor. Store at 4°C. Observe any expiration date specified.

Use: This antiserum is used in the exercise on the ring precipitin test.

Used in exercise 71B.

Reagent: Serum, goat, normal

Composition: Goat serum, demonstrated to lack detectable antibodies reactive to a particular antigen (here, bovine serum albumin) in serologic tests (here, the precipitin test)

Special Instructions: Commercially available from a variety of sources that specialize in the preparation of serological reagents. Handle aseptically; the serum has been sterilized by the commercial vendor. Store at 4°C.

Use: This serum is used as a negative control reagent in the exercise on the ring precipitin test.

Used in exercise 71B.

Reagent: Serum, human, Nonreactive (−)

Composition: Human serum, tested to be free of reaginic antibodies, nonreactive in the RPR Card Test.

Special Instructions: This serum is available commercially from several sources that specialize in the preparation of diagnostic serologic reagents. Store at − 20°C.

Use: This serum is used as a negative control in the exercise on the use of the RPR Card Test to screen for reaginic antibodies as a presumptive finding suggestive of syphilis.

Used in exercise 71D.

Reagent: Serum, human, Reactive (4 +)

Composition: Human serum, demonstrated to give a strong (4 +) reaction for reaginic antibodies in the RPR Card Test.

Special Instructions: This material is commercially available from several sources that specialize in the preparation of diagnostic serologic reagents. Store at − 20°C.

Use: This serum is used as a positive control in the exercise on the use of the RPR Card Test to screen for reaginic antibodies as a presumptive finding for syphilis.

Used in exercise 71D.

Reagent: Skim milk, sterile

Composition: Tube skim milk in 3.0 ml amounts, add closure and autoclave at 116°C for 20 minutes to sterilize.

Use: Used to provide casein for skim milk agar plates in the test for casein hydrolysis. Prewarm one tube of sterile skim milk, for each plate, to 50–55°C. Aseptically mix 3.0 ml of sterile skim milk with 20.0 ml of sterile molten (cooled to 50–55°C) plate count agar in each plate and allow to solidify.

Used in exercises 23B, 55B.

Reagent: Sodium hydroxide, 1.0 N

Composition:

NaOH	40.00	g
Distilled water, make up to	1000.0	ml

Dissolve the sodium hydroxide in about 100 ml of water. Cool the solution, add it to a 1 liter volumetric flask and bring the volume to the mark with distilled water. Mix this reagent thoroughly and store it in a polyethylene bottle.

Special Instructions: Avoid direct contact with sodium hydroxide pellets or concentrated solutions. Sodium hydroxide pellets are very hygroscopic, so weigh them quickly. Prepare this reagent in a well-ventilated chemical hood. Cool the solution during preparation, since a considerable amount of heat develops. Store the reagent in a cool area, away from containers of strong mineral acid reagents (particularly volatile acids such as concentrated hydrochloric acid).

Use: This alkaline solution is used to adjust the pH of trypticase soy broth (TSB) to a wide range of levels for the tests of the effects of environmental pH on microbial growth. This reagent is also used with pyrogallic acid to create a low oxygen zone above inoculated agar deep cultures so as to encourage growth of anaerobes. Finally, this reagent is used in the modified Whiteside test to alkalinize raw milk and permit its analysis for signs (and extent) of mastitis.

Used in exercises 17, 19D, 61B.

Reagent: Sodium hydroxide, 0.1N

Composition:
NaOH, 1.0 N	100.0	ml
Distilled water	900.0	ml

Mix thoroughly and store this reagent in a polyethylene bottle.

Special Instructions: Avoid direct contact with sodium hydroxide pellets or concentrated solutions.

Use: Used in the titration of acid production, as lactic acid, in sauerkraut fermentation.

Used in exercise 59A.

Reagent: Sulfamic acid

Composition:
Sulfamic acid	4.0	g
Distilled water	80.0	ml
Sulfuric acid, concentrated	20.0	ml

Carefully add the sulfuric acid to the water with continual mixing. Dissolve the sulfamic acid in the diluted sulfuric acid. Store at room temperature.

Special Instructions: Prepare in a well ventilated chemical hood. Avoid contact with concentrated acid solutions. Remember to add concentrated acid to water, not water to acid!

Use: This reagent is a substitute, for sulfanilic acid plus alpha-naphthylamine (or N,N-Dimethyl-1-naphthylamine), in the test to detect the formation of nitrate ions. When several drops are mixed with a grown nitrate broth culture, a positive finding of nitrite ions is indicated by the appearance (within about 5 minutes) of small bubbles rising from the depth of the culture.

Used in exercises 25C, 47C.

Reagent: Sulfanilic acid

Composition:
Prepare the solvent first (acetic acid, 5 N):
Glacial acetic acid	400.0	ml
Distilled water	1000.0	ml

Remember to slowly and carefully add concentrated acids to water (not water to acid!) with continual stirring.

Next, prepare the reagent:
Sulfanilic acid	8.0	g
Acetic acid, 5 N	1000.0	ml

Special Instructions: Prepare dilute acetic acid in a well ventilated chemical hood. Avoid inhalation of the corrosive fumes of glacial acetic acid.

Use: This reagent is used in a test to detect the presence of nitrite ions formed by microbial reduction of nitrate ions. When added with an equal volume of N,N-Dimethyl-1-naphthylamine to nitrate broth/agar cultures, a red color will develop within a few minutes if nitrite ions have been produced. The same reagents may be used to detect whether or not a culture has utilized nitrite ions in a nitrite agar/broth.

Used in exercises 25C, 72B.

Reagent: Sulfuric acid, 1:3, v/v

Composition:
Sulfuric acid, concentrated reagent	100.0	ml
Distilled water	200.0	ml

Carefully and slowly add the concentrated sulfuric acid to the distilled water with stirring.

Special Instructions: Prepare reagent in a well ventilated chemical hood. Wear protective clothing, gloves and goggles. Avoid physical contact and wash any accidentally affected areas immediately with copious amounts of tap water. Remember to add concentrated acid slowly to water with stirring, not water to acid!

Use: This reagent is mixed with Trommsdorf's reagent and ammonium broth culture in a spot plate to detect free nitrite ions as a sign of nitrifying activity.

Used in exercise 47B.

Reagent: Sulfuric acid, 1:1,000, v/v

Composition:
Sulfuric acid, concentrated, reagent grade	0.1	ml
Distilled water	100.0	ml

Special Instructions: Exercise caution in the use of concentrated sulfuric acid to prepare this reagent. It should be prepared in a well ventilated chemical hood and direct contact with concentrated sulfuric acid should be avoided.

Use: This dilute acidic solution may be used as the decolorizer in the stain procedure (Gohar, 1944) for the detection of metachromatic granules (volutin).

Used in exercises 11A, 12.

Reagent: TMG (thiomethylgalactoside), 10%, w/v, sterile

Composition:
TMG	100.0	g
Distilled water	1000.0	ml

Filter to sterilize.

Use: This sterile solution is used in the exercise on the activation of the lactose operon (an inducible enzyme system).

Used in exercise 38A.

Reagent: Toluene

Composition: Chemically pure toluene (also known as methyl-benzene) is a flammable, volatile, liquid with a benzene like odor and is only slightly soluble in water. Store only in a well closed container in a cool area away from flames.

Special Instructions: Poisonous, highly flammable member of the benzene family. Use with caution, only in a well-ventilated chemical hood. Avoid inhalation, ingestion, or contact with skin. Toluene ingestion has also caused macrocytic anemia; material is narcotic in high concentrations. Avoid exposure of chemical or its fumes to open flames or spark-producing devices.

Use: This organic fluid is used in the exercise on the activation of the lactose operon (an inducible enzyme system).

Used in exercise 38A.

Reagent: Trommsdorf's reagent

Composition:

Zinc chloride	20.0	g
Distilled water	100.0	ml

Boil to dissolve and add boiling solution to:

Starch	4.0	g
Distilled water	100.0	ml

Add 200 ml distilled water; then add:

Zinc iodide	2.0	g

Add distilled water to make 1000.0 ml of reagent; filter and store in tightly stoppered bottles.

Use: This reagent is used to test ammonium broth cultures for the appearance of nitrite ions, signifying nitrifying activity. When a drop of this reagent is mixed with a drop of sulfuric acid (1:3, v/v) and a drop of culture in a spot plate, the appearance of a blue-black color is indicative of nitrite ions.

Used in exercise 47B.

Reagent: Xylene

Composition: Xylene of commerce is commonly a mixture of the three isomers, ortho-, meta-, and para-xylene, with meta- predominating.

Special Instructions: Use only in a well ventilated chemical hood. Avoid contact with skin, ingestion or inhalation of its vapors. Although the extent of its toxicity is not well defined, it has been reported to be narcotic in high doses, a cause of photodermatitis in some individuals, and may be occasionally associated with chronic kidney damage. Since xylene is more dense than water, waste xylene should not be disposed of down sink drains, but to a special container.

Use: This organic solvent is used as the second reagent (or "de-colorizer") in the procedure to detect poly-beta-hydroxybutyric acid (PHB) inclusions.

Used in exercises 11B, 12.

Reagent: Zinc dust

Composition: Zinc, pure, powdered.

Special Instructions: Avoid inhalation of dust.

Use: This material is used to test for the persistence of nitrate ions in nitrate broth cultures which have been observed not to show a large gas bubble in the Durham tube (if present, this is evidence of denitrification-reduction of nitrate to gaseous end products such as nitrogen gas) or signs of nitrate reduction to nitrite. If nitrate ions are present in such a culture (or in the necessary sterile broth control—as they should be), gas bubbles will be observed after adding Zn.

Used in exercises 25C, 47C.

REPORT FOR EXERCISE 1. USE AND CARE OF THE MICROSCOPE

Name: _____ Date: _____ Lab Section: _____

A. Using the Bright Field Light Microscope

Bacteria

Protozoans

Algae

Fungi

Microscopic animals

B. Using the Oil Immersion Lens Directly

Bacteria

Protozoans

Algae

Fungi

Microscopic animals

REPORT FOR EXERCISE 2. MEASURING CELLS USING THE MICROSCOPE

Name: _____   Date: _____   Lab Section: _____

Report:

REPORT FOR EXERCISE 3. OBSERVATION OF LIVING MICROORGANISMS

Name: _____ Date: _____ Lab Section: _____

A. Making a Wet Mount

Report:

B. Making a Hanging Drop Mount

Report:

REPORT FOR EXERCISE 4. SPECIALIZED LIGHT MICROSCOPES

Name: _____ Date: _____ Lab Section: _____

A. Dark Field Light Microscope

Report:

B. Phase Contrast Light Microscope

Report:

C. Fluorescence Microscope

Report:

REPORT FOR EXERCISE 5. PREPARATION OF CULTURE MEDIA

Name: _____ Date: _____ Lab Section: _____

REPORT FOR EXERCISE 6. MICROORGANISMS IN THE LABORATORY ENVIRONMENT

Draw the shapes of some of the colonies appearing on your plates. Indicate the source of microorganisms.

SOURCE OF MICROORGANISMS	NUMBER OF COLONIES ON PLATE	DESCRIPTION OF COLONIES

REPORT FOR EXERCISE 7. PURE CULTURE TECHNIQUES

Name: _____ Date: _____ Lab Section: _____

A. Procedure for Making a Streak Plate

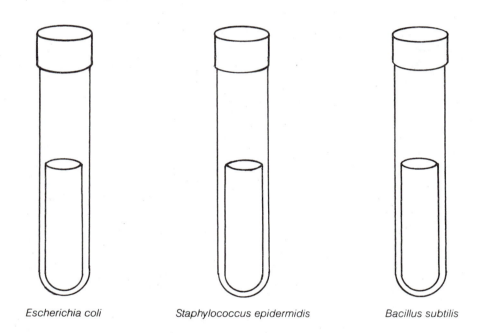

Escherichia coli   Staphylococcus epidermidis   Bacillus subtilis

B. Procedure for Subculturing a Pure Culture into Broth

Escherichia coli   Staphylococcus epidermidis   Bacillus subtilis

ORGANISM TESTED	GROWTH PATTERN SEEN
Escherichia coli	
Staphylococcus epidermidis	
Bacillus subtilis	

C. Procedure for Subculturing a Pure Culture onto an Agar Slant

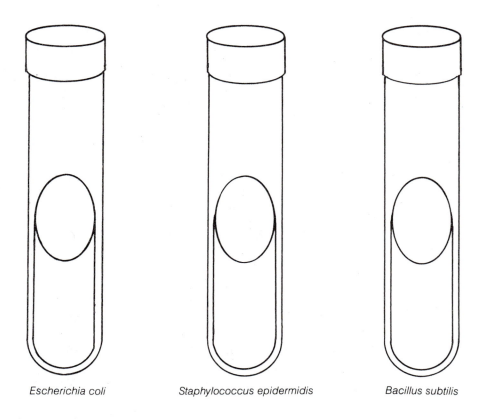

Escherichia coli   *Staphylococcus epidermidis*   *Bacillus subtilis*

ORGANISM TESTED	GROWTH PATTERN SEEN
Escherichia coli	
Staphylococcus epidermidis	
Bacillus subtilis	

D. Procedure for Aseptic Transfers Using Pipets

Report:

E. Selective and Differential Media

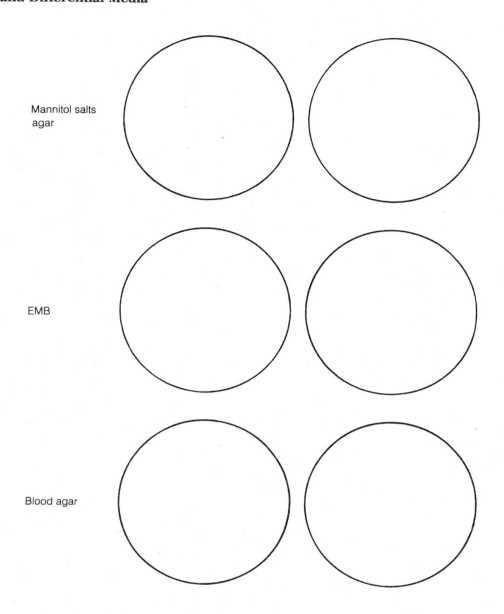

Mannitol salts
agar

EMB

Blood agar

ORGANISM	APPEARANCE ON CULTURE MEDIUM		
	MANNITOL-SALTS	EMB	BLOOD AGAR
Staphylococcus aureus			
Staphylococcus epidermidis			
Escherichia coli			
Proteus vulgaris			
Streptococcus lactis			
Streptococcus mitis			
Streptococcus faecalis			

REPORT FOR EXERCISE 8. POSITIVE AND NEGATIVE STAINING

Name: _____ Date: _____ Lab Section: _____

A. Procedure for Positive Staining

Report:

B. Procedure for Negative Staining

Report:

Name: _____ Date: _____ Lab Section: _____

A. Gram Stain

Report:

B. Acid-Fast Stain

Report:

Name: _____   Date: _____   Lab Section: _____

A. Capsule Stain

Report:

B. Endospore Stain

Report:

C. Flagella Stain

Report:

Name: _____ Date: _____ Lab Section: _____

A. Metachromatic (Volutin) Granules

Report:

B. Poly-β-Hydroxybutyric Acid (PHB)

Report:

C. Starch and Glycogen

Report:

REPORT FOR EXERCISE 12. MORPHOLOGICAL UNKNOWN

Name: _____ Date: _____ Lab Section: _____

REPORT FOR EXERCISE 13. DIRECT MICROSCOPIC COUNTS WITH A HEMOCYTOMETER

Name: _____ Date: _____ Lab Section: _____

DILUTION	CELLS/mm^2	CELLS/mm^3	CELLS/ml

REPORT FOR EXERCISE 14. VIABLE COUNTS

Name: _____ Date: _____ Lab Section: _____

A. Procedure for the Pour Plate Method

Type of sample: _____ Culture medium: _____

Temperature of incubation: _____

DILUTION OF SAMPLE	VOLUME PLATED	COLONIES PER PLATE	DILUTION FACTOR	CFU/ml OF SAMPLE
10^{-2}				
10^{-3}				
10^{-4}				

B. Procedure for the Spread Plate Method

DILUTION OF SAMPLE	VOLUME PLATED	COLONIES PER PLATE (CFU/0.1ml)	COLONIES/ml	DILUTION FACTOR	CFU/ml OF SAMPLE
10^{-1}	0.1 ml				
10^{-2}	0.1 ml				
10^{-3}	0.1 ml				

REPORT FOR EXERCISE 15. OPTICAL DENSITY MEASUREMENTS

Name: _____ Date: _____ Lab Section: _____

A. Procedure for Determining Population Size Using the Pour Plate Method

DILUTION PLATED	AMOUNT PLATED	CFU/PLATE 2	AVERAGE	CFU/ml OF SUSPENSION*
10^{-2}				
10^{-3}				
10^{-4}				

*Calculate results from average of plates with 30 to 300 colonies only.

B. Procedure for Determining the OD of the Culture

DILUTION	OD*	ESTIMATED CFU/ML (FROM EXERCISE 15-A)**
undiluted		
1:2		
1:4		
1:8		
1:16		
1:32		

*OD = optical density (absorbance)
**Estimate the number by multiplying the viable count by the dilution value (e.g., estimate = CFU/ml $\times$ ½).

C. Plotting the Data (O.D. vs C.F.U./ml)

ABSORBANCE (optical density)

CFU/ml

REPORT FOR EXERCISE 16. EFFECT OF TEMPERATURE ON MICROBIAL GROWTH

Name: _____ Date: _____ Lab Section: _____

ORGANISM	INCUBATION TEMPERATURE					
	4°C	18°C	37°C	45°C	55°C	60°C
Pseudomonas						
Escherichia						
Mycobacterium						
Bacillus						
Vibrio						

+ + = Heavy growth, + = Slight growth, − = No growth

REPORT FOR EXERCISE 17. EFFECT OF pH ON MICROBIAL GROWTH

Name: _____ Date: _____ Lab Section: _____

ORGANISM	pH									
	3	4	5	6	7	8	9	10	11	12
Lactobacillus										
Escherichia										
Pseudomonas										
Saccharomyces										
Penicillium										
Rhizopus										

+ + = Heavy growth, + = Slight growth, − = No growth

REPORT FOR EXERCISE 18. EFFECT OF OSMOTIC PRESSURE ON MICROBIAL GROWTH

Name: _____ Date: _____ Lab Section: _____

A. Effect of NaCl on Microorganisms

B. Effect of Sucrose on Microorganisms

ORGANISM	% NaCl							% SUCROSE					
	0.5	1	3	5	10	15	20	1	5	10	20	40	60
Escherichia													
Staphylococcus													
Saccharomyces													
Rhizopus													
Penicillium													

+ + = Heavy growth, + = Slight growth, − = No growth

REPORT FOR EXERCISE 19. EFFECT OF OXYGEN ON MICROBIAL GROWTH

Name: _____ Date: _____ Lab Section: _____

A. Growing Organisms in an Anaerobic GasPak Jar

ORGANISM	GASPAK JAR	NORMAL ATMOSPHERE
Micrococcus		
Clostridium		
Escherichia		
Streptococcus		
Neisseria		

+ + = Heavy growth, + = Slight growth, − = No growth

B. Growing Organisms in Agar Shake-Cultures

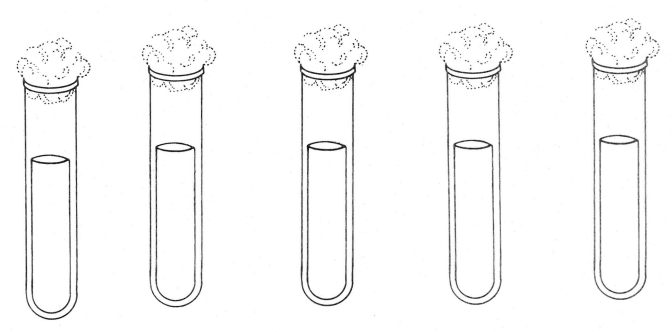

C. Growing Organisms in a Pyrogallic Acid-NaOH System

ORGANISM	PYROGALLIC ACID-NaOH	NORMAL ATMOSPHERE
Micrococcus		
Clostridium		
Escherichia		
Streptococcus		
Neisseria		

+ + = Heavy growth, + = Slight growth, − = No growth

D. Growing Organisms in a Candle Jar

ORGANISM	CANDLE JAR	NORMAL ATMOSPHERE
Micrococcus		
Clostridium		
Escherichia		
Streptococcus		
Neisseria		

+ + = Heavy growth, + = Slight growth, − = No growth

Name: _____ Date: _____ Lab Section: _____

3 M
NaCl

3 M
NaCl

Dark

Light

REPORT FOR EXERCISE 21. PHYSICAL METHODS OF CONTROL

Name: _____ Date: _____ Lab Section: _____

A. The Effectiveness of Hand Washing

HAND CONDITION	COLONIES PER PLATE			
	1st HAND	2nd HAND	3rd HAND	4th HAND
Contaminated				
Washed				

B. The Effectiveness of Heating

ORGANISM	UNINOCULATED CONTROL BROTH	BOILING (MINUTES)					
		1	5	10	15	20	30
Bacillus subtilis							
Escherichia coli							

+ + = Growth, + = Slight growth, − = No growth

C. The Effectiveness of Filtering

ORGANISM	UNINOCULATED CONTROL BROTH	PORE DIAMETER (μm)				
		10	5	1	0.45	0.3
Escherichia						

+ + = Growth, + = Slight growth, − = No growth

D. The Effectiveness of Ultraviolet Light

Colonies/plate

Time Irradiated (min)

REPORT FOR EXERCISE 22. CHEMICAL METHODS OF CONTROL

Name: _____ Date: _____ Lab Section: _____

A. Studying the Factors That Make an Antimicrobial Effective

SECOND PERIOD RESULTS

ORGANISM:		STAPHYLOCOCCUS EPIDERMIDIS														
TEMPERATURE:		0°C					18°C					45°C				
CHEMICAL CONCENTRATION:																
EXPOSURE TIME (MIN.):	0															
	5															
	10															
	15															
	20															

+ + = Growth, + = Slight growth, − = No growth

SECOND PERIOD RESULTS

ORGANISM:		BACILLUS SUBTILUS														
TEMPERATURE:		0°C					18°C					45°C				
CHEMICAL CONCENTRATION:																
EXPOSURE TIME (MIN.):	0															
	5															
	10															
	15															
	20															

+ + = Growth, + = Slight growth, − = No growth

Fill in the graphs below with (+) for growth or (−) for no growth.

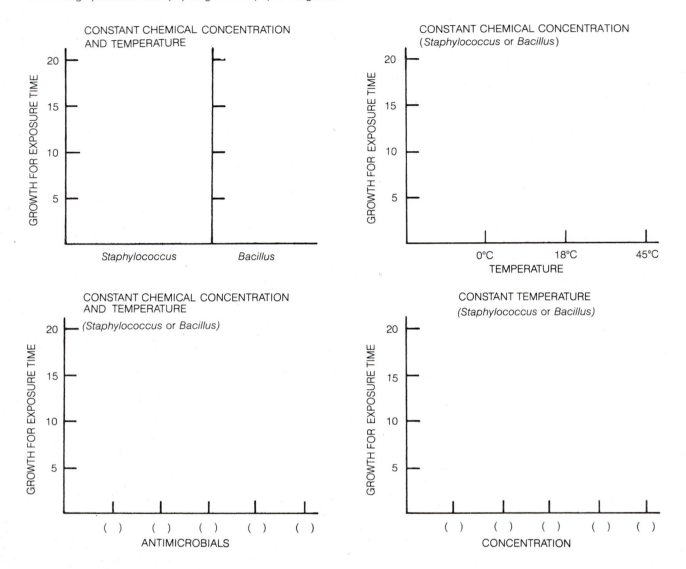

CONSTANT CHEMICAL CONCENTRATION AND TEMPERATURE

GROWTH FOR EXPOSURE TIME

20 15 10 5

Staphylococcus   Bacillus

CONSTANT CHEMICAL CONCENTRATION (Staphylococcus or Bacillus)

GROWTH FOR EXPOSURE TIME

20 15 10 5

0°C   18°C   45°C
TEMPERATURE

CONSTANT CHEMICAL CONCENTRATION AND TEMPERATURE
(Staphylococcus or Bacillus)

GROWTH FOR EXPOSURE TIME

20 15 10 5

( ) ( ) ( ) ( ) ( )
ANTIMICROBIALS

CONSTANT TEMPERATURE
(Staphylococcus or Bacillus)

GROWTH FOR EXPOSURE TIME

20 15 10 5

( ) ( ) ( ) ( ) ( )
CONCENTRATION

B. Determining the Phenol Coefficient

SECOND PERIOD RESULTS

ORGANISM:		STAPHYLOCOCCUS AUREUS												
CHEMICAL:		PHENOL				LYSOL DISINFECTANT								
DILUTION:		1:80	1:90	1:100	1:110	1:100	1:150	1:200	1:250	1:300	1:350	1:400	1:450	1:500
EXPOSURE TIME (MIN.):	5													
	10													
	15													

+ + = Growth, + = Slight growth, − = No growth

C. Testing the Bactericidal or Bacteriostatic Effect of Antibiotics and Drugs

SECOND PERIOD RESULTS

CHEMICAL:	SULFANILAMIDE (M)					
CONCENTRATION:	0.08	0.04	0.02	0.01	0.005	ZERO
GROWTH OF E. COLI						+ +

SECOND PERIOD RESULTS

CHEMICAL:	STREPTOMYCIN (µg/ml)					
CONCENTRATION:	400	200	100	50	25	ZERO
GROWTH OF E. COLI						+ +

THIRD PERIOD RESULTS

	BROTH FREE OF STREPTOMYCIN (µg/ml)					
CONCENTRATION OF CHEMICAL IN PREVIOUS BROTH:	400	200	100	50	25	ZERO
GROWTH OF E. COLI						+ +

+ + = Growth, + = Slight Growth, − = No Growth

SECOND PERIOD RESULTS

CHEMICAL:	PENICILLIN (UNITS/ml)					
CONCENTRATION:	200	100	50	25	10	ZERO
GROWTH OF E. COLI						+ +

THIRD PERIOD RESULTS

	BROTH FREE OF PENICILLIN (UNITS/ml)					
CONCENTRATION OF CHEMICAL IN PREVIOUS BROTH:	200	100	50	25	10	ZERO
GROWTH OF E. COLI						+ +

+ + = Growth, + = Slight Growth, − = No Growth

D. Determining the Selective Toxicity of Antibiotics and Drugs

SECOND PERIOD RESULTS

CHEMICALS	GROWTH	
	STAPHYLOCOCCUS AUREUS	SACCHAROMYCES CEREVISIAE
0.02M Sulfanilamide		
100 units/ml Penicillin		
10 units/ml Mycostatin		

+ + = Growth, + = Slight Growth, − = No Growth

REPORT FOR EXERCISE 23. HYDROLYSIS OF LARGE EXTRACELLULAR MOLECULES

Name: _____ Date: _____ Lab Section: _____

A. Hydrolysis of Polysaccharides

Results:

Escherichia coli

Bacillus subtilis

Organism:

Organism:

Organism:

Organism:

B. Hydrolysis of Proteins

Results:

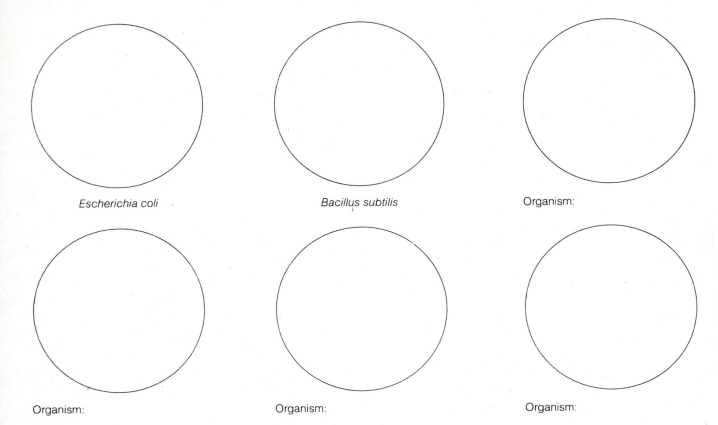

Escherichia coli

Bacillus subtilis

Organism:

Organism:

Organism:

Organism:

C. Hydrolysis of Nucleic Acids (DNA)

Results:

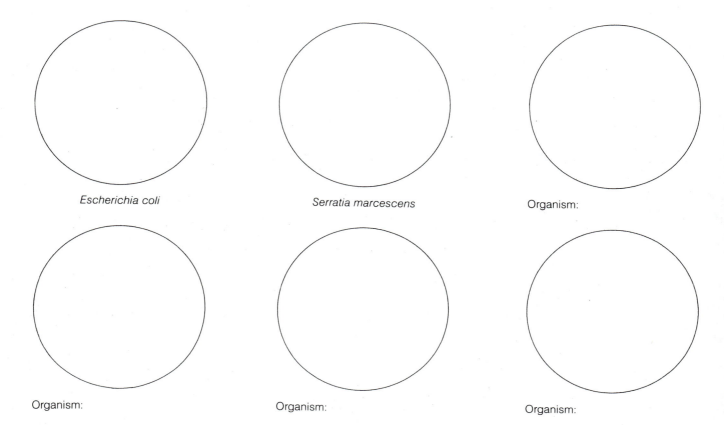

Escherichia coli

Serratia marcescens

Organism:

Organism:

Organism:

Organism:

D. Hydrolysis of Fats

Results:

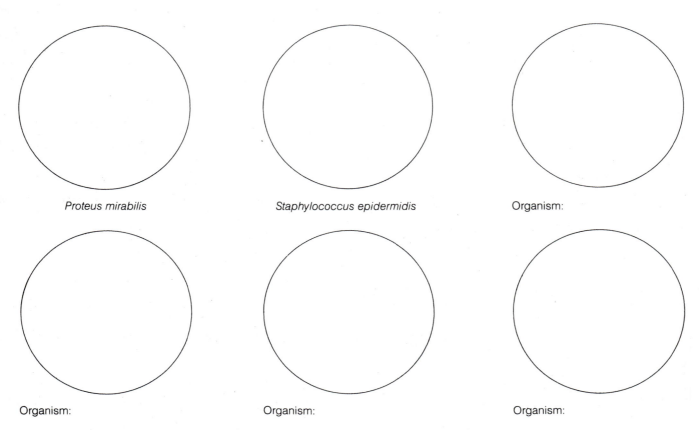

Proteus mirabilis	*Staphylococcus epidermidis*	Organism:
Organism:	Organism:	Organism:

ORGANISM TESTED	MACROMOLECULE			
	POLYSACCHARIDE	PROTEIN	DNA	FAT
Escherichia coli				
Bacillus subtilis				
Serratia marscescens				
Proteus mirabilis				
Staphylococcus epidermidis				

Name: _____   Date: _____   Lab Section: _____

A. Fermentation of Glucose, Lactose, Sucrose, Maltose, and Mannitol

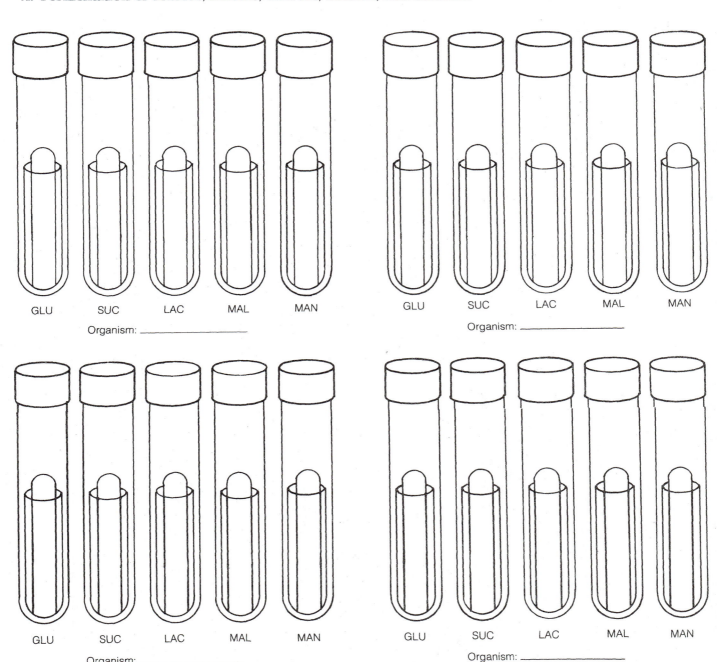

GLU   SUC   LAC   MAL   MAN

Organism: _____

ORGANISM TESTED	CARBOHYDRATE TESTED*				
	GLUCOSE	LACTOSE	SUCROSE	MALTOSE	MANNITOL
Escherichia coli					
Staphylococcus aureus					
Micrococcus luteus					
Saccharomyces cerevisiae					

*Record your results as follows: A = acid; G = gas; AG = acid and gas; B = alkaline reaction; + = no fermentation but growth; − = no growth.

B. Methyl Red Test

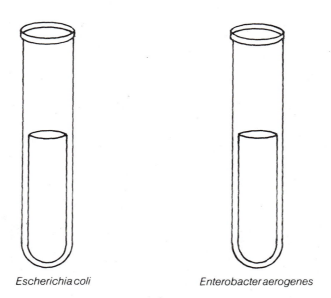

Escherichia coli   Enterobacter aerogenes

ORGANISM TESTED	COLOR AFTER ADDITION OF REAGENT
Escherichia coli	
Enterobacter aerogenes	

C. Voges-Proskauer Test

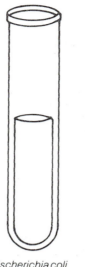

Escherichia coli

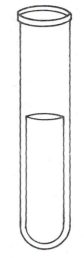

Enterobacter aerogenes

ORGANISM TESTED	COLOR AFTER ADDITION OF REAGENTS
Escherichia coli	
Enterobacter aerogenes	

Name: _____ Date: _____ Lab Section: _____

A. Oxidase Test

Escherichia coli

Pseudomonas fluorescens

ORGANISM TESTED	COLOR OF COLONY	OXIDASE REACTION
Pseudomonas fluorescens		
Escherichia coli		

B. Catalase Test

Staphylococcus epidermidis

Streptococcus faecalis

ORGANISM TESTED	COLONY APPEARANCE AFTER PEROXIDE ADDITION	CATALASE REACTION

C. Nitrate Reduction

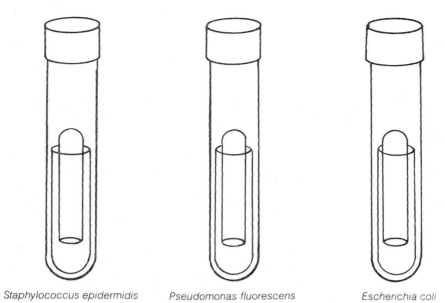

Staphylococcus epidermidis   Pseudomonas fluorescens   Escherichia coli

ORGANISM TESTED	APPEARANCE OF CULTURE AFTER α-NAPHTHYLAMINE	GAS IN DURHAM TUBE	INTERPRETATION OF RESULTS

REPORT FOR EXERCISE 26. UTILIZATION OF AMINO ACIDS

Name: _____ Date: _____ Lab Section: _____

A. Indole Production

Escherichia coli   Enterobacter aerogenes   Proteus vulgaris

ORGANISM TESTED	COLOR OF MEDIUM	COLOR OF LAYER	INTERPRETATION OF RESULTS
Escherichia coli			
Enterobacter aerogenes			
Proteus vulgaris			

B. Hydrogen Sulfide Production

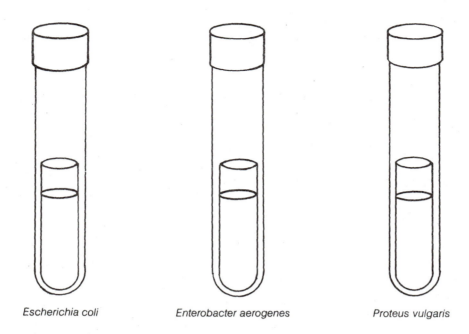

Escherichia coli   Enterobacter aerogenes   Proteus vulgaris

ORGANISM TESTED	COLOR OF MEDIUM BEFORE INOCULATION	COLOR OF MEDIUM FOLLOWING GROWTH	INTERPRETATION OF RESULTS
Escherichia coli			
Enterobacter aerogenes			
Proteus vulgaris			

C. Lysine Decarboxylase Test

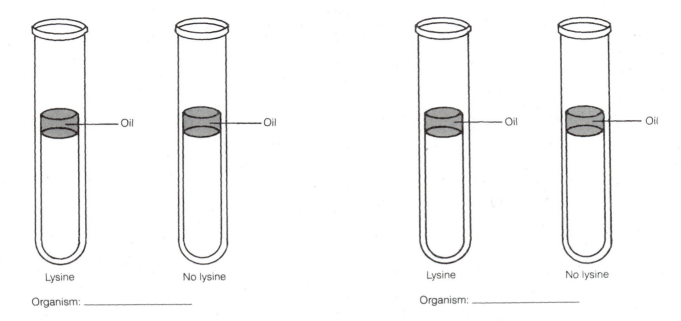

Organism: _____

Organism: _____

ORGANISM TESTED	COLOR OF LDC W/LYSINE	COLOR OF LDC W/O LYSINE	INTERPRETATION OF RESULTS
Enterobacter aerogenes			
Citrobacter freundii			

D. Phenylalanine Deamination

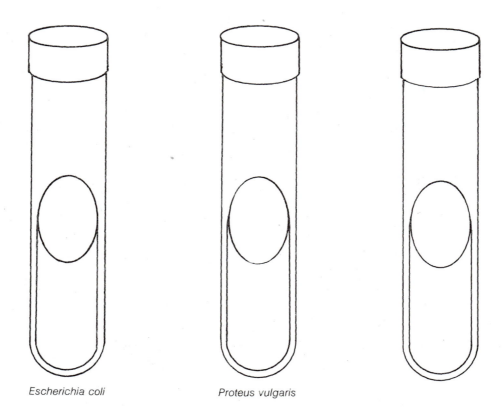

Escherichia coli   Proteus vulgaris

NAME OF ORGANISM	COLOR AFTER ADDING FERRIC CHLORIDE	INTERPRETATION OF RESULTS
Escherichia coli		
Proteus vulgaris		

REPORT FOR EXERCISE 27. UTILIZATION OF CITRATE, GELATIN, AND UREA

A. Citrate Utilization Test

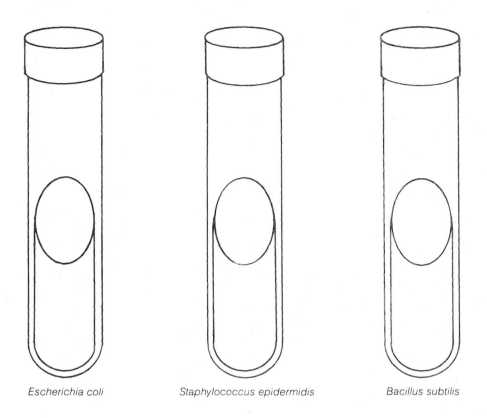

ORGANISM TESTED	COLOR OF AGAR FOLLOWING INCUBATION	INTERPRETATION OF RESULTS
Escherichia coli		
Staphylococcus epidermidis		
Bacillus subtilis		

B. Gelatin Hydrolysis

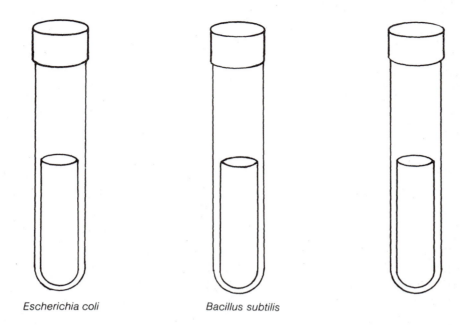

Escherichia coli   Bacillus subtilis

| ORGANISM TESTED | PHYSICAL STATE OF GELATIN | | INTERPRETATION OF RESULTS |
	BEFORE COOLING	AFTER COOLING	
Escherichia coli			
Bacillus subtilis			

C. Urea Hydrolysis

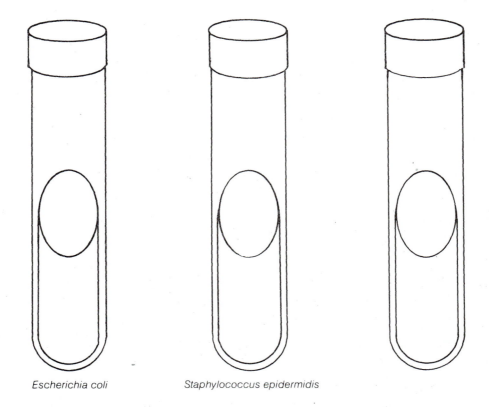

Escherichia coli   Staphylococcus epidermidis

ORGANISM TESTED	COLOR OF AGAR	INTERPRETATION OF RESULTS
Escherichia coli		
Staphylococcus epidermidis		

REPORT FOR EXERCISE 28. IDENTIFICATION OF AN UNKNOWN BACTERIUM

Name: _____   Date: _____   Lab Section: _____

TEST*	RESULTS		
	UNKNOWN		
Growth at 4°C (16)			
Growth at 20°C (16)			
Growth at 35°C (16)			
Growth at 45°C (16)			
Growth at 55°C (16)			
Growth at 60°C (16)			
Colony characteristics (6) form			
elevation			
margin			
consistency			
pigmentation			
Gram stain (9)			
Acid-fast stain (9)			
Cell morphology (3)			
Capsule (10)			
Endospores (10)			
Flagella stain (10)			
Motility (3)			
Glucose fermentation (24)†			
Lactose fermentation (24)			
Maltose fermentation (24)			
Sucrose fermentation (24)			
Mannitol fermentation (24)			

TEST*	RESULTS		
	UNKNOWN		
Methyl Red test (24)			
Voges-Proskauer test (24)			
Catalase test (25)			
Anaerobic growth (19)			
Oxidase test (25)			
Starch hydrolysis (23)			
Casein hydrolysis (23)			
DNA hydrolysis (23)			
Fat hydrolysis (23)			
Indole production (26)			
H_2S production (26)			
Nitrate reduction (25)			
Gelatin hydrolysis (27)			
Lysine decarboxylation (26)			
Phenylalanine deamination (26)			
Urea hydrolysis (27)			
Citrate utilization (27)			
Ammonification (47)			
Mannitol salts agar (7)			
Eosin Methylen blue agar (7)			
Blood agar (7)			

*Numbers in parenthesis indicate the number of the exercise in which the test is found.

†Report carbohydrate fermentation tests as follows: A = acid; G = gas; AG = acid and gas; 0 = no fermentation but growth; NG = no growth.

REPORT FOR EXERCISE 29. THE YEASTS AND MOLDS

Name: _____ Date: _____ Lab Section: _____

A. Examination of Cultures

ORGANISM EXAMINED	COLONY CHARACTERISTICS	MICROSCOPIC CHARACTERISTICS

B. Coverslip Preparation

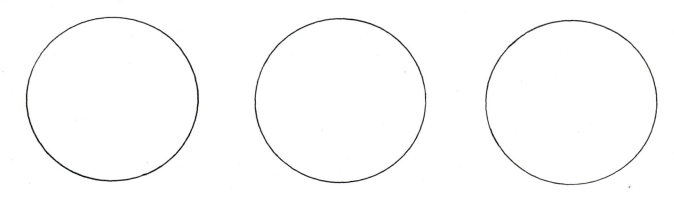

ORGANISM EXAMINED	MICROSCOPIC APPEARANCE	DESCRIPTION OF CHARACTERISTICS

C. Sexual Reproduction in *Rhizopus*

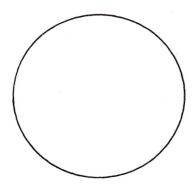

REPORT FOR EXERCISE 30. THE PROTOZOA

Name: _____ Date: _____ Lab Section: _____

A. Examination of Cultures

ORGANISM EXAMINED	SIZE	LOCOMOTION	REPRODUCTION	FEEDING	COMMENTS

B. Examination of Pond Water for Protozoa

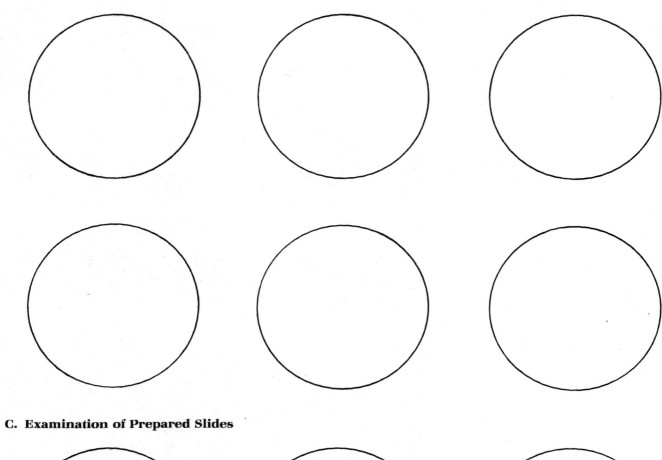

C. Examination of Prepared Slides

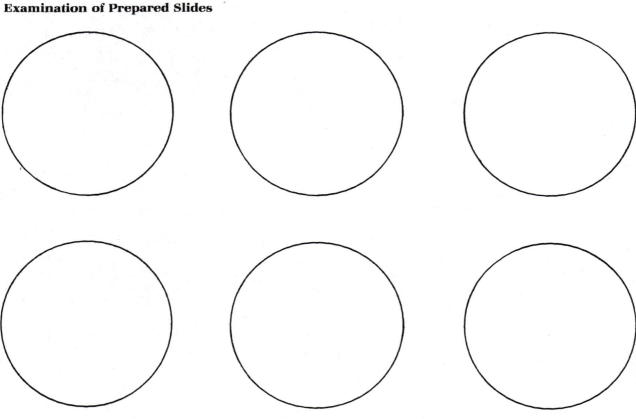

Name: _____ Date: _____ Lab Section: _____

A. Examination of Algal Cultures

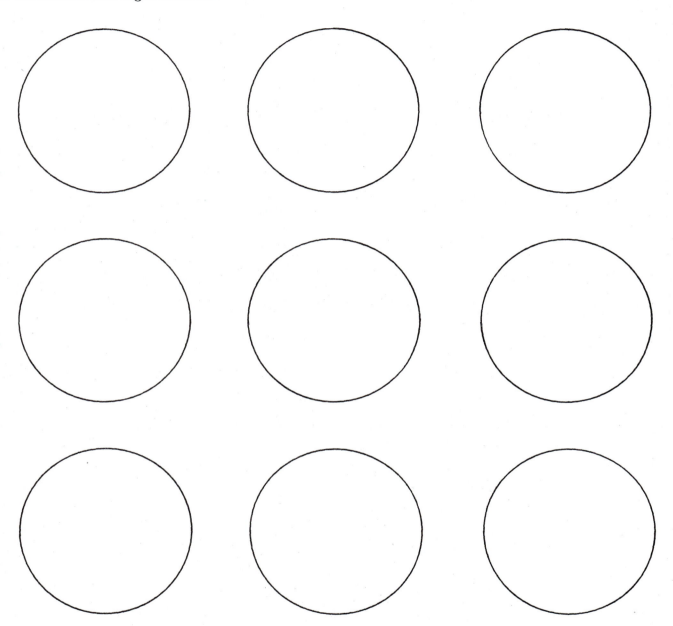

SPECIES OF ALGA	SHAPE	NUCLEUS	CHLOROPLAST	COLOR	MOTILITY	CELL WALL
Anabaena						
Oscillatoria						
Nostoc						
Spirogyra						
Chlamydomonas						
Euglena						
Scenedesmus						
Chlorella						
Ceratium						
Cyclotella						
Volvox						

B. Examination of Pond Water for Algae

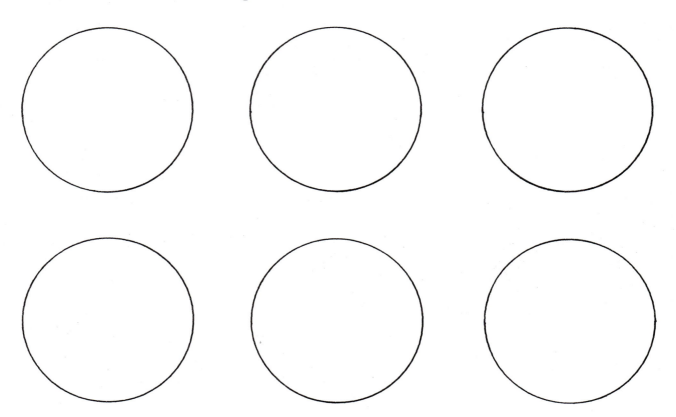

SAMPLE #	IDENTITY OF ORGANISM	CHARACTERISTICS NOTED
1		
2		
3		
4		
5		
6		
7		
8		

REPORT FOR EXERCISE 32. PHAGE ASSAY

Name: _____ Date: _____ Lab Section: _____

	DILUTION				
	10^{-5}	10^{-6}	10^{-7}	10^{-8}	10^{-9}
Plaques/0.1 ml					
Titer (PFU/ml)					

REPORT FOR EXERCISE 33. BACTERIOPHAGE ISOLATION

Name: _____ Date: _____ Lab Section: _____

	DILUTION*			
	3×10^{-1}	2×10^{-1}	10^{-1}	5×10^{-2}
PFU/ml				

*Take the number of plaques and divide by the dilution. For example, if there are 100 PFU on the plate where a 2×10^{-1} dilution was used, the number of PFU/ml of the phage solution is $100/2 \times 10^{-1} = 500$.

REPORT FOR EXERCISE 34. PLANT VIRUS ISOLATION

Name: _____ Date: _____ Lab Section: _____

LESIONS	DILUTION				
	10^0	10^{-1}	10^{-2}	10^{-3}	10^{-4}
LFU/leaf					
LFU/ml					

REPORT FOR EXERCISE 35. ONE-STEP GROWTH CURVE

Name: _____ Date: _____ Lab Section: _____

TIME (MINUTES) AFTER ADSORPTION	PLAQUE FORMING UNITS/ml*					
	DILUTIONS × REDUCTION					
	10^{-6}	10^{-7}	10^{-8}	10^{-9}	10^{-10}	10^{-11}
10						
15						
20						
25						
30						
35						
40						
45						

*Do not forget to account for the 0.1 ml used in the top agar. This represents another 10^{-1} reduction.

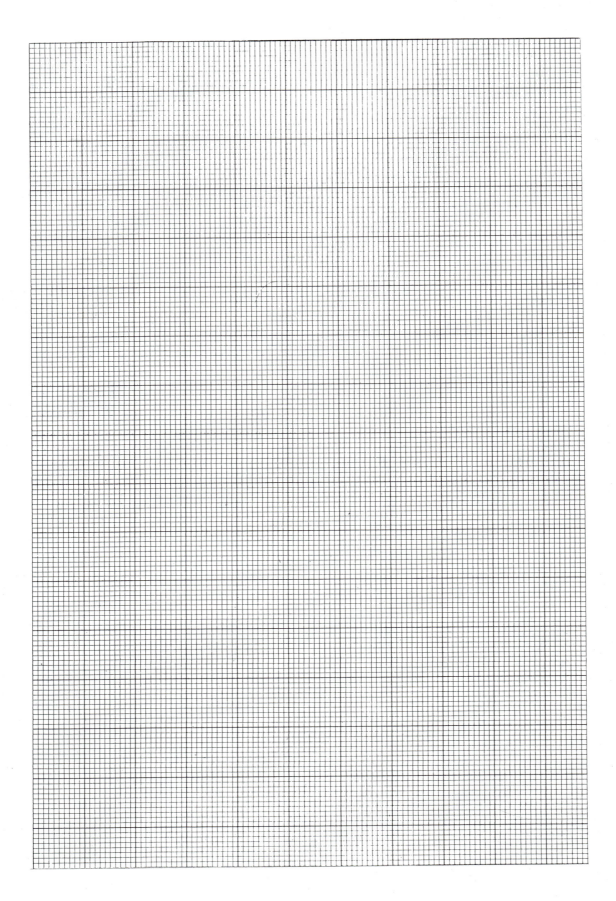

REPORT FOR EXERCISE 36. PHAGE TYPING OF *STAPHYLOCOCCUS AUREUS*

Name: _____   Date: _____   Lab Section: _____

S. Aureus	PHAGE TYPE					
	47	52A	71	81		
ATCC 27691						
ATCC 27693						
ATCC 27697						
Unknown						

+ = Lysis, − = No Lysis

Name: _____ Date: _____ Lab Section: _____

A. Illustrating Phenotypic Variation

SECOND PERIOD RESULTS		THIRD PERIOD RESULTS	
INCUBATION TEMP.	PIGMENTATION	INCUBATION TEMP.	PIGMENTATION
25°C		25°C	
		37°C	
37°C		25°C	
		37°C	

R = Red, C = Cream

B. Isolating a Mutant and Illustrating Genotypic Variation

SECOND PERIOD RESULTS		THIRD PERIOD RESULTS	
INCUBATION TEMP.	PIGMENTATION	INCUBATION TEMP.	PIGMENTATION
25°C	Pick Cream	25°C	
		37°C	
25°C	Pick Cream	25°C	
		37°C	
25°C	Pick Cream	25°C	
		37°C	
25°C	Pick Cream	25°C	
		37°C	
25°C	Pick Cream	25°C	
		37°C	

R = Red, C = Cream

C. Isolating Streptomycin-Resistant Mutants (Gradient-Plate Technique)

Second Period: Pattern of Growth

Third Period

Wild-Type   Mutant

D. Replica Plating

Fourth Period: Growth on replica plate with streptomycin

Fifth Period: Growth on replica plate with streptomycin

REPORT FOR EXERCISE 38. GENE REGULATION

Name: _____ Date: _____ Lab Section: _____

	INTENSITY OF THE COLOR*				
	MINUTES AFTER ADDING ONPG				
	15	20	25	30	35
Control					
LAC					
GLU					
GLU + cAMP					
LAC + GLU					
TMG					

*No color = −, Faint color = +, Intermediate intensity color = + +, Intense color = + + +

REPORT FOR EXERCISE 39. TRANSFORMATION

Name: _____ Date: _____ Lab Section: _____

SECOND PERIOD RESULTS

| | PLATE WITH LAWN | | | | CONTROL PLATE | | | |
| | LAWN + SPOTS | | | LAWN | SPOTS | | | CONTROL AREA |
	10^0	10^{-1}	10^{-2}	0.1 ml	10^0	10^{-1}	10^{-2}	—
Colonies								

THIRD PERIOD RESULTS

$$\frac{\text{number of colonies that grew}}{\text{number of colonies tested}} \times 100 =$$

Name: _____ Date: _____ Lab Section: _____

FOURTH PERIOD RESULTS

	MGTrp Plates			
	BACTERIAL CONTROLS	PHAGE CONTROLS	BACTERIA PLUS PHAGE DILUTIONS	
			10^0	10^{-1}
Cys$^+$				

FIFTH PERIOD RESULTS

	MG PLATES
Number of Trp$^+$Cys$^+$ Colonies	
Number of Cys$^+$ Colonies Picked	
Percentage of Trp$^+$Cys$^+$ Prototrophs	

REPORT FOR EXERCISE 41. CONJUGATION

Name: _____ Date: _____ Lab Section: _____

A. Mapping Mutations and Genes with F′ and F⁻ Mutants

SECOND AND THIRD PERIODS RESULTS

		F⁻ LAWN ON ML PLATES										
F′ SPOTS												

C = Complementation, R = Recombination, − = No Growth

B. Mapping Genes by Interrupting Matings between Hfr and F⁻ Mutants

SECOND PERIOD RESULTS

	PLATE																
MEDIA	MGlu + Str						MLac + Str + Leu						MGal + Str + Leu					
LOCUS	Leu						Lac						Gal					
TIME	10	15	20	25	30	40	10	15	20	25	30	40	10	15	20	25	30	40
COLONIES																		

REPORT FOR EXERCISE 42. AMES TEST

Name: _____ Date: _____ Lab Section: _____

SECOND PERIOD RESULTS

COLONY	CHEMICAL			
	CIGARETTE ASHES	MOTOR OILS	HAIR DYE	ETHYLENE OXIDE*
Distribution on Experimental Plate Count	◯	◯	◯	◯
Distribution on Control Plate Count	◯	◯	◯	◯

*Experimental plates sterilized with ethylene oxide, control plates (glass Petri plates) sterilized by autoclaving.

Name: _____ Date: _____ Lab Section: _____

TIME (MIN)	# VIABLE CELLS/ml	TURBIDITY (KLETT UNITS)
0		
30		
60		
90		
120		
150		
180		
210		
240		
270		
300		

Name: _____ Date: _____ Lab Section: _____

A. Competition

RESULTS: PURE CULTURE

TIME (MIN)	VIABLE *E. coli*/ml	VIABLE *S. aureus*/ml
0		
30		
60		
90		
120		
150		
180		
210		
240		
270		
300		

RESULTS: MIXED CULTURE

TIME (MIN)	VIABLE *E. coli*/ml	VIABLE *S. aureus*/ml
0		
30		
60		
90		
120		
150		
180		
210		
240		
270		
300		

B. Amensalism

Draw the results as you observed them on the respective plates of *E. coli* and *S. epidermidis* below.

Organisms in agar:

E. coli

S. epidermidis

Organisms streaked:

Pseudomonas streak

Bacillus streak

Penicillium streak

REPORT FOR EXERCISE 45. MICROBIAL SYNERGISM

Name: _____ Date: _____ Lab Section: _____

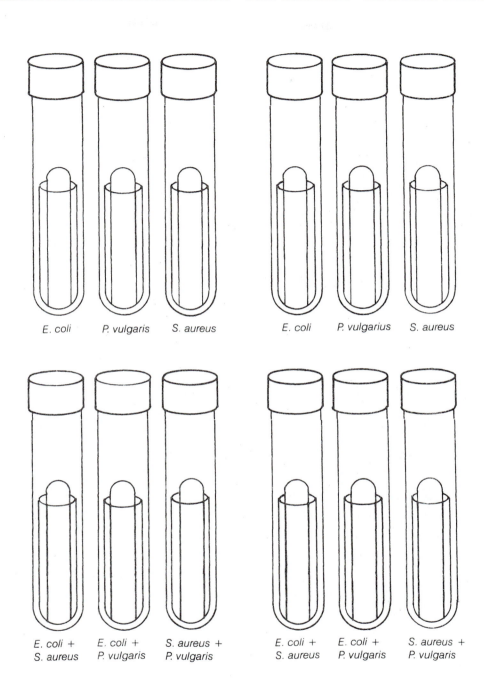

E. coli   P. vulgaris   S. aureus   E. coli   P. vulgarius   S. aureus

E. coli +   E. coli +   S. aureus +   E. coli +   E. coli +   S. aureus +
S. aureus   P. vulgaris   P. vulgaris   S. aureus   P. vulgaris   P. vulgaris

REPORT FOR EXERCISE 46. MICROBIAL POPULATIONS IN SOIL

Name: _____ Date: _____ Lab Section: _____

A. Isolating Microbial Populations

a. Total bacteria: _____
b. Fungi _____
c. Actinomycetes _____
d. Endospores _____

B. Isolation of an Endospore Former

Draw the dilution and plating series used in your experiment and determine the number of endospores/gram of soil.

# of endospores/gram of soil = _____

In the circles below, draw what you observed microscopically in the gram and endospore stains.

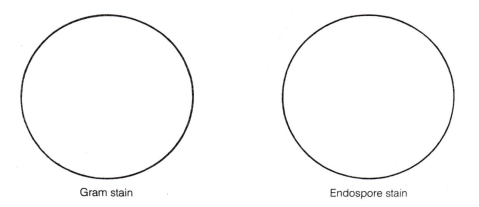

Gram stain   Endospore stain

C. Isolation of an Antibiotic Producer

Draw the results of the cross-streaks onto the drawing below showing the Petri dish of plate count agar containing the line inoculation of a potential antibiotic producer.

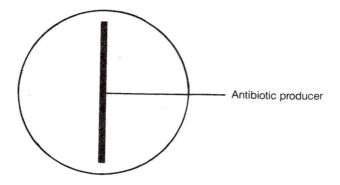

Antibiotic producer

REPORT FOR EXERCISE 47. THE NITROGEN CYCLE

Name: _____ Date: _____ Lab Section: _____

A. Ammonification

Indicate the results of the Nessler's test with *E. coli*, *P. fluorescens*, and the soil sample. Make sure you label correctly. Use the following symbols:

(−) = no yellowing any more intense than the control

\+ = slight yellow

\+ + = deeper yellow

\+ + + = brown precipitate

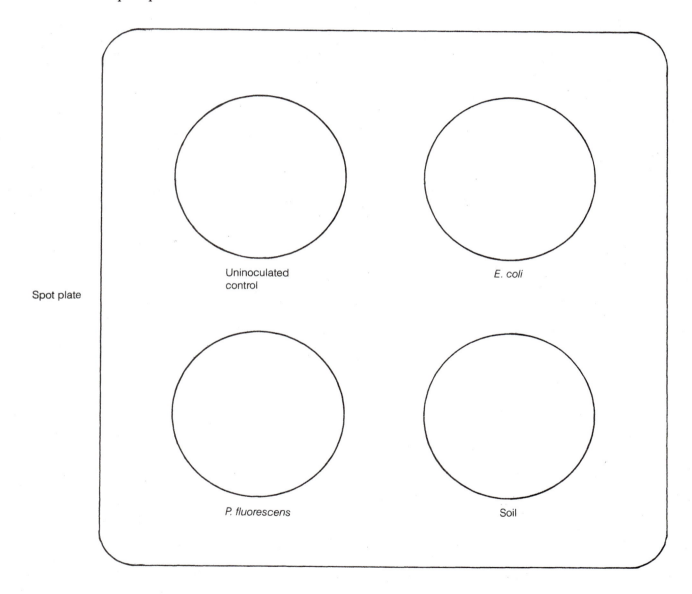

Spot plate

Uninoculated control

E. coli

P. fluorescens

Soil

B. Nitrification

AMMONIUM BROTH

TIME	NITRITE NITROGEN	RESIDUAL AMMONIUM	MORPHOLOGY, GRAM RTN.
1 week			
2 weeks			
3 weeks			

NITRITE BROTH

TIME	NITRITE NITROGEN	RESIDUAL NITRITE	MORPHOLOGY, GRAM RTN.
1 week			
2 weeks			
3 weeks			

C. Denitrification

NITRATE BROTH

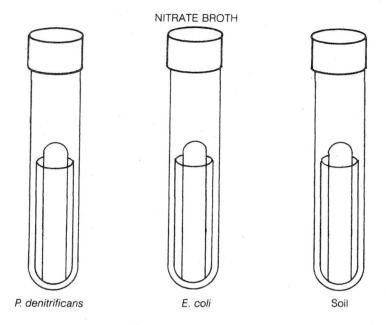

P. denitrificans   E. coli   Soil

TUBE	NO_3^- TO N_2	NO_3 TO NH_4^+	NO_3^- TO NO_2^-	$NO_3^- \neq$ (NO REACTION)
P. denitrificans				
E. coli				
soil				

Indicate results with a + or a −.

D. Nitrogen Fixation

GRAM STAINS

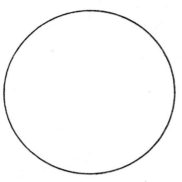

a. Root nodules

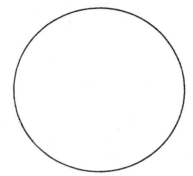

b. Nitrogen-free broth

REPORT FOR EXERCISE 48. THE SULFUR CYCLE

Name: _____ Date: _____ Lab Section: _____

A. The Winogradsky Column

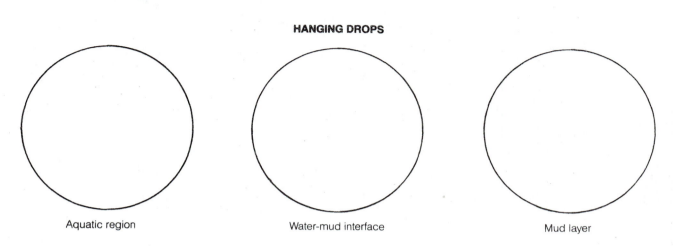

HANGING DROPS

Aquatic region   Water-mud interface   Mud layer

B. Examination of Photosynthetic Sulfur Bacteria

MODIFIED VAN NEIL'S MEDIUM

Serially diluted inoculum

Indicate the degree of turbidity in each bottle, noting any colors, and draw microscopic observations of hanging drops from the respective bottles.

HANGING DROPS

C. Isolation of *Thiobacillus*

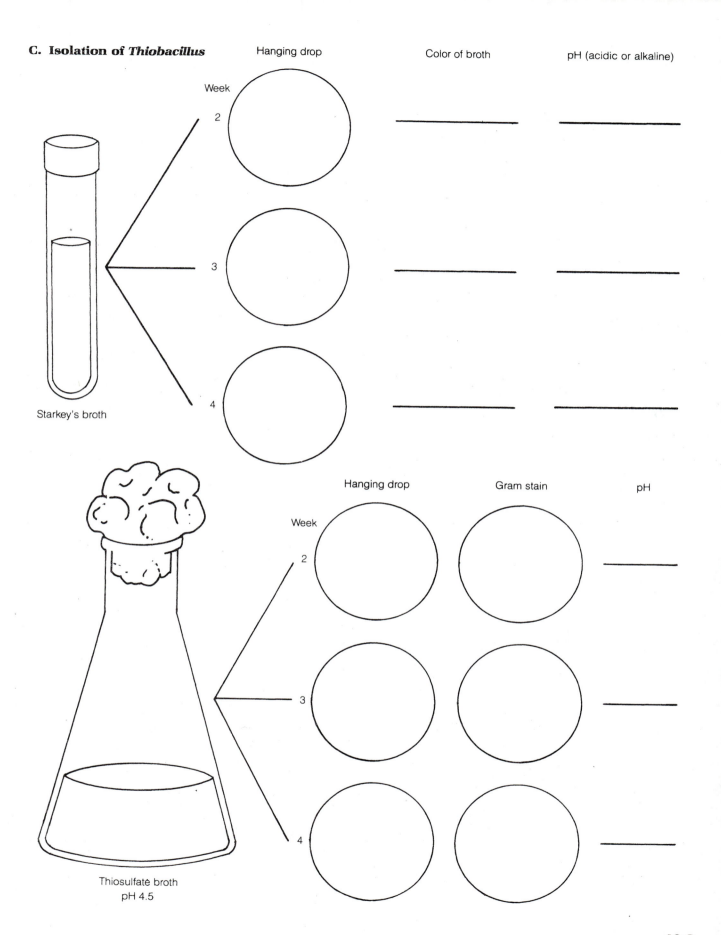

Hanging drop

Color of broth

pH (acidic or alkaline)

Week

2

3

4

Starkey's broth

Hanging drop

Gram stain

pH

Week

2

3

4

Thiosulfate broth
pH 4.5

Interpretation of results

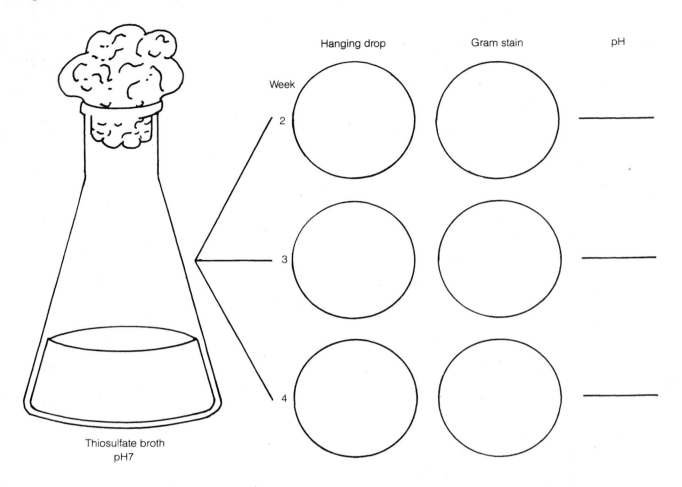

Thiosulfate broth
pH7

Week 2, 3, 4 — Hanging drop — Gram stain — pH

D. Isolation of Sulfate Reducers

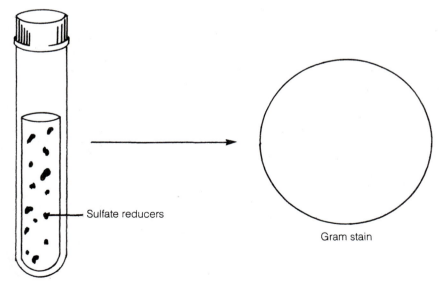

Sulfate reducers

Desulfovibrio agar

Gram stain

Name: _____   Date: _____   Lab Section: _____

Outline the reactions which occur in bioluminescence.

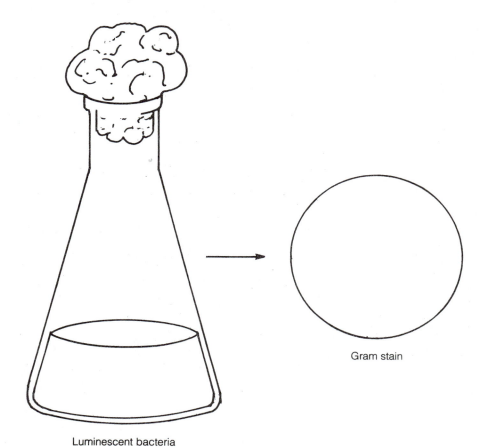

Luminescent bacteria

Gram stain

Name: _____ Date: _____ Lab Section: _____

MICROSCOPIC OBSERVATIONS

Day 5

Day 10

Field 1   Field 2   Field 3

REPORT FOR EXERCISE 51. MOST PROBABLE NUMBER METHOD FOR COLIFORM ANALYSIS

Name: _____ Date: _____ Lab Section: _____

TUBE NUMBER		1	2	3	4	5
Presumptive test (Lauryl tryptose broth)	24h					
	48h					
Confirmed test (BGBL broth)	24h					
	48h					
Fecal coliforms (EC broth)	24h					

Indicate in the boxes above (+) for gas, (−) for no gas, or nd for not done.

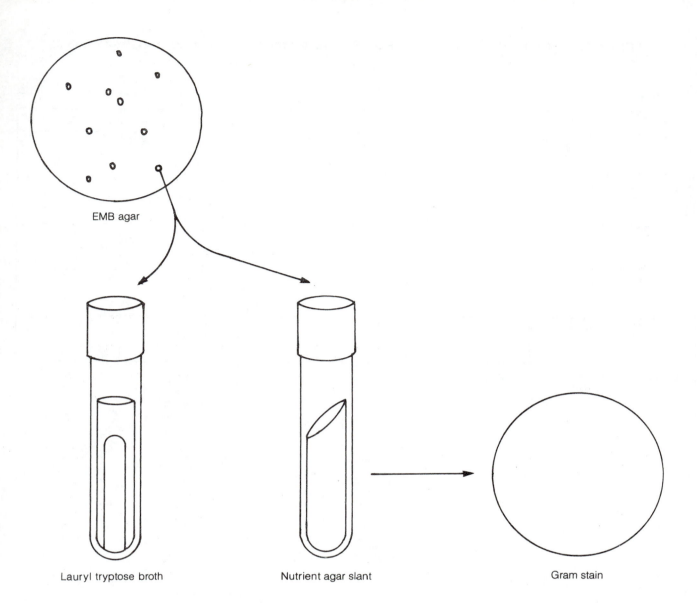

EMB agar

Lauryl tryptose broth

Nutrient agar slant

Gram stain

Completed test

Total Coliforms: _____ MPN/100 ml

Fecal Coliforms: _____ MPN/100 ml

REPORT FOR EXERCISE 52. MEMBRANE FILTER METHOD FOR COLIFORM ANALYSIS

Name: _____ Date: _____ Lab Section: _____

m-Endo

m-FC

Indicate the appearance of typical coliform colonies on each of the plates above.

m-Endo _____

m-FC _____

Results:

coliforms/100 ml _____

fecal coliforms/100 ml _____

REPORT FOR EXERCISE 53. NONCOLIFORM INDICATORS OF RECREATIONAL WATER QUALITY

Name: _____ Date: _____ Lab Section: _____

Asparagine broth
PRESUMPTIVE TEST

Acetamide broth
CONFIRMED TEST

Indicate the appearance of positive presumptive and confirmed tests in the *P. aeruginosa* MPN.

Results:

P. aeruginosa MPN/100 ml _____

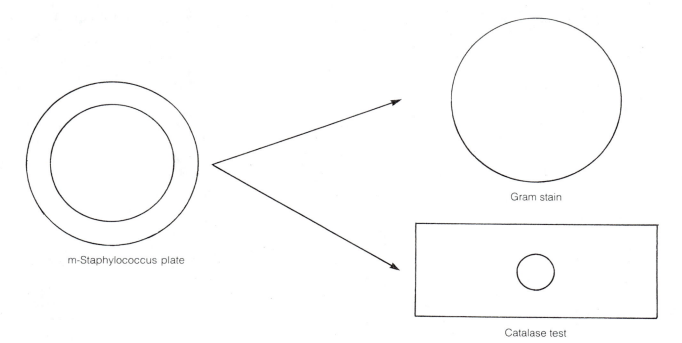

m-Staphylococcus plate

Gram stain

Catalase test

Results:

staphylococci/100 ml _____

REPORT FOR EXERCISE 54. ISOLATION OF ACTINOMYCETES FROM WATER AND WASTEWATER TREATMENT PLANTS

Name: _____ Date: _____ Lab Section: _____

Starch-casein agar

Microscopic observation
(100×)

Indicate the colonial appearance on the plate, and your observations at 100×,

Results:

actinomycetes/ml or gram _____

Name: _____ Date: _____ Lab Section: _____

A. Enumeration of Food Spoilage Organisms

Draw and label the dilution series used and the dilutions plated. Indicate the number of colonies on the countable plates.

Results:

APC/gram hamburger _____

B. Enumeration of Proteolytic Bacteria in Food

Draw and label the dilution series used and the dilutions plated. Indicate the number of proteolytic colonies on the countable plates.

Indicate your observations of the control organisms on the plates of skim milk agar below.

E. coli   *B. subtilis*

Results:

proteolytic bacteria/gram hamburger _____

C. Enumeration of Lipolytic Bacteria

Draw and label the dilution series used and the dilutions plated. Indicate the number of lipolytic colonies on the countable plates.

Indicate your observations of the control organisms on the plates of fat agar below.

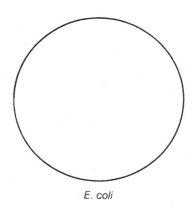

E. coli

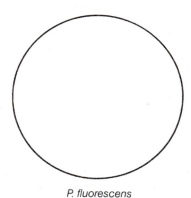

P. fluorescens

Results:

lipolytic bacteria/g hamburger _____

D. Enumeration of Molds and Yeasts from Fruit

Draw and label the dilution series used and the dilutions plated. Indicate the number of molds and yeasts on the countable plates.

Results:

molds and yeasts/gram hamburger _____

Name: _____ Date: _____ Lab Section: _____

Describe the appearance of the control organisms on phenol red mannitol salt agar.

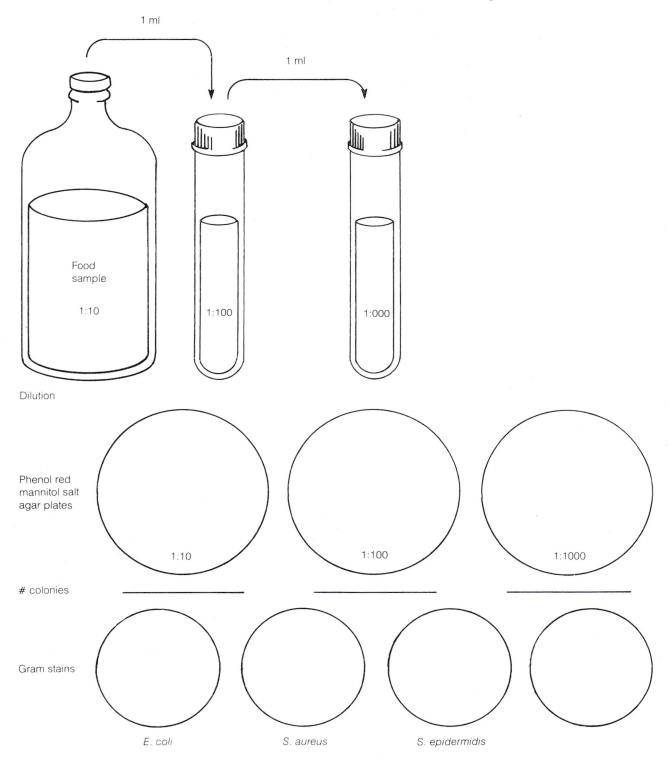

1 ml

1 ml

Food
sample

1:10

1:100

1:000

Dilution

Phenol red
mannitol salt
agar plates

1:10

1:100

1:1000

colonies _____ _____ _____

Gram stains

E. coli

S. aureus

S. epidermidis

Results:

staphylococci/g food _____

REPORT FOR EXERCISE 57. EVALUATING THE SANITARY QUALITY OF FOODS

Name: _____ Date: _____ Lab Section: _____

A. Coliform Analysis of Foods

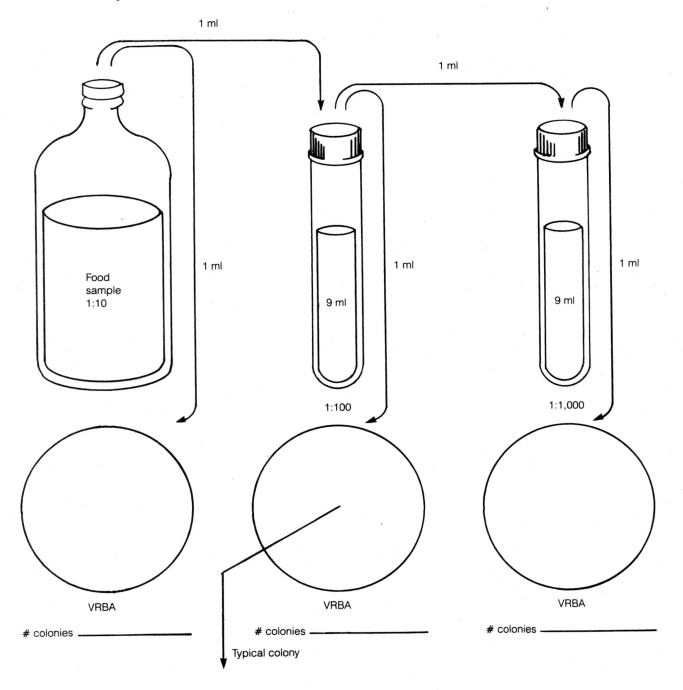

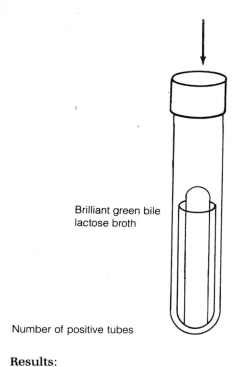

Brilliant green bile
lactose broth

Number of positive tubes

Results:

coliforms/g _____

B. Fecal Streptococcal Analysis of Foods

Draw a flow chart for your isolation and identification of fecal streptococci in foods. Include at each step the appearance of a positive reaction.

Results:

fecal streptocci/gram _____

Name: _____ Date: _____ Lab Section: _____

Results:

organisms/significant surface _____

REPORT FOR EXERCISE 59. MICROBES AND FOOD PRODUCTION

Name: _____ Date: _____ Lab Section: _____

A. Sauerkraut Production

TIME (DAYS)	pH	TITRATABLE ACIDITY	NUMBER CFU/ml	GRAM RXN/MORPHOLOGY	CATALASE REACTION

B. Vinegar Production

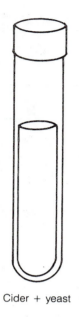

CONTROLLED PRODUCTION

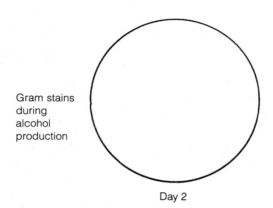

Gram stains
during
alcohol
production

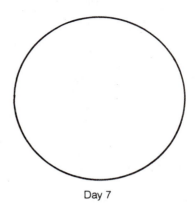

Day 2

Day 7

Cider + yeast

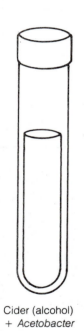

Gram stains
during
alcohol
oxidation

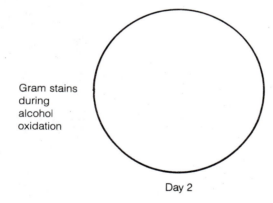

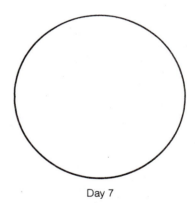

Day 2

Day 7

Cider (alcohol)
+ *Acetobacter*

NATURAL PRODUCTION

Cider + Raisins

Gram stains

Day 2

Day 14

Day 7

Day 21

C. Microorganisms in Sourdough Production

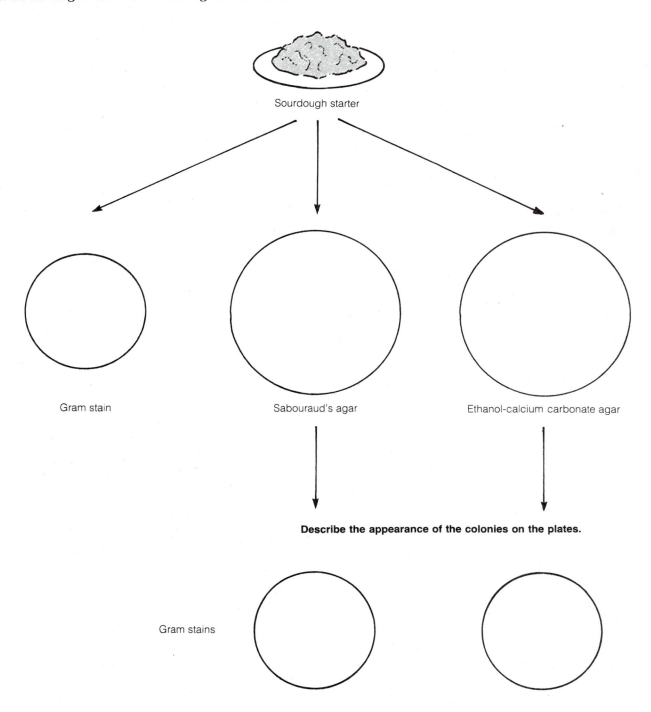

Sourdough starter

Gram stain

Sabouraud's agar

Ethanol-calcium carbonate agar

Describe the appearance of the colonies on the plates.

Gram stains

REPORT FOR EXERCISE 60. ANALYSIS OF THE SANITARY QUALITY OF MILK AND DAIRY PRODUCTS

Name: _____ Date: _____ Lab Section: _____

A. Coliform Analysis

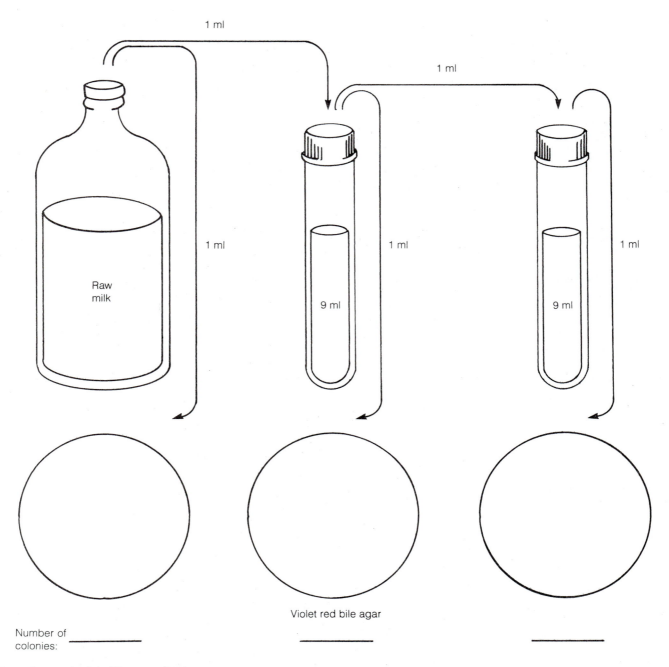

1 ml

1 ml

1 ml

1 ml

1 ml

1 ml

Raw milk

9 ml

9 ml

Violet red bile agar

Number of colonies: _____   _____   _____

Describe typical coliform colonies

Results:

\# coliforms/ml of milk _____

B. Dye Reduction Tests

Compare the tube of raw milk with the Munsell color standards shown in the color insert.

Results:

Sanitary quality of tested milk _____

C. Direct Microscopic Counts

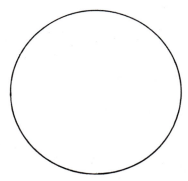

Typical field under oil

Determine the average number of clumps/ml using the following formula:

$$\text{average number of clumps per field} \times MF = CMC/ml$$

Results:

CMC/ml _____

Name: _____ Date: _____ Lab Section: _____

A. California Mastitis Test

Draw the appearance of the tested milk following addition of the CMT reagent.

Grade of milk: _____

B. Modified Whiteside Test

Draw the appearance of the tested milk after mixing with NaOH.

Grade of milk: _____

C. Direct Microscopic Somatic Cell Count

Determine the average number of somatic cells and leucocytes per ml using the following formula:

$$\text{average number of cells per field} \times \text{MF} = \text{somatic cells/ml}$$

Results:

Somatic cells/ml _____

Typical field under oil

D. Cultural Methods for Mastitis Milk

Record your observations of the growth (if any) on the blood agar plate below.

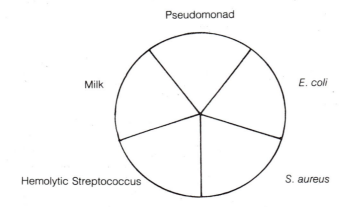

Record your observations of the growth (if any) on the EMB agar plate below.

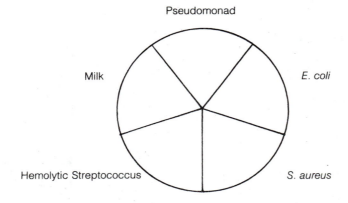

Record your observations of the growth (if any) on the PRMS agar below.

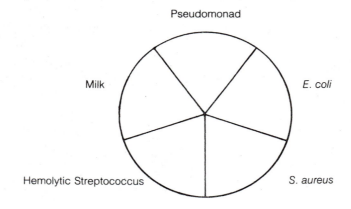

Name: _____ Date: _____ Lab Section: _____

Draw the dilution series and plating methods used on the raw milk sample for both coliform analysis and total plate count.

Results:

\# coliforms/ml _____

CFU/ml _____

Draw the dilution series and plating methods used on the pasteurized milk sample for both coliforms and total plate count.

Results:

\# coliforms/ml _____

CFU/ml _____

Name: _____ Date: _____ Lab Section: _____

A. Yogurt Production

Draw your observations in the spaces below.

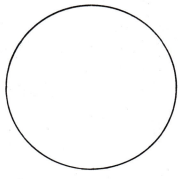

Gram stain of purchased yogurt

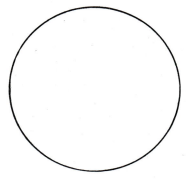

Gram stain of prepared yogurt

B. Microorganisms and Cheese Production

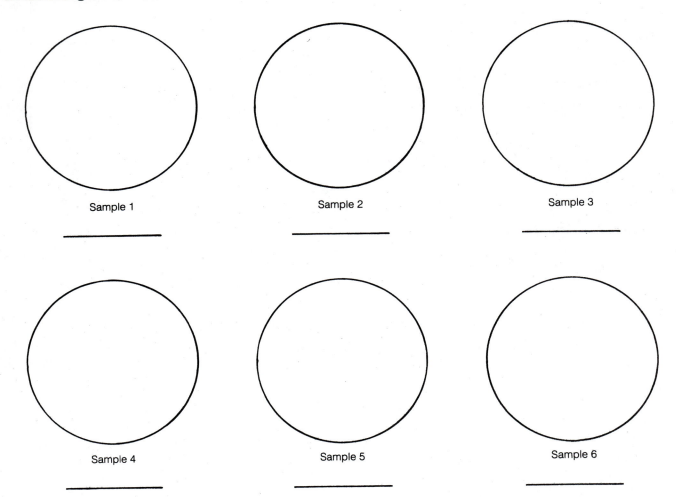

Sample 1

Sample 2

Sample 3

_____

_____

_____

Sample 4

Sample 5

Sample 6

_____

_____

_____

Draw the microorganisms observed in gram stains of cheese emulsions. Name the cheeses in the spaces provided.

Notes:

REPORT FOR EXERCISE 64. PLANT PATHOGENS

Name: _____ Date: _____ Lab Section: _____

A. Rusts and Smuts

Draw pictures of the following:

barberry leaves
with *P. graminis*

P. graminis

plants with
red rust

Urediospores

plants with
black rust

Teliospores

plants infected
with corn smut

Corn smut

B. Soft Rot

Carrot Soft Rot

	E. carotovora	*E. atroseptica*	Water
Appearance	_____	_____	_____
Odor	_____	_____	_____

Gram stains

Potato Soft Rot

	E. carotovora	*E. atroseptica*	Water
Appearance	_____	_____	_____
Odor	_____	_____	_____

Gram stains

C. Crown Gall

Draw a picture of the tomato plant showing the galls.

GRAM STAINS

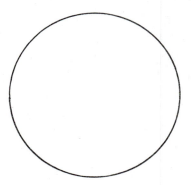

Agrobacterium
from crown gall pulp

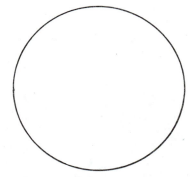

Agrobacterium
from colony on media

D. Tobacco Mosaic Virus

Draw a healthy tobacco plant and one with mosaic disease. Compare them.

REPORT FOR EXERCISE 65. MYCORRHIZAE

Name: _____ Date: _____ Lab Section: _____

PREPARED SLIDES

Mycorrhizal root

Orchid root

Phloxine stained mycorrhizae
examined microscopically

REPORT FOR EXERCISE 66. BIOINSECTICIDES

Name: _____ Date: _____ Lab Section: _____

Results:

TIME (DAYS)	VEGETABLE	NUMBER OF SURVIVORS	% SURVIVORS
1			
2			
3			
4			
5			

Prepare tables for the other vegetables used.

Name: _____ Date: _____ Lab Section: _____

Microscopic observation of tubes showing cellulose disintegration.

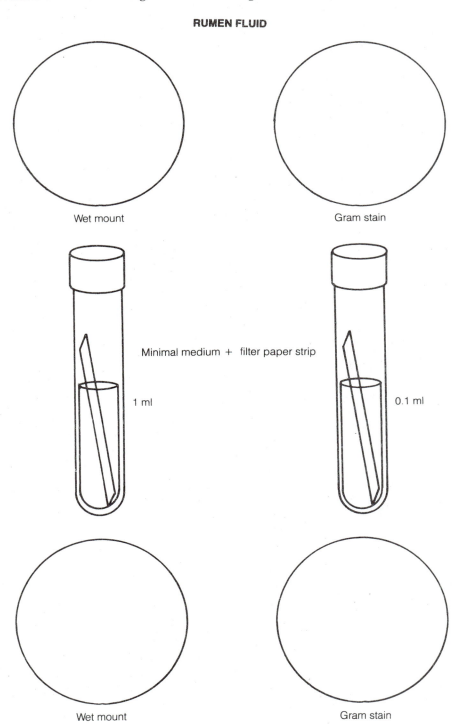

RUMEN FLUID

Wet mount

Gram stain

Minimal medium + filter paper strip

1 ml

0.1 ml

Wet mount

Gram stain

Name: _____ Date: _____ Lab Section: _____

A. The Oral Cavity and Cariogenic Bacteria

Record observations on the media.

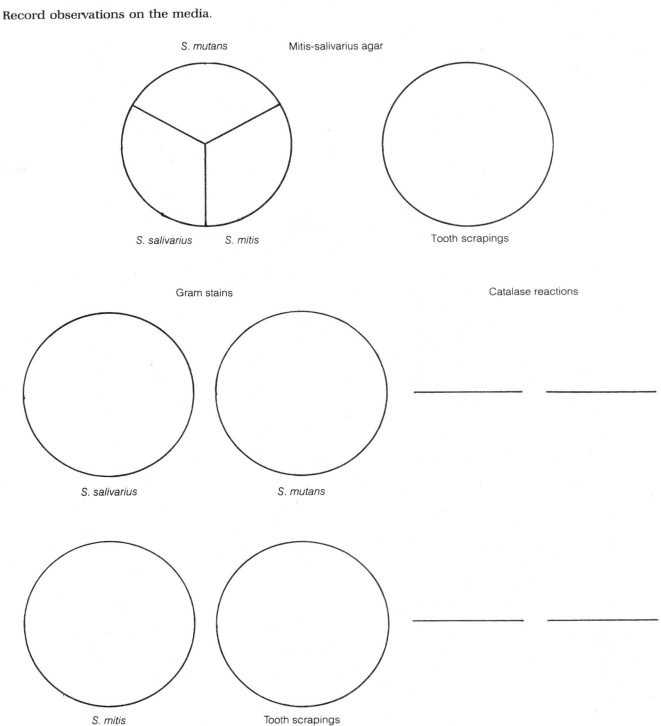

S. mutans   Mitis-salivarius agar

S. salivarius   S. mitis   Tooth scrapings

Gram stains   Catalase reactions

S. salivarius   S. mutans

S. mitis   Tooth scrapings

B. Skin Flora

Record your observations of the colonial appearance (and hemolytic activity on blood agar) on these media.

Skin swabs

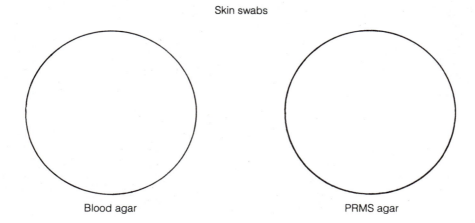

Blood agar   PRMS agar

Gram stains of predominant colonies

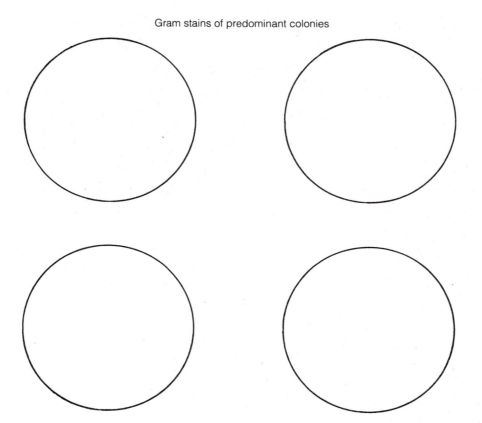

C. Throat Flora

Indicate the hemolytic reactions of predominant colonies in the space below.

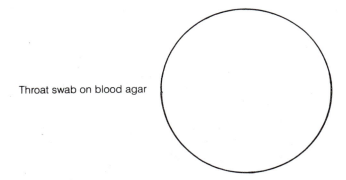

Throat swab on blood agar

Gram stains on predominant colonies

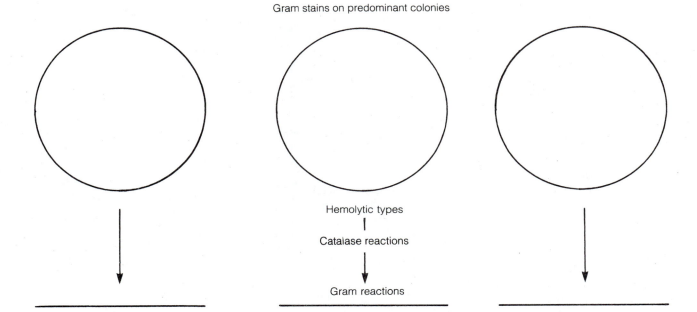

Hemolytic types

Catalase reactions

Gram reactions

REPORT FOR EXERCISE 69. ANTIBIOTIC SENSITIVITY TESTING USING THE KIRBY-BAUER PROCEDURE

Name: _____   Date: _____   Lab Section: _____

ANTIMICROBIAL AGENT	ORGANISM TESTED	ZONE DIAMETER (in mm)	INTERPRETATION*

*Interpret the results using Table 69-1 as a guide. Indicate whether the tested bacterium is resistant (R), susceptible (S), or of intermediate (I) susceptibility to the antimicrobial agent.

REPORT FOR EXERCISE 70. ABO BLOOD TYPING

Name: _____ Date: _____ Lab Section: _____

Anti-A

Anti-B

ANTISERUM ADDED	HEMAGGLUTINATION ((+) or (−))	BLOOD TYPE
Anti-A		
Anti-B		

Name: _____   Date: _____   Lab Section: _____

A. Slide Agglutination Test

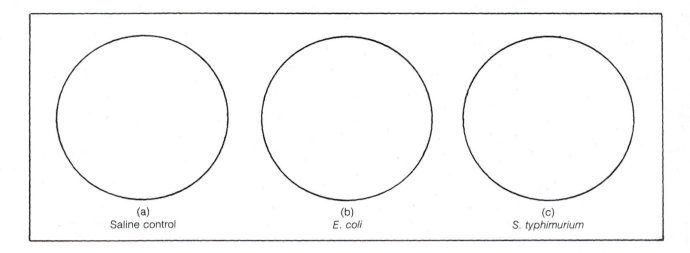

(a)
Saline control

(b)
E. coli

(c)
S. typhimurium

SAMPLE TESTED	AGGLUTINATION	INTERPRETATION
Phenol-normal saline		
E. coli		
S. typhimurium		

B. The Precipitin Test

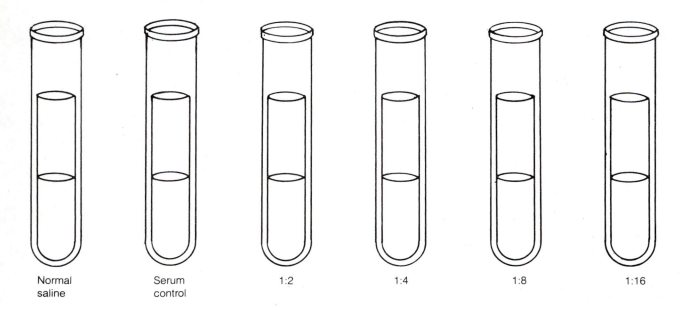

Normal saline   Serum control   1:2   1:4   1:8   1:16

REAGENT ADDED	TUBE NUMBER						RESULTS
	1	2	3	4	5	6	
Normal Saline							
Bovine Albumin							
Normal Serum							
Goat Antiserum							

C. Immunofluorescence

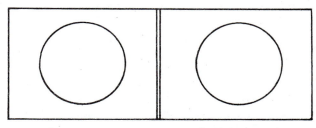

Streptococcus   Staphylococcus

ORGANISM TESTED	FLUORESCENCE		GRAM STAIN	
	RESULTS	APPEARANCE	RESULTS	APPEARANCE
Streptococcus faecalis				
Staphylococcus epidermidis				

D. The Rapid Plasma Reagin (RPR) Test for Syphilis

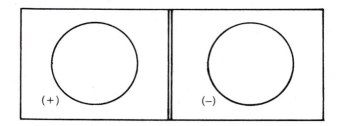

SPECIMEN #	APPEARANCE AFTER MIXING	RESULTS

Name: _____ Date: _____ Lab Section: _____

A. The Enterotube II System

IDENTIFICATION USING ID VALUE

TEST*	GLU	GAS	LYS	ORN	H₂S	IND	ADON	LAC	ARAB	SORB	DUL	PA	UREA	CIT
RESULTS														
TEST VALUE	2	+ 1	4	+ 2	+ 1		4	+ 2	+ 1		4	+ 2	+ 1		4	+ 2	+ 1

ID VALUE**

UNKNOWN # _____ IDENTIFICATION _____

 *The Voges-Proskauer test is used only as a confirmatory test
**The ID value can be obtained from the Enterotube II Computer Coding Manual. Ask your instructor for a copy.

IDENTIFICATION OF UNKNOWN USING TABLE 72-2 AS REFERENCE

TEST	GLU	GAS	LYS	ORN	H₂S	IND	ADON	LAC	ARAB	SORB	V-P	DUL	PA	UREA	CIT
RESULTS*															

*Consult table 72-2 to determine the identity of the unknown isolate.

UNKNOWN # _____ IDENTIFICATION _____

B. The API 20E System

TEST	ONPG	ADH	LDC	ODC	CIT	H₂S	URE	TDA	IND	VP	GEL	GLU	MAN	INO	SOR	RHA	SAC	MEL	AMY	ARA	OXI
TEST VALUE	1 +	2 +	4	1 +	2 +	4	1 +	2 +	4	1 +	2 +	4	1 +	2 +	4	1 +	2 +	4	1 +	2 +	4
RESULTS 5 HR.																					
RESULTS 24 HR.																					
PROFILE NUMBER																					

TEST	ONPG	ADH	LDC	ODC	CIT	H₂S	URE	TDA	IND	VP	GEL	GLU	MAN	INO	SOR	RHA	SAC	MEL	AMY	ARA	OXI
TEST VALUE	1 +	2 +	4	1 +	2 +	4	1 +	2 +	4	1 +	2 +	4	1 +	2 +	4	1 +	2 +	4	1 +	2 +	4
RESULTS 5 HR.																					
RESULTS 24 HR.																					
PROFILE NUMBER																					

INDEX

broth, 20, 35, 79
  gelatin, 116, 117
Nutritional factors, 24

Objective lens, 1
O.D. (optical density), 63, 69
Obligate
  aerobe, 80
  anaerobe, 80
Ocular
  lens, 1
  micrometer, calibration of, 11, 12
Oil immersion objective lens, 1, 4, 8–11
One-step growth curve, 146–150
Oomycetes, 124, 130
Operon, inducible, 159
Opportunistic pathogen, 250
Optical density (O.D.), 63, 69
Oral microorganisms (flora), 249, 250–251
Origin of transfer locus, 170
Organelle, 1
Ortho-nitrophenyl-beta-D
  galactoside (ONPG), 160–161
Oscillatoria, 85, 135
Osmophile, 78
Osmotolerant, 78
Osmotic pressure, 78
Oxidase (cytochrome oxidase) activity, test
  for, 108, 109, 119
Oxygen, effects on microbial growth, 74,
  80–84

Palladium catalyst, 80
Paramecium, 98, 132
Parasite, 209
Parasitism, 177
Parasporal bodies, 246
Parfocal, 8
Pasteur, Louis, 218, 249
Pasteurization, 223
  LTLT, 235
  HTST, 235
  UHT, 235
Pathogen, 131, 195, 242, 246, 250, 252
Pectinolytic, 242
Pediococcus, 218
Pellicle formation in broth cultures, 30, 220
Pelodictyon, 85
Penetration of cell by virus, 141
Penicillin, 37, 89, 95, 96, 185, 254
Penicillium, 77, 79, 124, 125, 179, 183, 223, 239
Pentose, 102
Pepper infusion, 14, 15, 16, 19
Peptide, 101
Peptidoglycan, 44
Peptococcus, 80
Peptone, 24, 184
Peptone glycerol phosphate, 192
Per cent transmission, 69
Periphytic, 193
Peritrichous flagella, 54
Peroxidase activity, 74, 80, 109
Petri dish (plate), 24
PFU (plaque forming unit), 143
pH
  common, indicators, Appendix B

control of, in culture media, 77
  definition of, 73
  effect on growth, 76–77
  measurement of, 73
Phage (see bacteriophage)
  assay, 143
  generalized transducing, 163
  isolation, 144, 145
  specialized transducing, 163
  typing, 142, 150–151
Phagocytosis, 49, 228
Phase contrast microscopy, 18–20
Phase plate, 18
PHB (poly-β-OH-butyrate)
  granules, 58–59
  inclusions, 58–59
Phenol
  coefficient, 93, 94
  -ic disinfectants, 88, 89
  normal, 258
  red, 104, 117, 177, 178, 211, 234, 251
Phenolphthalein, 219
Phenotypic variation, 153, 154–155
Phenylalanine deaminase, test for,
  114–115, 119
Phenylpyruvic acid, 114, 115
Phosphate-buffered saline, 263, E9
Phospholipids, 103
Phosphoresce, 20
Photoautotroph, 85, 134, 189
Photobacterium, 192
Photosynthesis
  bacterial, 85
  plant type, 85, 134–136
Phototroph, 85, 134, 189
Picornavirus, 140, 144
Pigment production, 154–155
Pileus, 127
Pilus, see also fimbria, 141, 153, 166
Pipette
  manipulation, 35–36
  precautions in use of, 35–36
  use of manual pipetting device (eg., Pro-
    pipette), 35–36
Pityrospum, 251
Plant pathology
  crown gall in, 244–245
  rusts and smuts in, 242–244
  soft rot in, 242
  tobacco mosaic virus in, 245–246
Plaque
  assay, 143
  forming units (PFU), 143
  viral, 143
Plasma, 264
Plasmid, 162, 166
Plasmodium, 133
Plasmolysis, 78
Plasmoptysis, 78
Plate count
  agar medium, 66, 70, 92, 100, 176, 181, 182,
    183, 206, 207, 216, 235
Pneumococcus, 49
Pneumonia, 49
Polar flagella, 54
Polio, 140, 144, 195
Poly-β-OH-butyrate (PHB)
  granules, 58, 59

inclusions, 58, 59
Polymyxin B, 89, 254
Polynucleotides
  digestion, test for, 101–102
  hydrolysis, test for, 101–102
Polypeptide
  digestion, test for, 100–101
  hydrolysis, test for, 100–101
Polysaccharide
  digestion, test for, 98–100
  hydrolysis, test for, 98–100
Pond water, 14, 15, 16, 19, 21, 132, 137, 188,
  193
Populations, 175
Positive stain, 39, 41, 42
Potassium chloride (KCl) broth, 146
Potassium carbonate, 161
Potassium hydroxide, 107, 266, 268, 273
Potato dextrose, acidified, 239
Pour plate technique, 67–68
Precipitin test, 258–262
Predation, 177
Preservation of
  food and dairy products, 28
Preservatives, 87
Presumptive test
  coliforms, 196
  P. aeruginosa, 200
Primary amoebic meningoencephalitis
  (PAM), 195
Primary stain, 45
Prodigiosin, 154
Profile Index Number, 272
Progesterone, 103
Prokaryote, 79, 187
Promotor, 171
Propiolactone, 89
Propionic acid, 89
  fermentation, 105
Propionobacterium, 105, 223, 238
Protease, 100
Proteolysis (see polypeptide digestion, hy-
  drolysis), 100, 101, 207–208
Proteolytic activity, test for, 101–102, 207–208
Proteus, 115
  P. mirabilis, 103, 119
  P. vulgaris, 37, 54, 111, 112, 115, 117, 119,
    180
Protista, 134
Protoslo, 132
Prototrophs, 153, 162, 170, 171
Protozoa, 123, 131–133, 175, 247
  classification of, 132
Pseudomonas
  P. aeruginosa, 108, 154, 195, 200, 253
  P. denitrificans, 187
  P. fluorescens, 14, 54, 75, 77, 109, 110, 119,
    154, 179, 183, 185, 208
Pseudoplasmodium, 130
Psychrophile, 75
pTi plasmid, 244
Puccinia, 124, 242, 243
Pulvinate colony, 29
Pure culture
  (aseptic culture, sterile) technique, 30–36
Purines, 102
Purple non-sulfur bacteria, 85
Purple sulfur bacteria, 85, 189, 190